Springer Series in Translational Stroke Research

Series Editor
John Zhang

For further volumes:
http://www.springer.com/series/10064

Jun Chen • Xiaoming Hu
Mary Stenzel-Poore • John H. Zhang
Editors

Immunological Mechanisms and Therapies in Brain Injuries and Stroke

Springer

Editors
Jun Chen
University of Pittsburgh School of Medicine
Pittsburgh, PA, USA

Xiaoming Hu
Department of Neurology
University of Pittsburgh School of Medicine
Pittsburgh, PA, USA

Mary Stenzel-Poore
Molecular Microbiology and Immunology
Oregon Health and Sciences University
Portland, OR, USA

John H. Zhang
Department of Physiology
Loma Linda University
Loma Linda, CA, USA

ISBN 978-1-4614-8914-6 ISBN 978-1-4614-8915-3 (eBook)
DOI 10.1007/978-1-4614-8915-3
Springer New York Heidelberg Dordrecht London

Library of Congress Control Number: 2013951660

Printed on acid-free paper

Springer is part of Springer Science+Business Media (www.springer.com)

Contents

Contributors

Yama Akbari University of California, Irvine, CA, USA

Ken Arai Massachusetts General Hospital, Harvard University, Charlestown, MA, USA

Jaroslaw Aronowski The University of Texas Medical School at Houston, Houston, TX, USA

Frances Rena Bahjat Department of Molecular Microbiology and Immunology, Oregon Health and Science University, Portland, OR, USA

Kyra J. Becker Neurology and Neurological Surgery, University of Washington School of Medicine, Seattle, WA, USA

Pak H. Chan Stanford University School of Medicine, Stanford, CA, USA

Jun Chen Department of Neurology, University of Pittsburgh, Pittsburgh, PA, USA

Sheng Chen Department of Physiology, Loma Linda University, Loma Linda, CA, USA

Sunghee Cho Neuroscience, Weill Cornell Medical College, Burke Medical Research Institute, White Plains, NY, USA

E. Sander Connolly Jr. Department of Neurosurgery, Columbia University, New York, NY, USA

Ulrich Dirnagl Charité Universitätsmedizin Berlin, Berlin, Germany

Mitchell S.V. Elkind Department of Epidemiology, Mailman School of Public Health, New York, NY, USA

Department of Neurology, Columbia University College of Physicians and Surgeons, New York, NY, USA

Gertrude H. Sergievsky Center, Columbia University, New York, NY, USA

Maria Febbraio University of Alberta, Edmonton, AB, Canada

David Fernández-López University of California San Francisco School of Medicine, San Francisco, CA, USA

Mutsumi Fujii Department of Physiology, Loma Linda University, Linda, CA, USA

Yu Gan Department of Neurology, University of Pittsburgh, Pittsburgh, PA, USA

Romergryko Geocadin The Johns Hopkins University School of Medicine, Baltimore, MD, USA

Raffaella Gesuete Oregon Health and Science University, Portland, OR, USA

D. Neil Granger Molecular and Cellular Physiology, Louisiana State University Health Sciences Center, Baton Rouge, LA, USA

Gaurav Gupta Department of Neurosurgery, Columbia University Medical Center, New York, NY, USA

John Hallenbeck Stroke Branch, NINDS, NIH, Bethesda, MD, USA

Xiaoming Hu Department of Neurology, University of Pittsburgh, Pittsburgh, PA, USA

Patricia D. Hurn The University of Texas System, Ausitn, TX, USA

Glen C. Jickling Department of Neurology, University of California, Davis, CA, USA

Joo Eun Jung Stanford University School of Medicine, Stanford, CA, USA

Damon Klebe Department of Physiology, Loma Linda University, Loma Linda, CA, USA

Jari Koistinaho A.I. Virtanen Institute for Molecular Sciences, University of Eastern Finland, Kuopio, Finland

Milla Koistinaho Medeia Therapeutics, Kuopio, Finland

Rehana Leak Pharmaceutical Sciences, Duquesne University, Pittsburgh, PA, USA

Peiying Li Department of Neurology, University of Pittsburgh, Pittsburgh, PA, USA

Eng H. Lo Massachusetts General Hospital, Harvard University, Charlestown, MA, USA

Jorge M. Luna Department of Epidemiology, Mailman School of Public Health, New York, NY, USA

Department of Neurology, Columbia University College of Physicians and Surgeons, New York, NY, USA

Department of Industrial and Operations Research, School of Engineering and Applied Sciences, Columbia University, New York, NY, USA

Carolina M. Maier Stanford University School of Medicine, Stanford, CA, USA

Takakuni Maki Department of Neurology, Kyoto University, Kyoto, Japan

Leilei Mao Department of Neurology, University of Pittsburgh, Pittsburgh, PA, USA

Michael McDowell Department of Neurosurgery, Columbia University Medical Center, New York, NY, USA

Maximilian Mulder Johns Hopkins Hospital, Baltimore, MD, USA

Purnima Narasimhan Stanford University School of Medicine, Stanford, CA, USA

Tatsuro Nishi Stanford University School of Medicine, Stanford, CA, USA

Halina Offner Neurology and Anesthesiology and Perioperative Medicine, Oregon Health and Science University, Portland, OR, USA

Ali Razmara Massachusetts General Hospital, Boston, MA, USA

Stephen F. Rodrigues Clinical and Toxicological Analyses, University of Sao Paulo, Sao Paulo, Brazil

Hiroyuki Sakata Stanford University School of Medicine, Stanford, CA, USA

Ji Hae Seo Massachusetts General Hospital, Harvard University, Charlestown, MA, USA

Nicholas Shea Department of Neurosurgery, Columbia University Medical Center, New York, NY, USA

Yejie Shi Department of Neurology, University of Pittsburgh, Pittsburgh, PA, USA

Yoshiteru Soejima Department of Physiology, Loma Linda University, Loma Linda, CA, USA

Mary P. Stenzel-Poore Oregon Health and Science University, Portland, OR, USA

Dandan Sun Department of Neurology, University of Pittsburgh, Pittsburgh, PA, USA

Alexander Thiel Department of Neurology and Neurosurgery, McGill University, Montreal, QC, Canada

Alexander Vakhmyanin Department of Physiology, Loma Linda University, Loma Linda, CA, USA

Zinaida S. Vexler Department of Neurology, University of California, San Francisco, CA, USA

Takuma Wakai Stanford University School of Medicine, Stanford, CA, USA

Changhong Xing Massachusetts General Hospital, Harvard University, Charlestown, MA, USA

Hui Yuan Department of Neurology, University of Pittsburgh, Pittsburgh, PA, USA

John H. Zhang Department of Physiology, Pharmacology, and Neurosurgery, Loma Linda University, Loma Linda, CA, USA

Xiurong Zhao Department of Neurology, The University of Texas Medical School at Houston, Houston, TX, USA

Chapter 1
Old Dogmas, Surprising Complexities, and Novel Therapeutic Targets

Ulrich Dirnagl

Abstract Immune responses to brain injury are more than a simple reaction to tissue damage. Brain and immune system are engaging in a tightly orchestrated communication, which may protect the brain, help recover lost function, or aggravate damage and impede repair. The bewildering complexity of these processes is reflected by the fact that for practically any cell type of the immune system evidence for beneficial as well as detrimental functions can be found in the literature. This introduction sets the stage for the chapters of this volume, which will summarize our current knowledge on the immunological mechanisms and therapies of brain injuries and stroke.

> *"Autopsies clearly demonstrate that the [apoplectic] brain is subject to inflammation and suppuration".*
>
> Translated from Richelmi, [1]

Introduction

That acute brain diseases, such as "apoplexy," can be accompanied by inflammation has been realized by physicians and pathologists already a long time ago. Today, we know that immune responses to brain injury are more than a simple reaction to tissue damage, at most responsible for clearing debris. We have come to realize that brain and immune system are engaging in a tightly orchestrated communication, which may protect the brain and even help recover lost function, but may also aggravate damage and impede repair. Indeed, not only resident immune cells of the brain, such as microglia, are involved in these responses, but also practically all cell types of the innate and adaptive immune system, which may home to the lesion, or act in the periphery. The bewildering complexity of the interaction of the two

U. Dirnagl (✉)
Charité Universitätsmedizin Berlin, Berlin, Germany
e-mail: ulrich.dirnagl@charite.de

J. Chen et al. (eds.), *Immunological Mechanisms and Therapies in Brain Injuries and Stroke*, Springer Series in Translational Stroke Research, DOI 10.1007/978-1-4614-8915-3_1,

"supersystems" [2] after brain lesion is reflected by the fact that for practically any cell type of the immune system evidence for beneficial as well as detrimental functions can be found in the literature, and the realization that despite many attempts targeting immune mechanisms has not been successful in large, randomized clinical trials of acute CNS diseases. With some general reflections, I would like to prepare the ground for the chapters of this volume, which summarize our current knowledge on the immunological mechanisms and therapies in brain injuries and stroke.

Nervous and Immune System: Precambrian Twins

Nervous and immune systems have coevolved over hundreds of millions of years. Both are engaged in the communication of the organism with the outside world. They share characteristics from a conceptual (memory, synapse, etc.) to a molecular level (e.g., identical signaling and guidance molecules). The most ancient and possibly most important task of the nervous system is to control movement and predation (or evasion from it), while the immune system protects against infection of the host by foreign organisms, parasitic, bacterial, or viral. Brain and immune system communicate intensely, sensing and controlling each other's state to maintain homeostasis. When things go wrong, however, primary diseases of the nervous system may harm the immune system, and vice versa. In fact, disorders of the immune system may lead to acute brain injury, as in atherothrombotic stroke, or acute brain injury, as in a stroke can cause brain inflammation, as well as immunodepression.

Nervous and Immune System: Friends, Foes, Then Friends Again?

For many decades, research into the pathophysiology of acute brain injury after ischemia or trauma was almost exclusively focused on the central nervous system. Involvement of the immune system was only considered insofar injury may lead to local inflammation, which involves not only brain cells, such as microglia and astrocytes, but also cells of the innate immune system which have homed to the lesion, such as granulocytes and monocytes. Early anti-inflammatory treatment in experimental models of stroke or brain trauma appears to be protective, although clinical trials were unable to confirm this effect in stroke patients. The interaction of adaptive immunity and the brain has traditionally been the domain of multiple sclerosis research, which has demonstrated that even the healthy brain is patrolled by T cells. Only recently, it was realized that cells of the adaptive immune system are players when the brain is acutely lesioned. Not only may the brain downregulate the peripheral immune system (innate and adaptive) after stroke [3], brain trauma [4], or spinal cord injury [5], cells of the adaptive immune system enter the brain where they may aggravate or contain damage, or potentially even partake in repair. These findings, as well as the notion that inflammation is intrinsically linked to wound healing in the periphery, have promoted the concept of the Janus-facedness of inflammation after brain injury; in its most simplistic version, acute cell death in the brain within hours leads to the activation of brain parenchymal and blood-borne immune cells and consequently the generation of toxic metabolites, such as free radicals, stressing the brain on top of the initial insult. After the acute phase, however, some proinflammatory cells shift their phenotype towards anti-inflammation (e.g., M1 → M2 polarization of macrophages), and other, primary anti-inflammatory and pro-regenerative cells (e.g., regulatory T cells, Tregs) take over, helping the brain to repair damage and recover function. This dichotomous concept is highly attractive, as it suggests that ill and beneficial effects of immune responses can be separated on a temporal scale. Anti-inflammation early on and modulation of inflammation towards "wound healing" later suggest itself as straightforward and promising therapeutic approaches. Various therapeutic agents (pharmacological and cellular) are ready to be tested, the only remaining challenge appears to develop and deploy noninvasive strategies (i.e., molecular imaging, such as TSPO-PET) to stratify patients to the right type of immune therapy.

Of Concepts and Misconceptions

But is it that simple? Can the outcome of an interaction between the two supersystems of the organism be either good or bad, can they be sometimes foes, and shortly thereafter friends again? There is nothing wrong in formulating reductionist biological concepts—they help to generate testable hypotheses in the face of overwhelming complexity. However, there is a risk that flawed concepts stick and may turn into dogmas. This has happened with several concepts which are relevant to our understanding of brain–immune interactions after injury. I will therefore briefly touch upon them.

The Immune Privilege of the Brain

The unique structure and function of the brain, the risk of erratic rewiring after damage and thus restricted capacity to regenerate, as well as its tight embedding into a bony structure limiting volume expansion necessitate protection against damage from inflammation. The organ has therefore developed tolerance against the introduction of antigens—the so-called immunological privilege [6]. However, this privilege is not absolute, and it is compartmentalized. An excellent treatment of this concept and misunderstandings associated with it can be found in Galea et al. [7]. In short, only the brain parenchyma has a tightly regulated immunosuppressive environment without an adaptive efferent arm of immunity. The ventricles (including choroid plexus and circumventricular organs), perivascular spaces, and meninges of the brain demonstrate responses of innate and adaptive immunity and antigen presentation very similar to peripheral sites. It should also be noted that the environment of the brain parenchyma rapidly loses its immunosuppressive capacity once inflammation has established itself after brain tissue damage. This is the result of blood–brain barrier breakdown, local production of chemoattractants and immunostimulants, and the appearance of dendritic or other antigen-presenting cells. Finally, although blood–brain barrier and relative immune privilege are linked, the one is not the primary consequence of the other. The immune privilege of the brain parenchyma results from a tightly regulated microenvironment and the lack of an efferent arm of adaptive immunity, rather than tight capillary endothelia.

The Blood–Brain Barrier and Leukocyte Trafficking

A "barrier" made of capillary tight junctions restricts the diffusion of molecules potentially disruptive for neurotransmission from the blood into the brain extracellular fluid and thus neuropil. However, this barrier is mostly restricted to the capillary bed, where no extravascular ("Virchow–Robin") space exists, as the basement membrane between endothelial cells and astrocytic endfeet of the glia limitans are fused—the so-called gliovascular or composite basement membrane. For an excellent treatment of the blood–brain barrier and a clarification of some prevalent misconceptions, the reader is referred to Bechmann et al. [8]. Importantly, leukocyte recruitment is a highly regulated process and does not normally involve the blood–brain barrier, as it occurs in postcapillary venules, where the cells first enter the Virchow–Robin space. Only a second step involving different molecular programs can take them into the neuropil, as they need to cross the basement membrane of the glia limitans. In other words, while solute movement in and out of the CNS is limited by properties of the endothelium, leukocyte migration is in addition hampered by extracellular matrix and membranes, which need to be actively degraded for passage [9, 10]. The lack of discrimination of the different barriers encountered by leukocytes in brain inflammation has confounded the literature. To understand the role of leukocytes in brain inflammation, we need to carefully locate and discriminate specific leukocyte subsets, such as neutrophils, monocytes, NK cells, T-cell subtypes, and B cells. A case in point is the dogma that polymorphonuclear leukocytes (PMNs) invade the brain parenchyma early after stroke, where their toxic products harm neurons. A recent study in experimental stroke and human neuropathological samples demonstrates, however, that after stroke the large majority of extravasated PMNs stay within the confines of the perivenular space and the meninges and do not gain access to the neuropil [11]. PMNs, therefore, appear to act at different sites than previously thought, which may at least partially explain the clinical failure of agents that block PMN infiltration and suggests alternative therapeutics targeting inflammation within the neurovascular unit.

Inflammation and Wound Healing

Tissue responses after brain lesions include resorption of debris, scar formation, and possibly attempted repair, and because of clear analogies have been compared to the tightly regulated process of wound healing in peripheral tissues, for example the skin. There the orchestrated response to injury includes elements of inflammation, such as leukocyte homing, in particular macrophage activity. Macrophages have been implicated in wound closure, reepithelialization, and angiogenesis. It should be noted, however, that even for wound healing in the periphery, the role of inflammation is still not clear. While some studies demonstrate disturbed wound healing by anti-inflammatory treatment or specific ablation of macrophages [12], others found normal wound healing (including angiogenesis) in the absence of inflammation [13]. It should be noted in this context that embryos demonstrate almost perfect wound healing without scarring, in the complete absence of inflammation [14]. Herz et al. [15] demonstrate that brain plasticity and repair after stroke can be fostered by anti-inflammatory therapy. Interestingly, however, Gliem et al. [16] found that bone marrow derived macrophages are critical for preventing hemorrhagic transformation of brain infarcts. Thored et al. [17] linked microglia accumulation to neurogenesis and repair after stroke. However, the same group went on to demonstrate that elimination of the microglia does not affect the neurogenic response [18]. This presents a nice illustration of the truism that correlation does not imply causation, which is unfortunately often overlooked, in particular regarding research on the relationship between brain and immune system (see below). The controversy surrounding the role of inflammation and repair or wound healing remains unresolved, and the reader is referred to the interesting debate in Crutcher et al. [19].

Correlation Versus Causation

Many experimental studies report pharmacologic manipulations, which lead to smaller infarcts via "anti-inflammatory" mechanisms. Claiming "anti-inflammatory action" quite often rests on the finding that giving the drug not only reduces damage but also many markers of inflammation, such as cytokines, influx of leukocytes, etc. Unfortunately, this conclusion is confounded by the problem that smaller infarcts (by whatever treatment) lead to a reduction of practically all mechanisms related to primary and secondary ischemic damage. For example, reducing infarct sizes by blocking the *N*-methyl-D-aspartate (NMDA) receptor (which is not found on cells of the immune system) via a reduction of tissue damage also leads to a reduction in secondary release of inflammatory cytokines or an influx of leukocytes into the affected hemisphere [20].

Resolution of Inflammation

Research on inflammation after brain lesion is highly focused on the mechanisms which induce and maintain inflammation, as well as its effects on tissue damage, protection, and repair. Surprisingly, little attention is devoted to the question how inflammation is terminated, and homeostasis is reestablished. This is achieved by an active, highly regulated process: resolution. In peripheral tissues, this process is well studied, and chemical mediators of resolution have been identified [21]. Resolution failure leads to chronic inflammation, with increased tissue injury and scarring. In the partially immune privileged CNS, resolution after injury may differ from other organs, and inflammation may in part be nonself-limiting [22]. Very little is known about resolution of inflammation after stroke and brain trauma, a field which deserves further inquiry as resolution agonists may be attractive therapeutics.

Open Issues and Future Challenges

Research of the last decades has clearly demonstrated that immunological responses to acute injury of the brain play an important role for tissue damage, protection, and repair. This research has also unraveled striking complexities in the interaction of brain and immune system: simple dichotomies, categorizing specific cells or responses as "good" or "bad" are no longer helpful [23]. We are beginning to understand the functional diversity of immune responses, which are highly context dependent. Numerous open issues remain (Table 1.1). The chapters of this volume explore these complex responses and the biological contexts in which they occur. They will also highlight a number of novel targets to inhibit secondary damage after stroke, brain trauma, or spinal cord injury. These targets include the induction or blockade of cytokines, subsets of cells of the innate and adaptive immune system, or pathways of communication between brain and immune system, such as the sympathetic and parasympathetic nervous systems.

Table 1.1 Exemplary open issues regarding immune responses to brain injuries

Can beneficial and detrimental effects of inflammation be disentangled? How can we noninvasively stratify patients to immunomodulatory therapies?
What is the role of specific types of T cells after brain injury? What is the role of antigen? If antigen presentation is needed, where does it occur? How can T cells damage neurons or regenerate neuronal function?
Does anti-inflammatory therapy affect the glial scar?
What are the sources of specific cytokines measured after brain injury in the blood?
Is immunodepression after brain injury an adaptive response? If so, does blocking it potentially exacerbate autoimmunity after brain injury?
Can adaptive immunity selectively be manipulated to protect or regenerate the brain?
How do comorbidities and aging affect immune responses after brain injury?
How do immune responses after stroke affect angio-, vasculo-, and neurogenesis?

Acknowledgment UD is supported by the European Union's Seventh Framework Program (FP7/2008-2013) under Grant Agreements 201024 and 202213 (European Stroke Network); the Deutsche Forschungsgemeinschaft (NeuroCure Cluster of Excellence, Exc 257); and the Bundesministerium für Bildung und Forschung (Center for Stroke Research Berlin, 01 EO 08 01).

References

1. Richelmi P (1811) Essai sur l'apoplexie, ou pathologie, séméiotique, hygiène et thérapeutique de cette maladie, considérée dans ses différentes espèces. Joseph-Francois Achard, Marseille. p 6 (footnote)
2. Tada T (1997) The immune system as a supersystem. Annu Rev Immunol 15:1–13
3. Prass K, Meisel C, Höflich C, Braun J, Halle E, Wolf T, Ruscher K, Victorov IV, Priller J, Dirnagl U, Volk HD, Meisel A (2003) Stroke-induced immunodeficiency promotes spontaneous bacterial infections and is mediated by sympathetic activation reversal by poststroke T helper cell type 1-like immunostimulation. J Exp Med 198(5):725–736
4. Woiciechowsky C, Asadullah K, Nestler D, Eberhardt B, Platzer C, Schöning B, Glöckner F, Lanksch WR, Volk HD, Döcke WD (1998) Sympathetic activation triggers systemic interleukin-10 release in immunodepression induced by brain injury. Nat Med 4(7):808–813
5. Riegger T, Conrad S, Schluesener HJ, Kaps HP, Badke A, Baron C, Gerstein J, Dietz K, Abdizahdeh M, Schwab JM (2009) Immune depression syndrome following human spinal cord injury (SCI): a pilot study. Neuroscience 158(3):1194–1199
6. Muldoon LL, Alvarez JI, Begley DJ, Boado RJ, Del Zoppo GJ, Doolittle ND, Engelhardt B, Hallenbeck JM, Lonser RR, Ohlfest JR, Prat A, Scarpa M, Smeyne RJ, Drewes LR, Neuwelt EA (2013) Immunologic privilege in the central nervous system and the blood-brain barrier. J Cereb Blood Flow Metab 33(1):13–21
7. Galea I, Bechmann I, Perry VH (2007) What is immune privilege (not)? Trends Immunol 28(1):12–18
8. Bechmann I, Galea I, Perry VH (2007) What is the blood-brain barrier (not)? Trends Immunol 28(1):5–11
9. Engelhardt B, Sorokin L (2009) The blood-brain and the blood-cerebrospinal fluid barriers: function and dysfunction. Semin Immunopathol 31(4):497–511
10. Sorokin L (2010) The impact of the extracellular matrix on inflammation. Nat Rev Immunol 10(10):712–723
11. Enzmann G, Mysiorek C, Gorina R, Cheng YJ, Ghavampour S, Hannocks MJ, Prinz V, Dirnagl U, Endres M, Prinz M, Beschorner R, Harter PN, Mittelbronn M, Engelhardt B, Sorokin L (2013) The neurovascular unit as a selective barrier to polymorphonuclear granulocyte (PMN) infiltration into the brain after ischemic injury. Acta Neuropathol 125(3):395–412
12. Mirza R, DiPietro LA, Koh TJ (2009) Selective and specific macrophage ablation is detrimental to wound healing in mice. Am J Pathol 175(6):2454–2462
13. Martin P, D'Souza D, Martin J, Grose R, Cooper L, Maki R, McKercher SR (2003) Wound healing in the PU.1 null mouse–tissue repair is not dependent on inflammatory cells. Curr Biol 13(13):1122–1128
14. Martin P, Dickson MC, Millan FA, Akhurst RJ (1993) Rapid induction and clearance of TGFß1 in response to wounding in the mouse embryo. Dev Genet 14:225–238
15. Herz J, Reitmeir R, Hagen SI, Reinboth BS, Guo Z, Zechariah A, ElAli A, Doeppner TR, Bacigaluppi M, Pluchino S, Kilic U, Kilic E, Hermann DM (2012) Intracerebroventricularly delivered VEGF promotes contralesional corticorubral plasticity after focal cerebral ischemia via mechanisms involving anti-inflammatory actions. Neurobiol Dis 45(3):1077–1085
16. Gliem M, Mausberg AK, Lee JI, Simiantonakis I, van Rooijen N, Hartung HP, Jander S (2012) Macrophages prevent hemorrhagic infarct transformation in murine stroke models. Ann Neurol 71(6):743–752

17. Thored P, Heldmann U, Gomes-Leal W, Gisler R, Darsalia V, Taneera J, Nygren JM, Jacobsen SE, Ekdahl CT, Kokaia Z, Lindvall O (2009) Long-term accumulation of microglia with proneurogenic phenotype concomitant with persistent neurogenesis in adult subventricular zone after stroke. Glia 57(8):835–849
18. Heldmann U, Mine Y, Kokaia Z, Ekdahl CT, Lindvall O (2011) Selective depletion of Mac-1-expressing microglia in rat subventricular zone does not alter neurogenic response early after stroke. Exp Neurol 229(2):391–398
19. Crutcher KA, Gendelman HE, Kipnis J, Perez-Polo JR, Perry VH, Popovich PG, Weaver LC (2006) Debate: "is increasing neuroinflammation beneficial for neural repair?". J Neuroimmune Pharmacol 1(3):195–211
20. Dirnagl U (2010) Chapter 19: Complexities, confounders, and challenges in experimental stroke research: a checklist for researchers and reviewers. In: Dirnagl U (ed) Rodent models of stroke, vol 47, Springer protocols neuromethods. Springer, New York, NY, pp 263–277
21. Schwab JM, Chiang N, Arita M, Serhan CN (2007) Resolvin E1 and protectin D1 activate inflammation-resolution programmes. Nature 447(7146):869–874
22. Prüss H, Kopp MA, Brommer B, Gatzemeier N, Laginha I, Dirnagl U, Schwab JM (2011) Non-resolving aspects of acute inflammation after spinal cord injury (SCI): indices and resolution plateau. Brain Pathol 21(6):652–660. doi:10.1111/j.1750-3639.2011.00488.x
23. Dirnagl U (2004) Inflammation in stroke: the good, the bad, and the unknown. In: Dirnagl U, Elger B (eds) Neuroinflammation in stroke, vol 47, Ernst schering research foundation workshop series. Springer, New York, NY, pp 88–99

Chapter 2
The Critical Roles of Immune Cells in Acute Brain Injuries

Peiying Li, Yu Gan, Leilei Mao, Rehana Leak, Jun Chen, and Xiaoming Hu

Abstract Acute brain injuries elicit prompt and robust immune responses characterized by the activation of local glial cells and mobilization of peripheral leukocytes. The activation of immune cells originally aims to clear the brain of cellular debris and promote brain repair; however, the immune system can also propel and propagate neuronal cell death when overactivated. Understanding the function of each type of immune cells in the acute brain injuries and their mechanisms of action promises to unveil effective immunomodulatory therapies that beneficially regulate post-injury immune responses. In this chapter, we discuss in detail how immune cells are recruited and/or activated in the injured brain and how they contribute to the evolvement of brain damage.

Introduction

A pivotal role of immune responses in the pathogenesis of acute brain injuries has emerged in recent years. Once an injury occurs, the brain and the immune system influence each other in specific and profound ways. Bidirectional or reciprocal neuro-immune communication presumably evolved to clear the brain of infections and dead cellular debris. However, in addition to its essential role in protecting the organism from harmful microbes and the necrotic spillage of intracellular contents, the immune system can also propel and propagate neuronal cell death when overactivated. In order to elicit activation of the immune system, injured neurons and other central

P. Li • Y. Gan • L. Mao • J. Chen • X. Hu (✉)
Department of Neurology, University of Pittsburgh, Pittsburgh, PA, USA
e-mail: hux2@upmc.edu

R. Leak
Pharmaceutical Sciences, Duquesne University, Pittsburgh, PA, USA

J. Chen et al. (eds.), *Immunological Mechanisms and Therapies in Brain Injuries and Stroke*, Springer Series in Translational Stroke Research, DOI 10.1007/978-1-4614-8915-3_2,

nervous system (CNS) cells release ATP and neuronal antigens. Other small molecules, such as cytokines, chemokines, and adhesion molecules, also convey an activation signal to the immune system. In addition, the autonomic nervous system (ANS) releases neurotransmitters such as acetylcholine (Ach) and norepinephrine from the compromised brain. The innate and adaptive immune systems promptly respond to these signals and participate in the development of secondary inflammatory brain injury. Therefore, revealing the paths of communication between the brain and immune system promises to unveil effective immunomodulatory therapies that break the vicious positive cycle of neuroimmune activation. In this chapter, we discuss the intricate dialogue between the brain and immune system after acute brain injuries, with an emphasis on acute ischemic stroke. How immune cells such as microglia and dendritic cells (DCs) are activated and how monocytes, neutrophils, natural killer (NK) cells, T cells, and B cells infiltrate deep into the brain and contribute to neuroinflammatory damage will all be discussed in detail.

The Response of the Innate Immune System Following Cerebral Injury

The innate immune system is the first line of defense that recognizes and responds to pathogens or injuries in a nonspecific manner. Unlike the adaptive immune system, it does not provide long-lasting protection to the host. The cells involved in the innate immune response to the cerebral injury include macrophages, neutrophils, DCs, NK cells, and γδT cells (Fig. 2.1).

Bone Marrow Monocytes

Microglial Activation

Microglia are a specialized type of macrophage residing in the CNS. Unlike other CNS cells, which are derived from neuroectoderm, microglia are derived from bone marrow monocytes and enter the CNS after birth. Therefore, microglia inextricably link the CNS with the immune system.

Microglial activation is the initial step in CNS inflammatory responses induced by acute brain injuries. For instance, activation of microglia after cerebral hemorrhage occurs much earlier than infiltration of neutrophils in and around the hematoma; the former occurs within 1 h, whereas the latter occurs after 4–5 h [1]. In response to insults, microglia, which are normally ramified in appearance, become activated and assume an amoeboid morphology. Meanwhile, the expression of surface markers, such as MHC-II, Iba-1, CD11, etc., is also dysregulated. Both the morphology and the surface markers of microglia are virtually indistinguishable from infiltrating macrophages/monocytes, reflecting their common heritage.

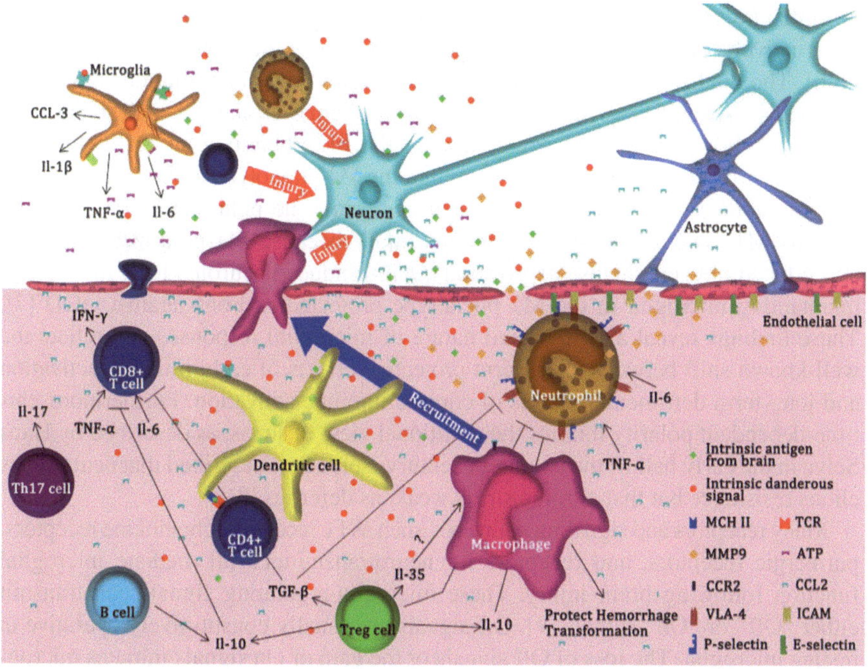

Fig. 2.1 Innate and adaptive immune response after stroke. In response to ischemic brain injury, a variety of danger signals are released from neurons, microglia, and astrocytes. These signals include ATP, cytokines, and chemokines, among other molecules. These signals activate microglia and induce their activation and further release of inflammatory cytokines, such as TNF-α and IL-6. Activated microglia can also release chemokines, such as CCL2 and CCL3, to recruit leukocyte infiltration. The CCL2 and CCR2 axis is responsible for the recruitment of macrophages. The infiltration of macrophages has been implicated as both detrimental to ischemic injury and protective against hemorrhagic transformation. In the ischemic brain, inflamed endothelial cells express adhesion molecules, such as ICAM-1, P-selectin, and E-selectin, which recruit leukocyte adhesion and infiltration. Following stimulation by IL-6 and TNF-α, neutrophils are activated and release MMP-9, which degrades matrix proteins of the blood–brain barrier and stimulates peripheral leukocyte infiltration. Brain-derived antigens can be processed by antigen-presenting cells, such as dendritic cells, and presented by MHC molecules on the cell surface. T-cell receptors on the surface of T cells can recognize the MHC and brain–antigen complex. Subsequently, the adaptive immune system is activated. $CD4^+$ T-helper cells, $CD8^+$ cytotoxic T cells, B cells, and Th17 cells all have been reported to contribute to the pathogenesis of neuroinflammation and neuronal cell death following stroke. Regulatory T cells (Tregs) and a special subset of regulatory B cells have been demonstrated to be neuroprotective through their release of IL-10. IL-35 and TGF-β are two other anti-inflammatory soluble factors which can be secreted by Tregs. Tregs can also inhibit MMP-9 production from neutrophils in a cell–cell contact manner

Microglial activation is not a uniform and monotonous state, as the morphological and gene expression changes associated with microglial activation vary enormously with the nature, strength, and duration of the stimulus [2]. In particular, in vitro stimulation with lipopolysaccharide (LPS) and IFN-γ promotes the differentiation of "classically activated" M1 microglia that typically release destructive pro-inflammatory mediators [3]. In contrast, interleukin (IL)-4 and IL-10 induce an "alternatively activated" M2 phenotype that possesses neuroprotective properties [4–6]. The dualistic roles of distinctly polarized microglial populations have been reported in several CNS injuries, including ischemic stroke [7] and spinal cord injury [4]. For instance, microglia assume the M2 phenotype at early stages of ischemic stroke but gradually transform into the M1 phenotype at the sites of injury. In vitro studies reveal that ischemic neurons prime the polarization of microglia towards M1. This is thought to enhance injury because M1-polarized microglia exacerbated oxygen glucose deprivation (OGD)-induced neuronal loss. In contrast, maintaining microglia in the M2 phenotype protected neurons against OGD [7]. These findings reveal a Janus-faced nature of microglial responses and reflect the well-known shift between short-term protective effects of early immune activation and long-term detrimental effects of chronic immune activation. Furthermore, this time-dependent polarity mimics the traditional view of stress, as described by Hans Selye in the early half of the twentieth century. Selye first described that acute stress elicits resistance but that chronic stress weakens defenses [8].

Many receptors and signaling pathways, such as Fc receptors, chemokine receptors, purinergic receptors, and receptors for neurotransmitters, all mediate microglial function following brain injury. These microglia-regulating signals can transmit either "Off" or "On" signals [9]. Off signals are usually constitutively operative in resting microglia. The loss of Off signals or the gain of On signals initiates microglial activation. This context-dependent activation orchestrates the complex and variable microglial responses to stresses.

Monocyte/Macrophage Infiltration

Monocytes are produced by bone marrow from monoblasts and mature into different types of macrophages or DCs. It is well established that peripheral monocyte/macrophage cells are attracted to infarct areas after cerebral ischemia reperfusion [10]. A recent study showed that the shift of spleen monocytes to a less pro-inflammatory state attenuated infarct volume after transient focal ischemia, suggesting a role for peripheral monocytes in the pathogenesis of cerebral ischemia [11]. Similarly, within 24 h of stroke, immature monocytes infiltrated into the infarct border zone and differentiated into mature phagocytes within the lesion compartment [12]. However, infiltrating macrophages are morphologically and functionally similar to reactive resident microglia that transform into a phagocytic phenotype [13, 14]. To distinguish the roles and distributions of microglia and peripheral macrophages in ischemic brain injury, some studies used bone marrow chimeric mice generated by transplanting green fluorescent protein (GFP) transgenic bone marrow into irradiated wild-type recipients [13–16]. Although the infiltration of hematogenous macrophages into the brain was shown to occur 1–2 days after focal cerebral ischemia, their number was much lower than the resident population of activated microglia [13, 17]. Peripheral macrophages were most abundant in the ischemic brain tissue 3–7 days after transient focal cerebral ischemia and decreased thereafter [14, 18]. Anatomical studies revealed that, at 7 days after transient ischemia, many process-bearing ramified blood-derived macrophages were distributed in the peri-infarct area, while phagocytic cells were detected in the core area of the infarction [13]. The majority of phagocytes in the infarct area were derived from local microglia, which were rapidly activated 1 day after ischemia and reached peak numbers within 2 days [15]. Taken together, most studies indicate that local microglial activation precedes and predominates over peripheral monocyte/macrophage infiltration for the first few days following cerebral ischemic injury, whereas blood-derived macrophages contribute to the delayed post-ischemic inflammation and brain injury. Mechanistically, monocyte chemoattractant protein-1 (MCP-1) and its receptor CC chemokine receptor 2 (CCR2) are involved in the migration of hematogenous inflammatory cells, including macrophages, following cerebral ischemia. Either MCP-1- or CCR-2-deficient mice exhibit reduced numbers of infiltrating macrophages as well as neutrophils [19]. More recently, it was shown that MCP-1/CCR-2 double deficiency virtually abrogates the recruitment of hematogenous macrophages [20], indicating a predominant role of the MCP-1/CCR-2 axis in chemotaxis of monocyte/macrophages after cerebral ischemia. MCP-1 is also implicated in macrophage recruitment into the damaged parenchyma after TBI. For example, sustained elevation of MCP-1 was detected in the cerebrospinal fluid (CSF) of severe TBI patients for 10 days after trauma and in cortical homogenates of C57Bl/6 mice subjected to closed head injury (CHI), peaking at 4–12 h [21].

Neutrophils

The infiltration of neutrophils into lesioned areas of the ischemic brain has been demonstrated not only in animal models [13, 18] but in stroke patients as well [22]. Neutrophil accumulation in injured parenchyma correlated with infarct expansion [22]. Higher peripheral neutrophil counts were associated with more severe stroke outcomes [23]. In rodent models of transient focal ischemia, depletion of neutrophils by treatment with anti-neutrophil monoclonal antibody (RP3) significant reduced the size of infarct as well as the formation of brain edema [24, 25], indicating an important role of neutrophil infiltration in post-ischemic brain injury.

Recent data support the notion that neutrophils infiltrate into the injured brain more rapidly than other types of peripheral inflammatory cells. This pattern in the brain closely mimics the leukocyte response to peripheral injuries, where neutrophils lead the way out of blood vessels into damaged tissue and are known as the first responders. Neutrophil extravasation into damaged peripheral tissues is followed by monocyte/macrophages, also called the second responders. In contrast to monocytes, which are most abundant in the injured parenchyma 3–7 days after ischemia, neutrophils are observed within a few hours and peak at 1–3 days after cerebral ischemia in animal models of stroke [26]. In stroke patients, neutrophil recruitment was also demonstrated to occur within 24 h of symptom onset [22]. Cellular adhesion molecules, including ICAM-1 and P-selectin, appear to be involved in the transendothelial migration of neutrophils into the brain similar to their role in peripheral tissues. Either ICAM-1-deficient or P-selectin-deficient mice exhibit less neutrophil infiltration as well as smaller infarct volumes following acute stroke compared with wild-type mice [27–29]. The MCP-1/CCR-2 chemotactic axis is also thought to contribute to post-ischemic neutrophil infiltration, as deficiency in either one or both significantly decreases the numbers of neutrophils recruited to the site of injury [19, 20]. These findings reveal many commonalities between peripheral and central immune responses.

Recruitment of neutrophil granulocytes is also characteristic of the early inflammatory response following human TBI. Neutrophil recruitment has been shown to increase over the first 24 h after experimental TBI and is dependent on both leukocyte CD11/CD18 and ICAM-1. Neutrophil recruitment is also involved in edema formation, cell death, and tissue loss following TBI in mice [30]. Recently, Victor Friedrich and his colleagues implicated neutrophils in early microvascular injury after subarachnoid hemorrhage (SAH) and showed that treatments which reduce neutrophil activity can increase survival after SAH and limit microvascular injury [31]. These studies support the notion that the neuroimmune responses to stroke and TBI are somewhat mechanistically similar and play the same negative roles in pathological outcome.

Infiltrating neutrophils, as well as microglia and macrophages, are able to release toxic amounts of nitric oxide (NO) via the inducible nitric oxidase (iNOS) isoform. The release of this free radical gas may contribute to post-ischemic brain damage [10] via multiple mechanisms including activated matrix metalloproteinase-9 (MMP-9) [32]. Many studies have highlighted the involvement of MMPs in ischemic pathophysiology. For example, MMP-9 is strongly implicated in the disruption of the blood–brain barrier (BBB) and hemorrhagic transformation following ischemic injury both in rodent models [33, 34] and in stroke patients [35, 36]. MMP-9 is elevated after stroke both in plasma and brain tissue [37] and participates in the neuroinflammatory response to stroke [38]. Importantly, current research suggests that neutrophils might be the main source of MMP-9 [37]. Neutrophils greatly contribute to elevations in MMP-9 following cerebral ischemia [39], which, in turn, causes BBB breakdown, promotes hematogenous leukocyte infiltration, and ultimately results in neuronal injury [40]. These results indicate that neutrophil inhibition might be a reasonable peripheral target for stroke treatment in the clinic.

Natural Killer Cells

Natural killer (NK) cells are cytotoxic lymphocytes critical to the innate immune system. An early study indicated the presence of $CD4^-/CD8^+$ NK cells at the infarct edge following permanent focal cerebral ischemia [41]. However, a recent study showed no significant change in the numbers of infiltrating NK cells 3 days after transient focal cerebral ischemia, although there was infiltration of neutrophils, macrophages, T lymphocytes, and even NKT cells [42]. In contrast, there is a significant decrease in blood NK cell counts [43], as well as peripheral NK cytotoxicity [44] in stroke patients. However, no change in NK cell counts was observed in gut-associated lymphoid tissue after experimental stroke in animal models [45]. Together, these studies indicate a role for NK cells in stroke-induced systemic immunosuppression rather than in the pathogenesis of post-ischemic brain injury.

γδT Cells

γδT cells represent a minor subset of T cells that possess a nonclassic T-cell receptor. The T-cell receptor of γδT cells is composed of two glycoprotein chains, one γ chain and one δ chain. Because they do not require antigen processing and antigen presentation by MHC, γδT cells are considered part of innate immunity. Many studies have demonstrated that γδT cells play a sentinel role in the early host response against many infectious agents [46, 47]. Direct evidence demonstrating their involvement in the progression of ischemic brain damage comes from the observation that depletion of γδT cells significantly ameliorated ischemia–reperfusion injury [48]. It was also shown that IL-17 is the key mediator of γδT cell-mediated delayed brain damage following cerebral ischemia–reperfusion [48]. These results suggest that γδT cells might be a rational therapeutic target for mitigating secondary inflammation-mediated damage following cerebral ischemia. However, it was also shown that the pro-inflammatory response of peripheral γδT cells was acutely diminished in stroke patients and that the subsequent increases in IFN-γ and perforin expression by γδT cells correlated well with clinical improvement [44]. Therefore, therapeutic strategies targeting γδT cells will have to be carefully timed to avoid undesired immunosuppression. This is not an unusual requirement, because all immunosuppressive therapies will have to ultimately ensure that the treatment is not accompanied by undesired side-effects such as increased vulnerability to microbial infections.

The Link Between the Innate and Adaptive Immune Response: APCs

Recently, a novel concept, the immunological synapse, was proposed to describe the structure formed by cytoskeletal molecules that assemble at the zone of contact between antigen-presenting cells (APCs) and T cells. The immunological synapse is associated with surface and cytoplasmic signaling [49]. Studies have shown that several hours of interaction between APCs and T cells were necessary for T-cell activation [50]. Biophysical analyses and two-photon microscopy studies confirmed immunological synapse formation at the site of robust interactions between APCs and T cells [50, 51].

There are professional APCs, including DCs, macrophages, B cells, and certain activated epithelial cells, and nonprofessional APCs, such as glial cells in the brain and vascular endothelial cells. Following stroke, autoimmune responses directed against neuroepitopes, which are not recognized by T cells under normal conditions, may be induced by cerebral ischemic injury. DCs, which have the broadest range of antigen presentation, might play an important role in the primary immune response against neuroepitopes after stroke [52].

There are two lineages of DCs: mDCs (myeloid DCs), which respond to bacteria and fungi and release IL-12, and plasmacytoid DCs (pDCs), which release IFN-α upon viral infection [53]. The precursor cells of both lineages can be detected in blood where these immature DCs patrol the circulation and invade tissue in response to a local infection. Upon migration into tissue, they pick up antigens and acquire the ability to stimulate T cells and subsequently induce the adaptive immune response against antigens not previously encountered [54], such as the cerebral antigens sequestered in the CNS under normal conditions.

A rapid increase in the levels of DCs in the ischemic brain was first observed in a rat model of permanent focal cerebral ischemia [55]. A subsequent study analyzed in detail the temporal dynamics of brain immune cell accumulation in a mouse model of transient focal ischemia [42]. This study revealed a significant increase in DCs in the ipsilateral brain at 1 day after cerebral ischemia/reperfusion, followed by a 20-fold increase on day 3 and a 12-fold increase on day 7 [42]. As there is already a DC population residing in the healthy brain, Felger et al. used radiation chimeras (wild-type hosts restored with CD11c/EYFP transgenic bone marrow or CD11c/EYFP transgenic hosts restored with wild-type bone marrow) to distinguish between brain original and peripheral DC [56]. Their results demonstrated DC infiltration in the ischemic hemisphere beginning 1 day after transient focal ischemia, and further indicated a role of original brain DCs in post-ischemic neuroinflammation [56]. On the other hand, a significant decrease in circulating DC precursors [57], consistent with a reduced costimulatory efficacy of circulating cells [58], was observed after human or experimental stroke. These results suggest that peripheral DCs are recruited from blood into the ischemic brain, which probably either triggers post-ischemic cerebral immune reactions or results in systemic immunosuppression. However, the role of DCs (both peripheral and brain original DCs) as potent mediators of inflammation in stroke has not been sufficiently investigated.

The Adaptive Immune Response

Unlike the innate immune system, adaptive immunity refers to antigen-specific defense mechanisms that require a relatively long time (several days) to become protective. Once the adaptive immune response has been established, more specific defense can be achieved. With the ability to remember specific antigens, the adaptive immune system can mount immediate, strong attacks each subsequent time that it encounters the same antigen. This powerful system of defense appeared first in jawed vertebrates and is phylogenetically more modern than the ancient innate immune system. The cells of the adaptive immune system include the T and B lymphocytes, both of which are related to the post-injury inflammatory response. Below we discuss the involvement of both cell types in the immune response following acute brain injuries.

T Lymphocytes

T lymphocytes play a central role in cell-mediated adaptive immunity. Accumulating evidence has verified the involvement of T lymphocytes in post-ischemic neuroinflammation and brain damage. Several earlier studies indicated that the infiltration of T lymphocytes into ischemic brain occurred relatively late (3–4 days) post-ischemia, following $CD11b^+$ microglia/macrophages and $Ly6G^+$ neutrophils [59, 60]. However, there is now mounting evidence showing that T-cell accumulation in the post-ischemic brain actually begins within the first 24 h of reperfusion in rodent models [61–63]. Treatment with anti-α4-integrin antibody blocked T lymphocyte recruitment into the post-ischemic brain and reduced infarct volume after transient cerebral ischemia in rats [64, 65]. In addition, mice lacking RANTES (also called CCL5), which plays a critical role in recruiting and activating T lymphocytes, exhibited smaller infarct volumes than wild-type littermates [66]. The most important evidence for the damaging effects of T lymphocytes arise from studies using lymphocyte-deficient mice. Smaller infarct volumes and improved functional outcomes after transient cerebral ischemia were consistently shown in severe combined immunodeficiency (SCID) mice [67] or $Rag1^{-/-}$ mice that lack mature B and T lymphocytes [62, 68]. Moreover, the lymphocyte deficiency-mediated neuroprotection was almost completely abolished when wild-type $CD3^+$ T lymphocytes were transplanted into $Rag1^{-/-}$ mice [62, 68], confirming a detrimental pathophysiologic role for circulating T lymphocytes in brain ischemia/reperfusion injury. Interestingly, a recent study showed that the detrimental effect of T cells neither depends on antigen recognition nor T-cell receptor (TCR) costimulation in the early stages of ischemic stroke, suggesting that classic adaptive immune pathways may be not involved in T-cell-mediated early ischemia/reperfusion injury. Recently, studies of TBI suggest that scavenging of reactive oxygen species (ROS) at the endothelial level dramatically reduced the infiltration of activated T lymphocytes [69]. Thus, targeting T lymphocyte trafficking to the injured brain at the microvascular level is a novel neuroprotective concept in TBI and warrants further exploration [69].

In recent years, there has been an increase in research on the roles of specific T-cell subtypes in brain injuries. Several subtypes, such as $CD8^+$ cytotoxic T cells and $CD4^+$ helper T cells, have been implicated in the pathogenesis of brain injury [62]. However, not all T-cell subtypes have detrimental effects after ischemic stroke. For example, neutralizing regulatory T cells with anti-CD25 antibody significantly increased infarct volume and neurological dysfunction, implicating an essential role for regulatory T cells in limiting post-ischemic brain injury [70]. Below, we will discuss the different effects of T-cell subtypes after brain injuries.

$CD8^+$ Cytotoxic T Cells and $CD4^+$ Helper T Cells

The unique expression profiles of specific cell-membrane proteins can be used to subtype different T-cell groups. The $CD3^+$ T lymphocytes mainly encompass both $CD8^+$ cytotoxic T cells and $CD4^+$ helper T cells in approximately equal proportions [71]. As the name suggests, the $CD8^+$ cytotoxic T cells are capable of directly inducing the death of intercellular pathogen-infected somatic cells or tumor cells either through the release of cytotoxins, perforin, and granzymes, or via cell–surface interactions such as Fas–FasL pathway [72, 73]. The $CD8^+$ cytotoxic T cells can also produce several cytokines, including IFN-γ and TNF-α, which serve to promote immune and inflammatory response [74]. In a rat TBI model, significant $CD8^+$ cell accumulation was observed 3 days post-injury. The $CD8^+$ cells were strictly distributed in the pan-necrotic areas and around the pan-necrotic perimeter [75]. The accumulation of activated antigen-specific T cells at traumatic injury sites, in addition to antigen-containing areas, could amplify local inflammatory processes in the CNS [75].

Although both $CD4^+$ and $CD8^+$ T cells were recruited to the ischemic hemisphere following stroke, there are greater numbers of $CD4^+$ helper T cells than $CD8^+$ cytotoxic T cells [42, 76]. The $CD4^+$ helper T cells have no cytotoxic activity themselves, but instead help to activate and direct other immune cells including $CD8^+$ cytotoxic T cells [77]. They can be further differentiated into several types according to their cytokine secretion profiles. Th1 and Th2 cells are two classic subgroups of helper T cells. Importantly, either $CD4^+$ or $CD8^+$ T-cell-deficient mice exhibited significant smaller infarct volumes than wild-type controls [62]. Mice deficient in these T-cell subtypes showed comparable reductions in neurological deficits at 24 h after ischemia, although the difference did not reach statistical significance [62]. This was confirmed by a recent study showing that selective antibody-mediated depletion of $CD4^+$ or $CD8^+$ T lymphocytes significantly reduced infarct volumes in mice with permanent cerebral ischemia [78]. Therefore, these findings demonstrate a detrimental role for both $CD4^+$ helper and $CD8^+$ cytotoxic T cells in the development of brain injury following stroke.

Stroke induces similar changes in peripheral $CD4^+$ helper and $CD8^+$ cytotoxic T cells. There is a stroke-induced reduction in numbers of both T-cell subtypes in spleen, thymus, lymph nodes, and gut-associated lymphoid tissue [45, 79–82]. This phenomenon may be partially due to an increased apoptosis of either $CD4^+$ or $CD8^+$ T cells in lymphatic organs following stroke and is mediated by the SNS [82]. Moreover, there is a shift from Th1 to Th2 cytokine production [82, 83], which also may suppress pro-inflammatory immune responses [84]. These findings indicate that T-cell subtypes are involved in stroke-induced immunosuppression.

Regulatory T Cells

The regulatory T cell (Treg), identified by expression of CD25 and the transcription factor FoxP3, is an important $CD4^+$ T-cell subtype that accounts for about

10 % of peripheral $CD4^+$ T cells [85]. The mechanisms by which Tregs exert their immunosuppressive function are still not clear; however, Tregs generally possess a diverse arsenal of inhibitory mechanisms, ranging from secretion of inhibitory cytokines, such as TGF-β, IL-10 [70], and a novel anti-inflammatory cytokine, IL-35 [86, 87], to the expression of inhibitory surface molecules, such as CTLA-4 and GITR [88]. With their vigorous immunosuppressive function, Tregs play critical roles in the maintenance of immunologic self-tolerance and negative control of a number of physiological and pathological immune responses [89]. For example, Tregs may be a critical factor in controlling experimental autoimmune encephalomyelitis and Parkinson's disease. Expansion of Tregs by prophylactic infusion of immunoglobulin prevented the development of experimental autoimmune encephalomyelitis and the protection was associated with increases in peripheral Treg number and function. Depletion of Tregs abrogated the protection of intravenous immunoglobulin, which strongly suggested that Tregs mediated the protection [90]. Adoptive transfer of Tregs to MPTP-intoxicated mice attenuated Th17 cell-mediated nigrostriatal dopaminergic neurodegeneration and provided significant protection of the nigrostriatal system [91, 92].

In the context of cerebral ischemia, the activation of regulatory T cells appears to be one of the intrinsic mechanisms that the body naturally uses to restrict the cerebral inflammation induced by ischemic stroke. Stroke induces a significant increase in peripheral regulatory T cells several days after the onset of stroke in patients [93] and in experimental models [94]. With regard to cell function, it was recently reported that the suppressive effect of Tregs in mouse and human was unaltered after stroke [58]. Thus, an increase in regulatory T-cell numbers, rather than enhanced function of each cell, may be responsible for the endogenous immunosuppression after stroke.

Expansion of endogenous Tregs might underlie the mechanism of protection afforded by MBP tolerance and E-selectin tolerance [95, 96]. This notion is supported by the following findings. Rats tolerant of MBP were less likely to develop Th1 responses compared to ovalbumin-tolerant rats following experimental stroke [95]. Mucosal tolerance to E-selectin protected against stroke in spontaneously hypertensive rats [96]. The protective effect of endogenous Tregs after stroke occurs by limiting secondary cerebral inflammatory responses and was further confirmed by a recent study showing that depletion of endogenous Tregs profoundly increased delayed brain damage and neurological dysfunction after brain ischemia [70]. Controversial evidence exists showing that Treg depletion failed to affect brain infarct volume [97]. However, our recent study confirmed neuroprotective effects of Tregs in experimental stroke. We showed that delayed post-stroke transplantation of Tregs robustly reduced brain damage and improved long-term neurological outcomes [98]. The neuroprotective effect of endogenous Tregs was previously shown to be associated with their ability to attenuate the activation of resident and invading inflammatory cells via releasing IL-10 [70]. In our study, we demonstrated that adoptively transferred Tregs reduced inflammatory responses both intrinsic and extrinsic to the central nervous system in rodent models of transient focal cerebral ischemia. Moreover, Tregs provided neurovascular protection against stroke by inhibiting peripheral neutrophil-derived MMP-9. These findings all suggest that Tregs might be promising candidates for cell-based therapies targeting post-stroke inflammatory dysregulation and neurovascular disruption [98].

B Lymphocytes

Currently, there are few studies investigating the role of B lymphocytes in cerebral ischemia/reperfusion injury. In various peripheral organs ranging from the intestines, heart, kidney, to skeletal muscles, the B lymphocyte has been found to be predominantly pathogenic following ischemia/reperfusion injury [99]. However, current data suggest the opposite role for B cells in cerebral ischemic injury. Although T- and B-cell-deficient mice, either $Rag^{-/-}$ [62] or SCID [67], exhibited significantly reduced infarct sizes and neurologic damage when subjected to transient MCAO, mice lacking only B cells failed to show improvement against ischemic injury [62]. Recently, another study showed B-cell deficiency worsened histological damage and functional outcomes after transient cerebral ischemia [100]. Adoptive transfer of B cells to B-cell-deficient mice reduced ischemic infarct size and improved neurological deficits [100]. The authors further identified IL-10-secreting regulatory B cells as a major regulatory cell type in stroke. Notably, the secretion of IL-10 was responsible for the B-cell-mediated neuroprotection [100]. These data suggest that the role of B lymphocytes might be protective rather than pathogenic in cerebral ischemia/reperfusion injury.

Conclusion

In summary, an exquisitely coordinated crosstalk between the CNS and peripheral immune system regulates the fate of the animal after acute brain injuries. In addition to the conventional idea that the brain signals danger to the periphery, recent studies also reveal multiple novel neural circuits through which brain injury evokes a robust immune response. Brain injury-induced immune responses used to be considered innate or antigen nonspecific. Recent work on $CD8^+$ cytotoxic T cells, $CD4^+$ helper T cells, B cells, and regulatory T cells have refreshed scientific thought by revealing that adaptive immunity also plays pivotal roles in secondary brain injury (Fig. 2.1). With these new findings in neuroimmune communication, therapeutic candidates to modulate immune responses and thereby improve stroke outcome may lie on the horizon. Moreover, it seems reasonable to propose that immunomodulation through immune cell transplantation will improve brain injuries more profoundly than immunosuppressive drugs because immune cell transplantation harnesses the protective power of the body's natural defenses. Further research on immunomodulatory therapies that leverage our current knowledge of brain injuries is highly warranted.

References

1. Wang J, Dore S (2007) Heme oxygenase-1 exacerbates early brain injury after intracerebral haemorrhage. Brain 130(Pt 6):1643–52
2. Ransohoff RM, Perry VH (2009) Microglial physiology: unique stimuli, specialized responses. Annu Rev Immunol 27:119–45
3. Ding AH, Nathan CF, Stuehr DJ (1988) Release of reactive nitrogen intermediates and reactive oxygen intermediates from mouse peritoneal macrophages. Comparison of activating cytokines and evidence for independent production. J Immunol 141(7):2407–12
4. Kigerl KA et al (2009) Identification of two distinct macrophage subsets with divergent effects causing either neurotoxicity or regeneration in the injured mouse spinal cord. J Neurosci 29(43):13435–44
5. Goerdt S et al (1999) Alternative versus classical activation of macrophages. Pathobiology 67(5–6):222–6
6. Durafourt BA et al (2012) Comparison of polarization properties of human adult microglia and blood-derived macrophages. Glia 60(5):717–27
7. Hu X et al (2012) Microglia/macrophage polarization dynamics reveal novel mechanism of injury expansion after focal cerebral ischemia. Stroke 43(11):3063–70
8. Selye H (1998) A syndrome produced by diverse nocuous agents, 1936. J Neuropsychiatry Clin Neurosci 10(2):230–1
9. Biber K et al (2007) Neuronal 'On' and 'Off' signals control microglia. Trends Neurosci 30(11):596–602
10. Amantea D et al (2009) Post-ischemic brain damage: pathophysiology and role of inflammatory mediators. FEBS J 276(1):13–26
11. Bao Y et al (2010) A role for spleen monocytes in post-ischemic brain inflammation and injury. J Neuroinflammation 7:92
12. Gliem M et al (2012) Macrophages prevent hemorrhagic infarct transformation in murine stroke models. Ann Neurol 71(6):743–52
13. Tanaka R et al (2003) Migration of enhanced green fluorescent protein expressing bone marrow-derived microglia/macrophage into the mouse brain following permanent focal ischemia. Neuroscience 117(3):531–9
14. Schilling M et al (2003) Microglial activation precedes and predominates over macrophage infiltration in transient focal cerebral ischemia: a study in green fluorescent protein transgenic bone marrow chimeric mice. Exp Neurol 183(1):25–33
15. Schilling M et al (2005) Predominant phagocytic activity of resident microglia over hematogenous macrophages following transient focal cerebral ischemia: an investigation using green fluorescent protein transgenic bone marrow chimeric mice. Exp Neurol 196(2):290–7
16. Schilling M et al (2009) Effects of monocyte chemoattractant protein 1 on blood-borne cell recruitment after transient focal cerebral ischemia in mice. Neuroscience 161(3):806–12
17. Kokovay E, Li L, Cunningham LA (2006) Angiogenic recruitment of pericytes from bone marrow after stroke. J Cereb Blood Flow Metab 26(4):545–55
18. Breckwoldt MO et al (2008) Tracking the inflammatory response in stroke in vivo by sensing the enzyme myeloperoxidase. Proc Natl Acad Sci USA 105(47):18584–9
19. Schilling M et al (2009) The role of CC chemokine receptor 2 on microglia activation and blood-borne cell recruitment after transient focal cerebral ischemia in mice. Brain Res 1289:79–84
20. Schuette-Nuetgen K et al (2012) MCP-1/CCR-2-double-deficiency severely impairs the migration of hematogenous inflammatory cells following transient cerebral ischemia in mice. Exp Neurol 233(2):849–58
21. Semple BD et al (2010) Role of CCL2 (MCP-1) in traumatic brain injury (TBI): evidence from severe TBI patients and CCL2-/- mice. J Cereb Blood Flow Metab 30(4):769–82
22. Price CJ et al (2004) Cerebral neutrophil recruitment, histology, and outcome in acute ischemic stroke: an imaging-based study. Stroke 35(7):1659–64

23. Kim J et al (2012) Different prognostic value of white blood cell subtypes in patients with acute cerebral infarction. Atherosclerosis 222(2):464–7
24. Shiga Y et al (1991) Neutrophil as a mediator of ischemic edema formation in the brain. Neurosci Lett 125(2):110–2
25. Matsuo Y et al (1994) Correlation between myeloperoxidase-quantified neutrophil accumulation and ischemic brain injury in the rat. Effects of neutrophil depletion. Stroke 25(7): 1469–75
26. Jin R, Yang G, Li G (2010) Inflammatory mechanisms in ischemic stroke: role of inflammatory cells. J Leukoc Biol 87(5):779–89
27. Connolly ES Jr et al (1996) Cerebral protection in homozygous null ICAM-1 mice after middle cerebral artery occlusion. Role of neutrophil adhesion in the pathogenesis of stroke. J Clin Invest 97(1):209–16
28. Connolly ES Jr et al (1997) Exacerbation of cerebral injury in mice that express the P-selectin gene: identification of P-selectin blockade as a new target for the treatment of stroke. Circ Res 81(3):304–10
29. Soriano SG et al (1996) Intercellular adhesion molecule-1-deficient mice are less susceptible to cerebral ischemia-reperfusion injury. Ann Neurol 39(5):618–24
30. Gerson A et al (1988) Umbilical arterial systolic/diastolic values in normal twin gestation. Obstet Gynecol 72(2):205–8
31. Bauer NJ et al (1988) Purification, characterization, and kinetic mechanism of S-adenosyl-L-methionine:macrocin O-methyltransferase from Streptomyces fradiae. J Biol Chem 263(30): 15619–25
32. Gu Z et al (2002) S-nitrosylation of matrix metalloproteinases: signaling pathway to neuronal cell death. Science 297(5584):1186–90
33. Romanic AM et al (1998) Matrix metalloproteinase expression increases after cerebral focal ischemia in rats: inhibition of matrix metalloproteinase-9 reduces infarct size. Stroke 29(5): 1020–30
34. Asahi M et al (2000) Role for matrix metalloproteinase 9 after focal cerebral ischemia: effects of gene knockout and enzyme inhibition with BB-94. J Cereb Blood Flow Metab 20(12):1681–9
35. Rosell A et al (2006) Increased brain expression of matrix metalloproteinase-9 after ischemic and hemorrhagic human stroke. Stroke 37(6):1399–406
36. Rosell A et al (2008) MMP-9-positive neutrophil infiltration is associated to blood-brain barrier breakdown and basal lamina type IV collagen degradation during hemorrhagic transformation after human ischemic stroke. Stroke 39(4):1121–6
37. Morancho A et al (2010) Metalloproteinase and stroke infarct size: role for anti-inflammatory treatment? Ann N Y Acad Sci 1207:123–33
38. Gidday JM et al (2005) Leukocyte-derived matrix metalloproteinase-9 mediates blood-brain barrier breakdown and is proinflammatory after transient focal cerebral ischemia. Am J Physiol Heart Circ Physiol 289(2):H558–68
39. Justicia C et al (2003) Neutrophil infiltration increases matrix metalloproteinase-9 in the ischemic brain after occlusion/reperfusion of the middle cerebral artery in rats. J Cereb Blood Flow Metab 23(12):1430–40
40. Ajmo CT Jr et al (2009) Blockade of adrenoreceptors inhibits the splenic response to stroke. Exp Neurol 218(1):47–55
41. Schroeter M et al (1994) Local immune responses in the rat cerebral cortex after middle cerebral artery occlusion. J Neuroimmunol 55(2):195–203
42. Gelderblom M et al (2009) Temporal and spatial dynamics of cerebral immune cell accumulation in stroke. Stroke 40(5):1849–57
43. Hug A et al (2009) Infarct volume is a major determiner of post-stroke immune cell function and susceptibility to infection. Stroke 40(10):3226–32
44. Peterfalvi A et al (2009) Impaired function of innate T lymphocytes and NK cells in the acute phase of ischemic stroke. Cerebrovasc Dis 28(5):490–8
45. Schulte-Herbruggen O et al (2009) Differential affection of intestinal immune cell populations after cerebral ischemia in mice. Neuroimmunomodulation 16(3):213–8

46. Kasper LH et al (1996) Induction of gammadelta T cells during acute murine infection with Toxoplasma gondii. J Immunol 157(12):5521–7
47. Sciammas R et al (1997) T cell receptor-gamma/delta cells protect mice from herpes simplex virus type 1-induced lethal encephalitis. J Exp Med 185(11):1969–75
48. Shichita T et al (2009) Pivotal role of cerebral interleukin-17-producing gammadelta T cells in the delayed phase of ischemic brain injury. Nat Med 15(8):946–50
49. Rodriguez-Fernandez JL, Riol-Blanco L, Delgado-Martin C (2010) What is an immunological synapse? Microbes Infect 12(6):438–45
50. Davis DM (2009) Mechanisms and functions for the duration of intercellular contacts made by lymphocytes. Nat Rev Immunol 9(8):543–55
51. Hosseini BH et al (2009) Immune synapse formation determines interaction forces between T cells and antigen-presenting cells measured by atomic force microscopy. Proc Natl Acad Sci U S A 106(42):17852–7
52. Steinman RM, Banchereau J (2007) Taking dendritic cells into medicine. Nature 449(7161): 419–26
53. Liu YJ (2001) Dendritic cell subsets and lineages, and their functions in innate and adaptive immunity. Cell 106(3):259–62
54. Banchereau J et al (2000) Immunobiology of dendritic cells. Annu Rev Immunol 18:767–811
55. Kostulas N et al (2002) Dendritic cells are present in ischemic brain after permanent middle cerebral artery occlusion in the rat. Stroke 33(4):1129–34
56. Felger JC et al (2010) Brain dendritic cells in ischemic stroke: time course, activation state, and origin. Brain Behav Immun 24(5):724–37
57. Yilmaz A et al (2010) Transient decrease in circulating dendritic cell precursors after acute stroke: potential recruitment into the brain. Clin Sci (Lond) 118(2):147–57
58. Hug A et al (2011) Reduced efficacy of circulating costimulatory cells after focal cerebral ischemia. Stroke 42(12):3580–6
59. Campanella M et al (2002) Flow cytometric analysis of inflammatory cells in ischemic rat brain. Stroke 33(2):586–92
60. Stevens SL et al (2002) The use of flow cytometry to evaluate temporal changes in inflammatory cells following focal cerebral ischemia in mice. Brain Res 932(1–2):110–9
61. Jander S et al (1995) Lymphocytic infiltration and expression of intercellular adhesion molecule-1 in photochemically induced ischemia of the rat cortex. J Cereb Blood Flow Metab 15(1):42–51
62. Yilmaz G et al (2006) Role of T lymphocytes and interferon-gamma in ischemic stroke. Circulation 113(17):2105–12
63. Brait VH et al (2010) Mechanisms contributing to cerebral infarct size after stroke: gender, reperfusion, T lymphocytes, and Nox2-derived superoxide. J Cereb Blood Flow Metab 30(7): 1306–17
64. Becker K et al (2001) Antibody to the alpha4 integrin decreases infarct size in transient focal cerebral ischemia in rats. Stroke 32(1):206–11
65. Relton JK et al (2001) Inhibition of alpha4 integrin protects against transient focal cerebral ischemia in normotensive and hypertensive rats. Stroke 32(1):199–205
66. Appay V, Rowland-Jones SL (2001) RANTES: a versatile and controversial chemokine. Trends Immunol 22(2):83–7
67. Hurn PD et al (2007) T- and B-cell-deficient mice with experimental stroke have reduced lesion size and inflammation. J Cereb Blood Flow Metab 27(11):1798–805
68. Kleinschnitz C et al (2010) Early detrimental T-cell effects in experimental cerebral ischemia are neither related to adaptive immunity nor thrombus formation. Blood 115(18):3835–42
69. Clausen F et al (2007) T lymphocyte trafficking: a novel target for neuroprotection in traumatic brain injury. J Neurotrauma 24(8):1295–307
70. Liesz A et al (2009) Regulatory T cells are key cerebroprotective immunomodulators in acute experimental stroke. Nat Med 15(2):192–9
71. Seder RA, Ahmed R (2003) Similarities and differences in CD4+ and CD8+ effector and memory T cell generation. Nat Immunol 4(9):835–42

72. Okruhlicova L et al (2002) L-arginine reduces structural remodeling in the diabetic rat myocardium. Methods Find Exp Clin Pharmacol 24(4):201–7
73. Russell JH, Ley TJ (2002) Lymphocyte-mediated cytotoxicity. Annu Rev Immunol 20:323–70
74. Phillips S et al (2010) CD8(+) T cell control of hepatitis B virus replication: direct comparison between cytolytic and noncytolytic functions. J Immunol 184(1):287–95
75. Ling C et al (2006) Traumatic injury and the presence of antigen differentially contribute to T-cell recruitment in the CNS. J Neurosci 26(3):731–41
76. Saino O et al (2010) Immunodeficiency reduces neural stem/progenitor cell apoptosis and enhances neurogenesis in the cerebral cortex after stroke. J Neurosci Res 88(11):2385–97
77. Santana MA, Rosenstein Y (2003) What it takes to become an effector T cell: the process, the cells involved, and the mechanisms. J Cell Physiol 195(3):392–401
78. Liesz A et al (2011) Inhibition of lymphocyte trafficking shields the brain against deleterious neuroinflammation after stroke. Brain 134(Pt 3):704–20
79. Liesz A et al (2009) The spectrum of systemic immune alterations after murine focal ischemia: immunodepression versus immunomodulation. Stroke 40(8):2849–58
80. Gendron A et al (2002) Temporal effects of left versus right middle cerebral artery occlusion on spleen lymphocyte subsets and mitogenic response in Wistar rats. Brain Res 955(1–2): 85–97
81. Martin A et al (2008) Imaging changes in lymphoid organs in vivo after brain ischemia with three-dimensional fluorescence molecular tomography in transgenic mice expressing green fluorescent protein in T lymphocytes. Mol Imaging 7(4):157–67
82. Prass K et al (2003) Stroke-induced immunodeficiency promotes spontaneous bacterial infections and is mediated by sympathetic activation reversal by poststroke T helper cell type 1-like immunostimulation. J Exp Med 198(5):725–36
83. Theodorou GL et al (2008) T helper 1 (Th1)/Th2 cytokine expression shift of peripheral blood CD4+ and CD8+ T cells in patients at the post-acute phase of stroke. Clin Exp Immunol 152(3):456–63
84. Abbas AK, Murphy KM, Sher A (1996) Functional diversity of helper T lymphocytes. Nature 383(6603):787–93
85. Sakaguchi S (2005) Naturally arising Foxp3-expressing CD25+CD4+ regulatory T cells in immunological tolerance to self and non-self. Nat Immunol 6(4):345–52
86. Collison LW et al (2007) The inhibitory cytokine IL-35 contributes to regulatory T-cell function. Nature 450(7169):566–9
87. Niedbala W et al (2007) IL-35 is a novel cytokine with therapeutic effects against collagen-induced arthritis through the expansion of regulatory T cells and suppression of Th17 cells. Eur J Immunol 37(11):3021–9
88. Vignali DA, Collison LW, Workman CJ (2008) How regulatory T cells work. Nat Rev Immunol 8(7):523–32
89. Sakaguchi S et al (2006) Foxp3+ CD25+ CD4+ natural regulatory T cells in dominant self-tolerance and autoimmune disease. Immunol Rev 212:8–27
90. Ephrem A et al (2008) Expansion of CD4+CD25+ regulatory T cells by intravenous immunoglobulin: a critical factor in controlling experimental autoimmune encephalomyelitis. Blood 111(2):715–22
91. Reynolds AD et al (2010) Regulatory T cells attenuate Th17 cell-mediated nigrostriatal dopaminergic neurodegeneration in a model of Parkinson's disease. J Immunol 184(5):2261–71
92. Reynolds AD et al (2007) Neuroprotective activities of CD4+CD25+ regulatory T cells in an animal model of Parkinson's disease. J Leukoc Biol 82(5):1083–94
93. Yan J et al (2009) Immune activation in the peripheral blood of patients with acute ischemic stroke. J Neuroimmunol 206(1–2):112–7
94. Offner H et al (2006) Splenic atrophy in experimental stroke is accompanied by increased regulatory T cells and circulating macrophages. J Immunol 176(11):6523–31
95. Gee JM et al (2008) Induction of immunologic tolerance to myelin basic protein prevents central nervous system autoimmunity and improves outcome after stroke. Stroke 39(5): 1575–82

96. Ishibashi S et al (2009) Mucosal tolerance to E-selectin promotes the survival of newly generated neuroblasts via regulatory T-cell induction after stroke in spontaneously hypertensive rats. J Cereb Blood Flow Metab 29(3):606–20
97. Ren X et al (2011) CD4+FoxP3+ regulatory T-cells in cerebral ischemic stroke. Metab Brain Dis 26(1):87–90
98. Li P et al (2013) Adoptive regulatory T-cell therapy protects against cerebral ischemia. Annals of neurology [published online ahead of print November 24, 2012]
99. Linfert D, Chowdhry T, Rabb H (2009) Lymphocytes and ischemia-reperfusion injury. Transplant Rev (Orlando) 23(1):1–10
100. Ren X et al (2011) Regulatory B cells limit CNS inflammation and neurologic deficits in murine experimental stroke. J Neurosci 31(23):8556–63

Chapter 3
Vascular Inflammation in Ischemic Stroke: Adhesion Receptors Controlling Leukocyte–Endothelial Interactions

Stephen F. Rodrigues and D. Neil Granger

Abstract The contribution of leukocytes to the pathogenesis of ischemic stroke has been extensively studied and thoroughly documented. In this chapter, different aspects of leukocyte involvement in the lesion formation caused by ischemic stroke are highlighted, including the inflammatory agents that mediate leukocyte recruitment to the site of injury, the primary leukocyte populations that contribute to tissue damage, and the adhesion receptors that control leukocyte–endothelial cell interactions in post-ischemic brain. Agents that interfere with leukocyte recruitment in the brain are also addressed as potential therapeutic interventions for ischemic stroke.

Ischemic Stroke

Ischemic stroke is a condition in which an area of the brain becomes poorly perfused as a consequence of partial or total blockade of an artery. A blood clot is a common cause of artery blockade. Blood clots can be formed in diseases such as atherosclerosis, atrial fibrillation, and heart attack. Due to the high oxygen and nutrient requirements of brain tissue, blood blockage results in ATP depletion in the neurons and consequent lack of ionic gradients across the cellular membranes, resulting in calcium and water influx and neurotransmitter release. Cytotoxic edema, excitotoxicity, and activation of intracellular enzymes then occur, leading to cellular damage and inflammation.

S.F. Rodrigues
Clinical and Toxicological Analyses, University of Sao Paulo, Sao Paulo, Brazil

D.N. Granger (✉)
Molecular and Cellular Physiology, Louisiana State University Health Sciences Center, Baton Rouge, LA, USA
e-mail: dgrang@lsuhsc.edu

J. Chen et al. (eds.), *Immunological Mechanisms and Therapies in Brain Injuries and Stroke*, Springer Series in Translational Stroke Research, DOI 10.1007/978-1-4614-8915-3_3,

Ischemic stroke is the most common kind of stroke, accounting for near 90 % of all strokes. Stroke is the second leading cause of long-term disability in industrialized countries and the second leading cause of death worldwide [1]. More than 16 million people had a stroke in 2005 and almost 40 % of them died as a result of the stroke [2]. For the survivors, 50 % have serious disabilities and require long-term care. The cost of stroke is estimated to be approximately 72 billion dollars in the USA [3]. The treatment for ischemic stroke in the USA and Canada is largely limited to intravenous recombinant tissue plasminogen activator (rtPA); however, a significant limitation of this treatment is that it must be administered within 3 h after the onset of symptoms [4, 5]. Furthermore, reperfusion of the ischemic tissue can potentiate the activation of parenchyma cells, blood cells, and endothelial cells, resulting in an inflammatory response in the brain. While events that lead to neuronal death can occur rapidly after ischemia, inflammation in the area neighboring the ischemic region (called the penumbra) develops more slowly, and this area may be amenable to therapeutic intervention [127].

There is growing evidence that leukocytes play a role in mediating the tissue damage associated with reperfusion of ischemic tissue [6]. The large increase in the leukocyte number in the brain tissue after reperfusion is consequence of the upregulation of cell adhesion molecules (CAMs) on both leukocytes and endothelial cells [7]. Leukocyte recruitment is a highly coordinated and sequential process. Neutrophils reach the inflammatory site first, and they are followed by mononuclear leukocytes (especially lymphocytes), which predominate in the inflammatory area 1 day after ischemic injury [8]. In this chapter, we address different aspects of the leukocyte involvement in the lesion formation caused by ischemic stroke, including the inflammatory agents that mediate leukocyte recruitment to the site of injury, the primary leukocyte populations that contribute to tissue damage, and the adhesion receptors that control leukocyte–endothelial cell adhesive interactions. Finally, we describe results of some studies that have focused on adhesion receptors as new potential therapeutic targets that are directed towards diminishing the damage caused by ischemic stroke.

Role of Leukocytes on Ischemic Stroke

The inflammatory phenotype which takes place after an ischemia–reperfusion (I/R) episode includes leukocyte infiltration [9]. Activation of resident cells, such as endothelial cells, microglia/macrophages, and astrocytes, in response to I/R injury leads to the release of a variety of mediators that attract and activate leukocytes. Activated leukocytes express a high density of adhesion receptors that facilitate adhesion to endothelial cells and their accumulation at the ischemic site [7]. Initially, the endothelial cell/leukocyte contact is weak, resulting in a phenomenon called leukocyte rolling. This binding becomes stronger at a later time, resulting in firm adhesion of the leukocyte. The rolling process is mediated by a family of adhesion molecules called selectins. P- and/or E-selectin expressed on endothelial cells mediate leukocyte rolling while firm adhesion is mediated by integrins such as β2-integrins on leukocytes, and immunoglobulin-like domain molecules such as ICAM-1 that is expressed on cerebral microvascular endothelial cells.

Several reports in the literature describe an increased presence of leukocytes in brain tissue both in mice [10–12] and rats subjected to ischemic stroke [13–16]. The same phenomenon has been demonstrated in humans [17, 18]. The importance of leukocyte infiltration for the ischemic lesion is highlighted by experiments wherein animals depleted of leukocytes exhibit a diminished injury response [19, 20], and prevention of leukocyte–endothelial cell adhesion affords protection against I/R-caused tissue injury [20]. Reactive oxygen species, proteases, and inflammatory cytokines are among the leukocyte-derived mediators that can cause I/R-induced tissue damage [21]. However, there are other reports that do not confirm a role for leukocytes in the genesis of stroke. Few leukocytes in the ischemic area [22], the occurrence of neutrophil accumulation after peak tissue damage [23, 24], and an absence of protection on infarct size following neutropenia [22] are major lines of evidence that argue against a role for neutrophils in the pathogenesis of cerebral ischemia. The contradictory results may be explained by different approaches used in each study, for example, differences in the duration and extent of the ischemic insult, duration of reperfusion, the maintenance of body temperature during the I/R protocol, methods used to quantify leukocyte infiltration and tissue damage, and anesthetics and/or analgesic agents used during/following the protocol. Despite the controversy, new reports continue to appear that describe a protective role of some subsets of leukocytes that could defend the brain tissue against more profound brain damage following a stroke [25]. Polymorphonuclear (PMN) leukocytes, mononuclear leukocytes, and infiltrating peripheral dendritic cells have all been implicated in the pathogenesis of cerebral ischemia.

PMN Leukocytes

Neutrophils reach the inflammatory area within minutes to hours after an initial ischemic insult [26, 27]. Mice rendered neutropenic with antineutrophil serum show a blunted number of migrated leukocytes to the ischemic area 4 h after focal I/R, showing neutrophils greatly account to the leukocyte invasion at that time [11]. At later times (24 h to days after injury), neutrophils account for a smaller proportion of the total leukocyte population at the ischemic site [28]. Other studies suggest that neutrophils are still important several days after initial lesion formation [29–31]. It is noteworthy that several reports demonstrate a correlation between PMN leukocyte infiltration after a stroke and the severity of the brain tissue injury and the worsening of the neurological deficits [17, 32, 33]. Hence, strategies directed towards reducing neutrophil infiltration may be beneficial from improving the final outcome after stroke.

Mononuclear Leukocytes

The infiltration of mononuclear leukocytes appears to surpass the accumulation of PMNs days to weeks after the ischemic insult [26, 34]. The mononuclear leukocytes that migrate to the ischemic area are mainly monocytes and T lymphocytes as shown by immunocytochemistry [35] and flow cytometry [31]. These leukocytes have been implicated in the outcome of I/R in the brain. CD4+ and CD8+ T lymphocytes were shown to contribute to the greater damage caused by temporary middle cerebral artery occlusion (MCAO) in mice at 24 h after injury compared to 4 h [28]. On the other hand, specific subsets of T lymphocytes such as CD4(+)CD25(+) forkhead box P3 (Foxp3)(+) regulatory T lymphocytes [T(reg) cells] have lately been regarded as protective against the neuronal damage caused by stroke [25, 36]. This result has proved to be controversial since others have found no influence of these cells on stroke outcome [37]. B lymphocytes have also been shown to limit inflammation in mice after stroke [38]. More studies are needed to clarify the role of each leukocyte population and its temporal profile during a stroke episode and to better define the feasibility of targeting specific leukocyte populations to treat ischemic stroke.

Inflammatory Mediators Contributing to Leukocyte Recruitment

There are several signaling events and mediators that occur or are produced/released locally after a stroke that can attract leukocytes to the site of an ischemic infarct. These include interferon gamma (IFN-γ), reactive oxygen species (ROS), CD40/CD40L interaction, Notch signaling, and monocyte chemoattractant protein-1 (MCP-1). Some of their effects during ischemia in brain are described below and illustrated in Fig. 3.1.

IFN-γ—A series of inflammatory mediators seems to contribute to the leukocyte recruitment following a stroke. IFN-γ, or type II interferon, is a cytokine that is critical for innate and adaptive immunity and for tumor control. IFN-γ is produced by natural killer (NK) cells, natural killer T (NKT) cells, CD4+ Th1, and CD8+ cytotoxic T lymphocyte (CTL) effector cells. A reduced number of adherent leukocytes in the cerebral microvasculature has been reported after I/R in IFN-γ KO mice, implicating a role for IFN-γ in leukocyte recruitment after MCAO [28]. Similarly, Rag-$1^{-/-}$ mice, which are deficient in both CD4+ and CD8+ T cells, show an attenuated recruitment of adherent leukocytes after cerebral I/R. This protection is completely reversed in Rag-$1^{-/-}$ mice reconstituted with T cells harvested from wild-type (WT) mice, but only partially reversed if the Rag-$1^{-/-}$ mice are reconstituted with T cells derived from IFN-γ KO mice [28]. This observation suggests that a component of the IFN-γ produced in response to cerebral I/R is produced either locally or by other circulating cells.

Reactive Oxygen Species

Superoxide and other ROS are produced after an I/R episode in different tissues, including in the brain. ROS have been implicated as mediators of the increased expression of several endothelial cell adhesion receptors (e.g., P-selectin) that participate in leukocyte recruitment into the post-ischemic brain. NADPH oxidase appears to be a major source of ROS production after stroke [11]. Indirectly, ROS can promote leukocyte–endothelial cell adhesion by inactivating nitric oxide, an endogenous inhibitor of leukocyte adhesion in postcapillary vessels [39]. This contention is supported by the observation that nitric oxide donors inhibit leukocyte adhesion in the cerebral microcirculation [40]. Thus, ROS can act both directly and indirectly to attract leukocytes to the ischemic site after stroke.

CD40/CD40L

CD40 is a membrane glycoprotein belonging to the tumor necrosis factor receptor superfamily and is expressed on different cell populations like lymphocytes, monocytes/macrophages, platelets, dendritic cells, endothelial cells, and neuronal cells [41]. CD40/CD40L interactions play an important role during inflammation since they induce cellular adhesion molecules [42] and tissue factor in endothelial cells and enhance the production of pro-inflammatory cytokines [43, 44]. The CD40/CD40L interaction appears to contribute to the leukocyte recruitment and ischemic damage during stroke since CD40 or CD40 ligand (CD40L) deficiency attenuates the recruitment of adherent leukocytes in the cerebral microcirculation after I/R, which is accompanied by a parallel reduction in blood–brain barrier (BBB) permeability and infarct volume [45]. That reduction in the number of adherent leukocytes can be mediated by platelets once CD40L on platelets can interact with CD40 on endothelial cells to induce P-selectin expression on endothelial cells, thereby resulting in the recruitment of leukocytes. CD40/CD40L signaling appears to be an important mechanism for leukocyte trafficking in the microcirculation after cerebral I/R injury.

Notch Signaling

Notch is a cell surface receptor that participates in a variety of physiological events such as angiogenesis, neurogenesis, and leukocyte recruitment [46]. Mutant mice with downregulated Notch signaling [47, 128], as well as wild-type mice treated with inhibitors of the Notch-activating enzyme, gamma secretase, exhibit an attenuated recruitment of leukocytes in the brain after cerebral I/R [48].

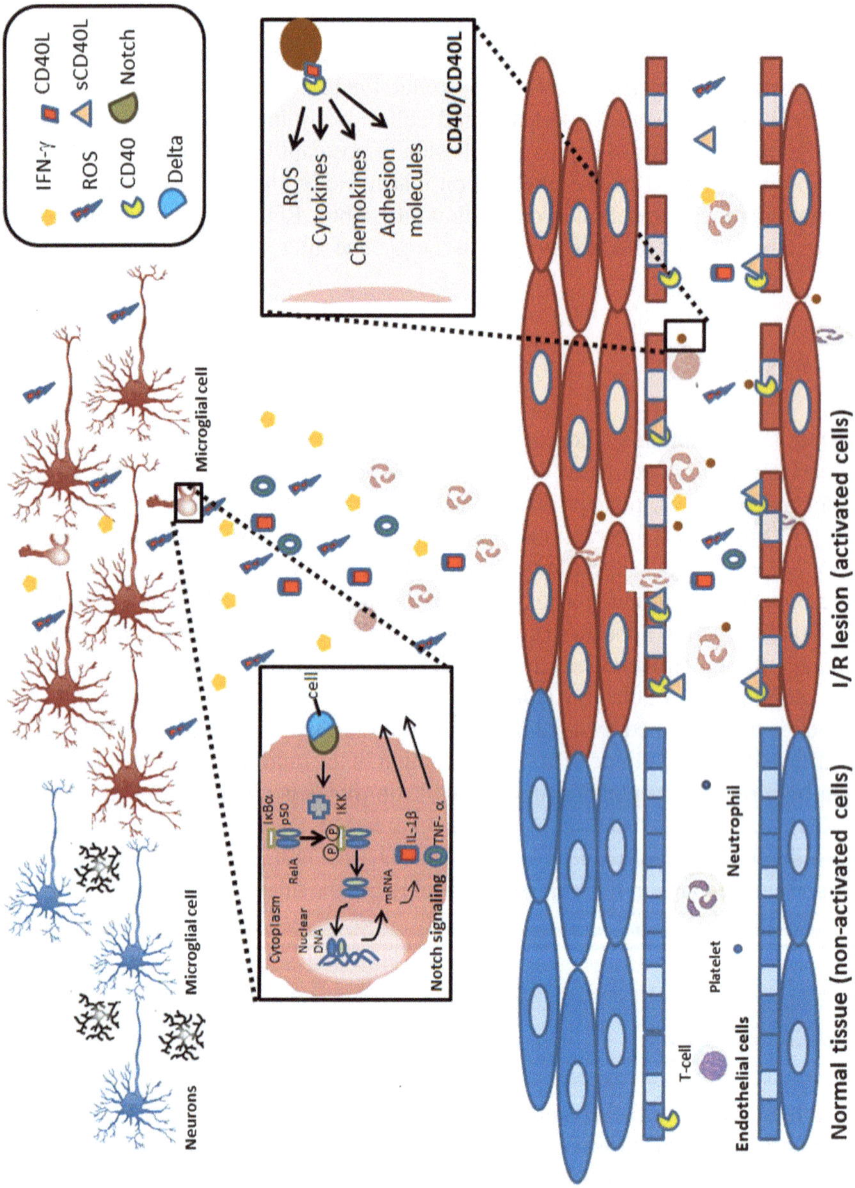
IFN-γ
CD40L
ROS
sCD40L
CD40
Notch
Delta
ROS
Cytokines
Chemokines
Adhesion molecules
CD40/CD40L
Microglial cell
Neurons
Microglial cell
Cytoplasm
Nuclear
DNA
RelA
p50
IκBα
IKK
mRNA
IL-1β
TNF-α
cell
Notch signaling
T-cell
Platelet
Neutrophil
Endothelial cells
Normal tissue (non-activated cells)
I/R lesion (activated cells)

Fig. 3.1 Mediators released following ischemia and reperfusion that lead to leukocyte activation and migration to the site of lesion development in the brain after stroke. In absence of blood flow disturbance, leukocytes are kept in the center of the bloodstream. Local and circulating cells remain nonactivated. There is no release of inflammatory mediators and the expression of endothelial cell adhesion molecules is low with few physical interactions between leukocyte and endothelial cells (represented in the *left side* of the figure, cells in *blue*). Following I/R, local cells including neurons, microglial cells, endothelial cells, and smooth muscle cells become activated, releasing and/or expressing a series of proteins/molecules such as IFN-γ, ROS, CD40, CD40L, sCD40L, and Notch that are inflammatory and/or chemotactic, resulting in enhanced leukocyte traffic to the site of the ischemic lesion (*right side* of the figure, cells in *brown*). Once neutrophils do not express either CD40 or CD40L, an indirect signaling can respond for the CD40/CD40L-mediated neutrophil recruitment. This can involve enhanced production of inflammatory mediators such as ROS, cytokines, chemokines, and adhesion molecules as a result of the CD40/CD40L interactions between other blood cell populations (e.g., platelets and T lymphocytes) and/or binding of circulating sCD40L with CD40 on endothelial cells, when activated, becoming hyperadhesive to neutrophils (*right square*). The Notch signaling involves cell–cell interactions where ligand–proteins binding to the extracellular domain induce proteolytic cleavage and release of the intracellular domain, which activates NF-κB and modify gene expression, including those genes that are going to be translated into inflammatory mediators such as IL-1β and TNF-α (*left square*). *CD40L* CD40 ligand, *IκBα* inhibitor of nuclear factor kappa-B subunit alpha, *IKK* IkappaB kinase, *IL-1β* interleukin 1β, *IFN-γ* interferon γ, *ROS* reactive oxygen species, *TNF-α* tumor necrosis factor α

Monocyte Chemoattractant Protein-1

This cytokine, also known as CCL2, belongs to the CC chemokine family. MCP-1 recruits monocytes, memory T cells, and dendritic cells to the sites of inflammation [49] and has been implicated in pathogeneses of several diseases, such as psoriasis, rheumatoid arthritis, atherosclerosis, glomerulonephritis, and neuroinflammatory processes [50, 51]. MCP-1 has been proposed to play a pivotal role in triggering the inflammatory reaction elicited by ischemia in the brain. It has been shown that MCP-1 overexpression leads to larger infarcts and more blood cell recruitment, while MCP-1-deficient mice develop smaller infarcts. Both macrophage and neutrophil infiltration are significantly reduced in MCP-1-deficient animals, compared to wild-type mice [129].

Adhesion Receptors Controlling Leukocyte–Endothelial Cell Interactions After I/R

The cerebral microcirculation exhibits unique features compared to other vascular beds relative to the leukocyte–endothelial cell interactions. In addition to the high venular shear rates [52], the lower basal expression levels of endothelial cell adhesion molecules appear to contribute to the fewer physical interactions between leukocytes and endothelial cells noted in normal (noninflamed) cerebral venules, compared to venules in other tissues [53]. A very low basal expression of both P- and E-selectin [54, 55] is detected in the cerebral microvasculature. Upon stimulation, however, an increased expression of endothelial cell adhesion molecules is observed in cerebral vessels. Interleukin-1β enhances the expression of ICAM-1 and VCAM-1 [56, 57], and TNF-alpha increases E- and P-selectin expression [58] in the brain. Following I/R, several different adhesion receptors contribute to the recruitment of leukocytes in the brain, with each adhesion receptor contributing in a time-dependent manner (Fig. 3.2).

Selectins

E-selectin expression is increased in cerebral microvessels at 2 h, peaks at 6–12 h, and is still visualized 24 h within the ischemic lesion in rats [55, 59], and treatment with CY-1503, an analog of sialyl Lewis(x) [SLe(x)], significantly reduces the infarct volume and neutrophil infiltration [60]. CY-1503 can act not only by blocking the E-selectin on endothelial cells but also L-selectin on leukocytes or P-selectin on endothelial cells or platelets. Due to the nonspecific mechanism of action of the CY-1503, a single specific selectin cannot be implicated in the neuronal damage

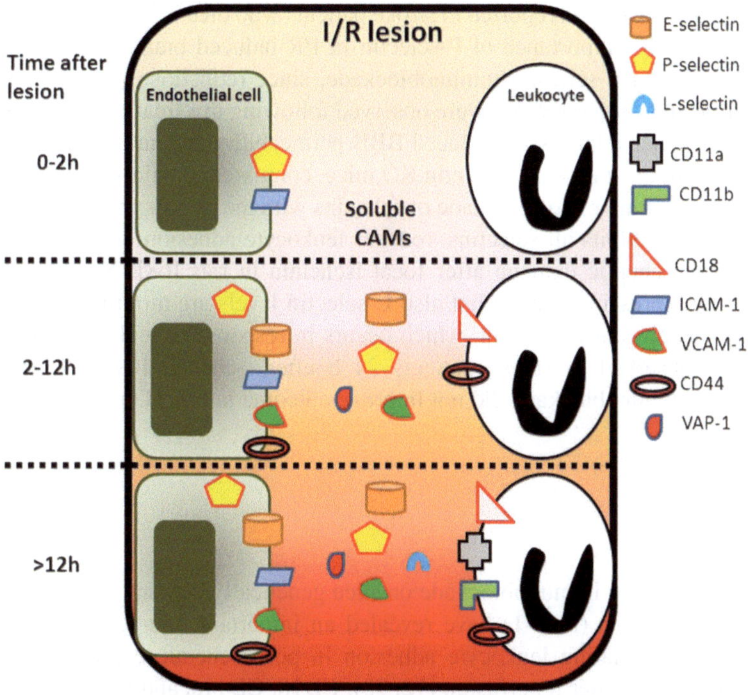

Fig. 3.2 Kinetics of cell adhesion molecule expression on endothelial cells, leukocytes and in the circulation (soluble CAMs) after ischemia and reperfusion (I/R). The increased P-selectin and ICAM-1 expression occurs within 2 h after occlusion of the middle cerebral artery and persist for several hours/days. Enhanced expression of E-selectin, VCAM-1, and CD44 on endothelial cells and CD18 and CD44 on leukocytes become evident 2 h after injury and the increased expression is maintained for several hours. A soluble version of these proteins (CAMs) also appears in the circulation. A longer time (more than 12 h) is necessary to increase the expression of CD11a and CD11b on leukocytes. *CAMs* cell adhesion molecules, *ICAM-1* intercellular adhesion molecule 1, *I/R* ischemia and reperfusion, *VAP-1* vascular adhesion protein 1, *VCAM-1* vascular cell adhesion molecule-1

following a stroke. However, studies targeted E-selectin using highly specific monoclonal antibodies have revealed reduced infarct size [61], but whether this is accompanied by and/or results from a reduction in the number of recruited leukocytes remains unclear. Increased E-selectin expression may be important for the leukocyte-induced damage caused by stroke in nonhuman primates and in humans. Increased soluble E-selectin levels are observed in the circulation 72 h after stroke in baboons [62] and 8 h up to 1 day after the beginning of the symptoms in stroke patients [63]. Another selectin, P-selectin, exhibits an increased expression as early as 15 min after occlusion of the middle cerebral artery, peaks 6 h later [55], and remains elevated up to 72 h after injury [64] in the brain of mice and rats. Increases in circulating soluble P-selectin [62] and in the expression of P-selectin in brain microvessels have been reported in stroke patients who died 3.5 h to 9 days after the stroke [65]. The importance of P-selectin in I/R-induced brain damage has been confirmed using P-selectin immunoblockade, since reductions in both leukocyte infiltration and cerebral injury were observed following this treatment [66]. In addition to a smaller infarct size, reduced BBB permeability and an improved survival rate were demonstrated in P-selectin KO mice, compared to wild-type control mice [10, 67]. In a similar way, blockade of selectins with the polysaccharide fucoidan, a polymer that inhibits all selectins, reduces leukocyte adhesion in cerebral vessels and blunts ischemic damage after focal ischemia in rats [68]. Not only P- and E-selectin expression change but also L-selectin levels are modified after stroke. In fact, shedding of L-selectin, which occurs in response to leukocyte activation, has been observed in stroke patients 24 h after ischemia [69, 70]. However, L-selectin immunoblockade did not improve outcome in MCAO mice [71].

Integrins

Studies employing immunoblockade or mice genetically deficient in either CD11a (LFA-1) or CD11b (Mac-1) have revealed an important role for these adhesion receptors in mediating leukocyte adhesion in post-ischemic cerebral venules at 24 h, but not 4 h, after reperfusion [72, 73]. CD18, CD11a, and CD11b expression are enhanced in mice after transient MCAO. Stroke patients and patients with transient ischemic attack also show elevated integrin expression at 12 h and up to 72 h after ischemia [74–76]. Interfering with the functionality of CD11a or CD11b reduces infarct volume and mortality after MCAO [77, 78].

Immunoglobulin-Like Domain Cell Adhesion Molecules

An increased expression of ICAM-1 in the cerebral vasculature has been detected after ischemia in different species, including baboons [79], rats [35], and humans [80]. The increased expression is detected as early as 1 h after reperfusion and it remains elevated up to 48 h after stroke [14, 15, 79, 81–84]. Blockade of ICAM-1 blunts the PMN leukocyte accumulation and ischemic lesion size after MCAO in rats, suggesting that ICAM-1-mediated leukocyte adhesion contributes to the damage caused by I/R in this species [23, 24, 85]. ICAM-1 knockout mice showed reduced leukocyte adhesion, smaller infarcts, improved cerebral flow, and reduced mortality after cerebral I/R [30]. The involvement of ICAM-1 on the I/R-induced lesion seems to be important in humans as well since induction of ICAM-1 in human cerebromicrovascular endothelial cells (HCEC) by ischemia-like conditions leads to enhanced neutrophil/HCEC adhesion [56]. VCAM-1 expression is also increased as early as 4 h after I/R injury both in mice and in rats, which parallels the increased leukocyte trafficking to the infarcted area [86–88]. Increases in soluble

VCAM-1 levels are also observed in the circulation of patients with acute stroke at 4 h and up to 5 days after the onset of symptoms [63] and in microvessels of patients who died 5 days to 3 months after stroke onset [89]. However, anti-VCAM-1 antibodies have shown no effectiveness in protecting patients against the deleterious effects of stroke [90]. Thus, strategies directed to reducing the influence of ICAM-1, but not VCAM-1, may prove useful in minimizing the tissue damage associated with ischemic stroke.

Others

CD73

Recent studies have revealed other cell membrane proteins that may mediate the leukocyte diapedesis elicited by ischemic stroke. CD73 is an ecto-5′ nucleotidase that catalyzes the terminal phosphohydrolysis of AMP and is expressed on the surface of glial cells, on cells of the choroid plexus, and on leukocytes. One of the main functions of the CD73 is to tighten epithelial barriers. Mice lacking CD73 were observed to have larger cerebral infarct volumes and more leukocyte accumulation after cerebral I/R. These responses were more evident in chimeric mice lacking CD73 in nonmyeloid tissue rather than in leukocytes. These observations reveal a potentially novel target for therapeutic intervention in ischemic stroke [91].

CD47

CD47 is a cell surface glycoprotein that helps mediate neutrophil transmigration across blood vessels. Ischemia does not upregulate the brain levels of CD47 in mice. However, CD47 knockout mice exhibit reductions in both infarct volume and tissue swelling at the lesion site after I/R. These effects occurred in parallel with a reduction of the extravasation of neutrophils into the brain parenchyma [92].

CD44

The CD44 antigen is a cell surface glycoprotein involved in cell–cell interactions, cell adhesion, and transmigration. Increased CD44 expression is observed 6 h and persists for up to 72 h after permanent MCAO in mice [93, 130]. Furthermore, significant reductions in ischemic infarct volume and neurological function were observed in CD44-deficient mice after transient and permanent MCAO [93]. These observations, coupled to the established role of CD44 in leukocyte trafficking, underscore the potential importance of this molecule in the pathogenesis of ischemic stroke.

VAP-1

Soluble vascular adhesion protein-1 (VAP-1) concentration is elevated postmortem in the serum of stroke patients [94]. VAP-1 is a semicarbazide-sensitive amine oxidase (SSAO) which functions as an endothelial adhesion molecule and participates in the process of transmigration of inflammatory cells into the ischemic brain. Moreover, VAP-1 may worsen ischemic brain injury due to its enzymatic function by producing toxic metabolites [94]. More studies are needed to better elucidate the role of VAP-1 in stroke.

Leukocyte Mediators Contributing to the I/R-Induced Cerebral Damage

Most of the leukocyte-derived mediators that cause damage to the brain tissue after ischemia remain unknown although recent studies have revealed some of them. These include high-mobility group protein B1 (HMGB1), interleukin 23 (IL-23), interleukin 17, regulated on activation, normal T-cell expressed and secreted (RANTES), metalloproteinase 9 (MMP-9), and neutrophil elastase. Leukocytes can also release mediators that blunt the brain damage caused by ischemia. Interleukin 10 (IL-10) is a good example of this type of mediator. These pro and anti-inflammatory mediators are described below and illustrated in Fig. 3.3.

HMGB1

High-mobility group protein B1 acts as a cytokine, induces pro-inflammatory mediators, such as inducible nitric oxide (NO) synthase, cyclooxygenase-2 (COX-2), interleukin (IL)-1β, and tumor necrosis factor (TNF)-α and causes neuronal apoptosis [95]. It is produced by inflammatory cells such as macrophages and monocytes and activated neurons. Fujioka et al. [96] recently reported increased plasma HMGB1 levels in MCAO mice, which corresponded with larger brain infarcts. Blocking HMGB1 has been shown to reduce infarct volume after cerebral artery occlusion [97]. The HMGB1-induced cerebral injury after ischemia seems to be related to Toll-like receptor-4 activation since it was demonstrated that TLR4-deficient mice presented reduced cerebral ischemia–reperfusion injury as well as downregulation of inflammatory cytokines, compared to WT animals [98] and intracerebroventricular injection of rhHMGB1 in TLR4(+/+) mice caused significantly more injury after cerebral ischemia–reperfusion than the control group [99]. Interestingly, the coexpression of HMGB1 and MPO was noted in ischemic brain, suggesting that neutrophils may be releasing HMGB1 locally and, as a result, causing larger lesions after ischemia.

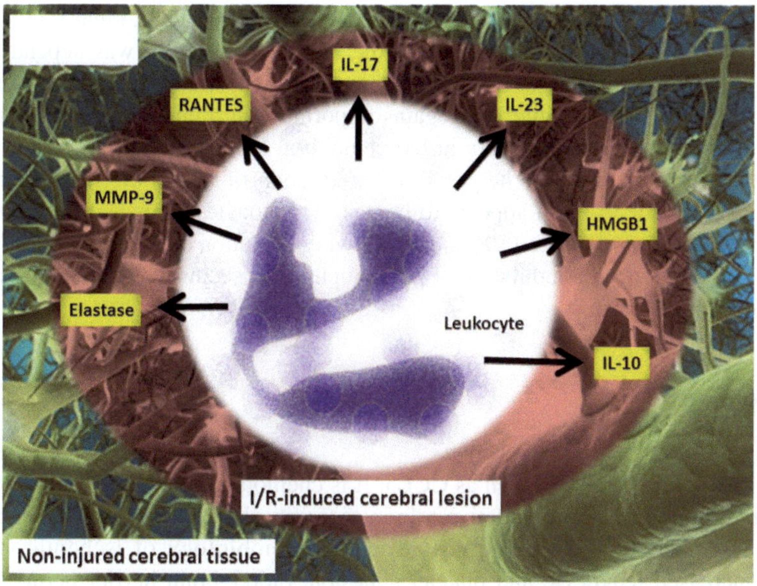

Fig. 3.3 Leukocyte-derived inflammatory mediators that contribute to cerebral tissue damage after ischemia and reperfusion. Acting as pro-inflammatory mediators are HMGB1, IL-23, IL-17, RANTES, MMP-9, and neutrophil elastase. Downregulating inflammation is IL-10. *HMGB1* high-mobility group protein B1, *IL-10* interleukin 10, *IL-17* interleukin 17, *IL-23* interleukin-23, *I/R* ischemia and reperfusion, *MMP-9* matrix metallopeptidase-9, *RANTES* regulated and normal T-cell expressed and secreted

IL-23

This is a heterodimeric cytokine consisting of two subunits: p40 and p19. The cytokine is considered to act as a pro-inflammatory agent and it appears to be released by infiltrating macrophages after cerebral ischemia. IL-23 is followed by peroxiredoxin release into the extracellular space by necrotic brain cells, resulting in further neural cell death. Peroxiredoxin inactivation suppresses both IL-23 expression and infarct volume [100].

IL-17

This cytokine acts as a potent inflammatory mediator in delayed-type reactions by increasing chemokine production in various tissues, including brain, to recruit monocytes and neutrophils to the site of inflammation. IL-17 is often produced by T-helper cells and is induced by IL-23, which result in destructive tissue damage in delayed-type reactions. In fact, it was observed that expression of IL-23, derived mostly from infiltrated macrophages, increases 24 h after I/R, and that IL-17 levels were raised 72 h later, suggesting that the induction of IL-17 was dependent on IL-23 [101]. It appears that IL-23 has greater pathophysiologic relevance immediately after I/R injury, while IL-17 gains importance at a later stage, when further apoptotic neuronal death occurs in the penumbra. However, contrary to the commonly held view of CD4+ helper T-cell production of IL-17, the main source of IL-17 during I/R in brain appears to be γδT lymphocytes. This is consistent with reports describing improved I/R injury following depletion of T lymphocytes. Thus, IL-17 production and modulation of γδT lymphocytes seems to be important targets for control of stroke.

RANTES

This chemokine is also known as CCL5. It is chemotactic for T cells, eosinophils, and basophils and plays an active role in recruiting leukocytes into inflammatory sites. RANTES has been implicated in the pathobiology of ischemic stroke. The increased leukocyte and platelet adhesion, blood–brain barrier permeability, and tissue infarction elicited in WT and control chimeric WT>WT mice after MCAO and reperfusion are significantly blunted in RANTES(−/−) mice. This protective effect is also observed in RANTES(−/−)>WT mice, suggesting that blood cell-produced RANTES contributes to the cerebral injury after ischemia [102].

MMP-9

This metalloproteinase is involved in the breakdown of extracellular matrix in normal physiological processes, such as embryonic development, reproduction, and tissue remodeling. Colocalization of MMP-9 and neutrophils has been described both within and at the periphery of an ischemic infarct detected 24 h after focal ischemia [103]. After 5 days, MMP-9 appears to stain with the macrophages present within the infarcted brain. Furthermore, MMP-9 blockade in wild-type mice as well as MMP-9$^{-/-}$ mice present with a significantly reduced infarct size, similar to that observed in bone marrow chimeras lacking leukocytic MMP-9 but not in chimeras with MMP-9-containing leukocytes [104]. These observations suggest that neutrophil and macrophage-released MMP-9 is important for the genesis of an infarct following ischemic stroke.

Neutrophil Elastase

This enzyme, also called ELA2 (elastase 2), is a serine proteinase secreted by neutrophils and macrophages during inflammation. It destroys bacteria but can cause damage to host tissue as well [105]. Neutrophil elastase degrades structural matrix proteins (e.g., elastin, collagens, laminins, and fibronectin), and endothelial junction proteins resulting in increased vascular permeability and leukocyte diapedesis [106, 107]. Release of neutrophil elastase appears to harm cerebral tissue during acute cerebral ischemia. Ikegame et al. [108] reported that most cerebral microvessels in I/R-induced infarcted region are destroyed and that sivelestat, a selective neutrophil elastase inhibitor, was effective in preserving the microvessels in the boundary zone. Sivelestat treatment also reduces brain edema, vascular permeability, and neurological deficit after acute focal ischemia [108].

IL-10

This anti-inflammatory cytokine is also known as human cytokine synthesis inhibitory factor (CSIF). It was recently observed that IL-10-released by B cells limits damage caused by ischemia in the brain. Ren et al. [38] reported greater brain damage caused by ischemia in B-cell-deficient μMT(−/−) mice, compared to WT controls and these MCAO-induced changes were completely prevented in B-cell-restored μMT(−/−) mice after transfer of highly purified WT GFP(+) B cells. However, this protection was no longer observed when B cells from IL-10(−/−) mice were transferred into μMT(−/−) mice. In a similar experiment, Frenkel et al. [109] showed that not only B cells but IL-10 released by CD4+ T cells protects the brain against the deleterious effects of ischemia. Overall, these findings implicate IL-10 as an important lymphocyte-derived mediator of protection against brain injury after cerebral I/R.

Adhesion Receptors as New Targets for Reducing Tissue Damage After Ischemic Stroke: Experimental and Clinical Trials

Targeting Selectins

Experimental Data

The clinical utility of P-selectin immunoblockade as a therapeutic strategy for reducing the brain damage after I/R is lessened by reports indicating that the protection noted in experimental models using antibody pretreatment is not evident in studies that administer the antibody after reperfusion [66, 110]. The same limitation has been described in nonhuman primates treated with a humanized monoclonal antibody against both E- and P-selectin (HuEP5C7) [62]. Although improved brain function was reported, there are no studies that employ the antibody after onset of the stroke. On the other hand, an antibody directed against E-selectin proved to be effective even when given 3 h after the beginning of the brain ischemia, suggesting that E-selectin may be a good target for prevention of brain damage following I/R [61]. A limitation of selectin blockers is that they appear to be more effective in preventing the tissue damage resulting from the combination of ischemia and reperfusion, rather than ischemia alone [111].

Another novel strategy for targeting selectins to reduce stroke-induced tissue injury in rats involves the nasal instillation of E-selectin. Takeda et al. [112] reported that nasal instillation of E-selectin potently inhibits the development of ischemic and hemorrhagic strokes in spontaneously hypertensive stroke-prone rats with untreated hypertension. Among the beneficial effects of E-selectin instillation are decreased infarction volume, increased numbers of Tregs in ischemic brain, reduced expression of tumor necrosis factor on blood vessels and an increase in the number of newly generated neuroblasts or neurons in the brain [113].

Clinical Trial

A phase I clinical trial is ongoing to define the maximum tolerated dose of recombinant E-selectin that can be instilled intranasally (ClinicalTrials.gov identifier: NCT00069069).

Targeting Integrins

Experimental Data

Similar to what has been observed with selectin targeting, blockade of integrins is therapeutically more efficient after tissue reperfusion [77]. Antagonism of neutrophil migration by treatment with a CD11/CD18 integrin blocking antibody (Hu23F2G) injected 20 min after occlusion reduces ischemic injury in a rabbit model of transient experimental stroke [114]. IB4, a monoclonal antibody directed against the CD18 leukocyte adhesion protein, was reported to reduce edema in ischemic brains of rabbits [115] and to improve reflow in cerebral blood vessels of baboons [116]. Matsuo et al. [33] observed the same beneficial results in rats using antibodies against CD11a (WT1 antibody) or CD18 (WT3 antibody) given before ischemia or immediately after reperfusion. In a similar way, treatment of pigs with a monoclonal antibody to the leukocyte adhesion glycoprotein complex CD11/CD18 severely attenuated both leukocyte adherence and the increase in vascular permeability [131]. It was also demonstrated that in experimental models of stroke, a combination of anti-b2 integrin (CD11/CD18) and thrombolytic therapy (rtPA) extends the therapeutic window for usage of rtPA, with a better outcome than the additive effects of these agents [83, 84]. In fact, treatment with UK-279,276, a selective CD11b/CD18 antagonist, in combination with rtPA at 2 or 4 h significantly ($P<0.01$) reduced infarct volume and enhanced recovery of neurological function, compared to controls [117]. The thrombolytic therapy was extended from 3–4 h after concomitant treatment with an integrin antagonist.

Clinical Trials

Leukarrest® (Hu23F2G)—This is a humanized anti-Mac-1 antibody that was previously tested in rabbits following MCAO and showed reduced infarct size and neuronal damage [114]. In a phase III clinical study, *Leukarrest*® from Icos (Bothell, WA) was halted while tested in stroke patients because it did not reach the success criteria [118] and no public information regarding the outcomes, safety issues, or relative number of severe adverse effects is available.

UK-279,276—This is a small recombinant glycoprotein that binds to CD11b integrin of Mac-1 and reduces infarct size in rats after stroke [119]. In a phase II clinical study, the UK-279,276 test was finished due to the absence of good results in UK-279,276-treated stroke patients. Only a slight improvement was observed when UK-279,276 was combined with rtPA treatment [120].

Rovelizumab®— It is a humanized monoclonal leukointegrin antibody developed by ICOS as a potential treatment for several diseases, including stroke. The company evaluated rovelizumab® in patients with ischemic stroke, in a double-blind, dose-escalating, placebo-controlled phase II trial in the US. Rovelizumab® was given 12 h after beginning of the symptoms and showed no significant difference in severe adverse effects compared to placebo treatment, with no immunogenicity observed [121].

Targeting Immunoglobulin-Like Domain Cell Adhesion Molecules

Experimental Data

ICAM-1 immunoblockade (1A29 antibody) has been reported to abrogate the damaging effects of I/R in the brain tissue of rats [33, 122]. Furthermore, antibodies directed against ICAM-1 are even more effective in reducing lesion area in a rabbit cerebral embolism stroke model when used concomitantly with tissue-type plasminogen activator, compared to the responses noted with either agent alone [123]. Thus, the work in animal models of stroke suggests that, when used either alone or in combination with thrombolytic therapy, ICAM-1 antibodies can be beneficial for treatment of ischemic stroke.

Clinical Trial

Enlimomab® (R6.5)—This is an anti-ICAM-1 murine monoclonal antibody tested in stroke patients in a phase III study. This study revealed that Enlimomab®

administration leads to higher mortality, infarct volume, and side effects in highly compromised patients [124]. This negative outcome is the likely result of using an antibody of murine origin, which can lead to an immunological response after treatment. In fact, tests made in rats showed murine anti-ICAM-1 antibodies can activate leukocytes secondary to the activation of complement [125, 126]. A verdict on the utility of anti-ICAM-1 therapy in ischemic stroke must await trials that test humanized antibodies.

Conclusion

Despite the massive amount of evidence showing benefits of leukocyte blocking in alleviating the tissue damage and cerebral dysfunction caused by cerebral ischemia in different animal species, similar outcomes have not been reproduced in humans. A variety of reasons have been offered to explain the absence of harmonization of anti-adhesion therapy outcome in animal models and stroke patients. It remains unclear whether the answer lies with the anti-adhesion therapy per se and an absence of importance of inflammation in human stroke or whether it reflects limitations of animals studies that do not allow for an accurate recapitulation of human stroke pathogenesis. Future advances in stroke research and in the discovery of effective therapeutic interventions for human stroke will rely on the continued improvement and refinement of technologies and approaches for assessing the responses of the human and animal brain to ischemic stroke. Since stroke continues to kill nearly six million people each year, the need for successful translation of successes in the laboratory to the clinical setting remains urgent.

References

1. Lopez AD, Mathers CD, Ezzati M, Jamison DT, Murray CJ (2006) Global and regional burden of disease and risk factors, 2001: systematic analysis of population health data. Lancet 367:1747–57
2. Strong K, Mathers C, Bonita R (2007) Preventing stroke: saving lives around the world. Lancet Neurol 6:182–7
3. Murphy J (2003) Pharmacological treatment of acute ischemic stroke. Crit Care Nurs Q 26(4):276–82
4. Adams HP Jr, Brott TG, Crowell RM, Furlan AJ, Gomez CR, Grotta J, Helgason CM, Marler JR, Woolson RF, Zivin JA (1994) Guidelines for the management of patients with acute ischemic stroke: a statement for healthcare professionals from a special writing group of the Stroke Council, American Heart Association. Circulation 90:1588–601
5. Khaja AM, Grotta JC (2007) Established treatments for acute ischaemic stroke. Lancet 27:319–30
6. Liesz A, Zhou W, Mracskó É, Karcher S, Bauer H, Schwarting S, Sun L, Bruder D, Stegemann S, Cerwenka A, Sommer C, Dalpke AH, Veltkamp R (2011) Inhibition of lymphocyte trafficking shields the brain against deleterious neuroinflammation after stroke. Brain 134:704–20
7. del Zoppo GJ (2010) Acute anti-inflammatory approaches to ischemic stroke. Ann N Y Acad Sci 1207:143–8
8. Yilmaz G, Granger DN (2010) Leukocyte recruitment and ischemic brain injury. Neuromolecular Med 12(2):193–204

9. Ziegler G, Freyer D, Harhausen D, Khojasteh U, Nietfeld W, Trendelenburg G (2011) Blocking TLR2 in vivo protects against accumulation of inflammatory cells and neuronal injury in experimental stroke. J Cereb Blood Flow Metab 31:757–66
10. Connolly ES Jr, Winfree CJ, Prestigiacomo CJ, Kim SC, Choudhri TF, Hoh BL, Naka Y, Solomon RA, Pinsky DJ (1997) Exacerbation of cerebral injury in mice that express the P-selectin gene: identification of P-selectin blockade as a new target for the treatment of stroke. Circ Res 81:304–10
11. Ishikawa M, Cooper D, Arumugam TV, Zhang JH, Nanda A, Granger DN (2004) Platelet-leukocyteendothelial cell interactions after middle cerebral artery occlusion and reperfusion. J Cereb Blood Flow Metab 24:907–15
12. Ren X, Akiyoshi K, Vandenbark AA, Hurn PD, Offner H (2011) Programmed death-1 pathway limits central nervous system inflammation and neurologic deficits in murine experimental stroke. Stroke 42:2578–83
13. Barone FC, Schmidt DB, Hillegass LM, Price WJ, White RF, Feuerstein GZ, Clark RK, Lee EV, Griswold DE, Sarau HM (1992) Reperfusion increases neutrophils and leukotriene B4 receptor binding in rat focal ischemia. Stroke 23:1337–47
14. Choi IY, Lee JC, Ju C, Hwang S, Cho GS, Lee HW, Choi WJ, Jeong LS, Kim WK (2011) A3 adenosine receptor agonist reduces brain ischemic injury and inhibits inflammatory cell migration in rats. Am J Pathol 179:2042–52
15. Choi JS, Park J, Suk K, Moon C, Park YK, Han HS (2011) Mild hypothermia attenuates intercellular adhesion molecule-1 induction via activation of extracellular signal-regulated kinase-1/2 in a focal cerebral ischemia model. Stroke Res Treat 2011:846716
16. Ritter LS, Orozco JA, Coull BM, McDonagh PF, Rosenblum WI (2000) Leukocyte accumulation and hemodynamic changes in the cerebral microcirculation during early reperfusion after stroke. Stroke 31:1153–61
17. Akopov SE, Simonian NA, Grigorian GS (1996) Dynamics of polymorphonuclear leukocyte accumulation in acute cerebral infarction and their correlation with brain tissue damage. Stroke 27:1739–43
18. Buck BH, Liebeskind DS, Saver JL, Bang OY, Yun SW, Starkman S, Ali LK, Kim D, Villablanca JP, Salamon N, Razinia T, Ovbiagele B (2008) Early neutrophilia is associated with volume of ischemic tissue in acute stroke. Stroke 39:355–60
19. Chen H, Chopp M, Bodzin G (1992) Neutropenia reduces the volume of cerebral infarct after transient middle cerebral artery occlusion in the rat. Neurosci Res Commun 11:93–9
20. Iadecola C, Alexander M (2001) Cerebral ischemia and inflammation. Curr Opin Neurol 14:89–94
21. Amantea D, Nappi G, Bernardi G, Bagetta G, Corasaniti MT (2009) Post-ischemic brain damage: pathophysiology and role of inflammatory mediators. FEBS J 276:13–26
22. Hayward NJ, Elliott PJ, Sawyer SD, Bronson RT, Bartus RT (1996) Lack of evidence for neutrophil participation during infarct formation following focal cerebral ischemia in the rat. Exp Neurol 139:188–202
23. Zhang RL, Chopp M, Chen H, Garcia JH (1994) Temporal profile of ischemic tissue damage, neutrophil response, and vascular plugging following permanent and transient (2H) middle cerebral artery occlusion in the rat. J Neurol Sci 125(1):3–10
24. Zhang RL, Chopp M, Li Y, Zaloga C, Jiang N, Jones ML, Miyasaka M, Ward PA (1994) Anti-ICAM-1 antibody reduces ischemic cell damage after transient middle cerebral artery occlusion in the rat. Neurology 44:1747–51
25. Brait VH, Arumugam TV, Drummond GR, Sobey CG (2012) Importance of T lymphocytes in brain injury, immunodeficiency, and recovery after cerebral ischemia. J Cereb Blood Flow Metab 32:598–611
26. Clark RK, Lee EV, White RF, Jonak ZL, Feuerstein GZ, Barone FC (1994) Reperfusion following focal stroke hastens inflammation and resolution of ischemic injured tissue. Brain Res Bull 35:387–92
27. Kriz J (2006) Inflammation in ischemic brain injury: timing is important. Crit Rev Neurobiol 18:145–57

28. Yilmaz G, Arumugam TV, Stokes KY, Granger DN (2006) Role of T lymphocytes and interferon-gamma in ischemic stroke. Circulation 113:2105–12
29. Breckwoldt MO, Chen JW, Stangenberg L, Aikawa E, Rodriguez E, Qiu S, Moskowitz MA, Weissleder R (2008) Tracking the inflammatory response in stroke in vivo by sensing the enzyme myeloperoxidase. Proc Natl Acad Sci USA 105:18584–9
30. Connolly ES Jr, Winfree CJ, Springer TA, Naka Y, Liao H, Yan SD, Stern DM, Solomon RA, Gutierrez-Ramos JC, Pinsky DJ (1996) Cerebral protection in homozygous null ICAM-1 mice after middle cerebral artery occlusion. Role of neutrophil adhesion in the pathogenesis of stroke. J Clin Invest 97:209–16
31. Stevens SL, Bao J, Hollis J, Lessov NS, Clark WM, Stenzel-Poore MP (2002) The use of flow cytometry to evaluate temporal changes in inflammatory cells following focal cerebral ischemia in mice. Brain Res 932:110–9
32. Matsuo Y, Onodera H, Shiga Y, Nakamura M, Ninomiya M, Kihara T, Kogure K (1994) Correlation between myeloperoxidase-quantified neutrophil accumulation and ischemia brain injury in the rat. Stroke 25:1469–75
33. Matsuo Y, Onodera H, Shiga Y, Shozuhara H, Ninomiya M, Kihara T, Tamatani T, Miyasaka M, Kogure K (1994) Role of cell adhesion molecules in brain injury after transient middle cerebral artery occlusion in the rat. Brain Res 656:344–52
34. Barone FC, Hillegass LM, Tzimas MN, Schmidt DB, Foley JJ, White RF, Price WJ, Feuerstein GZ, Clark RK, Griswold DE et al (1995) Time related changes in myeloperoxidase activity and leukotriene B4 receptor binding reflect leukocyte influx in cerebral focal stroke. Mol Chem Neuropathol 24:13–30
35. Jander S, Kraemer M, Schroeter M, Witte OW, Stoll G (1995) Lymphocytic infiltration and expression of intercellular adhesion molecule-1 in photochemically induced ischemia of the rat cortex. J Cereb Blood Flow Metab 15:42–51
36. Liesz A, Suri-Payer E, Veltkamp C, Doerr H, Sommer C, Rivest S, Giese T, Veltkamp R (2009) Regulatory T cells are key cerebroprotective immunomodulators in acute experimental stroke. Nat Med 15:192–9
37. Ren X, Akiyoshi K, Vandenbark AA, Hurn PD, Offner H (2011) CD4+FoxP3+ regulatory T-cells in cerebral ischemic stroke. Metab Brain Dis 26:87–90
38. Ren X, Akiyoshi K, Dziennis S, Vandenbark AA, Herson PS, Hurn PD, Offner H (2011) Regulatory B cells limit CNS inflammation and neurologic deficits in murine experimental stroke. J Neurosci 31:8556–63
39. Kubes P, Suzuki M, Granger DN (1991) Nitric oxide: an endogenous modulator of leukocyte adhesion. Proc Natl Acad Sci USA 88:4651–5
40. Gidday JM, Park TS, Shah AR, Gonzales ER (1998) Modulation of basal and postischemic leukocyte-endothelial adherence by nitric oxide. Stroke 29:1423–9
41. Tan J, Town T, Mori T, Obregon D, Wu Y, DelleDonne A, Rojiani A, Crawford F, Flavell RA, Mullan M (2002) CD40 is expressed and functional on neuronal cells. EMBO J 21:643–52
42. Omari KM, Dorovini-Zis K (2003) CD40 expressed by human brain endothelial cells regulates $CD4^+$ T cell adhesion to endothelium. J Neuroimmunol 134:166–78
43. Henn V, Slupsky JR, Grafe M, Anagnostopoulos I, Forster R, Muller-Berghaus G, Kroczek RA (1998) CD40 ligand on activated platelets triggers an inflammatory reaction of endothelial cells. Nature 391:591–4
44. Monaco C, Andreakos E, Young S, Feldmann M, Paleolog ET (2002) Cellmediated signaling to vascular endothelium: induction of cytokines, chemokines, and tissue factor. J Leukoc Biol 71:659–68
45. Ishikawa M, Vowinkel T, Stokes KY, Arumugam TV, Yilmaz G, Nanda A, Granger DN (2005) CD40/CD40 ligand signaling in mouse cerebral microvasculature after focal ischemia/reperfusion. Circulation 111:1690–6
46. Caolo V, Molin DG, Post MJ (2012) Notch regulation of hematopoiesis, endothelial precursor cells, and blood vessel formation: orchestrating the vasculature. Stem Cells Int 2012:805602

47. Chi Z, Zhang J, Tokunaga A, Harraz MM, Byrne ST, Dolinko A, Xu J, Blackshaw S, Gaiano N, Dawson TM, Dawson VL (2012) Botch promotes neurogenesis by antagonizing Notch. Dev Cell 22(4):707–20
48. Arumugam TV, Chan SL, Jo DG, Yilmaz G, Tang SC, Cheng A, Gleichmann M, Okun E, Dixit VD, Chigurupati S, Mughal MR, Ouyang X, Miele L, Magnus T, Poosala S, Granger DN, Mattson MP (2006) Gamma secretase-mediated Notch signaling worsens brain damage and functional outcome in ischemic stroke. Nat Med 12:621–3
49. Carr MW, Roth SJ, Luther E, Rose SS, Springer TA (1994) Monocyte chemoattractant protein 1 acts as a T-lymphocyte chemoattractant. Proc Natl Acad Sci USA 91:3652–6
50. Gerard C, Rollins BJ (2001) Chemokines and disease. Nat Immunol 2:108–15
51. Xia M, Sui Z (2009) Recent developments in CCR2 antagonists. Expert Opin Ther Pat 19:295–303
52. Liu L, Kubes P (2003) Molecular mechanisms of leukocyte recruitment: organ-specific mechanisms of action. Thromb Haemost 89:213–20
53. Barkalow FJ, Goodman MJ, Gerritsen ME, Mayadas TN (1996) Brain endothelium lack one of two pathways of P-selectin-mediated neutrophil adhesion. Blood 88:4585–93
54. Granger DN, Stokes KY (2005) Differential regulation of leukocyte-endothelial cell interactions. In: Aird WC (ed) Endothelial cells in health and disease. Taylor & Francis, Boca Raton, FL, pp 229–43
55. Zhang R, Chopp M, Zhang Z, Jiang N, Powers C (1998) The expression of P- and E-selectins in three models of middle cerebral artery occlusion. Brain Res 785(2):207–14
56. Stanimirovic D, Shapiro A, Wong J, Hutchison J, Durkin J (1997) The induction of ICAM-1 in human cerebromicrovascular endothelial cells (HCEC) by ischemia-like conditions promotes enhanced neutrophil/HCEC adhesion. J Neuroimmunol 76:193–205
57. Tamaru M, Tomura K, Sakamoto S, Tezuka K, Tamatani T, Narumi S (1998) Interleukin-1beta induces tissue- and cell type-specific expression of adhesion molecules in vivo. Arterioscler Thromb Vasc Biol 18(8):1292–303
58. Carvalho-Tavares J, Hickey MJ, Hutchison J, Michaud J, Sutcliffe IT, Kubes P (2000) A role for platelets and endothelial selectins in tumor necrosis factor-alphainduced leukocyte recruitment in the brain microvasculature. Circ Res 87:1141–8
59. Hu XM, Zhang Y, Zeng FD (2004) Effects of sodium beta-aescin on expression of adhesion molecules and migration of neutrophils after middle cerebral artery occlusion in rats. Acta Pharmacol Sin 25:869–75
60. Zhang RL, Chopp M, Zhang ZG, Phillips ML, Rosenbloom CL, Cruz R, Manning A (1996) E-selectin in focal cerebral ischemia and reperfusion in the rat. J Cereb Blood Flow Metab 16:1126–36
61. Huang J, Choudhri TF, Winfree CJ, McTaggart RA, Kiss S, Mocco J, Kim LJ, Protopsaltis TS, Zhang Y, Pinsky DJ, Connolly ES Jr (2000) Postischemic cerebrovascular E-selectin expression mediates tissue injury in murine stroke. Stroke 31:3047–53
62. Mocco J, Choudhri T, Huang J, Harfeldt E, Efros L, Klingbeil C, Vexler V, Hall W, Zhang Y, Mack W, Popilskis S, Pinsky DJ, Connolly ES Jr (2002) HuEP5C7 as a humanized monoclonal anti-E/P-selectin neurovascular protective strategy in a blinded placebo-controlled trial of nonhuman primate stroke. Circ Res 91:907–14
63. Fassbender K, Mössner R, Motsch L, Kischka U, Grau A, Hennerici M (1995) Circulating selectin- and immunoglobulin-type adhesion molecules in acute ischemic stroke. Stroke 26(8):1361–4
64. Kasahara Y, Taguchi A, Uno H, Nakano A, Nakagomi T, Hirose H, Stern DM, Matsuyama T (2010) Telmisartan suppresses cerebral injury in a murine model of transient focal ischemia. Brain Res 1340:70–80
65. Love S, Barber R (2001) Expression of P-selectin and intercellular adhesion molecule-1 in human brain after focal infarction or cardiac arrest. Neuropathol Appl Neurobiol 27:465–73
66. Suzuki H, Abe K, Tojo SJ, Kitagawa H, Kimura K, Mizugaki M, Itoyama Y (1999) Reduction of ischemic brain injury by anti-P-selectin monoclonal antibody after permanent middle cerebral artery occlusion in rat. Neurol Res 21(3):269–76

67. Jin AY, Tuor UI, Rushforth D, Kaur J, Muller RN, Petterson JL, Boutry S, Barber PA (2010) Reduced blood brain barrier breakdown in P-selectin deficient mice following transient ischemic stroke: a future therapeutic target for treatment of stroke. BMC Neurosci 11:12
68. Ruehl ML, Orozco JA, Stoker MB, McDonagh PF, Coull BM, Ritter LS (2002) Protective effects of inhibiting both blood and vascular selectins after stroke and reperfusion. Neurol Res 24:226–32
69. Htun P, Fateh-Moghadam S, Tomandl B, Handschu R, Klinger K, Stellos K, Garlichs C, Daniel W, Gawaz M (2006) Course of platelet activation and platelet-leukocyte interaction in cerebrovascular ischemia. Stroke 37:2283–7
70. Wei YS, Lan Y, Meng LQ, Nong LG (2011) The association of L-selectin polymorphisms with L-selectin serum levels and risk of ischemic stroke. J Thromb Thrombolysis 32:110–5
71. Yenari MA, Sun GH, Kunis DM, Onley D, Vexler V (2001) L-selectin inhibition does not reduce injury in a rabbit model of transient focal cerebral ischemia. Neurol Res 23:72–8
72. Arumugam TV, Salter JW, Chidlow JH, Ballantyne CM, Kevil CG, Granger DN (2004) Contributions of LFA-1 and Mac-1 to brain injury and microvascular dysfunction induced by transient middle cerebral artery occlusion. Am J Physiol Heart Circ Physiol 287:H2555–60
73. Chen H, Chopp M, Zhang RL, Bodzin G, Chen Q, Rusche JR, Todd RF 3rd (1994) Anti-CD11b monoclonal antibody reduces ischemic cell damage after transient focal cerebral ischemia in rat. Ann Neurol 35:458–63
74. Fiszer U, Korczak-Kowalska G, Palasik W, Korlak J, Górski A, Członkowska A (1998) Increased expression of adhesion molecule CD18 (LFA-1beta) on the leukocytes of peripheral blood in patients with acute ischemic stroke. Acta Neurol Scand 97:221–4
75. Khan M, Sekhon B, Giri S, Jatana M, Gilg AG, Ayasolla K, Elango C, Singh AK, Singh I (2005) S-Nitrosoglutathione reduces inflammation and protects brain against focal cerebral ischemia in a rat model of experimental stroke. J Cereb Blood Flow Metab 25:177–92
76. Kim JS, Chopp M, Chen H, Levine SR, Carey JL, Welch KM (1995) Adhesive glycoproteins CD11a and CD18 are upregulated in the leukocytes from patients with ischemic stroke and transient ischemic attacks. J Neurol Sci 128:45–50
77. Prestigiacomo CJ, Kim SC, Connolly ES Jr, Liao H, Yan SF, Pinsky DJ (1999) CD18-mediated neutrophil recruitment contributes to the pathogenesis of reperfused but not nonreperfused stroke. Stroke 30:1110–7
78. Soriano SG, Coxon A, Wang YF, Frosch MP, Lipton SA, Hickey PR, Mayadas TN (1999) Mice deficient in Mac-1 (CD11b/CD18) are less susceptible to cerebral ischemia/reperfusion injury. Stroke 30:134–9
79. Okada Y, Copeland BR, Mori E, Tung MM, Thomas WS, del Zoppo GJ (1994) P-selectin and intercellular adhesion molecule-1 expression after focal brain ischemia and reperfusion. Stroke 25:202–11
80. Lindsberg PJ, Carpén O, Paetau A, Karjalainen-Lindsberg ML, Kaste M (1996) Endothelial ICAM-1 expression associated with inflammatory cell response in human ischemic stroke. Circulation 94:939–45
81. Chen H, Song YS, Chan PH (2009) Inhibition of NADPH oxidase is neuroprotective after ischemia-reperfusion. J Cereb Blood Flow Metab 29:1262–72
82. Ding YH, Young CN, Luan X, Li J, Rafols JA, Clark JC, McAllister JP 2nd, Ding Y (2005) Exercise preconditioning ameliorates inflammatory injury in ischemic rats during reperfusion. Acta Neuropathol 109:237–46
83. Zhang RL, Zhang ZG, Chopp M, Zivin JA (1999) Thrombolysis with tissue plasminogen activator alters adhesion molecule expression in the ischemic rat brain. Stroke 30:624–9
84. Zhang RL, Zhang ZG, Chopp M (1999) Increased therapeutic efficacy with rt-PA and anti-CD18 antibody treatment of stroke in the rat. Neurology 52:273–9
85. Vemuganti R, Dempsey RJ, Bowen KK (2004) Inhibition of intercellular adhesion molecule-1 protein expression by antisense oligonucleotides is neuroprotective after transient middle cerebral artery occlusion in rat. Stroke 35:179–84
86. Brait VH, Jackman KA, Walduck AK, Selemidis S, Diep H, Mast AE, Guida E, Broughton BR, Drummond GR, Sobey CG (2010) Mechanisms contributing to cerebral infarct size after

stroke: gender, reperfusion, T lymphocytes, and Nox2-derived superoxide. J Cereb Blood Flow Metab 30:1306–17

87. Hoyte LC, Brooks KJ, Nagel S, Akhtar A, Chen R, Mardiguian S, McAteer MA, Anthony DC, Choudhury RP, Buchan AM, Sibson NR (2010) Molecular magnetic resonance imaging of acute vascular cell adhesion molecule-1 expression in a mouse model of cerebral ischemia. J Cereb Blood Flow Metab 30:1178–87
88. Wang J, Zhao Y, Liu C, Jiang C, Zhao C, Zhu Z (2011) Progesterone inhibits inflammatory response pathways after permanent middle cerebral artery occlusion in rats. Mol Med Rep 4:319–24
89. Krupinski J, Kaluza J, Kumar P, Kumar S, Wang JM (1994) Role of angiogenesis in patients with cerebral ischemic stroke. Stroke 25:1794–8
90. Justicia C, Martín A, Rojas S, Gironella M, Cervera A, Panés J, Chamorro A, Planas AM (2006) Anti-VCAM-1 antibodies did not protect against ischemic damage either in rats or in mice. J Cereb Blood Flow Metab 26:421–32
91. Petrovic-Djergovic D, Hyman MC, Ray JJ, Bouis D, Visovatti SH, Hayasaki T, Pinsky DJ (2012) Tissue-resident ecto-5′ nucleotidase (CD73) regulates leukocyte trafficking in the ischemic brain. J Immunol 188(5):2387–98
92. Jin G, Tsuji K, Xing C, Yang YG, Wang X, Lo EH (2009) CD47 gene knockout protects against transient focal cerebral ischemia in mice. Exp Neurol 217:165–70
93. Wang X, Xu L, Wang H, Zhan Y, Puré E, Feuerstein GZ (2002) CD44 deficiency in mice protects brain from cerebral ischemia injury. J Neurochem 83:1172–9
94. Airas L, Lindsberg PJ, Karjalainen-Lindsberg ML, Mononen I, Kotisaari K, Smith DJ, Jalkanen S (2008) Vascular adhesion protein-1 in human ischaemic stroke. Neuropathol Appl Neurobiol 34:394–402
95. Kim JB, Sig Choi J, Yu YM, Nam K, Piao CS, Kim SW, Lee MH, Han PL, Park JS, Lee JK (2006) HMGB1, a novel cytokine-like mediator linking acute neuronal death and delayed neuroinflammation in the postischemic brain. J Neurosci 26:6413–21
96. Fujioka M, Nakano T, Hayakawa K, Irie K, Akitake Y, Sakamoto Y, Mishima K, Muroi C, Yonekawa Y, Banno F, Kokame K, Miyata T, Nishio K, Okuchi K, Iwasaki K, Fujiwara M, Siesjö BK (2012) ADAMTS13 gene deletion enhances plasma high-mobility group box1 elevation and neuroinflammation in brain ischemia-reperfusion injury. Neurol Sci 33:1107–15
97. Kim ID, Shin JH, Lee HK, Jin YC, Lee JK (2012) Intranasal delivery of HMGB1-binding heptamer peptide confers a robust neuroprotection in the postischemic brain. Neurosci Lett 525:179–83
98. Cao CX, Yang QW, Lv FL, Cui J, Fu HB, Wang JZ (2007) Reduced cerebral ischemia-reperfusion injury in Toll-like receptor 4 deficient mice. Biochem Biophys Res Commun 353:509–14
99. Yang QW, Li JC, Lu FL, Wen AQ, Xiang J, Zhang LL, Huang ZY, Wang JZ (2008) Upregulated expression of toll-like receptor 4 in monocytes correlates with severity of acute cerebral infarction. J Cereb Blood Flow Metab 28:1588–96
100. Shichita T, Hasegawa E, Kimura A, Morita R, Sakaguchi R, Takada I, Sekiya T, Ooboshi H, Kitazono T, Yanagawa T, Ishii T, Takahashi H, Mori S, Nishibori M, Kuroda K, Akira S, Miyake K, Yoshimura A (2012) Peroxiredoxin family proteins are key initiators of post-ischemic inflammation in the brain. Nat Med 18:911–7
101. Shichita T, Sugiyama Y, Ooboshi H, Sugimori H, Nakagawa R, Takada I, Iwaki T, Okada Y, Iida M, Cua DJ, Iwakura Y, Yoshimura A (2009) Pivotal role of cerebral interleukin-17-producing gammadeltaT cells in the delayed phase of ischemic brain injury. Nat Med 15:946–50
102. Terao S, Yilmaz G, Stokes KY, Russell J, Ishikawa M, Kawase T, Granger DN (2008) Blood cell-derived RANTES mediates cerebral microvascular dysfunction, inflammation, and tissue injury after focal ischemia-reperfusion. Stroke 39:2560–70
103. Romanic AM, White RF, Arleth AJ, Ohlstein EH, Barone FC (1998) Matrix metalloproteinase expression increases after cerebral focal ischemia in rats: inhibition of matrix metalloproteinase-9 reduces infarct size. Stroke 29:1020–30

104. Gidday JM, Gasche YG, Copin JC, Shah AR, Perez RS, Shapiro SD, Chan PH, Park TS (2005) Leukocyte-derived matrix metalloproteinase-9 mediates blood-brain barrier breakdown and is proinflammatory after transient focal cerebral ischemia. Am J Physiol Heart Circ Physiol 289:H558–68
105. Belaaouaj A, Kim KS, Shapiro SD (2000) Degradation of outer membrane protein A in Escherichia coli killing by neutrophil elastase. Science 289:1185–8
106. Hermant B, Bibert S, Concord E, Dublet B, Weidenhaupt M, Vernet T, Gulino-Debrac D (2003) Identification of proteases involved in the proteolysis of vascular endothelium cadherin during neutrophil transmigration. J Biol Chem 278:14002–12
107. Ishikawa N, Oda M, Kawaguchi M, Tsunezuka Y, Watanabe G (2003) The effects of a specific neutrophil elastase inhibitor (ONO-5046) in pulmonary ischemia-reperfusion injury. Transpl Int 16:341–6
108. Ikegame Y, Yamashita K, Hayashi S, Yoshimura S, Nakashima S, Iwama T (2010) Neutrophil elastase inhibitor prevents ischemic brain damage via reduction of vasogenic edema. Hypertens Res 33:703–7
109. Frenkel D, Huang Z, Maron R, Koldzic DN, Moskowitz MA, Weiner HL (2005) Neuroprotection by IL-10-producing MOG CD4+ T cells following ischemic stroke. J Neurol Sci 233:125–32
110. Goussev AV, Zhang Z, Anderson DC, Chopp M (1998) P-selectin antibody reduces hemorrhage and infarct volume resulting from MCA occlusion in the rat. J Neurosci 161:16–22
111. Morikawa E, Zhang SM, Seko Y, Toyoda T, Kirino T (1996) Treatment of focal cerebral ischemia with synthetic oligopeptide corresponding to lectin domain of selectin. Stroke 27:951–5
112. Takeda H, Spatz M, Ruetzler C, McCarron R, Becker K, Hallenbeck J (2002) Induction of mucosal tolerance to E-selectin prevents ischemic and hemorrhagic stroke in spontaneously hypertensive genetically stroke-prone rats. Stroke 33:2156–63
113. Ishibashi S, Maric D, Mou Y, Ohtani R, Ruetzler C, Hallenbeck JM (2009) Mucosal tolerance to E-selectin promotes the survival of newly generated neuroblasts via regulatory T-cell induction after stroke in spontaneously hypertensive rats. J Cereb Blood Flow Metab 29:606–20
114. Yenari MA, Kunis D, Sun GH, Onley D, Watson L, Turner S, Whitaker S, Steinberg GK (1998) Hu23F2G, an antibody recognizing the leukocyte CD11/CD18 integrin, reduces injury in a rabbit model of transient focal cerebral ischemia. Exp Neurol 153(2):223–33
115. Bednar MM, Wright SD, Raymond-Russell SJ, Kohut JJ, Gross CE (1996) IB4, a monoclonal antibody against the CD18 leukocyte adhesion protein, reduces intracranial pressure following thromboembolic stroke in the rabbit. Neurol Res 18:171–5
116. Mori E, del Zoppo GJ, Chambers JD, Copeland BR, Arfors KE (1992) Inhibition of polymorphonuclear leukocyte adherence suppresses no-reflow after focal cerebral ischemia in baboons. Stroke 23:712–8
117. Zhang L, Zhang ZG, Zhang RL, Lu M, Krams M, Chopp M (2003) Effects of a selective CD11b/CD18 antagonist and recombinant human tissue plasminogen activator treatment alone and in combination in a rat embolic model of stroke. Stroke 34:1790–5
118. Becker KJ (2002) Anti-leukocyte antibodies: LeukArrest (Hu23F2G) and Enlimomab (R6.5) in acute stroke. Curr Med Res Opin 18:18–22
119. Jiang N, Chopp M, Chahwala S (1998) Neutrophil inhibitory factor treatment of focal cerebral ischemia in the rat. Brain Res 788:25–34
120. Krams M, Lees KR, Hacke W, Grieve AP, Orgogozo JM, Ford GA, ASTIN Study Investigators (2003) Acute Stroke Therapy by Inhibition of Neutrophils (ASTIN): an adaptive dose-response study of UK-279,276 in acute ischemic stroke. Stroke 34(11):2543–8
121. Jones R (2000) Rovelizumab (ICOS Corp). IDrugs 3:442–6
122. Zhang RL, Chopp M, Jiang N, Tang WX, Prostak J, Manning AM, Anderson DC (1995) Anti-intercellular adhesion molecule-1 antibody reduces ischemic cell damage after transient but not permanent middle cerebral artery occlusion in the Wistar rat. Stroke 26:1438–42, discussion 1443

123. Bowes MP, Rothlein R, Fagan SC, Zivin JA (1995) Monoclonal antibodies preventing leukocyte activation reduce experimental neurologic injury and enhance efficacy of thrombolytic therapy. Neurology 45:815–9
124. Enlimomab Acute Stroke Trial Investigators (2001) Use of anti-ICAM-1 therapy in ischemic stroke: results of the Enlimomab Acute Stroke Trial. Neurology 57:1428–34
125. Furuya K, Takeda H, Azhar S, McCarron RM, Chen Y, Ruetzler CA, Wolcott KM, DeGraba TJ, Rothlein R, Hugli TE, del Zoppo GJ, Hallenbeck JM (2001) Examination of several potential mechanisms for the negative outcome in a clinical stroke trial of enlimomab, a murine anti-human intercellular adhesion molecule-1 antibody: a bedside-to-bench study. Stroke 32:2665–74
126. Vuorte J, Lindsberg PJ, Kaste M, Meri S, Jansson SE, Rothlein R, Repo H (1999) Anti-ICAM-1 monoclonal antibody R6.5 (Enlimomab) promotes activation of neutrophils in whole blood. J Immunol 162(4):2353–7
127. Barone FC, Feuerstein GZ (1999) Inflammatory mediators and stroke: new opportunities for novel therapeutics. J Cereb Blood Flow Metab 19(8):819–34
128. Arumugam TV, Cheng YL, Choi Y, Choi YH, Yang S, Yun YK, Park JS, Yang DK, Thundyil J, Gelderblom M, Karamyan VT, Tang SC, Chan SL, Magnus T, Sobey CG, Jo DG (2011) Evidence that gamma-secretase-mediated Notch signaling induces neuronal cell death via the nuclear factor-kappaB-Bcl-2-interacting mediator of cell death pathway in ischemic stroke. Mol Pharmacol 80(1):23–31
129. Schilling M, Strecker JK, Schäbitz WR, Ringelstein EB, Kiefer R (2009) Effects of monocyte chemoattractant protein 1 on blood-borne cell recruitment after transient focal cerebral ischemia in mice. Neuroscience 161(3):806–12
130. Wang H, Zhan Y, Xu L, Feuerstein GZ, Wang X (2001) Use of suppression subtractive hybridization for differential gene expression in stroke: discovery of CD44 gene expression and localization in permanent focal stroke in rats. Stroke 32(4):1020–7
131. Gidday JM, Park TS, Gonzales ER, Beetsch JW (1997) CD18-dependent leukocyte adherence and vascular injury in pig cerebral circulation after ischemia. Am J Physiol 272 (6 Pt 2):H2622–9

Chapter 4
Immune Cell-Derived Free Radicals in Acute Brain Injury

Purnima Narasimhan, Hiroyuki Sakata, Joo Eun Jung, Tatsuro Nishi, Takuma Wakai, Carolina M. Maier, and Pak H. Chan

Abstract Oxidative stress, which generates reactive oxygen species (ROS), plays an important role after acute brain injuries, including transient cerebral ischemia. Brain injuries like ischemic–reperfusion result in a surge of excess oxygen that leads to generation of free radicals. Free radicals are present at low levels in the normal state where they play a critical role in signaling pathways. Antioxidants help in maintaining the redox level in the cells, but during an insult this homeostasis is disturbed resulting in excessive ROS. Mitochondrial ROS are among the main intracellular ROS. Cerebral ischemia triggers inflammation in response to injury, which also leads to the generation of free radicals and eventually to neuronal cell death. Studies using genetically manipulated animals where antioxidant genes are overexpressed or knocked down show the key role that ROS play in ischemia. Oxidative stress affects the injured area in a multifaceted way. It activates apoptotic markers, inflammatory mediators including cytokines and chemokines, and transcriptional activators. Therefore, it has a significant function in cell death and survival signaling cascades. Several recent reports have demonstrated the various effects of ROS generation and its link to the inflammatory response after ischemia. In this chapter, we present an overview of these mechanisms that have been elucidated, focusing on the damaging effects of ROS and their crucial role in inflammation after stroke.

Abbreviations

cyt c	Cytochrome c
DAPI	4′,6 Diamidino-2-phenylindole
GPx	Glutathione peroxidase
HEt	Hydroethidine

P. Narasimhan • H. Sakata • J.E. Jung • T. Nishi • T. Wakai • C.M. Maier • P.H. Chan (✉)
Stanford University School of Medicine, Stanford, CA, USA
e-mail: phchan@stanford.edu

J. Chen et al. (eds.), *Immunological Mechanisms and Therapies in Brain Injuries and Stroke*, Springer Series in Translational Stroke Research, DOI 10.1007/978-1-4614-8915-3_4,

HO-1	Hemoxygenase-1
ICAM1	Intercellular adhesion molecule 1
IL	Interleukin
I/R	Ischemia/reperfusion
MCAO	Middle cerebral artery occlusion
MCP-1	Monocyte chemoattractant protein-1
MIP-1α	Macrophage inflammatory protein-1α
mNSS	Modified neurologic severity scores
NF-κB	Nuclear factor-kappa B
nNOS	Neuronal nitric oxide synthase
NO	Nitric oxide
non-PC	Non-preconditioned
NOX	NADPH oxidase
NQO1	NADH quinone oxidoreductase
NSCs	Neural stem cells
$O_2^{\bullet -}$	Superoxide anion
O.D.	Optical density
OH	Hydroxyl ion
ONOO	Peroxynitrite
PC	Preconditioned
ROS	Reactive oxygen species
s.d.	Standard deviation
siRNA	Small interfering RNA
SOD	Superoxide dismutase
STAT3	Signal transducer and activator of transcription 3
Tg	Transgenic
TNF-α	Tumor necrosis factor-α
TUNEL	Terminal deoxynucleotidyl transferase-mediated uridine 5′-triphosphate-biotin nick end labeling
Wt	Wild-type

Introduction

Free radicals are highly reactive molecules produced after various acute brain injuries, including stroke, which is the third major cause of death worldwide after heart disease and cancer. Oxidative stress is a major contributing factor to reperfusion injury that occurs after stroke. Many pro-oxidant enzymes, including nitric oxide synthase, cyclooxygenase, and xanthine oxidase, participate in oxidative injury in cerebral ischemia [9]. The most common free radicals are those that contain unpaired valence electrons in oxygen, such as superoxide and its derivatives hydrogen peroxide and hydroxyl ions. Other free radicals include reactive nitrogen species, such as nitric oxide, which reacts with superoxide to release peroxynitrite.

Low levels of pro-oxidants are critical signaling molecules that participate in normal neuronal and vascular functions, but when the homeostasis is perturbed due to acute injury, an increase in pro-oxidants can be detrimental [12]. Homeostasis maintains non-pathological levels of reactive oxygen species (ROS) as part of the body's natural antioxidant defenses. Several studies have reported the complexity of the mechanisms that lead to cell injury after ischemia/reperfusion (I/R). Both oxidative stress and inflammation are associated with the damage observed in the brain after stroke, but the mechanisms involved have not been elucidated. In this chapter, we focus on the damaging effects of free radicals in stroke and their relation to inflammatory mediators.

Ischemic Stroke and ROS

Acute ischemic stroke results from the occlusion of a cerebral artery due to thrombosis or embolism in the brain, which is then followed by reperfusion. I/R produces an oxygen surge that generates free radicals, eventually leading to neuronal damage. In the early period after ischemia, insufficient oxygen and glucose result in energy failure and cell damage. Furthermore, the sudden decrease in blood flow initiates disintegration of cell membranes, leading to neuronal death in the core of the ischemic region. However, in the surrounding cerebral area that has residual blood flow, neuronal cells can be salvaged depending on the duration of ischemia. This area surrounding the core is the penumbra, which refers to cells that are damaged but have not yet died. This suggests that targeting this region for new therapies could help rescue cells and reduce the extent of post-stroke disability. ROS generation and lipid peroxidation have been observed at both early and late time points after I/R. Different cell types in the brain react differently to ROS production. Neurons are more susceptible to cellular damage than astrocytes and endothelia. There is evidence that impaired mitochondrial function after ischemia results in a large release of ROS, in particular massive amounts of superoxide anions ($O_2^{\bullet-}$), in neurons and astrocytes. Subsequently, the free radicals trigger lipid peroxidation and release substances that promote and activate leukocytes and endothelial cell injury, with a secondary release of ROS and other proteases. Immune cells like neutrophils and microglia generate ROS as an inflammatory response. Damage continues in the penumbra, and over several days and weeks delayed brain damage occurs due to inflammation processes [2, 33, 48], eventually leading to neuronal cell death after I/R. Taken together, this growing evidence shows that free radicals play a major role in brain injury in a cell-specific and temporal manner.

Previous studies have shown that ROS are generated by mitochondria after I/R [9]. Additionally, NADPH oxidase (NOX) is activated after cerebral ischemia, resulting in $O_2^{\bullet-}$ generation, which then causes lipid and DNA damage [4, 6, 11].

Antioxidants

Endogenous antioxidants, such as superoxide dismutase (SOD), glutathione peroxidase (GPx), and catalase, act as natural defenses and help in the maintenance of low-concentration oxidants and redox homeostasis in tissue. Other antioxidants like Nrf2 play an important role in triggering the response of the second phase of ROS scavenging proteins. There are three main SODs, each having specific cellular localization and cofactors. SOD1 is a cytosolic enzyme requiring copper and zinc as cofactors for its activity. SOD2 is present in the mitochondria and requires manganese for enzyme activity. SOD3 is present mainly in the extracellular space, cerebral fluid, and vessels and its cofactors are copper and zinc. Studies using overexpressing and knockout SOD1 and SOD2 animals have yielded important evidence providing significant insights into neuroprotection after cerebral ischemia. ROS and multiple mechanisms have been identified in SOD1-blocked mitochondrial-mediated apoptosis and are involved in regulation of cell death and survival signal transduction pathways, including p53, p53-upregulated modulator of apoptosis and Bax [15, 17, 37].

Transcription regulation of SOD2 by signal transducer and activator of transcription 3 (STAT3), a novel transcription factor, shows evidence that deactivation of STAT3 leads to downregulation of the SOD2 gene, which results in overproduction of ROS under conditions of cerebral I/R, followed by neuronal cell death [25]. In these studies, STAT3 was downregulated at early postischemic reperfusion times. Interestingly, there was high phosphorylation of STAT3 in normal neuronal cells not subjected to ischemia (Fig. 4.1). This indicates that SOD2, a housekeeping gene that is highly inducible by various cellular stimuli, is critical for sustaining the defense system against oxidative stress.

Nrf2 is a transcription factor that regulates other antioxidant genes (phase II genes) whose proteins act to scavenge ROS through enzyme reactions. Nrf2 is normally bound to kelch-like erythroid cell-derived protein with CNC homology associated protein 1 in the cytoplasm [22]. When oxidative stress occurs, the complex dissociates and Nrf2 translocates to the nucleus. Activated Nrf2 binds to promoters of genes that specifically have the antioxidant-response element. These phase II genes include hemoxygenase-1, 1-ferritin, and GPx, which maintain the redox balance and influence the inflammatory response. Increasing Nrf2 activity after stroke reduces damage to the cortical penumbra 24 h after reperfusion [43]. Functional recovery improves with Nrf2 treatment up to 1 month after transient focal cerebral ischemia, suggesting Nrf2 plays a role in delayed apoptosis and inflammation [31]. Moreover, Nrf2-deficient mice are significantly more susceptible to ischemic brain injury and neurological deficits than wild-type (Wt) mice [43].

Studies done in our laboratory showed that neural stem cells (NSCs) are protected after transplantation in mice subjected to I/R injury when they are preconditioned with minocycline. This occurs via upregulation of Nrf2 and Nrf2-regulated antioxidant genes [42]. Transfecting the NSCs with Nrf2 small interfering RNA (siRNA) before transplantation abolishes minocycline-induced neuroprotection.

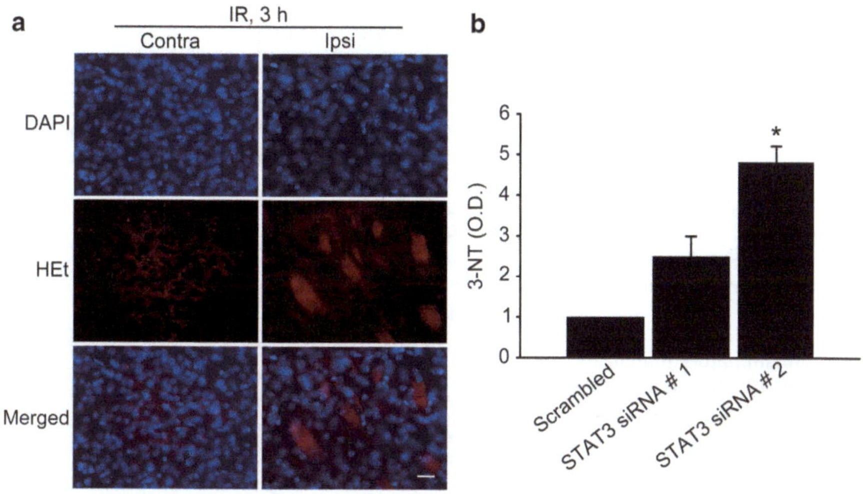

Fig. 4.1 (**a**) Inhibition of STAT3 by I/R increases generation of $O_2^{\bullet-}$. ROS production shown by hydroethidine (HEt, *red*) and 4′,6 diamidino-2-phenylindole (DAPI, *blue*) staining of ischemic regions (cortex) on the contralateral and ipsilateral sides. *Scale bar*, 20 μm. (**b**) STAT3 inhibition decreased SOD2 expression and increased protein nitrosylation. Summary graph depicting the changes in 3-nitrotyrosine (3-NT) relative to total protein loading in primary cortical neurons transfected with STAT3 siRNA. $^{*}P<0.05$ ($n=4$ per group). *OD* optical density [25]

This indicates that preconditioning with minocycline reprograms the NSCs to tolerate oxidative stress after I/R injury and to express higher levels of paracrine factors through Nrf2 upregulation. Furthermore, transplantation of these preconditioned NSCs significantly decreases infarct size and improves neurological performance.

GPx is a major antioxidant enzyme containing selenocysteine that scavenges ROS by reducing hydrogen peroxide and organic hydroperoxides to their corresponding alcohols. GPx-3 is the only member of the GPx family that is present in the extracellular space and plasma. GPx-3 deficiency results in significantly larger infarcts compared with Wt animals as a consequence of platelet activation and impaired endothelial function due to increased oxidative stress after I/R injury [24]. ROS generation facilitates recruitment of platelets to the growing platelet plug (thrombus). GPx-3-deficient mice treated with clopidogrel, a platelet inhibitor, have relatively small infarcts and better neurological scores, supporting the role of GPx-3 during oxidative stress, which contributes to platelet-dependent thrombosis and cerebral infarction. Ishibashi et al. [21] have shown that GPx-1-overexpressing mice have infarct volumes decreased by 48 % compared with Wt mice. Increased expression of GPx-1 in transgenic mice protects against I/R injury by modulating the inflammatory response and by decreasing the number of injured cells. There is an increase in transcriptional activation of cytokines and chemokines, including stress-response kinases that are targeted by GPx-1. Using gene transfer, overexpression of GPx was also shown to be neuroprotective after stroke, with increased survival of

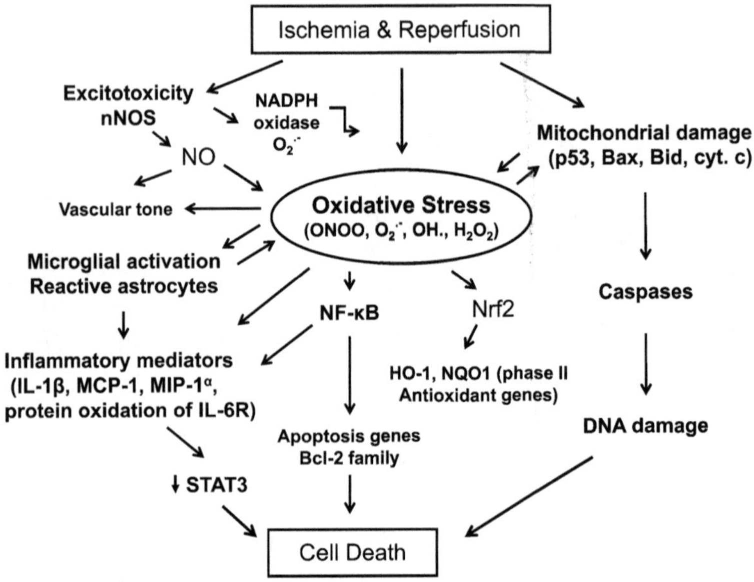

Fig. 4.2 Role of oxidative stress in the signaling pathways after I/R. Oxidative stress plays a major role in cerebral damage. In addition to mitochondrial-mediated apoptotic cell death, activation of inflammatory mediators leads to vascular dysfunction and neuronal cell death. *nNOS* neuronal nitric oxide synthase, *NO* nitric oxide, *ONOO* peroxynitrite, *OH* hydroxyl ion, *cyt c* cytochrome c, *HO-1* hemoxygenase-1, *NQO1* NADH quinone oxidoreductase

GPx-targeted striatal neurons via attenuation of apoptosis-related events with increased cytochrome c release [20]. These studies suggest that GPx is important in postischemic inflammation as well as in neuronal apoptotic signaling pathways.

ROS, Oxidative Stress, and Inflammation

In response to stroke, inflammation mediators and cytokines, in addition to free radicals, are generated along with activation of glia and microglia (Fig. 4.2). In fact, immune cells produce $O_2^{\bullet-}$ via NOX, which is comprised of membrane-bound subunits NOX-2 and p22 and cytoplasmic subunits p47, p67, and p40, as well as one of the small Rho guanosine triphosphate-binding proteins Rac1. Removal of NOX-4, another subunit, reduces brain infarction by more than 60 % at 24 h of reperfusion [29]. NOX is constitutively expressed in neurons, microglia, and astrocytes, in which it may function as a physiologic redox signal [4]. In NOX-2 knockout mice, there is a reduction in $O_2^{\bullet-}$ generated at 3 h of reperfusion after transient focal cerebral ischemia. Ablation of NOX-2 reduces oxidative stress [11]. This is ameliorated in SOD1-overexpressing mice, suggesting that NOX plays a role in oxidative stress and inflammation. Oxidative stress plays an important role in regulation of the

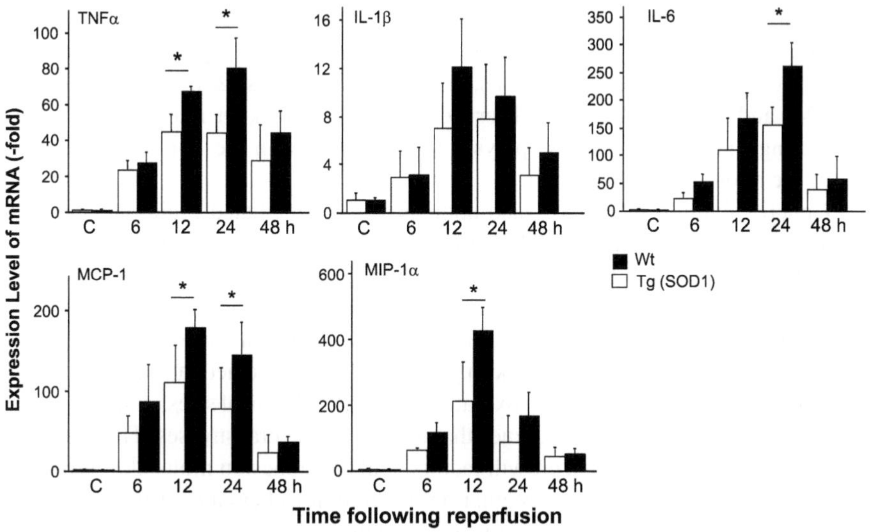

Fig. 4.3 Real-time quantitative reverse transcriptase polymerase chain reaction showing the time course of mRNA levels of proinflammatory cytokines (TNF-α, IL-1β, and IL-6) and β-chemokines (MCP-1 and MIP-1α). In general, the SOD1 transgenic (Tg) mice tended to have decreased mRNA expression compared with the Wt mice. Statistically significant differences were observed at 12 and 24 h of recirculation for TNF-α and MCP-1, at 12 h of recirculation for MIP-1α, and at 24 h of recirculation for IL-6. All data were standardized by glyceraldehyde 3-phosphate dehydrogenase and divided by controls (sham-operated animals). The *bars* show the mean ± s.d. ($^*P<0.05$) C, control [38]

postischemic neuroinflammatory process. Chen et al. [11] showed a greater reduction in expression of myeloperoxidase and intercellular adhesion molecule 1 (ICAM1) in NOX-2-knockout mice and apocynin-treated mice than in Wt mice at 24 h of reperfusion, suggesting that NOX inhibition can also protect against cerebral damage by alleviating inflammation processes such as neutrophil infiltration. Studies elucidating the molecules and pathways that are activated by NOX and that mediate changes in postischemic inflammation have shown the interaction between immune response and oxidative stress [5, 23, 27].

Activated microglia also play a key role in the inflammatory process after ischemia by releasing factors such as interleukin (IL)-1β, tumor necrosis factor-α (TNF-α), monocyte chemoattractant protein-1 (MCP-1), and macrophage inflammatory protein-1α (MIP-1α) [1]. Both MCP-1 and MIP-1α are induced after cerebral ischemia in rodents [49]. ROS facilitate MCP-1 and MIP-1α expression after ischemia [38]. Moreover, Nishi et al. [38] demonstrated that overexpression of the antioxidant enzyme SOD1 reduces MCP-1 and MIP-1α gene transcription and protein expression after brain ischemia (Fig. 4.3). The difference in protein expression between Wt and SOD1-overexpressing mice became very apparent at 24–48 h of reperfusion, which coincides with the peak of the inflammatory response in this transient focal ischemic model. MIP-1α is a well-known chemoattractant that modulates macrophage function [16] and triggers hydrogen peroxidase production in leukocytes. MIP-1α is mainly expressed in microglia/macrophages but has been observed in astrocytes [10], neurons, and endothelial cells [28]. This reduction in MCP-1 and MIP-1α was also observed in NOX-2-knockout mice [11], strongly indicating that oxidative stress plays a significant role in the immunomodulation response after cerebral ischemia.

Genetic ablation of NOX-2 also abolishes IL-1β-mediated brain damage after ischemia. After cerebral I/R, IL-1β is rapidly induced and exacerbates brain damage by inducing a strong inflammatory response [46]. It is thought that NOX-2 modulates the IL-1β effect by generating H_2O_2, which is well known as a critical redox-signaling intermediate [40]. NOX-2-derived H_2O_2 is also involved in the induction of the nuclear factor-kappa B (NF-κB) pathway after TNF-α activation [32]. Indeed, NF-κB is activated in response to oxidative stress and chemical stress and is involved in the induction of many of the genes that respond to several stresses. These include pro-inflammatory cytokines, growth factors, and cell adhesion molecules. NF-κB is a multiprotein complex and its regulation in cerebral ischemia is different depending on the nature of the damaging insult, the status of oxidative stress, and the time points after injury [13].

In SOD1-overexpressing mice, protein levels of IκBα, part of the NF-κB complex, do not change after transient focal cerebral ischemia. SOD1 prevents NF-κB activation and IκBα degradation. This finding suggests that transient loss of the IκB kinase α, β, and γ complex (part of NF-κB signaling) is mediated by ROS [44]. This supports other reports suggesting that NF-κB is regulated by the redox system [39]. NF-κB is thought to be triggered by ROS in leukocyte–endothelial interactions [48]. NF-κB activation elicits a cascade of events, including activation of adhesion molecules like ICAM-1, thereby mediating microvascular dysfunction and tissue injury [8, 51].

While some cytokines, such as IL-1, exacerbate cerebral injury after an ischemic insult, others like IL-6 and transforming growth factor-β provide neuroprotection and tissue repair. The role of IL-6 in stroke is controversial. Its expression significantly increases during the acute phase of cerebral ischemia and remains elevated in neurons and reactive microglia of the ischemic penumbra up to 14 days after ischemic insult [45]. Early studies in patients demonstrate a positive correlation between increased serum levels of IL-6 and larger infarct volumes and long-term poor outcomes [3, 47]. Recent reports support the idea that IL-6 has a neuroprotective effect in ischemic brain injuries [34, 45, 50]. Hirano et al. [19] and Brivanlou and Darnell [7] have reported that the presence of IL-6 results in the phosphorylation of STAT3, which then translocates to the nucleus and binds to specific promoters of target genes involved in cytoprotection and angiogenesis. The neuroprotective effect of exogenous IL-6 has been observed in PC12 cells in a hypoxia-reoxygenation paradigm [35]. Endogenous IL-6 transiently increases in the acute phase of cerebral ischemia and prevents neuronal cell death mediated by STAT3 activation [26, 50]. The mechanism by which IL-6 is regulated is described as the phosphatidylinositol 3-kinase/Akt and Ras/extracellular signal-regulated kinase pathways [36].

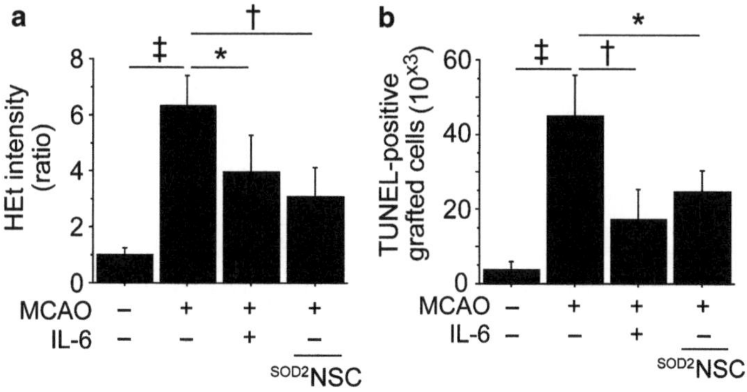

Fig. 4.4 Reduced grafted-cell death with IL-6 preconditioning in vivo. NSCs were transplanted into the brain 6 h after stroke. (**a**) Hydroethidine (HEt) signals increased in the nonpreconditioned NSCs under I/R injury, but this signal increase was reduced in the preconditioned NSCs and SOD2 NSCs ($n=4$). *MCAO* middle cerebral artery occlusion. (**b**) IL-6 preconditioning and SOD2 overexpression significantly reduced the number of terminal deoxynucleotidyl transferase-mediated uridine 5′-triphosphate-biotin nick end labeling (TUNEL)—positive-grafted cells in the ischemic brain. $^{*}P<0.05$, $^{\dagger}P<0.005$, and $^{\ddagger}P<0.001$ [41]

IL-6 acts by binding to IL-6R [18], which is composed of two major subunits, a ligand-binding α subunit (IL-6Rα) and a β subunit (gp130). Induction of gp130 results in phosphorylation of Janus-activating kinase and phosphorylation of tyrosine residues on gp130 providing docking sites for STAT3, which is recruited and phosphorylated. Phosphorylated STAT3 then translocates to the nucleus and binds to promoters of specific target genes, including SOD2. STAT3 is involved in regulation of the SOD2 gene in the mouse cerebral cortex and cortical neurons, thereby affecting the generation of ROS during oxidative stress by regulating SOD2 transcription, which is the cell's primary antioxidant defense [26]. IL-6 injected in mice subjected to cerebral ischemia and reperfusion protects neurons against oxidative stress. It restores signal transduction of STAT3-mediated SOD2 expression through the recovery of IL-6R association, which is blocked by I/R injury. Transfection with IL-6Rα or gp130 disrupts IL-6R function and significantly increases superoxide radical production via suppression of SOD2 transcription through blockage of STAT3 recruitment to the SOD2 promoter. This results in neuronal cell death caused by reduction in SOD2 expression and hence by suppression of the ROS defense system.

Protection offered by IL-6 has also been examined in NSCs. Indeed, preconditioning with IL-6 enhances the effectiveness of cell transplantation therapy after ischemic injury. Preconditioning with IL-6 protects transplanted NSCs from I/R injury. Preconditioned NSCs confer antioxidant properties by activating STAT3, which leads to induction of SOD2. Using STAT3 siRNA and SOD2 siRNA, IL-6-induced cytoprotection is abolished in NSCs subjected to oxygen–glucose deprivation and reoxygenation [41]. When NSCs are grafted into the ischemic penumbra 6 h after stroke, hydroethidine signals of NSCs preconditioned with IL-6 are significantly reduced compared with non-preconditioned NSCs (Fig. 4.4). Twenty-eight

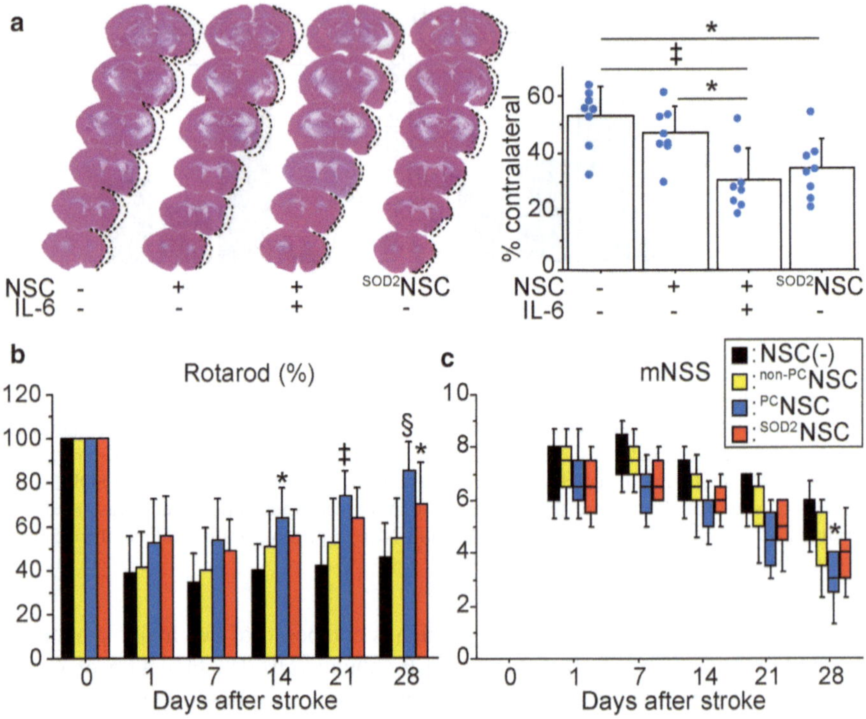

Fig. 4.5 Effects of preconditioned NSCs on infarct size and behavioral performance. NSCs were transplanted into the brain 6 h after stroke. (**a**) Cortical infarct size was significantly attenuated in the preconditioned NSC and SOD2 NSC groups compared with the nontransplanted control group ($n=4$) 28 days after stroke and transplantation. Transplantation of preconditioned NSCs resulted in the greatest functional recovery 28 days after stroke and transplantation ($n=8$). Indices of behavioral performance using the rotarod test (**b**) and modified neurologic severity scores (mNSS) (**c**). *Black bars* denote nontransplanted control group; *yellow bars* denote nonpreconditioned (non-PC) NSC group; *blue bars* denote preconditioned (PC) NSC group; and *red bars* denote SOD2 NSC group. The labels show *P*-value compared with the nontransplanted control group. $^{*}P<0.05$; $^{‡}P<0.005$; and $^{§}P<0.001$ [41]

days after stroke and transplantation, the NSCs migrate towards the ischemic lesion and a significantly greater number of IL-6 preconditioned NSCs survive compared with non-preconditioned NSCs. There is a significant reduction in infarct size as well as improvement in neurological performance as measured by rotarod test in the transplanted preconditioned NSC group compared with non-transplanted control and non-preconditioned NSC groups (Fig. 4.5). There is also a reduction in infarct size in the SOD2 NSC-transplanted group. Furthermore, preconditioned NSCs promote angiogenesis, as observed by increased blood vessel density 14 days after transplantation, through expression of STAT3-induced vascular endothelial growth factor. Transfection of STAT3 siRNA in IL-6 preconditioned NSCs abolishes angiogenesis and furthermore, there is no attenuation in infarct size or behavior.

In addition to the influence of inflammatory molecules on neurons during cerebral ischemia, a number of studies have reported on the effects of ROS in the microvasculature, including neutrophil and leukocyte infiltration and increased leukocyte–cerebrovascular endothelial cell interactions. Postischemic reperfusion is a markedly vulnerable time for the brain with a massive influx of ROS. In vascular cells, ROS provide signaling mechanisms affecting cerebral blood flow. Studies have demonstrated ROS action as endogenous vasodilators. Using arachidonic acid, dilation of feline cerebral arterioles is inhibited by the presence of SOD and catalase, an H_2O_2 scavenger [30]. SOD1 plays a role in affecting vascular tone where its genetic inhibition elicits a large increase in superoxide levels. This impairs nitric oxide-mediated vasodilation [14]. Evidence in recent years has also indicated that NOX represents a major source of ROS in cerebral arteries in the ischemic penumbra, which have detrimental effects in the perfusion of the penumbra, contributing to delayed neuronal damage.

In conclusion, numerous studies of cellular macromolecules have demonstrated that ROS are directly involved in oxidative damage such as lipid peroxidation, protein oxidation, protein nitrosylation/nitration, and nucleic acid damage, as well as modulation of the inflammatory response in ischemic tissue, all of which lead to cell death. The mechanisms of these oxidative stress-mediated processes have been elucidated by genetic manipulation of pro-oxidant or antioxidant enzymes, which has helped us to identify several key molecules that might have therapeutic potential against cerebral ischemic injury. Some of these molecules that have been promising in experimental studies have not held up during clinical trials. The discrepancies may be due to differences in the subjects, age, and comorbidity found in patients vs. experimental animal (rodent) studies. Stroke severity is another variable that is hard to translate from a rodent model to humans. The concept of targeting various antioxidant molecules with pleiotropic properties that block ROS generation as well as contain inflammation in the later stages and promote tissue repair would help in salvaging neurons and would also protect other cell types from oxidative injury. Further elucidation of mechanisms and the interrelationship between oxidative stress and inflammatory response would provide insights into the clinical treatment of stroke.

Acknowledgments This work was supported by NIH grants PO1 NS014543, RO1 NS025372, and RO1 NS038653, and by the James R. Doty Endowment.

References

1. Allan SM, Rothwell NJ (2001) Cytokines and acute neurodegeneration. Nat Rev Neurosci 2:734–44
2. Barone FC, Feuerstein GZ (1999) Inflammatory mediators and stroke: new opportunities for novel therapeutics. J Cereb Blood Flow Metab 19:819–34
3. Beamer NB, Coull BM, Clark WM, Hazel JS, Silberger JR (1995) Interleukin-6 and interleukin-1 receptor antagonist in acute stroke. Ann Neurol 37:800–5

4. Bedard K, Krause K-H (2007) The NOX family of ROS-generating NADPH oxidases: physiology and pathophysiology. Physiol Rev 87:245–313
5. Brait VH, Jackman KA, Walduck AK, Selemidis S, Diep H, Mast AE et al (2010) Mechanisms contributing to cerebral infarct size after stroke: gender, reperfusion, T lymphocytes, and Nox2-derived superoxide. J Cereb Blood Flow Metab 30:1306–17
6. Brennan AM, Suh SW, Won SJ, Narasimhan P, Kauppinen TM, Lee H et al (2009) NADPH oxidase is the primary source of superoxide induced by NMDA receptor activation. Nat Neurosci 12:857–63
7. Brivanlou AH, Darnell JE Jr (2002) Signal transduction and the control of gene expression. Science 295:813–8
8. Carvalho-Tavares J, Hickey MJ, Hutchison J, Michaud J, Sutcliffe IT, Kubes P (2000) A role for platelets and endothelial selectins in tumor necrosis factor-α–induced leukocyte recruitment in the brain microvasculature. Circ Res 87:1141–8
9. Chan PH (2001) Reactive oxygen radicals in signaling and damage in the ischemic brain. J Cereb Blood Flow Metab 21:2–14
10. Che X, Ye W, Panga L, Wu D-C, Yang G-Y (2001) Monocyte chemoattractant protein-1 expressed in neurons and astrocytes during focal ischemia in mice. Brain Res 902:171–7
11. Chen H, Song YS, Chan PH (2009) Inhibition of NADPH oxidase is neuroprotective after ischemia–reperfusion. J Cereb Blood Flow Metab 29:1262–72
12. Chrissobolis S, Faraci FM (2008) The role of oxidative stress and NADPH oxidase in cerebrovascular disease. Trends Mol Med 14:495–502
13. Denk A, Wirth T, Baumann B (2000) NF-κB transcription factors: critical regulators of hematopoiesis and neuronal survival. Cytokine Growth Factor Rev 11:303–20
14. Didion SP, Faraci FM (2002) Effects of NADH and NADPH on superoxide levels and cerebral vascular tone. Am J Physiol Heart Circ Physiol 282:H688–95
15. Endo H, Kamada H, Nito C, Nishi T, Chan PH (2006) Mitochondrial translocation of p53 mediates release of cytochrome *c* and hippocampal CA1 neuronal death after transient global cerebral ischemia in rats. J Neurosci 26:7974–83
16. Fahey TJ III, Tracey KJ, Tekamp-Olson P, Cousens LS, Jones WG, Shires GT et al (1992) Macrophage inflammatory protein 1 modulates macrophage function. J Immunol 148: 2764–9
17. Fujimura M, Morita-Fujimura Y, Noshita N, Sugawara T, Kawase M, Chan PH (2000) The cytosolic antioxidant copper/zinc-superoxide dismutase prevents the early release of mitochondrial cytochrome c in ischemic brain after transient focal cerebral ischemia in mice. J Neurosci 20:2817–24
18. Heinrich PC, Behrmann I, Müller-Newen G, Schaper F, Graeve L (1998) Interleukin-6-type cytokine signalling through the gp130/Jak/STAT pathway. Biochem J 334:297–314
19. Hirano T, Ishihara K, Hibi M (2000) Roles of STAT3 in mediating the cell growth, differentiation and survival signals relayed through the IL-6 family of cytokine receptors. Oncogene 19:2548–56
20. Hoehn B, Yenari MA, Sapolsky RM, Steinberg GK (2003) Glutathione peroxidase overexpression inhibits cytochrome *c* release and proapoptotic mediators to protect neurons from experimental stroke. Stroke 34:2489–94
21. Ishibashi N, Prokopenko O, Reuhl KR, Mirochnitchenko O (2002) Inflammatory response and glutathione peroxidase in a model of stroke. J Immunol 168:1926–33
22. Itoh K, Wakabayashi N, Katoh Y, Ishii T, Igarashi K, Engel JD et al (1999) Keap1 represses nuclear activation of antioxidant responsive elements by Nrf2 through binding to the amino-terminal Neh2 domain. Genes Dev 13:76–86
23. Jackson SH, Devadas S, Kwon J, Pinto LA, Williams MS (2004) T cells express a phagocyte-type NADPH oxidase that is activated after T cell receptor stimulation. Nat Immunol 5:818–27
24. Jin RC, Mahoney CE, Coleman Anderson L, Ottaviano F, Croce K, Leopold JA et al (2011) Glutathione peroxidase-3 deficiency promotes platelet-dependent thrombosis in vivo. Circulation 123:1963–73

25. Jung JE, Kim GS, Narasimhan P, Song YS, Chan PH (2009) Regulation of Mn-superoxide dismutase activity and neuroprotection by STAT3 in mice after cerebral ischemia. J Neurosci 29:7003–14
26. Jung JE, Kim GS, Chan PH (2011) Neuroprotection by interleukin-6 is mediated by signal transducer and activator of transcription 3 and antioxidative signaling in ischemic stroke. Stroke 42:3574–9
27. Kazama K, Anrather J, Zhou P, Girouard H, Frys K, Milner TA et al (2004) Angiotensin II impairs neurovascular coupling in neocortex through NADPH oxidase–derived radicals. Circ Res 95:1019–26
28. Kim JS, Gautam SC, Chopp M, Zaloga C, Jones ML, Ward PA et al (1995) Expression of monocyte chemoattractant protein-1 and macrophage inflammatory protein-1 after focal cerebral ischemia in the rat. J Neuroimmunol 56:127–34
29. Kleinschnitz C, Grund H, Wingler K, Armitage ME, Jones E, Mittal M et al (2010) Post-stroke inhibition of induced NADPH oxidase type 4 prevents oxidative stress and neurodegeneration. PLoS Biol 8(9):e1000479. doi:10.1371/journal.pbio.1000479
30. Kontos HA, Wei EP, Povlishock JT, Christman CW (1984) Oxygen radicals mediate the cerebral arteriolar dilation from arachidonate and bradykinin in cats. Circ Res 55:295–303
31. Lakhan SE, Kirchgessner A, Hofer M (2009) Inflammatory mechanisms in ischemic stroke: therapeutic approaches. J Transl Med 7:97
32. Li Q, Spencer NY, Oakley FD, Buettner GR, Engelhardt JF (2009) Endosomal Nox2 facilitates redox-dependent induction of NF-κB by TNF-α. Antioxid Redox Signal 11:1249–63
33. Lipton P (1999) Ischemic cell death in brain neurons. Physiol Rev 79:1431–568
34. Loddick SA, Turnbull AV, Rothwell NJ (1998) Cerebral interleukin-6 is neuroprotective during permanent focal cerebral ischemia in the rat. J Cereb Blood Flow Metab 18:176–9
35. Maeda Y, Matsumoto M, Hori O, Kuwabara K, Ogawa S, Yan SD et al (1994) Hypoxia/reoxygenation-mediated induction of astrocyte interleukin 6: a paracrine mechanism potentially enhancing neuron survival. J Exp Med 180:2297–308
36. Negoro S, Kunisada K, Fujio Y, Funamoto M, Darville MI, Eizirik DL et al (2001) Activation of signal transducer and activator of transcription 3 protects cardiomyocytes from hypoxia/reoxygenation-induced oxidative stress through the upregulation of manganese superoxide dismutase. Circulation 104:979–81
37. Niizuma K, Endo H, Nito C, Myer DJ, Chan PH (2009) Potential role of PUMA in delayed death of hippocampal CA1 neurons after transient global cerebral ischemia. Stroke 40: 618–25
38. Nishi T, Maier CM, Hayashi T, Saito A, Chan PH (2005) Superoxide dismutase 1 overexpression reduces MCP-1 and MIP-1α expression after transient focal cerebral ischemia. J Cereb Blood Flow Metab 25:1312–24
39. Oka S-I, Kamata H, Kamata K, Yagisawa H, Hirata H (2000) *N*-Acetylcysteine suppresses TNF-induced NF-κB activation through inhibition of IκB kinases. FEBS Lett 472:196–202
40. Park HS, Lee SH, Park D, Lee JS, Ryu SH, Lee WJ et al (2004) Sequential activation of phosphatidylinositol 3-kinase, βPix, Rac1, and Nox1 in growth factor-induced production of H_2O_2. Mol Cell Biol 24:4384–94
41. Sakata H, Narasimhan P, Niizuma K, Maier CM, Wakai T, Chan PH (2012) Interleukin 6-preconditioned neural stem cells reduce ischaemic injury in stroke mice. Brain 135: 3298–310
42. Sakata H, Niizuma K, Yoshioka H, Kim GS, Jung JE, Katsu M et al (2012) Minocycline-preconditioned neural stem cells enhance neuroprotection after ischemic stroke in rats. J Neurosci 32:3462–73
43. Shih AY, Li P, Murphy TH (2005) A small-molecule-inducible Nrf2-mediated antioxidant response provides effective prophylaxis against cerebral ischemia in vivo. J Neurosci 25:10321–35
44. Song YS, Lee Y-S, Chan PH (2005) Oxidative stress transiently decreases the IKK complex (IKKα, β, and γ), an upstream component of NF-κB signaling, after transient focal cerebral ischemia in mice. J Cereb Blood Flow Metab 25:1301–11

45. Suzuki S, Tanaka K, Suzuki N (2009) Ambivalent aspects of interleukin-6 in cerebral ischemia: inflammatory versus neurotrophic aspects. J Cereb Blood Flow Metab 29:464–79
46. Touzani O, Boutin H, LeFeuvre R, Parker L, Miller A, Luheshi G et al (2002) Interleukin-1 influences ischemic brain damage in the mouse independently of the interleukin-1 type I receptor. J Neurosci 22:38–43
47. Vila N, Castillo J, Dávalos A, Chamorro A (2000) Proinflammatory cytokines and early neurological worsening in ischemic stroke. Stroke 31:2325–9
48. Wang Q, Tang XN, Yenari MA (2007) The inflammatory response in stroke. J Neuroimmunol 184:53–68
49. Wang HK, Park UJ, Kim SY, Lee JH, Kim SU, Gwag BJ et al (2008) Free radical production in CA1 neurons induces MIP-1α expression, microglia recruitment, and delayed neuronal death after transient forebrain ischemia. J Neurosci 28:1721–7
50. Yamashita T, Sawamoto K, Suzuki S, Suzuki N, Adachi K, Kawase T et al (2005) Blockade of interleukin-6 signaling aggravates ischemic cerebral damage in mice: possible involvement of Stat3 activation in the protection of neurons. J Neurochem 94:459–68
51. Zhang RL, Chopp M, Jiang N, Tang WX, Prostak J, Manning AM et al (1995) Anti-intercellular adhesion molecule-1 antibody reduces ischemic cell damage after transient but not permanent middle cerebral artery occlusion in the Wistar rat. Stroke 26:1438–42

Chapter 5
The Complement Cascade in Acute Brain Injury

Michael McDowell, Nicholas Shea, Gaurav Gupta, and E. Sander Connolly Jr.

Abstract The prominent role of inflammatory pathways in acute brain injury has become increasingly clear in recent literature. The complement system represents a heterogeneous group of inflammatory molecules capable of being activated by numerous stimuli to a large number of ends. In this chapter, we review the mechanisms of the complement system, with emphasis on C5 and C3. We then present the leading theories of the conflicting role of the complement system in central nervous system disease and the current state of investigations attempting to modify injury through modulation of the complement system.

Introduction

Acute brain injury (ABI) represents one of the principal neurological emergencies, with the primary etiologies being traumatic, hemorrhagic, and ischemic in nature. Stroke is the fourth most common cause of death and a leading cause of disability in the USA [1]. It affects patients of all ages and backgrounds and results in profound losses in terms of quality years of life and economic resources [2, 3]. In the USA, there are 795,000 strokes per year, of which approximately 87 % are ischemic and 10 % are intracerebral hemorrhage (ICH) [4, 5]. Traumatic head injury severe

M. McDowell • N. Shea • G. Gupta • E.S. Connolly Jr. (✉)
Department of Neurosurgery, Columbia University Medical Center, New York, NY, USA
e-mail: esc5@columbia.edu

J. Chen et al. (eds.), *Immunological Mechanisms and Therapies in Brain Injuries and Stroke*, Springer Series in Translational Stroke Research, DOI 10.1007/978-1-4614-8915-3_5,

enough to result in brain injury has been estimated to have an incidence of 100–600/100,000, with 15 % of this population suffering from severe brain injury [6, 7].

The medical and surgical management of ABI is rooted in applying interventions designed to either decrease the primary insult to cerebral tissue or to minimize secondary injury due to detrimental physiologic responses to the original injury. In recent years, numerous potential mechanisms and markers of secondary ABI have been elucidated and investigated as potential approaches to curbing the severity ABI. A major topic of interest has been the complement system, which has long been known to be a central part of the innate immune system, with a role spanning various mechanisms of cell lysis, opsonization, chemotaxis, and inflammation [8].

In this chapter, we summarize the current understanding of the role of the complement system in each category of ABI, and the current level of success in modulating the pathway as a method of improving patient outcomes.

Review of the Current Knowledge of the Complement Pathways

The complement cascade is an elegant component of the innate immune system that was originally associated with bacterial killing but has demonstrated roles in mechanisms of inflammation, cell signaling, phagocytosis, adaptive immunity, and opsonization. Traditionally, the activation of the complement system has been via three well-demarcated pathways: the classical, alternate, and lectin pathways. The mechanisms of all three pathways are similar enough to allow for generalization into the activation phase, C3 convertase phase, C5 convertase phase, and the membrane attack complex (MAC) phase [8].

Each pathway begins with a unique activation phase. The Classical pathway utilizes IgG and IgM binding to the surfaces of foreign cells to stimulate serine protease activation of C1, allowing for the generation of one type C3 convertase by cleaving C2 and C4 via serine protease action into the components required. This convertase is also formed via the lectin pathway, which cleaves C2 and C4 via a complex of mannose-binding lectin and associated proteins, which forms in the presence of the mannose molecules that are often incorporated into the structure of pathogen surfaces. The C3 convertase can then cleave C3 into its active form, which is incorporated into the convertase to form C5 convertase. C3 can alternatively be used to create an alternative form of C3 convertase, allowing for a cascade effect. This alternative C3 convertase is essential in the Alternative pathway of complement. C3 is capable of spontaneous activation when associated with cell surfaces, something that is vigorously inhibited on host cells, allowing for the formation of C3 convertase and the progression to C5 convertase as described above [9]. The cleavage of C5 convertase allows for the initiation of the MAC. C5b binds to C6, C7, and C8 to form a "scaffold" on the cell membrane for the attraction of C9, which extends through the membrane. C9 polymerizes with numerous copies of itself, forming a ring-shaped pore linking the intracellular and extracellular spaces [9, 10].

Signaling

Another product of the complement pathways are signaling molecules, composed of portions of the complement molecules that were not used in the direct pathway after cleavage. As discussed below, these molecules play a significant role in the specific mechanisms of complement in acute brain injury. The two major molecules are C5a and C3a. C5a is a highly potent inflammatory molecule, as well as a chemokine. There is substantial evidence that C5a also promotes apoptosis through MAP kinase-mediated caspase mechanisms and potentially directly via a tumor necrosis factor-type receptor, resulting in release of cytochrome c from mitochondria [9, 11, 12]. C3a, in conjunction with C5a, is a major anaphylatoxin responsible for stimulation of smooth muscle contraction, mast cell degranulation, and histamine release, and increasing the permeability of capillary beds, precipitating edema, and increasing the rate of chemotaxis [13].

Interactions with the Coagulation Pathway

Recent advances in the understanding of complement have yielded a potential new pathway through interactions with the coagulation cascade that may someday explain many of the idiosyncrasies of complement in acute brain injury. Early data showed that there was potential serine protease cross-reactivity, and inhibitors of the complement pathway have also been shown to be potent inhibitors of the coagulation pathway in vivo [14]. It has been well established that fibrin deposition with a counterbalancing fibrinolytic response occurs in situations of severe injury [15]. Similar fibrin deposition has been seen secondary to elevated thrombin levels detected in areas of complement-mediated inflammation [16]. At the same time, portions of the coagulation cascade have also demonstrated the ability to regulate complement levels of activity. Data have suggested a correlation between early coagulopathy after acute injury and hyperactivation of the complement system, particularly elevation of C3 and C5a [17, 18]. Some data suggest that thrombin may be capable of activation of C3 and C5 independent of traditional mechanisms, whereas plasmin may inactivate these molecules similarly to thrombin [19]. The abundance of serine protease activity in each pathway provides biological plausibility for potential cross talk between the two mechanisms (Fig. 5.1) [18].

Risk Factors, Etiology, Pathophysiology, Prognosis, and Treatment for Ischemic Stroke

Major risk factors for ischemic stroke include hypertension, hypercholesterolemia, heart disease, atrial fibrillation, diabetes mellitus, and cigarette smoking [20]. Most strokes result from an embolus or local thrombosis occluding a major cerebral

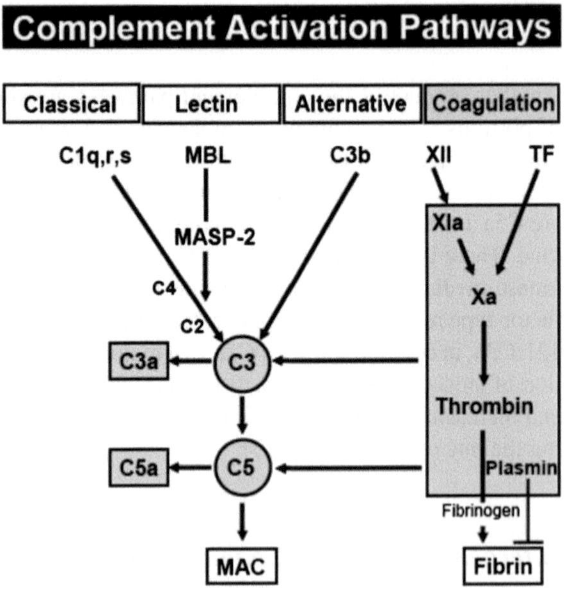

Fig. 5.1 Summary of complement activations pathways and the cross-regulation with the coagulation cascade (*MAC* membrane attack complex, *MBL* mannose-binding lectine, *MASP-2* mannose-associated serine protease-2, *TF* tissue factor). Reprinted with permission from Amara U, et al (2008) Interaction between the coagulation and complement system. Adv Exp Med Biol 632: 71–9

artery leading to infarction [21]. Secondary damage to the penumbra, the region around the irreversibly infarcted core, may occur despite prompt reperfusion ("reperfusion injury") [22]. The mechanisms mediating this process include the release of free radical species from damaged cells, cytotoxic edema, and inflammation [2, 23]. Weakening of the blood–brain barrier is maximal 3–6 h following ischemic onset [24] and promotes additional inflammation and edema, leading to increased intracranial pressure, mass effect, and potentially herniation [21, 25, 26].

Degree of edema is among the earliest markers of ensuing pathophysiology after ischemic stroke and a major predictor of survival beyond the first few hours [2]. Treatment for ischemic stroke includes a combination of recanalization therapies and supportive measures and requires achieving balance between preventing ischemic injury, minimizing reperfusion injury, and avoiding hemorrhage. Intravascular thrombolytic therapies administered within the first 3 h of stroke have been shown to be effective [27], but efficacy decreases as a function of time from symptom onset [28] and carries increased risk of fatal hemorrhage [27].

Risk Factors, Etiology, Pathophysiology, Prognosis, and Treatment for ICH

Hypertension is the main risk factor for and the most significant etiology of ICH [29]. Mass effect and midline shift caused by the hematoma physically disrupts neighboring tissues and brain structures and can result in secondary brain injury, neurological impairment, and, often, fatality [30–32]. Edema and inflammation, spurred on by activation of the complement cascade, follow shortly after hemorrhage and peak after several days [30, 33–36].

ICH volume and Glasgow Coma Scale grade on admission are the best predictors of death 30-day post-ictus [37, 38]. Edema is closely associated with poorer clinical outcomes [30, 32, 39, 40]. The current treatment for ICH consists of supportive measures, prevention of hematoma expansion [30], and maintenance of a safe intracranial pressure [38, 41]. Medical therapy entails glucose management, cerebral temperature control, and antiepileptic drugs in patients with seizures [38, 41].

Putative Mechanisms of Complement Activation and Complement-Mediated Injury in Acute Cerebrovascular Injury

Putative Mechanisms of Complement Activation and Complement-Mediated Injury in Ischemic Stroke

Complement activation and deposition on neurons is known to occur following cerebral ischemia [42–45]. The origin of this deposition is not currently well understood, but local synthesis of complement proteins may contribute. While complement components are primarily produced in the liver, astrocytes, microglia, oligodendrocytes, and neurons can express most of the complement proteins [46–51] and some types of brain cells have been shown to produce mRNA for the entire complement cascade in vitro [52–54].

Activation of the cascade in the setting of cerebral ischemia may occur through the MBL pathway or an autoimmune version of the classical pathway. In systemic organs, ischemia and reperfusion is believed to lead to exposure of self-antigens, which subsequently interact with circulating natural IgM antibodies and MBL to initiate the complement cascade [55]. A 2011 study by our laboratory demonstrates that MBL deficiency in mice reduces C3 cleavage as well as the accumulation of mononuclear cells in the ischemic region, providing evidence that such a process of activation may occur in ischemic stroke [56].

Once activated, complement contributes to the spread of the infarct into the surrounding penumbra regions. Cleavage of C3 and C5 produces the diffusible anaphylatoxins C3a and C5a, which serve as chemoattractants for leukocytes and may damage the blood–brain barrier further, allowing for propagation of inflammation and injury into healthy tissue. Initiation of inflammatory pathways leads to the release of proteases and free radicals by infiltrating neutrophils [57, 58] as well as increased leukocyte adhesion to capillary walls, which can result in microvascular failure [42]. Leukocytes also produce cytokines that further attract inflammatory cells and can potentially accumulate to toxic levels.

Putative Mechanisms of Complement Activation and Complement-Mediated Injury in ICH

Spontaneous intracerebral hemorrhage (ICH) is a disease associated with significant disability [37]. The estimated incidence of ICH is 10–20/100,000 people and typically presents after the fifth decade of life [59]. Commonly assessed risk factors include chronic hypertension, smoking, impaired coagulation, vascular malformations, and cerebral amyloid angiopathy [37]. Injury from mass effect during the initial bleed and subsequent expansion in up to 1/3 of patients disrupt tissue architecture and is the major source of early fatality or neurological impairment [30–32]. Post-ictal edema begins almost immediately upon hemorrhage and peaks mid-week [30–32, 60]. It is within the pathophysiology of edema and the underlying inflammation of nervous tissue that complement plays a significant role [33, 34].

The complement system is excluded for central nervous system tissues in normal physiology due to a lack of permeability across the blood–brain barrier. Following ICH, the complement system is present in local tissue as part of the original hematoma and secondarily through continued disruption of the blood–brain barrier [60]. Complement activation following ICH may depend upon prior activation of the coagulation cascade via normal hemostatic mechanisms. While not a traditionally accepted role, rapidly activated coagulation factors Xa, Xia, plasmin, and thrombin have been associated with cleavage of C3 and C5 via serine protease action into C3a, C3b, C5a, and C5b [18, 19, 61]. Much of this evidence stems from non-neurological models, but Gong et al. recently provided evidence that this extrinsic protease pathway is present in rat ICH models [62]. Additional evidence has been generated to support that thrombin may be a primary cause of blood–brain barrier disruption, early peri-hematomal edema, activation of complement in ICH [63–65].

C5 and C3 serve to promote cerebral injury via an anaphylatoxin-mediated mechanism (C5a and C3a) and a membrane attack complex (MAC)-mediated mechanism (C3b and C5b). The binding of C3a to its receptor has been found to increase granulocyte infiltration and edema when studied within ICH animal models [66]. The role of C5a is less clear in ICH specifically, but it is well known to be a powerful chemoattractant for neutrophils and is heavily involved in other types of acute brain injury [19, 67]. Further, C5a likely plays a causal role in increasing peri-hematomal edema [68]. C5 has the additional complexity of being activated in the traditional way via C3b-dependent pathways, or independent of C3 levels via the coagulation cascade [18, 19]. The exact mechanism in nervous tissue is still being elucidated, but these molecules assist in the upregulation and secretion of pro-inflammatory cytokines, including TNF-alpha, IL-1Alpha, and IL-1Beta, and stimulate leukocyte release of myeloperoxidase [13, 69, 70]. This milieu of inflammatory biomolecules results in secondary tissue injury and edema.

While C5a and C3a mediate neurological damage through secondary inflammation and edema, C5b and C3b are crucial regulators of direct MAC cell lysis. MAC formation is mediated by C5b forming the initial component of the complex, which subsequently attracts C6, C7, C8, and C9 to form the membrane pore. C5b is in term upregulated heavily by C3b-dependent C5 convertase, as well as potentially supplementary pathways such as via thrombin activation [63–65, 71, 72]. Activated MAC within the acute and resolving hematoma assist in hemolysis, causing a highly concentrated hemoglobin and iron level [73, 74]. Iron freed via upregulated heme oxygenase has a demonstrable effect on neural tissue via oxidative stress and caspase activation [65, 75, 76]. The data also suggest that hematoma development and resolution may play a role in late-onset edema starting several days after thrombin-related edema, approximately at the time of peak edema level [74]. Direct MAC destruction of parenchymal tissue has been associated with ABI, but its role in ICH is yet unclear [76].

Experimental Support for the Pathophysiological Role of Complement Activation in Acute Brain Injury

Complement Activation Is Implicated in the Pathophysiology of Other Central Nervous System Diseases

The complement cascade is implicated in the pathophysiology of a number of central nervous system diseases, including multiple sclerosis [77], Alzheimer's disease [47, 78–81], Pick's disease [47, 82, 83], Huntington's Disease [47], aneurysmal subarachnoid hemorrhage [84, 85], and age-related macular degeneration [86]. Perhaps most relevant to ischemic and hemorrhagic stroke, complement has been shown to be an important mediator of injury following acute brain trauma [54, 87–94].

Neurons Seem to Be Unusually Susceptible to Complement Activation

In vitro, neurons are capable of spontaneously activating the complement cascade, eventually leading to opsonization and lysis [95, 96]. This phenomenon may be due to a relative neuronal deficiency in key complement modulators such as membrane cofactor protein (MCP, CD46), membrane inhibitor of active lysis (MIRL, CD59), decay-activating factor (DAF, CD55), C4b-binding protein, and Factor H, all of which are more abundantly produced by astrocytes [51, 96].

Ischemic Stroke: Evidence that Complement Is Activated and Is Associated with Pathology

Early research with nonspecific complement inhibitors such as cobra venom factor (CVF) and C1 esterase inhibitor (C1-INH) demonstrated an important role for the complement cascade in the pathophysiology of ischemic brain injury. CVF has been shown to deplete systemic and brain complement in rat stroke models and result in reduced infarct size [44, 97] and diminished neurological deficits as measured by somatosensory evoked potentials (SSEPs) [98]. C1-INH treatment in rat and mouse models of transient focal brain ischemia/reperfusion (I/R) similarly results in reduced infarct volume, neutrophil accumulation, and neuronal damage [99–101] as well as significantly improved general and focal deficits [100].

A more specific complement inhibitor, soluble complement receptor type 1 (sCR1), promotes inactivation of C3b and methylamine-treated C4 (C4-ma), thereby blocking complement activation by multiple pathways [102]. In murine stroke models, sCR1 reduces neutrophil and platelet aggregation and significantly improves neurologic function as well as providing modest but not statistically significant reductions in cerebral infarct volume [45]. These effects are improved by sialylation of the sCR1 molecule to form a compound called sCR1-sLex [45]. Studies by our laboratory investigating sCR1 and sCR1-sLex treatment in nonhuman primates failed to demonstrate comparable neuroprotective effects, however [103, 104].

Further investigations have examined specific complement components. Cleavage of C3, the central and most abundant component of the complement cascade, appears to be a critical step in complement-related inflammatory tissue injury, although other components play a role as well [105, 106]. A 2006 study by our group examined C1q, C5, and C3 knockout murine models of focal cerebral ischemia and found that only the C3 strain was protected [107]. Furthermore, administration of a C3a receptor antagonist (C3aRA) to wild-type mice 45 min prior to ischemia resulted in neurological improvement and stroke volume reduction equivalent to that demonstrated in C3 knockout mice, indicating C3aR could represent a potential target of therapy in ischemic stroke [107].

Numerous studies have confirmed that complement is activated and deposited on neurons following cerebral ischemia in humans [43–45, 108]. Translation of results from laboratory models into human therapies has proven difficult, however. In 2001, the Enlimomab Acute Stroke Trial of an anti-ICAM-1 antibody was halted after Enlimomab-treated patients were found to experience worse outcomes than placebo-treated patients [109]. The mouse antibody utilized in this study was later shown to activate the complement cascade when incubated with blood from healthy human volunteers, providing a possible explanation for the adverse effects observed in the trial [110].

Complement activation following stroke in humans may also be distinct from murine models as a result of heterogeneity of activation of the different components over time. Work by our laboratory demonstrates that human plasma levels of complement proteins exhibit complex temporal variation following ischemic stroke. We found that C3a is elevated acutely after ischemia, C5a is elevated 7–14 days after ischemia, and soluble C5b-9 is depressed acutely after ischemia [108]. Additional evidence suggests that I/R injury in human ischemic stroke results not just from variable upregulation of complement proteins but also from variable downregulation of key complement regulators [111].

ICH: Evidence that Complement Is Activated and Is Associated with Pathology

The first direct evidence that complement activation plays a role in the pathophysiology of ICH was provided by Hua et al. [34, 35]. In a rat model of ICH, a sixfold increase in C9 deposition around the hematoma was identified 24 h after hemorrhage and peri-hematomal MAC formation was evident by 72 h. *N*-acetyl heparin, a complement inhibitor, significantly reduced brain edema in the ipsilateral basal ganglia at 24 h and 72 h after ICH. These results show that ICH leads to complement activation and that initiation of the MAC contributes to cerebral edema formation after ICH. Experiments by Xi and colleagues examining the effect of systemic depletion of complement proteins confirmed these findings [33, 68, 112]. Intraperitoneal administration of CVF in rats reduced brain edema at 24 h and 72 h. Cerebellar water was unaffected. TNF-α production was reduced 2 h after ICH. Peri-hematomal C9 deposition, C3d production, and the number of C5a- and myeloperoxidase-positive cells were significantly reduced by complement depletion. A similar study in rats using venom defibrase DF-521 batroxobin, another nonspecific complement inhibitor, likewise identified reduced expression of C3d and C9, as well as ICAM-1, in the peri-hematomal area after ICH [113].

Broad evidence of complement involvement in ICH pathogenesis prompted further investigation into the roles of specific complement proteins. The first such study by Nakamura and colleagues examined ICH in C5-deficient mice. This study produced surprising results. C5-deficient mice in the setting of ICH were found to have significantly increased brain water content at 3-day post-hemorrhage compared with C5-sufficient mice [114]. These paradoxic findings indicate that C5a may have a beneficial effect in this setting, a conclusion supported by previous work in models of hippocampal damage using transgenic C5a receptor (C5aR) knockout mice [115].

Given these findings regarding the potential protective role of C5a in ICH stroke and evidence suggesting a predominant role for C3 in the pathogenesis of cerebral I/R injury [107, 108, 116], subsequent work has focused on the role of C3 activation in ICH. Compared to C3-sufficient mouse ICH models, Yang et al. found C3-deficient mouse ICH models have less brain edema, lower hemeoxygenase-1 levels, and reduced microglia activation and neutrophil infiltration around the clot after ICH [117, 118]. C3-deficient mice also demonstrated improved neurological function compared to C3-sufficient mice. In 2008, using a new murine model of ICH developed in our laboratory, we demonstrated that a C3a receptor antagonist (C3aRA) attenuates brain injury after cerebral hemorrhage [66, 119]. Animals pretreated with C3aRA experienced significant amelioration of neurologic dysfunction, decreased brain water content, and reduced granulocyte infiltration relative to vehicle-treated animals at 72 h post-ICH. Animals treated with C3aRA at the delayed time point of 6 h following hemorrhage onset likewise showed reduced edema formation and significantly improved spatial memory and sensorimotor capacity.

Subsequent work has built upon this evidence, examining the time course of inflammation-related injury in ICH and exploring other potential synergistic targets. Zhang et al. found that inflammatory infiltrate (especially neutrophils) and cells immunopositive for expression of TNF-α, IL-6, and NF-κB were maximum at 48 h following ICH [36]. C3 expression peaked at 48–72 h. Garrett et al. compared mice treated with both C5aRA and C3aRA with animals treated with C5aRA alone and found that, while administration of only C5aRA provides some neuroprotective effects versus vehicle, combination C3aRA/C5aRA therapy post-ICH leads to synergistic improvements in neurologic function as measured by a Morris water-maze, a 28-point scale, and a corner test at 6, 12, 24, 48, and 72 h after ICH onset [120]. Mice treated with combination therapy also demonstrated reduced inflammatory cell infiltration and brain edema. In addition, C5aRA-treated mice and C3aRA/C5aRA-treated mice had a decreased ratio of granulocytes in the hemorrhagic versus nonhemorrhagic hemispheres relative to vehicle-treated animals. These results indicate that a combined approach of complement inhibition targeting both the C3a receptor and the C5a receptor may offer superior neuroprotection in hemorrhagic stroke.

The Beneficial Role of Complement in Recovery Following Acute Brain Injury

The role of complement in the pathophysiology of acute brain injury is not singularly harmful; indiscriminate inhibition of the complement cascade may interfere with physiologic mechanisms important for recovery following ischemic stroke and ICH.

It has been long established that complement is involved in opsonization and clearance of apoptotic cell debris [121–125]. Deficiencies in complement leading to accumulation of apoptotic fragments have been associated with the pathogenesis of several diseases including systemic lupus erythematosus and Alzheimer's disease [123, 125, 126]. Apoptotic fragments and other cellular debris that go uncleared undergo secondary necrotic cell death and release damaging intracellular substances into the surrounding extracellular tissue space [125]. Whereas phagocytosis of apoptotic cell fragments leads to disposal without an induction of inflammation [121, 125, 127], necrotic cell death is pro-inflammatory [128–132]. Furthermore, effective complement-mediated phagocytosis has been shown to actively shift the cytokine milieu from pro-inflammatory to anti-inflammatory [125, 128, 130, 133–136].

Complement components, particularly the anaphylatoxins C3a and C5a, also have beneficial actions in promoting tissue regeneration following ischemia by sensitizing damaged tissues to key growth factors [137–139]. Complement may regulate endogenous neurogenesis following ischemic and ICH stroke through such a process. Evidence indicates that neurogenesis occurs following acute brain injury [140, 141], that neural progenitor cells and immature neurons express both C3aR and C5aR [46], and that both C3a receptor antagonism and C3aR deficiency reduces their proliferation [46]. C3-deficient mice demonstrate diminished ischemia-induced neurogenesis and larger infarcts when assessed at delayed time points, indicating that chronic blockade of C3aR might suppress neurogenesis and could thereby negatively impact functional outcome [46]. Some more recent studies have failed to support such a clear role for complement in neurogenesis, however, and this issue remains an ongoing subject of investigation [142, 143].

Recent work by our group confirms that complement inhibition does not have wholly neuroprotective effects and further suggests that the balance between the detrimental and beneficial actions of complement shifts predictably as a function time. In Ducruet, 2011, we demonstrate that MBL deficiency in murine stroke models ameliorates reperfusion injury in the acute phase (24 h post-ischemia), but that this neuroprotective effect of complement inhibition is not sustained in the subacute phase (7 day post-ischemia). A study published this year by our group similarly indicates that, relative to the cohort treated at a delayed time point, C3aRA administration limited to the acute post-stroke period increases neurogenesis, improves neurological functional outcome, and reduces mortality [144].

Therapeutic Strategies for Complement Inhibition in Acute Brain Injury

The complement cascade is an appealing therapeutic target in the treatment of acute brain injury following ischemic stroke and ICH for a number of reasons. First, the outcomes achieved by current strategies evince the need for new approaches in general. Second, the limitations of surgical interventions and the recognition that the biological processes involved in the pathogenesis of these diseases extend beyond the initial injurious event and are dynamic over the ensuing time course indicate a need for pharmacological alternatives that have ongoing activity in the brain and can be adjusted over time. Third, the understanding that complement activation is a key feature of the secondary cerebral damage on which functional outcomes often hinge paired with the rapid expansion in the array of available immunomodulatory agents has targeted complement modulation specifically as the best possibility for new treatments in these devastating disease processes.

Given the centrality of C3 in the pathways of activation, the possibility for cascade-independent cleavage of C3 by thrombin and other proteases, and available evidence from laboratory models, therapeutic strategies targeting C3 currently appear to have the most promise. Inhibition of C3 cleavage would block inflammatory exacerbation by anaphylatoxins as well as downstream MAC-mediated lysis and associated iron toxicity. Research by our group and others has largely focused on antagonism of the C3a receptor, with promising initial results in murine models [46, 66, 107, 116, 120]. Additional evidence suggests the therapeutic benefits of C3aRA treatment can be augmented by administration of a C5aR antagonist [120].

It has become clear, however, that a number of complicating factors continue to impede the translation of these results to human stroke and ICH. First, new research has recognized that complement proteins, in particular C3a and C5a, have beneficial effects in the recovery process following ischemic and ICH stroke, both in stimulating neurogenesis and in shifting the cytokine milieu away from a pro-inflammatory state. As a result, inhibition of complement activation through antagonism of the C3a/C3aR axis could potentially impair neurorestorative processes. Second, recent investigations demonstrate that the complicated balance between the harmful and beneficial effects of complement activation is not homogenous over time. Specifically, evidence seems to indicate that the injurious actions of complement predominate in the acute phase immediately following ischemia or hemorrhage but that long-term inhibition of the complement cascade into the subacute phase interferes with complement-mediated reparative processes [46, 56, 144].

The gap between laboratory models of these diseases and their pathophysiology in humans was further evidenced by the failure of sCR1 and sCR1-sLex trials in nonhuman primates despite promising results in mice [103, 104]. To date, the only relevant drug trial in humans, the enlimomab acute stroke trial, was a notable failure [109].

Another approach to complement inhibition involves intravenous immunoglobulin (IVIG) therapy. IVIG is less specific in targeting complement than other experimental alternatives such as C3aRA, but it also therefore has the potential to act on multiple pro-inflammatory pathways at once, including pro-inflammatory cytokine production and leukocyte adhesion [145]. IVIG treatment targeting neuroinflammation through inhibition of complement has been shown to be beneficial in numerous neurological diseases [145–149]. In murine models of stroke, one study found IVIG reduced infarction size by 50–60 %, significantly improved functional outcome, and almost entirely eliminated mortality [150]. In addition, greater numbers of neurons were spared with only occasional cell loss observed within the ischemic region. The authors of this study posit that IVIG therefore could reduce the long-term neurological consequences of ischemic stroke in humans that result from neuron loss.

Conclusion

Recent revelations about the complexities of the role played by complement in acute brain injury and its temporal pattern of activity suggest, on the one hand, that the rift between our current understanding and the ultimate end-goal, a viable drug for use in humans, is perhaps wider than initially believed. On the other hand, the intricacies of complement activity in the setting of ischemic and ICH stroke, of which our understanding has grown dramatically in recent years, indicate that, with precise targeting and timing, nuanced and personalized pharmacological treatments that are both safe and produce significant improvements in neurologic outcome may be achievable in the near future.

Future research must elucidate further the various actions of the complement cascade, the different roles played by the many component proteins, and the temporal patterns of complement activity in the setting of acute brain injury. While C3a receptor antagonism still appears to be the most promising approach for a potential drug, greater understanding of the effects its blockade will produce in humans is needed. Meanwhile, alternative approaches like IVIG also should continue to be explored.

References

1. Miniño AM, et al (2011) Deaths: final data for 2008. In: National Vital Statistics Report. Center for Disease Control and Prevention
2. Dirnagl U, Iadecola C, Moskowitz MA (1999) Pathobiology of ischaemic stroke: an integrated view. Trends Neurosci 22(9):391–7
3. Heidenreich PA et al (2011) Forecasting the future of cardiovascular disease in the United States: a policy statement from the American Heart Association. Circulation 123(8):933–44
4. National Heart, Lung, and Blood Institute (2006) Incidence and prevalence: 2006 chart book on cardiovascular and lung diseases. National Institute of Health: National Heart, Lung, and Blood Institute, Bethesda, MD
5. Roger VL et al (2011) Heart disease and stroke statistics—2011 update: a report from the American Heart Association. Circulation 123(4):e18–209
6. Bazarian JJ et al (2005) Mild traumatic brain injury in the United States, 1998–2000. Brain Inj 19(2):85–91
7. Carroll LJ et al (2004) Prognosis for mild traumatic brain injury: results of the WHO Collaborating Centre Task Force on mild traumatic brain injury. J Rehabil Med 2004(43 Suppl):84–105
8. Walport MJ (2001) Complement. First of two parts. N Engl J Med 344(14):1058–66
9. Cole DS, Morgan BP (2003) Beyond lysis: how complement influences cell fate. Clin Sci (Lond) 104(5):455–66
10. Peitsch MC, Tschopp J (1991) Assembly of macromolecular pores by immune defense systems. Curr Opin Cell Biol 3(4):710–6
11. Lee A, Whyte MK, Haslett C (1993) Inhibition of apoptosis and prolongation of neutrophil functional longevity by inflammatory mediators. J Leukoc Biol 54(4):283–8
12. Riedemann NC et al (2002) C5a receptor and thymocyte apoptosis in sepsis. FASEB J 16(8):887–8
13. Takabayashi T et al (1996) A new biologic role for C3a and C3a desArg: regulation of TNF-alpha and IL-1 beta synthesis. J Immunol 156(9):3455–60
14. Davis AE 3rd (2004) Biological effects of C1 inhibitor. Drug News Perspect 17(7):439–46
15. Lampl L et al (1994) Blood coagulation parameters as prognostic factors in multiple trauma: can clinical values be an early diagnostic aid? Zentralbl Chir 119(10):683–9
16. Lidington EA, Haskard DO, Mason JC (2000) Induction of decay-accelerating factor by thrombin through a protease-activated receptor 1 and protein kinase C-dependent pathway protects vascular endothelial cells from complement-mediated injury. Blood 96(8):2784–92
17. Ganter MT et al (2007) Role of the alternative pathway in the early complement activation following major trauma. Shock 28(1):29–34
18. Amara U et al (2008) Interaction between the coagulation and complement system. Adv Exp Med Biol 632:71–9
19. Huber-Lang M et al (2006) Generation of C5a in the absence of C3: a new complement activation pathway. Nat Med 12(6):682–7
20. Hankey GJ, Norman PE, Eikelboom JW (2006) Medical treatment of peripheral arterial disease. JAMA 295(5):547–53
21. Deb P, Sharma S, Hassan KM (2010) Pathophysiologic mechanisms of acute ischemic stroke: an overview with emphasis on therapeutic significance beyond thrombolysis. Pathophysiology 17(3):197–218
22. Mocco J et al (2006) The complement system: a potential target for stroke therapy. Adv Exp Med Biol 586:189–201
23. Feuerstein GZ, Wang X, Barone FC (1998) Cerebrovascular disease: pathophysiology, diagnosis, and management. In: Ginsberg MD, Bogousslavsky J (eds), Blackwell, Malden, MA
24. Belayev L et al (1996) Quantitative evaluation of blood–brain barrier permeability following middle cerebral artery occlusion in rats. Brain Res 739(1–2):88–96
25. Klatzo I (1987) Pathophysiological aspects of brain edema. Acta Neuropathol 72(3):236–9

26. Fishman RA (1992) Cerebrospinal fluid in diseases of the nervous system, 2nd edn. Saunders, Philadelphia, p 431
27. Adams HP Jr et al (2007) Guidelines for the early management of adults with ischemic stroke: a guideline from the American Heart Association/American Stroke Association Stroke Council, Clinical Cardiology Council, Cardiovascular Radiology and Intervention Council, and the Atherosclerotic Peripheral Vascular Disease and Quality of Care Outcomes in Research Interdisciplinary Working Groups: the American Academy of Neurology affirms the value of this guideline as an educational tool for neurologists. Stroke 38(5):1655–711
28. Lees KR et al (2010) Time to treatment with intravenous alteplase and outcome in stroke: an updated pooled analysis of ECASS, ATLANTIS, NINDS, and EPITHET trials. Lancet 375(9727):1695–703
29. Fujii Y et al (1994) Hematoma enlargement in spontaneous intracerebral hemorrhage. J Neurosurg 80(1):51–7
30. Xi G, Keep RF, Hoff JT (2006) Mechanisms of brain injury after intracerebral haemorrhage. Lancet Neurol 5(1):53–63
31. Broderick JP et al (1990) Ultra-early evaluation of intracerebral hemorrhage. J Neurosurg 72(2):195–9
32. Zazulia AR et al (1999) Progression of mass effect after intracerebral hemorrhage. Stroke 30(6):1167–73
33. Xi G et al (2001) Systemic complement depletion diminishes perihematomal brain edema in rats. Stroke 32(1):162–7
34. Hua Y et al (2000) Complement activation in the brain after experimental intracerebral hemorrhage. J Neurosurg 92(6):1016–22
35. Hua Y et al (1999) Complement C9 accumulation, membrane attack complex (MAC) formation and cluster in up regulation following intracerebral hemorrhage. J Cereb Blood Flow Metab 19(suppl 1):S670
36. Zhang X et al (2006) Brain edema after intracerebral hemorrhage in rats: the role of inflammation. Neurol India 54(4):402–7
37. Broderick JP et al (1993) Volume of intracerebral hemorrhage. A powerful and easy-to-use predictor of 30-day mortality. Stroke 24(7):987–93
38. Broderick J et al (2007) Guidelines for the management of spontaneous intracerebral hemorrhage in adults: 2007 update: a guideline from the American Heart Association/American Stroke Association Stroke Council, High Blood Pressure Research Council, and the Quality of Care and Outcomes in Research Interdisciplinary Working Group. Circulation 116(16):e391–413
39. Ropper AH (1986) Lateral displacement of the brain and level of consciousness in patients with an acute hemispheral mass. N Engl J Med 314(15):953–8
40. Ropper AH, King RB (1984) Intracranial pressure monitoring in comatose patients with cerebral hemorrhage. Arch Neurol 41(7):725–8
41. Morgenstern LB et al (2010) Guidelines for the management of spontaneous intracerebral hemorrhage: a guideline for healthcare professionals from the American Heart Association/ American Stroke Association. Stroke 41(9):2108–29
42. Sughrue ME et al (2004) Anti-adhesion molecule strategies as potential neuroprotective agents in cerebral ischemia: a critical review of the literature. Inflamm Res 53(10):497–508
43. Schafer MK et al (2000) Complement C1q is dramatically up-regulated in brain microglia in response to transient global cerebral ischemia. J Immunol 164(10):5446–52
44. Cowell RM, Plane JM, Silverstein FS (2003) Complement activation contributes to hypoxic-ischemic brain injury in neonatal rats. J Neurosci 23(28):9459–68
45. Huang J et al (1999) Neuronal protection in stroke by an sLex-glycosylated complement inhibitory protein. Science 285(5427):595–9
46. Rahpeymai Y et al (2006) Complement: a novel factor in basal and ischemia-induced neurogenesis. EMBO J 25(6):1364–74
47. Gasque P et al (2000) Complement components of the innate immune system in health and disease in the CNS. Immunopharmacology 49(1–2):171–86

48. Nataf S, Levison SW, Barnum SR (2001) Expression of the anaphylatoxin C5a receptor in the oligodendrocyte lineage. Brain Res 894(2):321–6
49. Spiegel K, Emmerling M, Barnum S (1998) Strategies for inhibition of complement activation in the treatment of neurodegenerative diseases. In: Wood PL (ed) Neuroinflammation: mechanisms and management. Humana, Totowa, NJ, p 375
50. Thomas A et al (2000) Expression of a complete and functional complement system by human neuronal cells in vitro. Int Immunol 12(7):1015–23
51. D'Ambrosio AL, Pinsky DJ, Connolly ES (2001) The role of the complement cascade in ischemia/reperfusion injury: implications for neuroprotection. Mol Med 7(6):367–82
52. Gasque P et al (1993) Expression of the complement classical pathway by human glioma in culture. A model for complement expression by nerve cells. J Biol Chem 268(33): 25068–74
53. Gasque P et al (1992) Expression of complement components of the alternative pathway by glioma cell lines. J Immunol 149(4):1381–7
54. van Beek J, Elward K, Gasque P (2003) Activation of complement in the central nervous system: roles in neurodegeneration and neuroprotection. Ann N Y Acad Sci 992:56–71
55. Thrane AS, Skehan JD, Thrane PS (2007) A novel interpretation of immune redundancy and duality in reperfusion injury with important implications for intervention in ischaemic disease. Med Hypotheses 68(6):1363–70
56. Ducruet AF et al (2011) The neuroprotective effect of genetic mannose-binding lectin deficiency is not sustained in the sub-acute phase of stroke. Transl Stroke Res 2(4):588–99
57. Yano T et al (2003) Neuroprotective effect of urinary trypsin inhibitor against focal cerebral ischemia–reperfusion injury in rats. Anesthesiology 98(2):465–73
58. Shimakura A et al (2000) Neutrophil elastase inhibition reduces cerebral ischemic damage in the middle cerebral artery occlusion. Brain Res 858(1):55–60
59. Qureshi AI et al (2001) Spontaneous intracerebral hemorrhage. N Engl J Med 344(19): 1450–60
60. Xi G et al (2001) Mechanisms of edema formation after intracerebral hemorrhage: effects of extravasated red blood cells on blood flow and blood–brain barrier integrity. Stroke 32(12): 2932–8
61. Clark A et al (2008) Evidence for non-traditional activation of complement factor C3 during murine liver regeneration. Mol Immunol 45(11):3125–32
62. Gong Y et al (2005) Complement inhibition attenuates brain edema and neurological deficits induced by thrombin. Acta Neurochir Suppl 95:389–92
63. Kitaoka T et al (2002) Delayed argatroban treatment reduces edema in a rat model of intracerebral hemorrhage. Stroke 33(12):3012–8
64. Matsuoka H, Hamada R (2002) Role of thrombin in CNS damage associated with intracerebral haemorrhage: opportunity for pharmacological intervention? CNS Drugs 16(8):509–16
65. Hua Y et al (2007) Brain injury after intracerebral hemorrhage: the role of thrombin and iron. Stroke 38(2 Suppl):759–62
66. Rynkowski MA et al (2009) C3a receptor antagonist attenuates brain injury after intracerebral hemorrhage. J Cereb Blood Flow Metab 29(1):98–107
67. Lucchesi BR (1993) Complement activation, neutrophils, and oxygen radicals in reperfusion injury. Stroke 24(12):I41–7
68. Xi G et al (2002) Brain edema after intracerebral hemorrhage: the effects of systemic complement depletion. Acta Neurochir Suppl 81:253–6
69. Aronowski J, Hall CE (2005) New horizons for primary intracerebral hemorrhage treatment: experience from preclinical studies. Neurol Res 27(3):268–79
70. Okusawa S et al (1988) C5a stimulates secretion of tumor necrosis factor from human mononuclear cells in vitro. Comparison with secretion of interleukin 1 beta and interleukin 1 alpha. J Exp Med 168(1):443–8
71. Lee KR et al (1997) Mechanisms of edema formation after intracerebral hemorrhage: effects of thrombin on cerebral blood flow, blood–brain barrier permeability, and cell survival in a rat model. J Neurosurg 86(2):272–8

72. Muller-Eberhard HJ (1986) The membrane attack complex of complement. Annu Rev Immunol 4:503–28
73. Wu J et al (2003) Iron and iron-handling proteins in the brain after intracerebral hemorrhage. Stroke 34(12):2964–9
74. Xi G, Keep RF, Hoff JT (1998) Erythrocytes and delayed brain edema formation following intracerebral hemorrhage in rats. J Neurosurg 89(6):991–6
75. Wang X et al (2002) Hemoglobin-induced cytotoxicity in rat cerebral cortical neurons: caspase activation and oxidative stress. Stroke 33(7):1882–8
76. Wagner KR et al (2003) Heme and iron metabolism: role in cerebral hemorrhage. J Cereb Blood Flow Metab 23(6):629–52
77. Gasque P et al (1998) The receptor for complement anaphylatoxin C3a is expressed by myeloid cells and nonmyeloid cells in inflamed human central nervous system: analysis in multiple sclerosis and bacterial meningitis. J Immunol 160(7):3543–54
78. Head E et al (2001) Complement association with neurons and beta-amyloid deposition in the brains of aged individuals with Down syndrome. Neurobiol Dis 8(2):252–65
79. Matsuoka Y et al (2001) Inflammatory responses to amyloidosis in a transgenic mouse model of Alzheimer's disease. Am J Pathol 158(4):1345–54
80. Eikelenboom P et al (1989) Complement activation in amyloid plaques in Alzheimer's dementia. Virchows Arch B Cell Pathol Incl Mol Pathol 56(4):259–62
81. McGeer PL et al (1989) Activation of the classical complement pathway in brain tissue of Alzheimer patients. Neurosci Lett 107(1–3):341–6
82. Yasuhara O et al (1994) Expression of the complement membrane attack complex and its inhibitors in Pick disease brain. Brain Res 652(2):346–9
83. Singhrao SK et al (1996) Role of complement in the aetiology of Pick's disease? J Neuropathol Exp Neurol 55(5):578–93
84. Kasuya H, Shimizu T (1989) Activated complement components C3a and C4a in cerebrospinal fluid and plasma following subarachnoid hemorrhage. J Neurosurg 71(5 Pt 1):741–6
85. Mack WJ et al (2007) Early plasma complement C3a levels correlate with functional outcome after aneurysmal subarachnoid hemorrhage. Neurosurgery 61(2):255–60, discussion 260-1
86. Yates JR et al (2007) Complement C3 variant and the risk of age-related macular degeneration. N Engl J Med 357(6):553–61
87. Rancan M et al (2003) Central nervous system-targeted complement inhibition mediates neuroprotection after closed head injury in transgenic mice. J Cereb Blood Flow Metab 23(9):1070–4
88. Bellander BM et al (2001) Complement activation in the human brain after traumatic head injury. J Neurotrauma 18(12):1295–311
89. Kaczorowski SL et al (1995) Effect of soluble complement receptor-1 on neutrophil accumulation after traumatic brain injury in rats. J Cereb Blood Flow Metab 15(5):860–4
90. Keeling KL et al (2000) Local neutrophil influx following lateral fluid-percussion brain injury in rats is associated with accumulation of complement activation fragments of the third component (C3) of the complement system. J Neuroimmunol 105(1):20–30
91. Morganti-Kossmann M-C et al (2002) Inflammatory responses to traumatic brain injury: an overview for the new millennium. In: Rothwwell N, Lodick S (eds) Immune and inflammatory responses in the nervous system. Oxford University Press, Oxford, UK
92. Nataf S et al (1999) Complement anaphylatoxin receptors on neurons: new tricks for old receptors? Trends Neurosci 22(9):397–402
93. Stahel PF, Morganti-Kossmann MC, Kossmann T (1998) The role of the complement system in traumatic brain injury. Brain Res Brain Res Rev 27(3):243–56
94. Stahel PF et al (2001) Intrathecal levels of complement-derived soluble membrane attack complex (sC5b-9) correlate with blood–brain barrier dysfunction in patients with traumatic brain injury. J Neurotrauma 18(8):773–81
95. Gasque P et al (1996) Complement activation on human neuroblastoma cell lines in vitro: route of activation and expression of functional complement regulatory proteins. J Neuroimmunol 66(1–2):29–40

96. Singhrao SK et al (2000) Spontaneous classical pathway activation and deficiency of membrane regulators render human neurons susceptible to complement lysis. Am J Pathol 157(3):905–18
97. Figueroa E et al (2005) The administration of cobra venom factor reduces post-ischemic cerebral injury in adult and neonatal rats. Neurosci Lett 380(1–2):48–53
98. Vasthare US et al (1998) Complement depletion improves neurological function in cerebral ischemia. Brain Res Bull 45(4):413–9
99. Akita N et al (2001) The effect of C 1 esterase inhibitor on ischemia: reperfusion injury in the rat brain. No To Shinkei 53(7):641–4
100. De Simoni MG et al (2003) Neuroprotection by complement (C1) inhibitor in mouse transient brain ischemia. J Cereb Blood Flow Metab 23(2):232–9
101. De Simoni MG et al (2004) The powerful neuroprotective action of C1-inhibitor on brain ischemia-reperfusion injury does not require C1q. Am J Pathol 164(5):1857–63
102. Weisman HF et al (1990) Soluble human complement receptor type 1: in vivo inhibitor of complement suppressing post-ischemic myocardial inflammation and necrosis. Science 249(4965):146–51
103. Mocco J et al (2006) Preclinical evaluation of the neuroprotective effect of soluble complement receptor type 1 in a nonhuman primate model of reperfused stroke. J Neurosurg 105(4):595–601
104. Ducruet AF et al (2007) Pre-clinical evaluation of an sLe x-glycosylated complement inhibitory protein in a non-human primate model of reperfused stroke. J Med Primatol 36(6):375–80
105. Kilgore KS et al (1998) Attenuation of interleukin-8 expression in C6-deficient rabbits after myocardial ischemia/reperfusion. J Mol Cell Cardiol 30(1):75–85
106. Mack WJ et al (2006) Temporal pattern of C1q deposition after transient focal cerebral ischemia. J Neurosci Res 83(5):883–9
107. Mocco J et al (2006) Complement component C3 mediates inflammatory injury following focal cerebral ischemia. Circ Res 99(2):209–17
108. Mocco J et al (2006) Alterations in plasma complement levels after human ischemic stroke. Neurosurgery 59(1):28–33, discussion 28–33
109. Furuya K et al (2001) Examination of several potential mechanisms for the negative outcome in a clinical stroke trial of enlimomab, a murine anti-human intercellular adhesion molecule-1 antibody: a bedside-to-bench study. Stroke 32(11):2665–74
110. Vuorte J et al (1999) Anti-ICAM-1 monoclonal antibody R6.5 (Enlimomab) promotes activation of neutrophils in whole blood. J Immunol 162(4):2353–7
111. Pedersen ED et al (2009) In situ deposition of complement in human acute brain ischaemia. Scand J Immunol 69(6):555–62
112. Xi G, Keep RF, Hoff JT (2002) Pathophysiology of brain edema formation. Neurosurg Clin N Am 13(3):371–83
113. Wu G, Huang FP (2005) Effects of venom defibrase on brain edema after intracerebral hemorrhage in rats. Acta Neurochir Suppl 95:381–7
114. Nakamura T et al (2004) Intracerebral hemorrhage in mice: model characterization and application for genetically modified mice. J Cereb Blood Flow Metab 24(5):487–94
115. Mukherjee P, Pasinetti GM (2000) The role of complement anaphylatoxin C5a in neurodegeneration: implications in Alzheimer's disease. J Neuroimmunol 105(2):124–30
116. Ducruet AF et al (2008) C3a receptor modulation of granulocyte infiltration after murine focal cerebral ischemia is reperfusion dependent. J Cereb Blood Flow Metab 28(5):1048–58
117. Yang S et al (2006) The role of complement C3 in intracerebral hemorrhage-induced brain injury. J Cereb Blood Flow Metab 26(12):1490–5
118. Yang S et al (2006) Intracerebral hemorrhage in complement C3-deficient mice. Acta Neurochir Suppl 96:227–31
119. Rynkowski MA et al (2008) A mouse model of intracerebral hemorrhage using autologous blood infusion. Nat Protoc 3(1):122–8
120. Garrett MC et al (2009) Synergistic neuroprotective effects of C3a and C5a receptor blockade following intracerebral hemorrhage. Brain Res 1298:171–7

121. Mevorach D (1999) The immune response to apoptotic cells. Ann N Y Acad Sci 887:191–8
122. Mevorach D (2000) Opsonization of apoptotic cells. Implications for uptake and autoimmunity. Ann N Y Acad Sci 926:226–35
123. Mevorach D (2003) Systemic lupus erythematosus and apoptosis: a question of balance. Clin Rev Allergy Immunol 25(1):49–60
124. Mevorach D et al (1998) Complement-dependent clearance of apoptotic cells by human macrophages. J Exp Med 188(12):2313–20
125. Fishelson Z, Attali G, Mevorach D (2001) Complement and apoptosis. Mol Immunol 38(2–3):207–19
126. Wyss-Coray T et al (2002) Prominent neurodegeneration and increased plaque formation in complement-inhibited Alzheimer's mice. Proc Natl Acad Sci USA 99(16):10837–42
127. Savill J, Fadok V (2000) Corpse clearance defines the meaning of cell death. Nature 407(6805):784–8
128. Fadok VA (1999) Clearance: the last and often forgotten stage of apoptosis. J Mammary Gland Biol Neoplasia 4(2):203–11
129. Fadok VA et al (1998) The role of phosphatidylserine in recognition of apoptotic cells by phagocytes. Cell Death Differ 5(7):551–62
130. Fadok VA, Henson PM (1998) Apoptosis: getting rid of the bodies. Curr Biol 8(19):R693–5
131. Fadok VA et al (1998) CD36 is required for phagocytosis of apoptotic cells by human macrophages that use either a phosphatidylserine receptor or the vitronectin receptor (alpha v beta 3). J Immunol 161(11):6250–7
132. Elward K, Gasque P (2003) "Eat me" and "don't eat me" signals govern the innate immune response and tissue repair in the CNS: emphasis on the critical role of the complement system. Mol Immunol 40(2–4):85–94
133. Savill J (2000) Apoptosis in resolution of inflammation. Kidney Blood Press Res 23(3–5): 173–4
134. Fadok VA, Chimini G (2001) The phagocytosis of apoptotic cells. Semin Immunol 13(6):365–72
135. Savill J (1997) Apoptosis in resolution of inflammation. J Leukoc Biol 61(4):375–80
136. De Simone R et al (2003) Apoptotic PC12 cells exposing phosphatidylserine promote the production of anti-inflammatory and neuroprotective molecules by microglial cells. J Neuropathol Exp Neurol 62(2):208–16
137. Strey CW et al (2003) The proinflammatory mediators C3a and C5a are essential for liver regeneration. J Exp Med 198(6):913–23
138. Markiewski MM et al (2004) C3a and C3b activation products of the third component of complement (C3) are critical for normal liver recovery after toxic injury. J Immunol 173(2):747–54
139. DeAngelis RA, Markiewski MM, Lambris JD (2006) Liver regeneration: a link to inflammation through complement. Adv Exp Med Biol 586:17–34
140. Shen J et al (2008) Neurogenesis after primary intracerebral hemorrhage in adult human brain. J Cereb Blood Flow Metab 28(8):1460–8
141. Yang S et al (2008) Effects of thrombin on neurogenesis after intracerebral hemorrhage. Stroke 39(7):2079–84
142. Bogestal YR et al (2007) Signaling through C5aR is not involved in basal neurogenesis. J Neurosci Res 85(13):2892–7
143. Moriyama M et al (2011) Complement receptor 2 is expressed in neural progenitor cells and regulates adult hippocampal neurogenesis. J Neurosci 31(11):3981–9
144. Ducruet AF et al (2012) Complement inhibition promotes endogenous neurogenesis and sustained anti-inflammatory neuroprotection following reperfused stroke. PLoS One 7(6):e38664
145. Arumugam TV et al (2009) Neuroprotection in stroke by complement inhibition and immunoglobulin therapy. Neuroscience 158(3):1074–89
146. Hughes RA, Cornblath DR (2005) Guillain–Barre syndrome. Lancet 366(9497):1653–66
147. Archelos JJ, Fazekas F (2006) IVIG therapy in neurological disorders of childhood. J Neurol 253(Suppl 5):V80–6

148. Ringel I, Zettl UK (2006) Intravenous immunoglobulin therapy in neurological diseases during pregnancy. J Neurol 253(Suppl 5):V70–4
149. Stangel M, Pul R (2006) Basic principles of intravenous immunoglobulin (IVIg) treatment. J Neurol 253(Suppl 5):V18–24
150. Arumugam TV et al (2007) Intravenous immunoglobulin (IVIG) protects the brain against experimental stroke by preventing complement-mediated neuronal cell death. Proc Natl Acad Sci USA 104(35):14104–9

Chapter 6
Matrix Metalloproteinases as an Inflammatory Mediator in the Neurovascular Unit

Changhong Xing, Takakuni Maki, Ji Hae Seo, Ken Arai, and Eng H. Lo

Abstract Matrix metalloproteinases (MMPs) represent a critical set of mediators in the neurovascular unit after stroke. Responses in MMPs underlie both acute injury as well as delayed remodeling as tissue transitions from damage into repair. This chapter briefly reviews how MMPs may contribute to acute infarction, stroke recovery, gray versus white matter responses, and how cell–cell signaling in neuronal, glial, and vascular compartments can be interpreted in the overall context of inflammation within the perturbed neurovascular unit. Dissection of these underlying MMP pathways may eventually lead us to novel therapeutic approaches for target neuroprotection as well as new ways to search for biomarkers for mapping stroke recovery.

Introduction

Inflammation and activation of both innate and adaptive immunity are now recognized to significantly contribute to stroke pathophysiology. In the context of the neurovascular unit, inflammation may be expressed as a set of coordinated cell–cell and cell–matrix signals that allow the damaged neurovascular unit to respond to injury.

The importance of cell–cell signaling between all elements of the neurovascular unit in stroke, brain injury, and neurodegeneration has been extensively discussed [1–4]. Dysfunctional cross talk between neurons, glia, and vascular compartments

C. Xing • K. Arai • J.H. Seo • E.H. Lo (✉)
Massachusetts General Hospital, Harvard University, Charlestown, MA, USA
e-mail: Lo@helix.mgh.harvard.edu

T. Maki
Department of Neurology, Kyoto University, Kyoto, Japan

J. Chen et al. (eds.), *Immunological Mechanisms and Therapies in Brain Injuries and Stroke*, Springer Series in Translational Stroke Research, DOI 10.1007/978-1-4614-8915-3_6,

contributes to multiple aspects of acute pathophysiology in CNS disease. Impaired glutamate release–reuptake mechanisms in neurons and astrocytes can amplify excitotoxicity [5]. Perturbed signaling between cerebral endothelium, astrocytes, and pericytes can disrupt blood–brain barrier integrity [6]. Dysfunctional coupling between neuronal activation and vascular responses can promote deleterious spreading depression [7]. And ultimately, disordered signaling between all neurovascular and gliovascular elements can underlie the evolution of neuroinflammation and cell death [8]. This concept of the neurovascular unit may be important because, by understanding how complex multicellular events unfold, we may be able to move beyond a singular focus on "preventing neuron death" towards a more integrative approach where we attempt to rescue function within and signaling between all cell types in the entire neurovascular unit.

Increasingly, however, investigations into cell–cell signals of the neurovascular unit in stroke have suggested that these mechanisms may be extremely complex. A key challenge may be the recognition that many of the signals within the neurovascular unit may play biphasic roles as injury and disease evolves. What is damaging may become protective and what is protective may become damaging. Acute overactivation of the NMDA receptor is excitotoxic, but without well-regulated NMDA signaling during stroke recovery, neuroplasticity cannot take place. Inflammatory activation of microglia might be initially beneficial for tissue clean-up and repair, but prolonged microglia activation in Alzheimer's disease becomes highly neurotoxic. In this chapter, we discuss one of the best examples of this biphasic nature of inflammatory signals in the neurovascular unit, i.e., the extracellular family of proteases comprising matrix metalloproteinases (MMPs).

In stroke, a large body of preclinical data and accumulating clinical findings support a role of MMPs as both biomarker as well as target. When stroke or brain injury occurs, MMPs become dysregulated and may be a central cause of tissue damage. However, some caution may also be warranted since MMPs can play biphasic roles after brain injury—deleterious in the acute phase but potentially beneficial in delayed remodeling and recovery.

MMPs in Acute Brain Injury after Stroke

MMPs comprise a large family of extracellular zinc endopeptidases [9]. But in the context of stroke, the largest amount of data may exist for the gelatinases MMP-2 and MMP-9. In animal models of focal and global cerebral ischemia, MMPs are upregulated, and treatment with MMP inhibitors prevent neuronal cell death, decrease infarction, and improve outcomes [10–12]. Knockout mice that lack MMP-9 show significantly reduced brain cell death after cerebral ischemia or traumatic brain injury [13–16]. Conversely, transgenic mice that overexpress tissue inhibitors of metalloproteinase (TIMP) have better outcomes [17].

Mechanistically, the data in animal models fit well with the premise that high levels of MMPs can damage neurovascular matrix and cause BBB injury, edema, and hemorrhage [18]. Degradation of various basal lamina and tight junction proteins has been correlated with BBB leakage and blockade of MMPs reduce edema [13, 15]. Matrix proteolysis and BBB disruption was reduced in knockout mice lacking MMP-9 [15]. MMP activation and BBB leakage also appears to coincide with the generation of free radicals. And as neurovascular injury continues to evolve, recruitment of cytokines and vascular adhesion molecules add onto the accumulating tissue damage and may even further amplify MMPs and inflammation [19].

Beyond vascular leakage per se, MMP-mediated proteolysis of neurovascular matrix may also interfere with homeostatic signals between different cell types in the neurovascular unit. Resting matrix signaling via integrins is vital for normal cell function. Disruption of extracellular matrix by MMPs can induce anoikis in neurons and cerebral endothelial cells [20, 21]. In animal models, degradation of matrix correlates with cell death [22]. In a nonhuman primate model of focal cerebral ischemia, areas where matrix antigens are lost correspond to growing regions of collapsing penumbra and dying cores [23]. The importance of these matrix signals is further confirmed in fibronectin knockout mice in which neuronal apoptosis and brain damage are amplified after cerebral ischemia [24]. Broadly speaking, MMPs may degrade homeostatic cell–matrix integrin signals that then lead to anoikis-like cell death in a wide spectrum of CNS cell types. Hence, blocking MMPs may reduce neurovascular damage and infarction after stroke.

MMPs in Delayed Recovery after Stroke

MMPs play a deleterious role during acute stroke by augmenting BBB disruption, edema, hemorrhage, and brain injury. However, emerging data now suggest that MMPs may play biphasic roles. During the acute stages of stroke, MMPs are deleterious. But during delayed phases of stroke recovery, MMPs may play surprisingly beneficial roles [25, 26]. In part, this duality of MMP phenotype may be related to its original physiologic roles in normal development of brain morphology [27]. In developing brain, these proteases modify extracellular matrix to allow newborn cells to migrate and neurites and axons to extend and connect. Additionally, MMPs may also facilitate the actions of other signaling molecules. For example, MMP-9 may be an "angiogenic switch" by processing and releasing bioactive VEGF to promote vascular growth and/or remodeling [28]. MMP-9 has also been implicated in associative learning in the hippocampus. The broad-spectrum MMP inhibitor FN-439 interferes with long-term potentiation [29]. MMP-9 knockout mice display deficits in learning and memory [30].

During stroke recovery, the brain attempts to remodel. MMPs may be recruited as part of this endogenous recovery process. So blocking MMPs at the wrong place or wrong time may worsen outcomes. Following focal cerebral ischemia in mice, endogenous neurogenesis is amplified in the subventricular zone and newborn neuroblasts are diverted from their original rostral migratory stream towards damaged brain [31]. This process requires MMPs, and delayed blockade of MMPs disrupts neuroblast migration [32]. At 2 weeks after focal strokes in rats, a secondary upregulation of MMPs in peri-infarct cortex can be detected in astrocytes and endothelial cells [33]. Late blockade of MMPs in this model system is damaging since MMPs appear to mediate VEGF processing, compensatory angiogenesis, and stroke recovery [33]. Hence, MMP inhibitors can sometimes lead to beneficial reductions of acute edema, while resulting in impaired long-term recovery [34]. Similar biphasic properties of MMPs may also exist in spinal cord injury, where MMP-2 is increased together with reactive gliosis [35]. But genetic deletion of MMP-2 exacerbated white matter damage and decreased motor recovery [36]. Of course, not all MMP-mediated plasticity is guaranteed to be beneficial. How MMPs augment normal or abnormal rewiring in recovering brains after stroke or trauma remains to be fully elucidated.

MMPs and Inflammatory Demyelination

Although the neurovascular unit was originally proposed as a concept to assess neuronal function and disease, it has become clear that analogous cell–cell interactions take place in white matter. An "oligovascular unit" has been described and investigated that operationally comprises trophic interactions between axonal, glial, and vascular compartments [37]. Myelin proteins produced by oligodendrocytes can support axonal health and function [38]. Cerebral endothelium and white matter astrocytes can also secrete critical trophic factors such as IGF1 and FGF that support homeostasis and proliferation in oligodendrocyte precursor cells [39, 40]. Just as in neurovascular unit of gray matter, extracellular proteolysis by MMPs will also play a central role in the oligovascular unit after stroke.

MMPs were initially implicated in white matter injury because it was discovered that they could proteolytically cleave myelin basic protein [41]. In EAE animal models that mimic multiple sclerosis, activated microglia secrete MMPs that degrade the blood–brain barrier and myelin basic protein, thus leading to inflammatory demyelination [42]. After traumatic brain injury, similar events may also be detected in damaged white matter. MMPs become upregulated, especially MMP-9, which then results in degradation of myelin basic protein and disruption of axonal function [43]. These events may also impact on the recovery process. The extracellular matrix substrate laminin is known as a critical cofactor for neurite growth [44]. MMP-2 and MMP-9 can both recognize laminin, but the effects may be complex. Outright degradation of laminin should reduce this pro-neurite outgrowth factor, thus interfering with recovery. However, under some conditions, extracellular matrix proteolysis may conversely increase laminin bioavailability and promote neurite recovery in white matter after stroke and trauma [45].

In the context of cerebrovascular disease, white matter dysfunction may occur in various conditions of vascular dementia. A recent study has implicated a role for MMPs in this disease [46]. After prolonged cerebral hypoperfusion in mice, white matter tracts of the corpus callosum begin to demyelinate over the course of weeks to months. At the same time, there is neurovascular inflammation evidenced by BBB leakage, activation of microglia, and invasion of neutrophils. When MMP-9 levels were measured, a multicellular response was detected. During the initial 2–3 days post-onset, MMP-9 was upregulated in oligodendrocyte precursor cells. Thereafter, by 1–2 weeks, MMP-9 expression shifted into microglia and perturbed cerebral endothelium. In concert with this biphasic temporal profile, early blockade of MMPs in oligodendrocyte precursor cells significantly reduced blood–brain barrier leakage and subsequent demyelination and neutrophil infiltration. The biphasic profile of MMPs in this model of white matter vascular dementia may again support to demonstrate a biphasic role for inflammation in CNS disease. After white matter injury, an "overactivation" of oligodendrocyte precursors may initially exacerbate injury by damaging the blood–brain barrier. But in the delayed phase, oligodendrocyte precursors may remain essential to rebuild lost mature oligodendrocyte populations and restore myelination and white matter function.

MMPs as Target and Biomarker in Clinical Stroke

Because MMPs can degrade BBB integrity and function in the neurovascular unit, they have been proposed as potential biomarkers in clinical stroke and brain injury. Plasma levels of MMP-9 are elevated during acute stages of both ischemic and hemorrhagic stroke and appear to be correlated with poor neurological outcomes [47, 48]. In animal models of embolic stroke, tPA amplifies MMP-9 [49]. Emerging clinical data may be consistent with the experimental literature. Patients with higher plasma levels of MMP-9 may be more susceptible to hemorrhagic transformation following tPA thrombolysis for acute ischemic strokes [50, 51].

In addition to serving as positive protein signals in plasma, MMP responses have also been detected in genetic and brain compartments of stroke patients. After ischemia or brain injury, circulating blood cells show rapid alterations in gene expression. In particular, responses in MMP-9 genes are highly conserved [52, 53]. In the brain parenchyma itself, MMP-9 positive astrocytes colocalize with peri-hematoma edema [54]. After ischemic strokes, MMP-9 positive neutrophils appear to coincide with local disruptions in microvessels [55, 56]. Taken together, these signals are broadly consistent with data and mechanisms derived from experimental models.

Beyond their utility as biomarkers, MMPs have also been pursued as potential therapeutic targets. In animal models of focal cerebral ischemia, MMP inhibitors reduce infarct volumes when administered early during acute ischemic strokes. Consistent with its proposed mechanisms, MMP inhibitors appear to be especially effective in terms of reducing brain edema and hemorrhage. One aspect of this pathophysiology with particular clinical relevance may be the relationship between MMPs and tPA thrombolysis. tPA is known to bind several lipoprotein receptors in cerebral endothelium that can upregulate MMP-9 [57]. Therefore, it is possible that some of the hemorrhagic transformation complications seen in tPA-thrombolysis patients may be caused by an inadvertent increase in MMPs [50, 51]. An obvious question is whether clinically acceptable compounds can be used as MMP inhibitors in stroke thrombolysis? In this regard, minocycline has been proposed as a potential "re-purposed drug" to target MMP-9 [58]. In experimental clot-based models of focal stroke in hypertensive rats, minocycline plus tPA as combination stroke therapy seem to suppress MMP-9, decrease hemorrhagic transformation, and widen the therapeutic time window for safe and effective reperfusion [59]. Based in part on these experimental data, clinical trials have been started [60]. Initial findings are promising as minocycline appeared to dampen plasma MMP-9 biomarker levels as hypothesized [61]. Nevertheless, some caution might be warranted. As discussed earlier, MMP blockade may interfere with endogenous recovery after brain injury, and long-term use of minocycline worsened outcomes in an ALS clinical trial [62].

Although the majority of clinical MMP data has been collected in ischemic strokes, recent efforts extend the role of these proteases to hemorrhagic strokes as well. MMPs are upregulated in subarachnoid hemorrhage patients, and blood and CSF levels of MMP-9 may track vasospasm and clinical outcomes [63]. Mechanistically, MMPs contribute to early brain injury and may also process gelsolin that can further amplify neuroinflammation [64]. In experimental models of subarachnoid hemorrhage, MMP inhibition improves outcomes [65]. Whether these targets work clinically remains to be determined.

Besides acute ischemic and hemorrhagic strokes, MMPs may also contribute to the pathophysiology of "slower" cerebrovascular disorders such as vascular dementia. Similar to findings in experimental animal models, dysregulation of MMPs may occur in white matter injury after vascular dementia. MMP-9 in particular has been found to be upregulated in the CSF of vascular dementia patients [66]. More recently, it has been suggested that a panel of biomarkers including MMPs, myelin basic protein, tau, and amyloid-beta may help distinguish vascular dementia from Alzheimer's disease [67].

Proper functioning of the neurovascular unit and oligovascular unit requires integrated cell and matrix signaling between multicellular compartments. In this regard, MMPs can disrupt signaling or promote remodeling, depending on timing and context. Hence, MMPs may serve as useful biomarkers and targets in stroke and brain injury.

Conclusion

Inflammation and immune reactions play key roles in the pathophysiology of stroke and cerebrovascular disease. From a clinical standpoint, these mechanisms are important because they provide many potential targets for diagnostics and therapeutics. However, what is being increasingly recognized, is that inflammatory and immune responses are highly complex and are not always deleterious. Indeed, inflammation can be broadly interpreted as a set of evolutionarily conserved responses that underlie endogenous attempts at tissue repair. After stroke, multiple inflammatory signals are activated within the neurovascular unit that also have both deleterious as well as beneficial consequences. In this regard, MMPs may represent a stereotyped example of this concept. As extracellular proteases in the CNS, MMPs become upregulated in the damaged neurovascular unit. Initially, overactivated MMPs may injure the neurovascular unit and lead to BBB leakage, edema, hemorrhage, anoikis-like cell death, activation of microglia, and secondary infiltration of immune cells. But as the neurovascular unit continues to recover, MMPs may then switch into a repair mode whereby they help neurovascular plasticity and remodeling. Therefore, any attempt to define MMPs as an inflammatory target in stroke will have to carefully balance the transition between initial injury and subsequent repair.

Acknowledgments Supported in part by grants from the American Heart Association, NIH, and the Rappaport Foundation. Based in part on ideas previously discussed in Seo et al. (Curr Pharm Des 2012), Xing et al. (Neurol Res 2012), and Rosell et al. (Curr Opin Pharmacol 2008).

References

1. del Zoppo GJ (2009) Relationship of neurovascular elements to neuron injury during ischemia. Cerebrovasc Dis 27(Suppl 1):65–76
2. Iadecola C (2004) Neurovascular regulation in the normal brain and in Alzheimer's disease. Nat Rev Neurosci 5:347–60
3. Lo EH, Dalkara T, Moskowitz MA (2003) Mechanisms, challenges and opportunities in stroke. Nat Rev Neurosci 4:399–415
4. Lok J, Gupta P, Guo S et al (2007) Cell–cell signaling in the neurovascular unit. Neurochem Res 32:2032–45
5. Besancon E, Guo S, Lok J, Tymianski M, Lo EH (2008) Beyond NMDA and AMPA glutamate receptors: emerging mechanisms for ionic imbalance and cell death in stroke. Trends Pharmacol Sci 29:268–75
6. Zlokovic BV (2008) The blood–brain barrier in health and chronic neurodegenerative disorders. Neuron 57:178–201
7. Dreier JP (2011) The role of spreading depression, spreading depolarization and spreading ischemia in neurological disease. Nat Med 17:439–47
8. Iadecola C, Anrather J (2011) The immunology of stroke: from mechanisms to translation. Nat Med 17:796–808
9. Yong VW (2005) Metalloproteinases: mediators of pathology and regeneration in the CNS. Nat Rev Neurosci 6:931–44
10. Romanic AM, White RF, Arleth AJ, Ohlstein EH, Barone FC (1998) Matrix metalloproteinase expression increases after cerebral focal ischemia in rats: inhibition of matrix metalloproteinase-9 reduces infarct size. Stroke; A Journal of Cerebral Circulation 29:1020–30
11. Gasche Y, Fujimura M, Morita-Fujimura Y et al (1999) Early appearance of activated matrix metalloproteinase-9 after focal cerebral ischemia in mice: a possible role in blood–brain barrier dysfunction. J Cereb Blood Flow Metab: official Journal of the International Society of Cerebral Blood Flow and Metabolism 19:1020–8
12. Heo JH, Lucero J, Abumiya T, Koziol JA, Copeland BR, del Zoppo GJ (1999) Matrix metalloproteinases increase very early during experimental focal cerebral ischemia. J Cereb Blood Flow Metab: Official Journal of the International Society of Cerebral Blood Flow and Metabolism 19:624–33
13. Asahi M, Asahi K, Jung JC, del Zoppo GJ, Fini ME, Lo EH (2000) Role for matrix metalloproteinase 9 after focal cerebral ischemia: effects of gene knockout and enzyme inhibition

with BB-94. J Cereb Blood Flow Metab: Official Journal of the International Society of Cerebral Blood Flow and Metabolism 20:1681–9
14. Wang X, Jung J, Asahi M et al (2000) Effects of matrix metalloproteinase-9 gene knock-out on morphological and motor outcomes after traumatic brain injury. J Neurosci: The Official Journal of the Society for Neuroscience 20:7037–42
15. Asahi M, Wang X, Mori T et al (2001) Effects of matrix metalloproteinase-9 gene knock-out on the proteolysis of blood–brain barrier and white matter components after cerebral ischemia. J Neurosci: The Official Journal of the Society for Neuroscience 21:7724–32
16. Lee SR, Tsuji K, Lee SR, Lo EH (2004) Role of matrix metalloproteinases in delayed neuronal damage after transient global cerebral ischemia. J Neurosci: The Official Journal of the Society for Neuroscience 24:671–8
17. Tejima E, Guo S, Murata Y et al (2009) Neuroprotective effects of overexpressing tissue inhibitor of metalloproteinase TIMP-1. J Neurotrauma 26:1935–41
18. Cunningham LA, Wetzel M, Rosenberg GA (2005) Multiple roles for MMPs and TIMPs in cerebral ischemia. Glia 50:329–39
19. Rosenberg LJ, Lucas JH (1997) The effects of ciliary neurotrophic factor on murine spinal cord neurons subjected to dendrite transection injury. Brain Res 775:209–13
20. Gu Z, Kaul M, Yan B et al (2002) S-nitrosylation of matrix metalloproteinases: signaling pathway to neuronal cell death. Science 297:1186–90
21. Lee SR, Lo EH (2004) Induction of caspase-mediated cell death by matrix metalloproteinases in cerebral endothelial cells after hypoxia-reoxygenation. J Cereb Blood Flow Metab: Official Journal of the International Society of Cerebral Blood Flow and Metabolism 24:720–7
22. Gu Z, Cui J, Brown S et al (2005) A highly specific inhibitor of matrix metalloproteinase-9 rescues laminin from proteolysis and neurons from apoptosis in transient focal cerebral ischemia. J Neurosci: The Official Journal of the Society for Neuroscience 25:6401–8
23. Fukuda S, Fini CA, Mabuchi T, Koziol JA, Eggleston LL Jr, del Zoppo GJ (2004) Focal cerebral ischemia induces active proteases that degrade microvascular matrix. Stroke; A Journal of Cerebral Circulation 35:998–1004
24. Sakai T, Johnson KJ, Murozono M et al (2001) Plasma fibronectin supports neuronal survival and reduces brain injury following transient focal cerebral ischemia but is not essential for skin-wound healing and hemostasis. Nat Med 7:324–30
25. Zhao BQ, Tejima E, Lo EH (2007) Neurovascular proteases in brain injury, hemorrhage and remodeling after stroke. Stroke; A Journal of Cerebral Circulation 38:748–52
26. Rosell A, Lo EH (2008) Multiphasic roles for matrix metalloproteinases after stroke. Curr Opin Pharmacol 8:82–9
27. Canete Soler R, Gui YH, Linask KK, Muschel RJ (1995) MMP-9 (gelatinase B) mRNA is expressed during mouse neurogenesis and may be associated with vascularization. Brain Res Dev Brain Res 88:37–52
28. Bergers G, Brekken R, McMahon G et al (2000) Matrix metalloproteinase-9 triggers the angiogenic switch during carcinogenesis. Nat Cell Biol 2:737–44
29. Meighan SE, Meighan PC, Choudhury P et al (2006) Effects of extracellular matrix-degrading proteases matrix metalloproteinases 3 and 9 on spatial learning and synaptic plasticity. J Neurochem 96:1227–41
30. Nagy V, Bozdagi O, Matynia A et al (2006) Matrix metalloproteinase-9 is required for hippocampal late-phase long-term potentiation and memory. J Neurosci: The Official Journal of the Society for Neuroscience 26:1923–34
31. Arvidsson A, Collin T, Kirik D, Kokaia Z, Lindvall O (2002) Neuronal replacement from endogenous precursors in the adult brain after stroke. Nat Med 8:963–70
32. Lee SR, Kim HY, Rogowska J et al (2006) Involvement of matrix metalloproteinase in neuroblast cell migration from the subventricular zone after stroke. J Neurosci: The Official Journal of the Society for Neuroscience 26:3491–5
33. Zhao BQ, Wang S, Kim HY et al (2006) Role of matrix metalloproteinases in delayed cortical responses after stroke. Nat Med 12:441–5

34. Sood RR, Taheri S, Candelario-Jalil E, Estrada EY, Rosenberg GA (2008) Early beneficial effect of matrix metalloproteinase inhibition on blood–brain barrier permeability as measured by magnetic resonance imaging countered by impaired long-term recovery after stroke in rat brain. J Cereb Blood Flow Metab: Official Journal of the International Society of Cerebral Blood Flow and Metabolism 28:431–8
35. Goussev S, Hsu JY, Lin Y et al (2003) Differential temporal expression of matrix metalloproteinases after spinal cord injury: relationship to revascularization and wound healing. J Neurosurg 99:188–97
36. Hsu JY, McKeon R, Goussev S et al (2006) Matrix metalloproteinase-2 facilitates wound healing events that promote functional recovery after spinal cord injury. J Neurosci: The Official Journal of the Society for Neuroscience 26:9841–50
37. Arai K, Lo EH (2009) Experimental models for analysis of oligodendrocyte pathophysiology in stroke. Exp Transl Stroke Med 1:6
38. Nave KA (2010) Myelination and support of axonal integrity by glia. Nature 468:244–52
39. Arai K, Lo EH (2009) An oligovascular niche: cerebral endothelial cells promote the survival and proliferation of oligodendrocyte precursor cells. J Neurosci 29:4351–5
40. Arai K, Lo EH (2010) Astrocytes protect oligodendrocyte precursor cells via MEK/ERK and PI3K/Akt signaling. J Neurosci Res 88:758–63
41. Proost P, Van Damme J, Opdenakker G (1993) Leukocyte gelatinase B cleavage releases encephalitogens from human myelin basic protein. Biochem Biophys Res Commun 192:1175–81
42. Gijbels K, Proost P, Masure S, Carton H, Billiau A, Opdenakker G (1993) Gelatinase B is present in the cerebrospinal fluid during experimental autoimmune encephalomyelitis and cleaves myelin basic protein. J Neurosci Res 36:432–40
43. Wang X, Jung J, Asahi M et al (2000) Effects of matrix metalloproteinase-9 gene knock-out on morphological and motor outcomes after traumatic brain injury. J Neurosci 20:7037–42
44. Lander AD, Fujii DK, Reichardt LF (1985) Laminin is associated with the "neurite outgrowth-promoting factors" found in conditioned media. Proc Natl Acad Sci USA 82:2183–7
45. Costa S, Planchenault T, Charriere-Bertrand C et al (2002) Astroglial permissivity for neuritic outgrowth in neuron-astrocyte cocultures depends on regulation of laminin bioavailability. Glia 37:105–13
46. Seo JH, Miyamoto N, Hayakawa K et al (2013) Oligodendrocyte precursors induce early blood–brain barrier opening after white matter injury. J Clin Invest 123:782–6
47. Montaner J, Alvarez-Sabin J, Molina C et al (2001) Matrix metalloproteinase expression after human cardioembolic stroke: temporal profile and relation to neurological impairment. Stroke; A Journal of Cerebral Circulation 32:1759–66
48. Montaner J, Alvarez-Sabin J, Molina CA et al (2001) Matrix metalloproteinase expression is related to hemorrhagic transformation after cardioembolic stroke. Stroke; A Journal of Cerebral Circulation 32:2762–7
49. Sumii T, Lo EH (2002) Involvement of matrix metalloproteinase in thrombolysis-associated hemorrhagic transformation after embolic focal ischemia in rats. Stroke; A Journal of Cerebral Circulation 33:831–6
50. Montaner J, Molina CA, Monasterio J et al (2003) Matrix metalloproteinase-9 pretreatment level predicts intracranial hemorrhagic complications after thrombolysis in human stroke. Circulation 107:598–603
51. Ning M, Furie KL, Koroshetz WJ et al (2006) Association between tPA therapy and raised early matrix metalloproteinase-9 in acute stroke. Neurology 66:1550–5
52. Tang Y, Xu H, Du X et al (2006) Gene expression in blood changes rapidly in neutrophils and monocytes after ischemic stroke in humans: a microarray study. Journal of Cerebral Blood Flow and Metabolism : Official Journal of the International Society of Cerebral Blood Flow and Metabolism 26:1089–102
53. Barr TL, Conley Y, Ding J et al (2010) Genomic biomarkers and cellular pathways of ischemic stroke by RNA gene expression profiling. Neurology 75:1009–14

54. Tejima E, Zhao BQ, Tsuji K et al (2007) Astrocytic induction of matrix metalloproteinase-9 and edema in brain hemorrhage. J Cereb Blood Flow Metab: Official Journal of the International Society of Cerebral Blood Flow and Metabolism 27:460–8
55. Gidday JM, Gasche YG, Copin JC et al (2005) Leukocyte-derived matrix metalloproteinase-9 mediates blood–brain barrier breakdown and is proinflammatory after transient focal cerebral ischemia. Am J Physiol Heart Circ Physiol 289:H558–68
56. Rosell A, Cuadrado E, Ortega-Aznar A, Hernandez-Guillamon M, Lo EH, Montaner J (2008) MMP-9-positive neutrophil infiltration is associated to blood–brain barrier breakdown and basal lamina type IV collagen degradation during hemorrhagic transformation after human ischemic stroke. Stroke; A Journal of Cerebral Circulation 39:1121–6
57. Wang X, Lee SR, Arai K et al (2003) Lipoprotein receptor-mediated induction of matrix metalloproteinase by tissue plasminogen activator. Nat Med 9:1313–7
58. Fagan SC, Cronic LE, Hess DC (2011) Minocycline development for acute ischemic stroke. Transl Stroke Res 2:202–8
59. Murata Y, Rosell A, Scannevin RH, Rhodes KJ, Wang X, Lo EH (2008) Extension of the thrombolytic time window with minocycline in experimental stroke. Stroke; A Journal of Cerebral Circulation 39:3372–7
60. Fagan SC, Waller JL, Nichols FT et al (2010) Minocycline to improve neurologic outcome in stroke (MINOS): a dose-finding study. Stroke; A Journal of Cerebral Circulation 41:2283–7
61. Switzer JA, Hess DC, Ergul A et al (2011) Matrix metalloproteinase-9 in an exploratory trial of intravenous minocycline for acute ischemic stroke. Stroke; A Journal of Cerebral Circulation 42:2633–5
62. Gordon PH, Moore DH, Miller RG et al (2007) Efficacy of minocycline in patients with amyotrophic lateral sclerosis: a phase III randomised trial. Lancet Neurol 6:1045–53
63. Chou SH, Feske SK, Simmons SL et al (2011) Elevated peripheral neutrophils and matrix metalloproteinase 9 as biomarkers of functional outcome following subarachnoid hemorrhage. Transl Stroke Res 2:600–7
64. Chou SH, Lee PS, Konigsberg RG et al (2011) Plasma-type gelsolin is decreased in human blood and cerebrospinal fluid after subarachnoid hemorrhage. Stroke; A Journal of Cerebral Circulation 42:3624–7
65. Guo ZD, Zhang XD, Wu HT, Lin B, Sun XC, Zhang JH (2011) Matrix metalloproteinase 9 inhibition reduces early brain injury in cortex after subarachnoid hemorrhage. Acta Neurochir Suppl 110:81–4
66. Adair JC, Charlie J, Dencoff JE et al (2004) Measurement of gelatinase B (MMP-9) in the cerebrospinal fluid of patients with vascular dementia and Alzheimer disease. Stroke 35:e159–62
67. Bjerke M, Zetterberg H, Edman A, Blennow K, Wallin A, Andreasson U (2011) Cerebrospinal fluid matrix metalloproteinases and tissue inhibitor of metalloproteinases in combination with subcortical and cortical biomarkers in vascular dementia and Alzheimer's disease. J Alzheimers Dis 27:665–76

Chapter 7
Toll-Like Receptors in Ischemic Stroke and Other Acute Brain Injuries

Frances Rena Bahjat, Raffaella Gesuete, and Mary P. Stenzel-Poore

Abstract Each year a substantial number of Americans suffer from hypoxic injury to the brain due to diminished blood flow and few effective treatments are available. A fruitful area of current investigation involves toll-like receptors (TLRs), which are a family of highly conserved receptors that play a key role in the pathology of brain injury. Studies in animals deficient in specific TLRs as well as genetic data from patients with altered TLR biology suggest that the activation of TLRs exacerbates damage in the setting of ischemia. Paradoxically, the stimulation of TLRs prior to injury is known to induce a state of tolerance to subsequent ischemic injury or "preconditioning". Such preconditioning results in a profound neuroprotective effect and the mechanisms involved are under intense investigation. Understanding these divergent roles of TLRs in brain injury and neuroprotection offers great promise in the discovery of new therapeutic targets and the mitigation of ischemic brain injury in "at risk" patients through the use of prophylactic TLR stimulation as a therapeutic strategy. This chapter focuses on these two divergent roles of TLRs—one role that promotes and another that prevents ischemic injury in the brain in the context of stroke and other acute brain injuries.

F.R. Bahjat
Department of Molecular Microbiology and Immunology,
Oregon Health and Science University, Portland, OR, USA

R. Gesuete • M.P. Stenzel-Poore (✉)
Oregon Health and Science University, Portland, OR, USA
e-mail: poorem@ohsu.edu

J. Chen et al. (eds.), *Immunological Mechanisms and Therapies in Brain Injuries and Stroke*, Springer Series in Translational Stroke Research, DOI 10.1007/978-1-4614-8915-3_7,

Toll-Like Receptors: An Introduction

Damage to the brain following injury occurs through the activation of multiple pathological mechanisms (e.g., hypoxia, excitotoxicity, ionic imbalance, spreading depolarization, oxidative stress, endothelial dysfunction, inflammation, and apoptosis) [1]. These mechanisms develop over hours or even days after the initial injurious event and negatively affect the complex network of cellular interactions in the brain. Among these mechanisms, the inflammatory response plays a critical role. Stroke, cerebral ischemia, and traumatic brain injury each elicit an inflammatory response consisting of systemic and local responses that can result in significant pathology or even death [1, 2].

Inflammation can be observed systemically by peripheral immune activation and centrally by the activation of microglia, endothelial cells, and astrocytes in the central nervous system. A key factor in the activation and modulation of the inflammatory responses in the brain involves TLRs, a family of receptors initially characterized in mammals by their ability to recognize "nonself" antigens in the form of pathogen-associated molecular patterns (PAMPs) [3]. PAMPs are components derived from pathogens (e.g., bacteria, viruses, parasites, and fungi) [4]. TLRs constitute a major component of innate immunity, traditionally considered to be the nonspecific first line of immune defense. Structurally, TLRs are transmembrane proteins characterized by the extracellular presence of leucine-rich repeats that recognize PAMPs and by the intracellular presence of Toll-interleukin 1 (IL-1) receptor (TIR) domains that are responsible for downstream signal transduction [5]. Ten TLRs have been identified in humans, whereas 12 functional TLRs have been found in mice [5]. TLRs can form heterodimers or homodimers and some associate with co-receptors or other accessory molecules (e.g., MD2 and CD14 in the case of TLR4), which can be critical factors necessary for tailoring the outcome of TLR signaling. Certain TLRs localize to the plasma membrane (TLRs 1, 2, 4, 5, and 6), whereas others are located intracellularly on endosomes or lysosomes (e.g., TLRs 3, 7, 8, and 9) [5].

TLR stimulation induces a diverse array of intracellular signaling pathways that regulate the scale, duration, and nature of the inflammatory response elicited by a given ligand. TLR signaling initiates the recruitment and association of adaptor molecules characterized by a structurally conserved Toll/interleukin-1 receptor (TIR) domain. There are two principal adaptor molecules that associate with TLRs: myeloid differentiation factor-88 (MyD88) and TIR-domain-containing adaptor inducing IFNβ (TRIF) [6]. MyD88 associates with all TLRs, with the exception of TLR3, which exclusively associates with TRIF [7]. Among the TLRs, TLR4 is the only receptor that is known to associate with both MyD88 and TRIF [8]. Briefly, the association of MyD88 with an activated TLR results in the activation of the transcription factor nuclear factor kappa B (NFκB) and subsequent generation of primarily pro-inflammatory cytokines including interleukin (IL)-6 and tumor necrosis factor-alpha (TNF) [9]. In contrast, recruitment of TRIF typically leads to the generation of anti-inflammatory (e.g., IL-10 and TGFβ), antiviral, and type I interferon (IFN)-associated molecules including IFNβ [6]. These differences in signaling indicate a tailoring of the response for a given PAMP or pathogen (e.g., virus versus bacteria).

TLR Activation: Detection of Self and Nonself

Studies conducted in mice deficient in individual TLRs have elucidated distinct functions for each receptor, with regard to PAMP recognition [4]. The most common PAMPs recognized by TLRs include the bacterial cell wall components peptidoglycan (TLR2) and lipopolysaccharide (LPS, endotoxin) (TLR4), and the nucleic acids double-stranded RNA (TLR3), single-stranded RNA (TLR7), and unmethylated cytosine–guanosine-containing DNA oligonucleotides (CpG ODNs) (TLR9) [4, 10]. Depending upon their localization, PAMPs are recognized by TLRs present in the plasma membrane, or in the endosomal, lysosomal, and endolysosomal compartments inside the cell [4], thus providing an additional level of regulation.

In addition to pathogen or "nonself" detection, TLRs act as sentinels of cell or tissue damage. As a result, TLRs mediate inflammatory responses to aseptic tissue injury or "self". In the absence of pathogens, the activation of TLRs depends upon the recognition of certain molecules derived from host cells. These self-derived signals indicate the presence of an endogenous threat and convey the nature of the damage (e.g., mechanical injury, cellular necrosis, or hyperactivation) [11]. A variety of signals are released or modified during cellular damage or death, including damage-associated molecular patterns (DAMPs) as well as PAMP-binding or PAMP-sensitizing proteins [11, 12]. These molecules consist of intracellular proteins, extracellular matrix components, modified lipids, nucleic acids, and other soluble mediators that represent "self or modified self" and accumulate at the site of injury (Fig. 7.1). Some of these ligands are thought to activate TLRs directly while others are postulated to participate with other ligands, chaperones, or co-receptors to fine-tune TLR responses [13].

At least 24 distinct ligands have been reported to represent danger signals released or modified during injury and some constitute key effectors identified in the injured brain including chromatin-associated protein high mobility group protein B1 (HMGB1) [14], uric acid [15], heat shock proteins (HSP) [16], ATP [17], S100 proteins [18], heparin sulfate [19], and host-derived nucleic acids in the form of DNA [20], RNA [21], and microRNA [22] (Table 7.1). Antioxidant proteins called peroxiredoxins are the most recently identified DAMPs released from the ischemic brain during the late phase of injury [48] (Table 7.1). In contrast, HMGB1 is released early from damaged neurons and is detectable in the cerebral spinal fluid within hours of the ischemic event [50]. These data suggest that distinct roles may exist for the variety of DAMPs released at various stages of injury.

These endogenous molecules with their unique location and timing of appearance clearly contribute to TLR activation and suggest that danger signals play a critical role in the brain's overall response to injury. For example, the administration of HMGB1 directly into the brain resulted in exacerbated stroke severity and inflammation [51, 52], whereas specific blockade of HMGB1 ameliorated infarction in a rodent stroke model [53]. Another example is that of S100A8 and S100A14, putative ligands of TLR4 [49] whose targeted deletion resulted in reduced lesion volumes, brain edema, and less inflammatory infiltrates following stroke in mice [54]. These are but two examples of the detrimental effects of TLR activation via danger signals generated by brain injury.

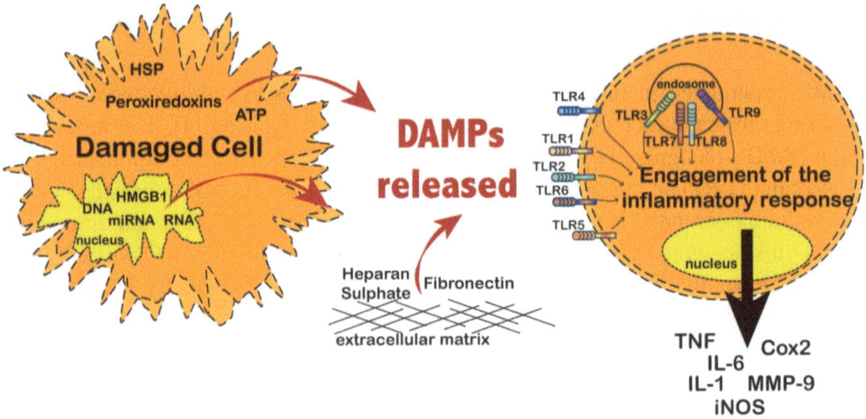

Fig. 7.1 Endogenous danger signals derived from dead or dying cells. Damaged or necrotic cells release products known as damage-associated molecular patterns (DAMPs). These DAMPs can activate TLR pathways. DAMPs are released either passively through damaged plasma membranes or actively by secretion from the damaged cell. In addition, disruption of the surrounding extracellular matrix releases components that act as DAMPs. Upon DAMP recognition, responding cells initiate pathways that culminate in a pro-inflammatory cascade that drives inflammation or repair. When exaggerated or prolonged, these inflammatory signals may cause significant pathology in the brain

Table 7.1 Selected danger-associated molecular pattern (DAMP) ligands and proposed TLR associations

Compartment	Ligand	TLR Association	References
Extracellular matrix derived	Fibronectin	TLR4	[23]
	Hyaluronan	TLR2, TLR4	[24, 25]
	Heparan sulfate	TLR4	[19]
	α-Crystallin	TLR4	[26]
Intracellular	ATP	TLR4	[27]
	Heat shock proteins	TLR2, TLR4	[28, 26, 29, 30]
	Uric acid	TLR2	[15, 31]
Nuclear	Chromatin:IgG complex	TLR9	[32]
	dsRNA	TLR3	[33]
	DNA	TLR9	[34]
	MicroRNA	TLR7	[22]
	HMGB1	TLR2, TLR4	[35]
	ssRNA	TLR7, TLR8	[36, 37]
Other	Amyloids	TLR2, TLR4	[38]
	β-defensin	TLR1/2, TLR2,TLR4	[39, 40, 41, 42]
	Eosinophil-derived neurotoxin	TLR2	[43]
	Fibrinogen	TLR4	[44]
	Low-density lipoproteins	TLR4	[45, 46, 47]
	Peroxiredoxins	TLR4	[48]
	S100A	TLR4	[49]

TLR Expression and Distribution in the CNS

As part of immune surveillance, TLRs are generally expressed by antigen-presenting cells such as B cells, dendritic cells, monocytes, and macrophages from blood, organs, or lymphoid tissues (Table 7.2). These important cell types function in innate and antigen-acquired immune responses designed to target specific pathogens or altered self. TLRs in the brain are constitutively expressed by microglia and astrocytes—cells that are considered to be key sentinels of innate CNS immunity. Other brain cells, not strictly involved in the CNS immunity, such as brain endothelial cells, oligodendrocytes, and neurons, also express TLRs [55, 69, 70]. Human neurons express TLRs 2, 3, 4, 7, 8, and 9, whereas microglia and astrocytes express TLRs 1–9 (Table 7.2). Human astrocytes highly express TLR3 as compared to other cells [57], suggesting a special role for this receptor in the brain. Other cells in the brain typically express a restricted number of TLRs including TLR2, TLR3 (oligodendrocytes and endothelial cells), and TLR4, TLR6, and TLR9 (endothelial cells) (Table 7.2). These cells increase their expression of TLRs in response to activating stimuli and subsequent TLR stimulation generates cytokine profiles reflective of the specific cell type as well as the TLR, co-receptors, or ligands [71]. Thus, the varied distribution of TLRs on cells in the brain suggests unique aspects of TLR biology in the regulation of the brain's response to injury.

TLRs in CNS Immunity

The CNS has traditionally been considered an immune privileged site completely separate from immunological processes in peripheral organs. However, accumulating evidence argues that the CNS has its own well-organized innate immune reaction to injury or infection [72, 73]. The primary function of these cells is to maintain and restore tissue homeostasis by regulating the balance between the protective and the potentially harmful effects of their activation following stimuli, such as acute brain injury [74]. Microglia are central to the process of immune regulation in the brain, as the immune response of the CNS depends upon phagocytic and scavenger receptors that distinguish foreign and host-derived molecules [75].

Microglia are the resident phagocytic cells of the CNS similar to macrophages found in other tissues [76]. Under basal conditions microglia, like macrophages, continuously survey the CNS microenvironment for the presence of infection or injury. Upon infection or CNS injury, microglia are rapidly activated leading to increased proliferation, motility, phagocytic activity, and the release of cytokines and reactive oxygen species [77]. Because microglia play a central role in innate immunity, recognizing both PAMPs and DAMPs, it is not surprising that they are implicated in a range of detrimental neuronal inflammatory processes associated with diseases of the CNS [78]. However, microglia when appropriately activated may also protect neurons from damage induced by resident or infiltrating cells [79–81].

Table 7.2 TLR expression on cell types in the blood, brain, and cerebrovasculature

Compartment	Cell type	TLR1	TLR2	TLR3	TLR4	TLR5	TLR6	TLR7	TLR8	TLR9	TLR10	TLR11
Blood	APCs[a]	+	+	+	+	+	+	+	+	+	+	+
	Lymphocytes	+	+									
Cerebrovasculature	Endothelial cell		+	+	+		+			+		
Brain	Astrocytes	+	+	+	+	+	+	+	+	+		
	Microglia	+	+	+	+	+	+	+	+	+		
	Oligodendrocytes		+	+								
	Neurons		+	+	+			+	+	+		

[a]APCs denote antigen-presenting cells including dendritic cells, B cells, and monocytes

Constantin et al. [55], Cristofaro and Opal [56], Farina et al. [57], Guo and Schluesener [58], Hanke and Kielian [59], Hornung et al. [60], Lafon et al. [61], Leow-Dyke et al. [62], Muzio et al. [63], Nishimura and Naito [64], Takeda and Akira [65], Tsan and Gao [66], Wang et al. [67], Zarember and Godowski [68]

The effects induced by the stimulation of a particular TLR can differ according to the specific cell type involved. For example, microglia and astrocytes can respond differently upon engagement of a given TLR. Microglia secrete cytokines and chemokines (e.g., TNF, IL-6, IL-10, IL-12, CXCL-10, and IFNβ) in response to the stimulation of TLR2, TLR3, and TLR4 with their cognate ligands, whereas astrocytes produce small amounts of IL-6 to all but TLR3 stimulation [82]. Upon activation, adult human astrocytes were shown to express TLR3 followed by induction of neuroprotective mediators, anti-inflammatory cytokines (e.g., IL-10), as well as regulators of growth, differentiation, and migration. This activation of TLR3 and the downstream processes resulted in a functional inhibition of neuronal cell death in human organotypic brain slice cultures [83].

A recent study showed that stimulation of neurons with LPS in vitro induced JNK activation and high CXCL1 (KC) chemokine with limited pro-inflammatory cytokine secretion when compared to mixed glial cultures [62]. However, when cocultured with endothelial cells, LPS-treated neurons induced the expression of cellular adhesion molecules (ICAM-1 and VCAM-1) and increased the infiltration of neutrophils [62]. As with peripheral immune responses, the activation of CNS immune effector mechanisms can have both toxic and protective effects depending upon the circumstances and cellular context [84].

TLRs in the Brain: At the Crossroads Between Injury and Protection

Detrimental Role of TLRs in Ischemic Stroke

Acute brain injury is associated with a myriad of inflammatory responses, both systemically and locally. For example, acute stroke stimulates an inflammatory cascade in the occluded vessel, the arterial wall, and the brain parenchyma [85]. The stroke-induced damage of those compartments leads to the release of DAMPs into the extracellular environment including extracellular matrix components (hyaluronan, heparan sulfate, and fibronectin), modified lipids, nucleic acids (microRNA, RNA, and DNA), and intracellular proteins (heat shock proteins, gp96, HMGB1, and uric acid) (Fig. 7.1, Table 7.1). These DAMPs activate the innate and adaptive arms of the immune system and further stimulate the inflammatory cascade [86, 87].

Studies in animals deficient in specific TLRs and genetic data from patients with altered TLR biology suggest that the activation of TLRs exacerbates damage in the setting of ischemia [86, 88–90]. The involvement of TLRs, in particular TLR4 and TLR2, in ischemic brain injury has been studied extensively. The expression of TLR4 was upregulated on glial cells in vitro in response to hypoxic conditions and in vivo after ischemia [89, 91]. Microglia cultured in vitro and exposed to hypoxia show increased mRNA and protein levels of TLR4 [91] and TLR4 mRNA upregulation in neurons was found as early as 1 h after cerebral ischemia in vivo [92, 93].

More extensive genomic analyses conducted in a mouse stroke model showed that TLRs and TLR pathway genes were regulated by ischemia–reperfusion injury. Specifically, genes encoding TLRs 1, 2, 4, 7, and 13 were upregulated in the brain of animals that had been subjected to a stroke, whereas TLRs 3, 6, 8, and 9 showed no differential regulation compared to control animals (Stenzel-Poore unpublished data and [94]). A similar, although not identical, set of genes were upregulated in the peripheral blood following stroke (Stenzel-Poore unpublished data).

Importantly, cortical neuronal cultures derived from mice deficient in TLR4 were protected from glucose deprivation [92]. In vivo studies also revealed that TLR4 deficiency conferred protection against permanent middle cerebral artery occlusion (pMCAO) or transient MCAO (tMCAO), as demonstrated by improved behavioral outcomes and smaller infarct volumes [89, 95–97]. The activation of TLR downstream effectors was also reduced following stroke in TLR4-deficient animals (e.g., reduced IκB phosphorylation, NFκB activity, and secretion of pro-inflammatory cytokines TNF and IL-6 [96, 98]). Other major mediators of brain damage such as inducible nitric oxide synthase (iNOS), cyclooxygenase-2 (COX-2), and matrix metalloproteinase-9 (MMP9) were also reduced in TLR4-deficient mice [89, 95]. TLR4 has also been shown to contribute to injury associated with intracerebral hemorrhage using a mouse model of perihematomal inflammation [99]. Collectively, these studies indicate that TLR4 signaling exacerbates brain injury in experimental models of ischemic and hemorrhagic stroke.

TLR2 appears to play a similar role to TLR4 in cerebral ischemic damage [92, 100–103]. After cerebral ischemia in mice TLR2 mRNA was upregulated in the brain [92] and neuronal cultures from TLR2 knockout mice showed reduced cell death compared to wild-type cells after energy deprivation [101]. In addition, brain damage and neurological deficits induced by MCAO were significantly reduced in TLR2-deficient mice compared with wild-type control mice [101]. TLR2 was the most highly induced TLR in the ipsilateral brain hemisphere following injury when compared to TLR4 and TLR9 [103]. In these studies, TLR2 protein was expressed mainly in microglia in post-ischemic brain tissue but also in selected endothelial cells, neurons, and astrocytes. In addition, genes having known pro-inflammatory (NFκB and IRAK) and pro-apoptotic (FADD, CASP1, and CASP8) functions were also observed. Similarly, brain TLR2 mRNA was upregulated acutely (6–24 h) following hypoxic–ischemic brain injury of neonatal mice and TLR2 deficiency was protective [104]. Therefore, the induction of TLR2 and downstream signaling are potentially important contributors to the damage resulting from focal cerebral ischemia in experimental models.

In a more recent study, TLR2 has been shown to mediate leukocyte and microglial infiltration and neuronal death, as these processes were attenuated upon TLR2 inhibition [102]. However, another study showed that in the absence of TLR2, animals were protected at early time points (~3 days) following stroke, yet subsequent observations revealed an exacerbation of the ischemic lesion at 7 and 14 days after stroke. In that study, TLR2 deficiency resulted in reduced microglia/macrophage activation acutely after stroke along with a reduced capacity of resident microglia to proliferate. The authors concluded that an early reduction in microglial response was associated with initially smaller lesions and that a long-term decrease in microglia/macrophage activation and proliferation led to increased neuronal apoptosis and delayed exacerbation of ischemic lesions in TLR2 null mice [105].

Collectively, this evidence implicates TLR4 and TLR2 as critical mediators of injury induced by stroke; thus, these two receptors are potential targets for therapeutic manipulation. While much is known regarding the putative roles of TLR2 and TLR4 in brain injury, more recent data suggest that other TLRs may play similar roles inducing divergent processes. The presence of let-7, an extracellular microRNA, which has recently been identified as a ligand for TLR7, induced neurodegeneration through neuronal TLR7 signaling in vivo. While a role for let-7 in human stroke has not yet been defined, studies showed that let-7f was expressed in microglia in the ischemic brain and antagonists of let-7 promoted neuroprotection in female rats [106]. In addition, high levels of let-7 have been detected in the cerebrospinal fluid of Alzheimer's patients [22]; thus, it may be fruitful to examine this TLR ligand in other CNS disorders including stroke.

Role of TLRs in Other Brain Injuries

TLRs have recently been implicated in a host of other brain pathologies including traumatic brain injury (TBI) and neurodegenerative diseases. TBI results from damage due to external forces and the severity of injury depends upon the nature of the precipitating event, as well as the location, intensity, and duration of the injurious forces acting on the brain. TBI can lead to a diverse array of histopathological changes, including but not limited to hemorrhagic contusion, intracerebral hemorrhage, subarachnoid hemorrhage, and widespread white matter damage [107]. A primary biomechanical injury occurs following TBI that is followed by a secondary wave of injury occurring within hours and lasting days after the initial insult [1]. The extent of the secondary injury (e.g., neurotransmitter release, free-radical generation, calcium-mediated damage, gene activation, mitochondrial dysfunction, and inflammation) is the major determinant of the extent of overall brain damage and functional deficits [108]. Inflammatory response to TBI involves glial activation, recruitment and activation of immune cells, activation of the complement cascade, upregulation of pro-inflammatory cytokines, chemokines, and continued release of endogenous DAMPs upon neuronal cell death [109, 110]. As such, TLR2 and TLR4 play pivotal roles in the injury associated with TBI [111, 112]. TBI induces a significant upregulation of TLR1, TLR2, and TLR9 as well as TLR4 and MyD88 [111, 112] in the brain and initiates the infiltration of TLR-expressing circulating immune cells into the CNS. TLR2 is induced mainly in neurons while TLR9 is upregulated in astrocytes following TBI [111]. TLR2, TLR4, and MyD88 were determined by immunohistochemistry to be localized to lesion regions and subcortical matter in astrocytes and macrophage/microglia, whereas the endogenous danger signal, HSP70, was found in glia at the site of injury and in peri-lesional neurons [112]. In this study, TLRs and downstream effectors were detected within 1–2 days after TBI and positive cells accumulated for up to 4 days following injury [112]. Importantly, genomic studies in a rodent TBI model have demonstrated the upregulation of two TLRs (TLR1 and TLR2), five TLR adaptor/interacting proteins (CD14, MD-1, HSPA1a, PGRP, and Ticam2) and 13 downstream target genes encoding proteins elicited by TLR pathways (Ccl2, Csf3, IL1a, IL1b, IL1r1, IL6, IL-10, TNFa, Tnfrsf1a, Cebpb, Clec4e, Ptgs2, and Cxcl10) [111].

A potential role for TLRs in the exacerbation of TBI-induced brain damage was evident from studies in TLR-deficient mice. Mice lacking TLR2 function showed reduced neurological deficit, apoptosis, and brain edema acutely following TBI. In addition, decreased expression of inflammatory cytokines such as TNF-α, IL-1β, and IL-6 was observed after TBI in these mice, as compared with their wild-type counterparts. These results argue that TLR2 signaling may exacerbate TBI-induced secondary brain injury, possibly by affecting key inflammatory processes in the cortex [113]. Similar to experimental TBI, humans demonstrate a robust inflammatory response characterized by intracerebral leukocyte infiltration and upregulation of inflammatory cytokines following a cerebral contusion [114–116]. Similarly, traumatic injury in the spinal cord (SCI) was shown to increase the expression of TLRs and TLR signaling mediators in the brain, although in this case, the absence of TLR2 or TLR4 function exacerbated injury suggesting a protective role for these receptors in SCI [117].

TLRs also play a role in the exacerbation of the damage caused by retinal ganglion cell (RGC) axotomy, a clinically relevant model of subacute neuronal degeneration [97]. TLR4-deficient and TLR2-deficient mice are protected from neuronal injury in this model, similar to stroke and TBI [86, 97]. Neurodegeneration can also occur secondary to alcohol abuse due to the ensuing neuroinflammatory response. Consumption of ethanol can elicit cytokines and other pro-inflammatory mediators in the brain that results in cell damage [118]. TLR4 plays a pivotal role in alcohol-associated brain damage [119]. Ethanol activates TLR4 signaling in astrocytes [120], microglia, and macrophages [121] in wild-type mice; however, mice deficient in TLR4 show reduced TLR signaling, reduced pro-inflammatory and apoptotic mediators [e.g., cytokines, caspases, cyclooxygenase-2 (COX-2), and reduced inducible nitric oxide synthase (iNOS)] in the cortex in response to chronic ethanol [119]. These data support the contention that TLR4 signaling contributes to glial activation in ethanol-induced brain damage and confirm the important role of TLRs in the development of brain damage regardless of the nature of the injury. Collectively, all of these data highlight the clinical potential of anti-inflammatory or TLR-modulatory strategies in the treatment of a multitude of brain injuries.

Neuroprotection Induced by TLR Stimulation

TLR activation has been demonstrated to potentiate inflammatory responses that exacerbate ischemic damage [71] and initiate anti-inflammatory or cytoprotective responses designed to ameliorate inflammation and promote repair [83]. Paradoxically, TLRs, when stimulated prior to injury, can induce a state of tolerance to a subsequent ischemic challenge [122–128]. The specific induction of any one of these effects depends upon the cell type, the particular TLR or stimulus, the timing of activation with respect to injury, the intensity and duration of signal, and the composition of the surrounding milieu (e.g., proteins, metabolites, and adjacent cells). Of particular relevance to ischemic injury is the finding that suboptimal stimulation of TLRs days prior to ischemia provides robust neuroprotection, a phenomenon known as delayed preconditioning [129]. TLR4-induced tolerance to cerebral ischemia was first demonstrated with low-dose systemic administration of LPS, a cell wall component of Gram-negative bacteria. LPS treatment prior to injury caused spontaneously hypertensive rats to become tolerant to subsequent ischemic brain damage induced by MCAO [128]. Subsequent studies in a mouse model of stroke and a porcine model of deep hypothermic circulatory arrest also demonstrated LPS-mediated tolerance to brain ischemia [130]. Neuroprotection induced by LPS is time and dose dependent. For example, tolerance to ischemic injury occurs within 1 day following LPS administration and extends to 7 days, but is lost by 14 days [131].

Similar regimens of preconditioning with low doses of TLR agonists of TLR2, TLR3, TLR4, TLR7, and TLR9 have been shown to protect against ischemic injury in rodent stroke models [122, 123, 125, 127, 128]. Typically, preconditioning with a low dose of a TLR agonist 3 days prior to ischemic injury was associated with reduced infarct volumes in mice. A pivotal proof of concept study performed by the Stenzel-Poore laboratory showed that nonhuman primates subjected to cerebral ischemia exhibit a similar time and dose-dependent neuroprotection induced 3 days following treatment with CpG ODNs (TLR9 agonists) [132]. Thus, TLR-induced neuroprotection shows considerable promise in humans at risk of ischemic brain injury.

Genomic Underpinnings of TLR-Induced Neuroprotection

TLR-induced neuroprotection in the brain, as demonstrated for other tissues, seems to be related to an early low-level activation of TLR signaling that results in a subsequent "tolerance" characterized by a hyporesponsive state [133, 134]. Efficacy using TLR agonist preconditioning is generally correlated with the early induction of systemic and brain pro-inflammatory mediators within hours of systemic administration. This is followed by the induction of interferon-related genes in the context of subsequent ischemic brain injury [135]. Using microarray analysis performed on multiple distinct preconditioning paradigms we found that the preconditioning-induced neuroprotection is associated with genomic reprogramming characterized by attenuated NFκB-mediated pro-inflammatory signaling and increased expression of IFN-associated genes [135, 136]. Our studies have identified a network of genes associated with preconditioning-induced neuroprotection against ischemic injury [135]. This gene network is comprised predominantly of genes that are transcriptionally associated with interferon regulatory factor (IRF) activity. Thus, we proposed that preconditioning reprograms injury-induced TLR signaling to drive substantial TRIF-IRF-mediated signaling.

A recent analysis performed using the genomic signatures elicited by various neuroprotectants of diverse mechanisms supports our findings in that common signatures were observed that also included interferon-associated genes [137]. Further, Pappas et al. showed that treatment with either IFNβ or IFNα was sufficient to increase neuronal survival in vitro following exposure to an excitotoxic insult [137]. These data suggest the possibility that IFNs arising after injury as a result of delayed TLR preconditioning may also be providing direct protection to neurons undergoing excitotoxic injury. Additionally, LPS can precondition against traumatic brain injury in mice, resulting in IFNβ expression and increased IL-6 [138]. Taken together, mouse genomic and functional studies suggest the intriguing possibility that TLR stimulation prior to stroke may reprogram or alter the outcome of post-injury TLR signaling and effectively reduce neuronal excitotoxic injury.

Cellular Targets of Neuroprotection

The identity of the initial target cell for TLR activation is not yet clear. Most experimental models use systemic administration of various TLR agonists, which are generally considered to be unable to cross the blood–brain barrier (BBB). It is possible that TLR agonists induce robust neuroprotection via actions on peripheral cells that subsequently signal indirectly leading to preservation of brain tissue in the setting of injury. In the context of preconditioning, TLR agonists induce circulating pro-inflammatory cytokines when administered systemically to mice or primates [139, 140]. Rapid acute expansion of neutrophils and concomitant reduction in other peripheral blood mononuclear cells also occur post-administration [139]. These systemic inflammatory cues (e.g., prostaglandins) may act as inflammatory messengers across the BBB [141] to prepare the brain for potentially injurious conditions.

Alternatively, TLR agonists could act directly on CNS cells to reprogram their response to ischemia, provided they gain access to the brain by crossing the BBB or by active transport. As discussed above, neurons and glial cells express TLRs and respond to many TLR ligands (Table 7.2). Specifically, ischemia modeled in vitro using cultured murine primary neuronal cells indicated that pretreatment with TLR ligands can directly protect cells in mixed cortical cultures from ischemic injury. Primary cortical cells pretreated with a TLR agonist prior to exposure to oxygen–glucose deprivation were protected from injury [125, 127, 131]. These primary cortical cell cultures were comprised of astrocytes, neurons, and microglia, all of which express TLRs and thus represent potential target cell populations. Due to the variable presence and activation of each TLR on these individual cell types, it is difficult to determine precisely which effectors are at play in this system. Recent data from our laboratory revealed that the neuroprotective effect of CpG preconditioning was absent in mice deficient in TLR9 expression on either hematopoietic cells or other parenchymal cells including the brain [142]. These findings argue that TLR9-induced neuroprotection requires expression in non-hematopoietic as well as hematopoietic cells in the circulation, thereby suggesting an important interplay between the peripheral and central responses to produce the neuroprotective effect of CpG preconditioning.

Brain ischemia activates perivascular inflammation and causes an increase in BBB permeability that results in increased leukocyte infiltration, brain edema, and tissue damage [143]. Endothelial cells are an integral part of the BBB that make up the interface between the blood and brain, which positions them strategically for TLR-induced neuroprotection. TLR agonists may act directly on brain endothelial cells to induce protection by reducing BBB dysfunction and inflammatory-cell infiltration, and ultimately leading to less damage. Using an in vitro BBB model consisting of a coculture of primary murine brain microvessel endothelial cells (BMEC) and primary mixed astrocytes and microglia cells, we and others have shown that preconditioning with TLR ligands can attenuate OGD-induced BBB dysfunction [122, 144]. For instance, the administration of poly-ICLC 1 day prior to exposure to OGD prevented the drastic reduction in transendothelial resistance (TEER) typically induced (Fig. 7.2a) and restored permeability coefficient (Pe) transport values (Fig. 7.2b). In addition, treatment preserved tight junction (TJ) structure (Fig. 7.2c). This indicates poly-ICLC preconditioning maintains both the function and integrity of the BBB endothelial cells in the setting of in vitro ischemic injury. Similar findings were obtained using Pam3CSK4, a TLR2 agonist using a similar in vitro BBB system [122].

We have observed that IFNβ plays a key role in the protective effect induced by preconditioning via poly-ICLC treatment [144]. Stimulation of cells with poly-ICLC induced astrocyte/microglial cell production of IFNβ prior to OGD, a response essential for the protective effects seen on BBB integrity and function. We further showed that the IFNβ produced by glial cells as well as type I IFN receptors on the endothelium were both required to maintain the integrity of the BBB in response to OGD. The significance of these findings is further highlighted by in vivo studies indicating that poly-ICLC preconditioning depends upon type I IFN signaling to protect the brain against ischemic injury [144]. These data correlate with findings in other neurological diseases. For instance, multiple sclerosis is an autoimmune-mediated nervous system disorder for which recombinant IFN-β-1-α (sold under trade names Avonex™, Betaseron™, Extavia™, and Rebif™) remains one of the most recognized worldwide therapeutic treatments. While the precise mechanism of action of IFNβ therapy in MS remains elusive, its immunomodulatory effects are likely to be a primary mechanism of efficacy [145].

There are at least three potential sites of action for TLRs in preconditioning-induced neuroprotection (Fig. 7.3). TLR agonists can have a systemic effect on hematopoietic cells, a direct effect on CNS cells or they can act on the endothelial cells that represent the physical interface between the two compartments. It is also likely that based on the specific TLR agonist being used, more than one mechanism could be involved in the neuroprotective effect observed in vivo.

Conclusion

Stroke is the leading cause of disability and the third leading cause of death in the USA today and is projected to be the leading cause of death by the year 2050 [146]. Despite intensive efforts by researchers to identify interventions that lessen

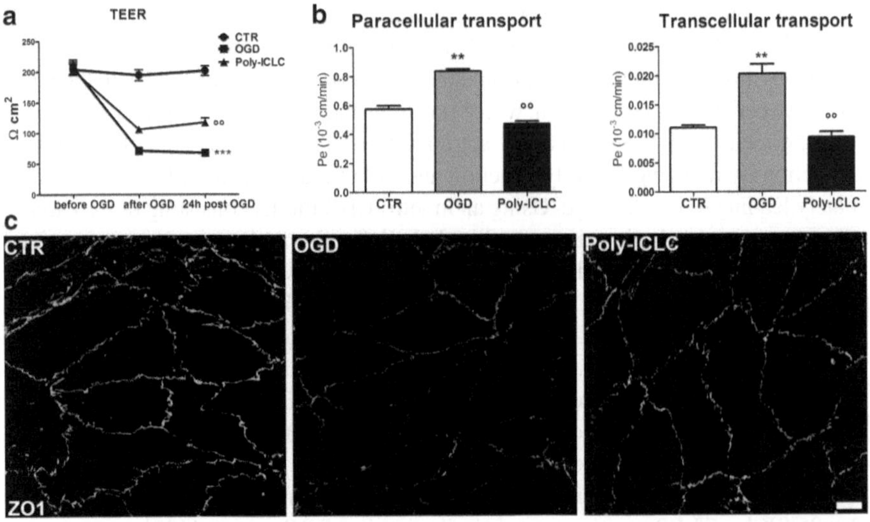

Fig. 7.2 Poly-ICLC preconditioning reduces damage and preserves the function of the BBB following OGD. Poly-ICLC preconditioning significantly prevented the drastic reduction of TEER values (**a**) and the increase of endothelial permeability coefficient (Pe) values for both paracellular and transcellular transport (**b**) induced by OGD. Poly-ICLC preconditioning also prevented the OGD-induced disorganization of the tight junction ZO-1. (**c**). Qualitative analysis of ZO-1 immunofluorescence showed an intense and continuous staining on the cell borders (**c**, *left panel*). OGD induced a decrease in immunostaining intensity, loss of the continuous junctional staining pattern, and a global disorganization of the proteins (**c**, *middle panel*). Poly-ICLC-treated cells showed a preserved staining intensity and distribution of ZO-1 following OGD (**c**, *right panel*). Data reflect mean±SEM. $^{**}P<0.01$ and $^{***}P<0.001$ vs CTR; $^{\circ\circ}P<0.01$ vs OGD, two-way ANOVA followed by Bonferroni test for TEER and one-way ANOVA followed by Bonferroni test for Pe. *Scale bar*=10 μm. Adapted from [144]

stroke-induced brain injury, few therapies exist for the 800,000 Americans affected by stroke each year. An equal number of Americans are at high risk of cerebral ischemia secondary to cardiovascular or cerebrovascular medical procedures (e.g., cardiac bypass grafting, cardiac valve replacement, carotid endarterectomy, aortic repair, endovascular coiling or clipping, peripheral vascular surgery, or resection of head and neck tumors). Depending upon the procedure as many as two-third of patients will demonstrate ischemic brain lesions detectable by diffusion-weighted magnetic resonance imaging (DWI) and some will have cognitive deficits [147–149]. Importantly, the true long-term clinical impact of these ischemic lesions identified in asymptomatic patients remains undefined. Many of these patients will experience peri- or postoperative stroke [150–153]. Moreover, traumatic brain injury (TBI) affects approximately 1.7 million Americans each year, one-third of which are children under the age of 14 [154]. In the USA, TBI contributes to ~30 % of all injury-related deaths. Given the prevalence and severity of acute brain injuries, there is compelling need to improve the development of novel therapeutics, which requires an improved understanding of the pathology that underlies these injuries.

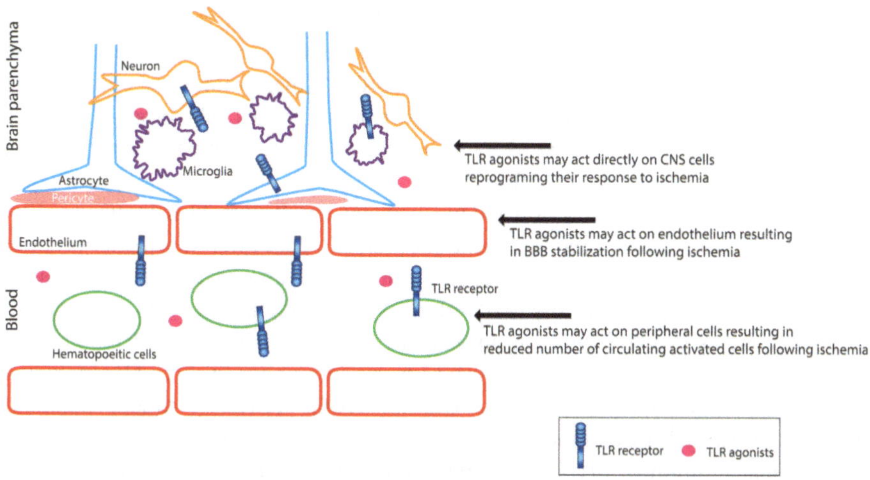

Fig. 7.3 Potential sites of action for TLR preconditioning-induced neuroprotection. TLR preconditioning via systemic TLR agonists administration prior to a harmful event may occur directly on peripheral cells to induce a cascade culminating in neuroprotection. This could result in decreased number of circulating activated cells and reduced leukocytes infiltration that in turn would contribute to decreased brain damage. Alternatively, TLR agonists may cross the BBB and directly act on CNS cells resulting in neuroprotection by reprogramming their response to subsequent ischemia. TLR agonists may also act directly on BBB at the interface between blood and brain resulting in increased BBB stabilization following ischemia. Adapted from [125]

Collectively, experimental data in rodents and nonhuman primates argue that TLR activation represents a key initiating event in acute brain injuries and thus represents a potential therapeutic target. Clinically, the expression levels of TLR7, TLR8, TLR2, and TLR4 were independently associated with poor outcome and increased levels of inflammatory mediators, while the expression levels of TLR4 and TLR8 were additionally associated with size of infarcted tissue [88, 155]. Genetic polymorphisms in the human genes encoding TLR2 [156] and TLR4 [157] have also been associated with ischemic stroke in some studies. While it is unproven whether modulation of TLR-induced inflammation in humans will lead to clinical improvement in stroke, current data from animal models suggest that inflammatory signals drive potent secondary injury. As such, manipulation of TLR signaling and blockade of endogenous and exogenous TLR ligand signaling remain promising clinical strategies. Importantly, therapeutic strategies aimed at blocking the inflammatory response in humans with brain injury have produced mixed results, which may indicate that selective modulation of the inflammatory response, rather than global suppression would be a more valuable clinical therapeutic strategy. For instance, downregulation of the deleterious side of inflammation and upregulation of protective molecules could provide clinical benefit. Thus, understanding the protective pathways of the neuroinflammatory response is a priority in planning future therapies aimed at its manipulation [110]. Novel therapies that target post-injury processes mediated by TLRs or downstream mediators also offer promise as important adjunct treatments that synergize with conventional pharmacotherapy (e.g., thrombolytics and clot removal devices) or enhance the efficacy of existing partially effective therapeutics. As with many human conditions, brain injury is multifactorial; thus, a single therapeutic, a so-called magic bullet, is not likely to exist for stroke or other brain injuries. Therefore, innovative combination strategies are most likely to achieve widespread benefit for patients but will require arduous preclinical efforts.

While the precise site of action and mechanisms involved in TLR-induced neuroprotection are unknown, preclinical and mechanistic data offer substantial promise for the clinical development of preventative therapeutic approaches directed against ischemic injury. Several agents under clinical development represent good candidates for use as prophylactic neuroprotectants (e.g., poly-ICLC and CpG ODNs). Three TLR agonists have been approved for use in humans by the FDA including *Bacillus Calmette–Guérin Calmette–Guérin* (BCG, TLR2/4), monophosphoryl lipid A (MPL, TLR4), and the small molecule agonist Imiquimod (TLR7/8) [158]. An attenuated strain of *Mycobacterium bovis* known as BCG is mainly used as a vaccine against tuberculosis, but it is also known to be an immunotherapeutic for bladder carcinoma. Cervarix®, a vaccine against human papillomavirus-16 and -18, has been formulated with monophosphoryl lipid A (MPL), a TLR4 ligand derived from the LPS of *Salmonella minnesota*. Lastly, Imiquimod is a well-characterized synthetic imidazoquinoline often used topically in patients with actinic keratosis, superficial basal cell carcinoma, and external genital warts. Given their favorable safety profile, these and related agents that similarly target TLRs could be useful in the multitude of high-risk populations impacted by brain ischemia. In addition, the use of prophylactic therapies that promote resistance to ischemic injury is an invaluable platform for the discovery of novel acute therapeutic targets.

The use of TLR agonists as an antecedent therapy could be a novel and powerful strategy to protect patients at risk of ischemic injury to the brain. Nearly one million US patients annually will undergo cardiovascular and cerebrovascular procedures that increase the risk of brain ischemia from embolic events associated with these medical procedures. Preclinical validation of preconditioning has produced extremely positive outcomes in nonhuman primate models of stroke involving severe cerebral ischemic injury. Clinical testing in humans is the next critical step to demonstrate whether this approach offers benefit by reducing ischemic brain injuries associated with these common vital surgical procedures.

References

1. Werner C, Engelhard K (2007) Pathophysiology of traumatic brain injury. Br J Anaesth 99:4–9
2. Jin R, Yang G, Li G (2010) Inflammatory mechanisms in ischemic stroke: role of inflammatory cells. J Leukoc Biol 87:779–789
3. Janeway CA Jr, Medzhitov R (2002) Innate immune recognition. Annu Rev Immunol 20:197–216
4. Akira S, Uematsu S, Takeuchi O (2006) Pathogen recognition and innate immunity. Cell 124:783–801

5. Kawai T, Akira S (2010) The role of pattern-recognition receptors in innate immunity: update on toll-like receptors. Nat Immunol 11:373–384
6. Yamamoto M, Sato S, Mori K, Hoshino K, Takeuchi O, Takeda K, Akira S (2002) Cutting edge: a novel toll/IL-1 receptor domain-containing adapter that preferentially activates the IFN-beta promoter in the toll-like receptor signaling. J Immunol 169:6668–6672
7. Yamamoto M, Sato S, Hemmi H, Hoshino K, Kaisho T, Sanjo H, Takeuchi O, Sugiyama M, Okabe M, Takeda K, Akira S (2003) Role of adaptor TRIF in the MyD88-independent toll-like receptor signaling pathway. Science 301:640–643
8. Tanimura N, Saitoh S, Matsumoto F, Akashi-Takamura S, Miyake K (2008) Roles for LPS-dependent interaction and relocation of TLR4 and TRAM in TRIF-signaling. Biochem Biophys Res Commun 368:94–99
9. Takeda K, Akira S (2004) TLR signaling pathways. Semin Immunol 16:3–9
10. Akira S, Takeda K, Kaisho T (2001) Toll-like receptors: critical proteins linking innate and acquired immunity. Nat Immunol 2:675–680
11. Erridge C (2010) Endogenous ligands of TLR2 and TLR4: agonists or assistants? J Leukoc Biol 87:989–999
12. Miyake Y, Yamasaki S (2012) Sensing necrotic cells. Adv Exp Med Biol 738:144–152
13. Lloyd-Jones KL, Kelly MM, Kubes P (2008) Varying importance of soluble and membrane CD14 in endothelial detection of lipopolysaccharide. J Immunol 181:1446–1453
14. Lotze MT, Tracey KJ (2005) High-mobility group box 1 protein (HMGB1): nuclear weapon in the immune arsenal. Nat Rev Immunol 5:331–342
15. Kono H, Chen CJ, Ontiveros F, Rock KL (2010) Uric acid promotes an acute inflammatory response to sterile cell death in mice. J Clin Invest 120:1939–1949
16. Quintana FJ, Cohen IR (2005) Heat shock proteins as endogenous adjuvants in sterile and septic inflammation. J Immunol 175:2777–2782
17. Bours MJ, Swennen EL, Di Virgilio F, Cronstein BN, Dagnelie PC (2006) Adenosine 5′-triphosphate and adenosine as endogenous signaling molecules in immunity and inflammation. Pharmacol Ther 112:358–404
18. Hofmann MA, Drury S, Fu C, Qu W, Taguchi A, Lu Y, Avila C, Kambham N, Bierhaus A, Nawroth P, Neurath MF, Slattery T, Beach D, McClary J, Nagashima M, Morser J, Stern D, Schmidt AM (1999) RAGE mediates a novel proinflammatory axis: a central cell surface receptor for S100/calgranulin polypeptides. Cell 97:889–901
19. Johnson GB, Brunn GJ, Kodaira Y, Platt JL (2002) Receptor-mediated monitoring of tissue well-being via detection of soluble heparan sulfate by toll-like receptor 4. J Immunol 168:5233–5239
20. Burckstummer T, Baumann C, Bluml S, Dixit E, Durnberger G, Jahn H, Planyavsky M, Bilban M, Colinge J, Bennett KL, Superti-Furga G (2009) An orthogonal proteomic-genomic screen identifies AIM2 as a cytoplasmic DNA sensor for the inflammasome. Nat Immunol 10:266–272
21. Kariko K, Ni H, Capodici J, Lamphier M, Weissman D (2004) mRNA is an endogenous ligand for toll-like receptor 3. J Biol Chem 279:12542–12550
22. Lehmann SM, Kruger C, Park B, Derkow K, Rosenberger K, Baumgart J, Trimbuch T, Eom G, Hinz M, Kaul D, Habbel P, Kalin R, Franzoni E, Rybak A, Nguyen D, Veh R, Ninnemann O, Peters O, Nitsch R, Heppner FL, Golenbock D, Schott E, Ploegh HL, Wulczyn FG, Lehnardt S (2012) An unconventional role for miRNA: let-7 activates toll-like receptor 7 and causes neurodegeneration. Nat Neurosci 15:827–835
23. Okamura Y, Watari M, Jerud ES, Young DW, Ishizaka ST, Rose J, Chow JC, Strauss JF 3rd (2001) The extra domain A of fibronectin activates toll-like receptor 4. J Biol Chem 276:10229–10233
24. Taylor KR, Trowbridge JM, Rudisill JA, Termeer CC, Simon JC, Gallo RL (2004) Hyaluronan fragments stimulate endothelial recognition of injury through TLR4. J Biol Chem 279:17079–17084
25. Termeer C, Benedix F, Sleeman J, Fieber C, Voith U, Ahrens T, Miyake K, Freudenberg M, Galanos C, Simon JC (2002) Oligosaccharides of hyaluronan activate dendritic cells via toll-like receptor 4. J Exp Med 195:99–111

26. Roelofs MF, Boelens WC, Joosten LA, Abdollahi-Roodsaz S, Geurts J, Wunderink LU, Schreurs BW, van den Berg WB, Radstake TR (2006) Identification of small heat shock protein B8 (HSP22) as a novel TLR4 ligand and potential involvement in the pathogenesis of rheumatoid arthritis. J Immunol 176:7021–7027
27. Harada K, Hide I, Seki T, Tanaka S, Nakata Y, Sakai N (2011) Extracellular ATP differentially modulates toll-like receptor 4-mediated cell survival and death of microglia. J Neurochem 116:1138–1147
28. Ohashi K, Burkart V, Flohe S, Kolb H (2000) Cutting edge: heat shock protein 60 is a putative endogenous ligand of the toll-like receptor-4 complex. J Immunol 164:558–561
29. Vabulas RM, Ahmad-Nejad P, da Costa C, Miethke T, Kirschning CJ, Hacker H, Wagner H (2001) Endocytosed HSP60s use toll-like receptor 2 (TLR2) and TLR4 to activate the toll/interleukin-1 receptor signaling pathway in innate immune cells. J Biol Chem 276:31332–31339
30. Asea A, Rehli M, Kabingu E, Boch JA, Bare O, Auron PE, Stevenson MA, Calderwood SK (2002) Novel signal transduction pathway utilized by extracellular HSP70: role of toll-like receptor (TLR) 2 and TLR4. J Biol Chem 277:15028–15034
31. Shi Y, Evans JE, Rock KL (2003) Molecular identification of a danger signal that alerts the immune system to dying cells. Nature 425:516–521
32. Leadbetter EA, Rifkin IR, Hohlbaum AM, Beaudette BC, Shlomchik MJ, Marshak-Rothstein A (2002) Chromatin-IgG complexes activate B cells by dual engagement of IgM and toll-like receptors. Nature 416:603–607
33. Alexopoulou L, Holt AC, Medzhitov R, Flavell RA (2001) Recognition of double-stranded RNA and activation of NF-kappaB by toll-like receptor 3. Nature 413:732–738
34. Hemmi H, Takeuchi O, Kawai T, Kaisho T, Sato S, Sanjo H, Matsumoto M, Hoshino K, Wagner H, Takeda K, Akira S (2000) A toll-like receptor recognizes bacterial DNA. Nature 408:740–745
35. Yu M, Wang H, Ding A, Golenbock DT, Latz E, Czura CJ, Fenton MJ, Tracey KJ, Yang H (2006) HMGB1 signals through toll-like receptor (TLR) 4 and TLR2. Shock 26:174–179
36. Heil F, Hemmi H, Hochrein H, Ampenberger F, Kirschning C, Akira S, Lipford G, Wagner H, Bauer S (2004) Species-specific recognition of single-stranded RNA via toll-like receptor 7 and 8. Science 303:1526–1529
37. Diebold SS, Kaisho T, Hemmi H, Akira S, Reis e Sousa C (2004) Innate antiviral responses by means of TLR7-mediated recognition of single-stranded RNA. Science 303:1529–1531
38. Tukel C, Wilson RP, Nishimori JH, Pezeshki M, Chromy BA, Baumler AJ (2009) Responses to amyloids of microbial and host origin are mediated through toll-like receptor 2. Cell Host Microbe 6:45–53
39. Biragyn A, Coscia M, Nagashima K, Sanford M, Young HA, Olkhanud P (2008) Murine beta-defensin 2 promotes TLR-4/MyD88-mediated and NF-kappaB-dependent atypical death of APCs via activation of TNFR2. J Leukoc Biol 83:998–1008
40. Biragyn A, Ruffini PA, Leifer CA, Klyushnenkova E, Shakhov A, Chertov O, Shirakawa AK, Farber JM, Segal DM, Oppenheim JJ, Kwak LW (2002) Toll-like receptor 4-dependent activation of dendritic cells by beta-defensin 2. Science 298:1025–1029
41. Funderburg N, Lederman MM, Feng Z, Drage MG, Jadlowsky J, Harding CV, Weinberg A, Sieg SF (2007) Human -defensin-3 activates professional antigen-presenting cells via toll-like receptors 1 and 2. Proc Natl Acad Sci USA 104:18631–18635
42. Hertz CJ, Wu Q, Porter EM, Zhang YJ, Weismuller KH, Godowski PJ, Ganz T, Randell SH, Modlin RL (2003) Activation of toll-like receptor 2 on human tracheobronchial epithelial cells induces the antimicrobial peptide human beta defensin-2. J Immunol 171:6820–6826
43. Yang D, Chen Q, Su SB, Zhang P, Kurosaka K, Caspi RR, Michalek SM, Rosenberg HF, Zhang N, Oppenheim JJ (2008) Eosinophil-derived neurotoxin acts as an alarmin to activate the TLR2-MyD88 signal pathway in dendritic cells and enhances Th2 immune responses. J Exp Med 205:79–90
44. Smiley ST, King JA, Hancock WW (2001) Fibrinogen stimulates macrophage chemokine secretion through toll-like receptor 4. J Immunol 167:2887–2894

45. Miller YI, Choi SH, Wiesner P, Bae YS (2012) The SYK side of TLR4: signalling mechanisms in response to LPS and minimally oxidized LDL. Br J Pharmacol 167:990–999
46. Miller YI, Viriyakosol S, Worrall DS, Boullier A, Butler S, Witztum JL (2005) Toll-like receptor 4-dependent and -independent cytokine secretion induced by minimally oxidized low-density lipoprotein in macrophages. Arterioscler Thromb Vasc Biol 25:1213–1219
47. Walton KA, Hsieh X, Gharavi N, Wang S, Wang G, Yeh M, Cole AL, Berliner JA (2003) Receptors involved in the oxidized 1-palmitoyl-2-arachidonoyl-sn-glycero-3-phosphorylcholine-mediated synthesis of interleukin-8. A role for toll-like receptor 4 and a glycosylphosphatidylinositol-anchored protein. J Biol Chem 278:29661–29666
48. Shichita T, Hasegawa E, Kimura A, Morita R, Sakaguchi R, Takada I, Sekiya T, Ooboshi H, Kitazono T, Yanagawa T, Ishii T, Takahashi H, Mori S, Nishibori M, Kuroda K, Akira S, Miyake K, Yoshimura A (2012) Peroxiredoxin family proteins are key initiators of post-ischemic inflammation in the brain. Nat Med 18:911–917
49. Vogl T, Tenbrock K, Ludwig S, Leukert N, Ehrhardt C, van Zoelen MA, Nacken W, Foell D, van der Poll T, Sorg C, Roth J (2007) Mrp8 and Mrp14 are endogenous activators of toll-like receptor 4, promoting lethal, endotoxin-induced shock. Nat Med 13:1042–1049
50. Qiu J, Nishimura M, Wang Y, Sims JR, Qiu S, Savitz SI, Salomone S, Moskowitz MA (2008) Early release of HMGB-1 from neurons after the onset of brain ischemia. J Cereb Blood Flow Metab 28:927–938
51. Agnello D, Wang H, Yang H, Tracey KJ, Ghezzi P (2002) HMGB-1, a DNA-binding protein with cytokine activity, induces brain TNF and IL-6 production, and mediates anorexia and taste aversion. Cytokine 18:231–236
52. O'Connor KA, Hansen MK, Rachal Pugh C, Deak MM, Biedenkapp JC, Milligan ED, Johnson JD, Wang H, Maier SF, Tracey KJ, Watkins LR (2003) Further characterization of high mobility group box 1 (HMGB1) as a proinflammatory cytokine: central nervous system effects. Cytokine 24:254–265
53. Liu K, Mori S, Takahashi HK, Tomono Y, Wake H, Kanke T, Sato Y, Hiraga N, Adachi N, Yoshino T, Nishibori M (2007) Anti-high mobility group box 1 monoclonal antibody ameliorates brain infarction induced by transient ischemia in rats. FASEB J 21:3904–3916
54. Ziegler G, Prinz V, Albrecht MW, Harhausen D, Khojasteh U, Nacken W, Endres M, Dirnagl U, Nietfeld W, Trendelenburg G (2009) Mrp-8 and -14 mediate CNS injury in focal cerebral ischemia. Biochim Biophys Acta 1792:1198–1204
55. Constantin D, Cordenier A, Robinson K, Ala'Aldeen DA, Murphy S (2004) Neisseria meningitidis-induced death of cerebrovascular endothelium: mechanisms triggering transcriptional activation of inducible nitric oxide synthase. J Neurochem 89:1166–1174
56. Cristofaro P, Opal SM (2006) Role of toll-like receptors in infection and immunity: clinical implications. Drugs 66:15–29
57. Farina C, Krumbholz M, Giese T, Hartmann G, Aloisi F, Meinl E (2005) Preferential expression and function of toll-like receptor 3 in human astrocytes. J Neuroimmunol 159:12–19
58. Guo LH, Schluesener HJ (2007) The innate immunity of the central nervous system in chronic pain: the role of toll-like receptors. Cell Mol Life Sci 64:1128–1136
59. Hanke ML, Kielian T (2011) Toll-like receptors in health and disease in the brain: mechanisms and therapeutic potential. Clin Sci (Lond) 121:367–387
60. Hornung V, Rothenfusser S, Britsch S, Krug A, Jahrsdorfer B, Giese T, Endres S, Hartmann G (2002) Quantitative expression of toll-like receptor 1–10 mRNA in cellular subsets of human peripheral blood mononuclear cells and sensitivity to CpG oligodeoxynucleotides. J Immunol 168:4531–4537
61. Lafon M, Megret F, Lafage M, Prehaud C (2006) The innate immune facet of brain: human neurons express TLR-3 and sense viral dsRNA. J Mol Neurosci 29:185–194
62. Leow-Dyke S, Allen C, Denes A, Nilsson O, Maysami S, Bowie AG, Rothwell NJ, Pinteaux E (2012) Neuronal toll-like receptor 4 signaling induces brain endothelial activation and neutrophil transmigration in vitro. J Neuroinflammation 9:230

63. Muzio M, Bosisio D, Polentaurutti N, D'amico G, Stoppacciaro A, Manicinelli R, Cv V, Penton-Rol R, Ruco LP, Allavena P, Mantovani A (2000) Differential expression and regulation of toll-like receptors (TLR) in human leukocytes: selective expression of TLR3 in dendritic cells. J Immunol 164:5998–6004
64. Nishimura M, Naito S (2005) Tissue-specific mRNA expression profiles of human toll-like receptors and related genes. Biol Pharm Bull 28:886–892
65. Takeda K, Akira S (2005) Toll-like receptors in innate immunity. Int Immunol 17:1–14
66. Tsan MF, Gao B (2004) Endogenous ligands of toll-like receptors. J Leukoc Biol 76:514–519
67. Wang YC, Lin S, Yang QW (2011) Toll-like receptors in cerebral ischemic inflammatory injury. J Neuroinflammation 8:134
68. Zarember KA, Godowski PJ (2002) Tissue expression of human toll-like receptors and differential regulation of toll-like receptor mRNAs in leukocytes in response to microbes, their products, and cytokines. J Immunol 168:554–561
69. Bsibsi M, Ravid R, Gveric D, Noort JMV (2002) Broad expression of toll-like receptors in the human central nervous system. J Neuropathol Exp Neurol 61:1013–1021
70. Carty M, Bowie AG (2011) Evaluating the role of toll-like receptors in diseases of the central nervous system. Biochem Pharmacol 81:825–837
71. Arumugam TV, Okun E, Tang SC, Thundyil J, Taylor SM, Woodruff TM (2009) Toll-like receptors in ischemia-reperfusion injury. Shock 32:4–16
72. Kielian T (2006) Toll-like receptors in central nervous system glial inflammation and homeostasis. J Neurosci Res 83:711–730
73. Nguyen MD, Julien JP, Rivest S (2002) Innate immunity: the missing link in neuroprotection and neurodegeneration? Nat Rev Neurosci 3:216–227
74. Wyss-Coray T, Mucke L (2002) Inflammation in neurodegenerative disease–a double-edged sword. Neuron 35:419–432
75. Griffiths MR, Gasque P, Neal JW (2009) The multiple roles of the innate immune system in the regulation of apoptosis and inflammation in the brain. J Neuropathol Exp Neurol 68:217–226
76. Kraft AD, Harry GJ (2011) Features of microglia and neuroinflammation relevant to environmental exposure and neurotoxicity. Int J Environ Res Public Health 8:2980–3018
77. Yang I, Han SJ, Kaur G, Crane C, Parsa AT (2010) The role of microglia in central nervous system immunity and glioma immunology. J Clin Neurosci 17:6–10
78. Lehnardt S (2010) Innate immunity and neuroinflammation in the CNS: the role of microglia in toll-like receptor-mediated neuronal injury. Glia 58:253–263
79. Neumann J, Gunzer M, Gutzeit HO, Ullrich O, Reymann KG, Dinkel K (2006) Microglia provide neuroprotection after ischemia. FASEB J 20:714–716
80. Neumann J, Sauerzweig S, Ronicke R, Gunzer F, Dinkel K, Ullrich O, Gunzer M, Reymann KG (2008) Microglia cells protect neurons by direct engulfment of invading neutrophil granulocytes: a new mechanism of CNS immune privilege. J Neurosci 28:5965–5975
81. Walter L, Neumann H (2009) Role of microglia in neuronal degeneration and regeneration. Semin Immunopathol 31:513–525
82. Jack CS, Arbour N, Manusow J, Montgrain V, Blain M, McCrea E, Shapiro A, Antel JP (2005) TLR signaling tailors innate immune responses in human microglia and astrocytes. J Immunol 175:4320–4320
83. Bsibsi M, Persoon-Deen C, Verwer RW, Meeuwsen S, Ravid R, Van Noort JM (2006) Toll-like receptor 3 on adult human astrocytes triggers production of neuroprotective mediators. Glia 53:688–695
84. Streit WJ (2002) Microglia as neuroprotective, immunocompetent cells of the CNS. Glia 40:133–139
85. Chen GY, Nunez G (2010) Sterile inflammation: sensing and reacting to damage. Nat Rev Immunol 10:826–837
86. Babcock AA, Wirenfeldt M, Holm T, Nielsen HH, Dissing-Olesen L, Toft-Hansen H, Millward JM, Landmann R, Rivest S, Finsen B, Owens T (2006) Toll-like receptor 2 signaling in response to brain injury: an innate bridge to neuroinflammation. J Neurosci 26:12826–12837

87. Rubartelli A, Lotze MT (2007) Inside, outside, upside down: damage-associated molecular-pattern molecules (DAMPs) and redox. Trends Immunol 28:429–436
88. Brea D, Blanco M, Ramos-Cabrer P, Moldes O, Arias S, Perez-Mato M, Leira R, Sobrino T, Castillo J (2011) Toll-like receptors 2 and 4 in ischemic stroke: outcome and therapeutic values. J Cereb Blood Flow Metab 31:1424–1431
89. Caso JR, Pradillo JM, Hurtado O, Lorenzo P, Moro MA, Lizasoain I (2007) Toll-like receptor 4 is involved in brain damage and inflammation after experimental stroke. Circulation 115:1599–1608
90. Hua F, Ma J, Ha T, Kelley JL, Kao RL, Schweitzer JB, Kalbfleisch JH, Williams DL, Li C (2009) Differential roles of TLR2 and TLR4 in acute focal cerebral ischemia/reperfusion injury in mice. Brain Res 1262:100–108
91. Ock J, Jeong J, Choi WS, Lee WH, Kim SH, Kim IK, Suk K (2007) Regulation of toll-like receptor 4 expression and its signaling by hypoxia in cultured microglia. J Neurosci Res 85:1989–1995
92. Tang SC, Arumugam TV, Xu X, Cheng A, Mughal MR, Jo DG, Lathia JD, Siler DA, Chigurupati S, Ouyang X, Magnus T, Camandola S, Mattson MP (2007) Pivotal role for neuronal toll-like receptors in ischemic brain injury and functional deficits. Proc Natl Acad Sci 104:13798–13803
93. Yang QW, Li JC, Lu FL, Wen AQ, Xiang J, Zhang LL, Huang ZY, Wang JZ (2008) Upregulated expression of toll-like receptor 4 in monocytes correlates with severity of acute cerebral infarction. J Cereb Blood Flow Metab 28:1588–1596
94. Tang Y, Lu A, Sharp FR (2001) Global gene expression profiling reveals mechanisms of brain injury following focal ischemia, intracerebral hemorrhage, seizures and hypoglycemia. Soc Neurosci Abstr 27:203.1
95. Caso JR, Pradillo JM, Hurtado O, Leza JC, Moro MA, Lizasoain I (2008) Toll-like receptor 4 is involved in subacute stress-induced neuroinflammation and in the worsening of experimental stroke. Stroke 39:1314–1320
96. Hua F, Ma J, Ha T, Xia Y, Kelley J, Williams DL, Kao RL, Browder IW, Schweitzer JB, Kalbfleisch JH, Li C (2007) Activation of toll-like receptor 4 signaling contributes to hippocampal neuronal death following global cerebral ischemia/reperfusion. J Neuroimmunol 190:101–111
97. Kilic U, Kilic E, Matter CM, Bassetti CL, Hermann DM (2008) TLR-4 deficiency protects against focal cerebral ischemia and axotomy-induced neurodegeneration. Neurobiol Dis 31:33–40
98. Cao CX, Yang QW, Lv FL, Cui J, Fu HB, Wang JZ (2007) Reduced cerebral ischemia-reperfusion injury in toll-like receptor 4 deficient mice. Biochem Biophys Res Commun 353:509–514
99. Sansing LH, Harris TH, Welsh FA, Kasner SE, Hunter CA, Kariko K (2011) Toll-like receptor 4 contributes to poor outcome after intracerebral hemorrhage. Ann Neurol 70:646–656
100. Abe T, Shimamura M, Jackman K, Kurinami H, Anrather J, Zhou P, Iadecola C (2010) Key role of CD36 in Toll-like receptor 2 signaling in cerebral ischemia. Stroke 41:898–904
101. Lehnardt S, Lehmann S, Kaul D, Tschimmel K, Hoffmann O, Cho S, Krueger C, Nitsch R, Meisel A, Weber JR (2007) Toll-like receptor 2 mediates CNS injury in focal cerebral ischemia. J Neuroimmunol 190:28–33
102. Ziegler G, Freyer D, Harhausen D, Khojasteh U, Nietfeld W, Trendelenburg G (2011) Blocking TLR2 in vivo protects against accumulation of inflammatory cells and neuronal injury in experimental stroke. J Cereb Blood Flow Metab 31:757–766
103. Ziegler G, Harhausen D, Schepers C, Hoffmann O, Rohr C, Prinz V, Konig J, Lehrach H, Nietfeld W, Trendelenburg G (2007) TLR2 has a detrimental role in mouse transient focal cerebral ischemia. Biochem Biophys Res Commun 359:574–579
104. Stridh L, Smith PL, Naylor AS, Wang X, Mallard C (2011) Regulation of toll-like receptor 1 and −2 in neonatal mice brains after hypoxia-ischemia. J Neuroinflammation 8:45
105. Bohacek I, Cordeau P, Lalancette-Hebert M, Gorup D, Weng YC, Gajovic S, Kriz J (2012) Toll-like receptor 2 deficiency leads to delayed exacerbation of ischemic injury. J Neuroinflammation 9:191

106. Selvamani A, Sathyan P, Miranda RC, Sohrabji F (2012) An antagomir to microRNA Let7f promotes neuroprotection in an ischemic stroke model. PLoS One 7:e32662
107. Graham DI (1996) Neuropathology of head injury. In: Na-Rayan RK, Willberger J, Povlishock JT (eds) Neurotrauma. McGraw Hill, New York, NY
108. Maas AI, Stocchetti N, Bullock R (2008) Moderate and severe traumatic brain injury in adults. Lancet Neurol 7:728–741
109. Sims JE, Smith DE (2010) The IL-1 family: regulators of immunity. Nat Rev Immunol 10:89–102
110. Ziebell JM, Morganti-Kossmann MC (2010) Involvement of pro- and anti-inflammatory cytokines and chemokines in the pathophysiology of traumatic brain injury. Neurotherapeutics 7:22–30
111. Hua F, Wang J, Ishrat T, Wei W, Atif F, Sayeed I, Stein DG (2011) Genomic profile of toll-like receptor pathways in traumatically brain-injured mice: effect of exogenous progesterone. J Neuroinflammation 8:42
112. Zhang Z, Zhang ZY, Wu Y, Schluesener HJ (2012) Immunolocalization of toll-like receptors 2 and 4 as well as their endogenous ligand, heat shock protein 70, in rat traumatic brain injury. Neuroimmunomodulation 19:10–19
113. Yu ZQ, Zha JH (2012) Genetic ablation of toll-like receptor 2 reduces secondary brain injury caused by cortical contusion in mice. Ann Clin Lab Sci 42:26–33
114. Hans VH, Kossmann T, Joller H, Otto V, Morganti-Kossmann MC (1999) Interleukin-6 and its soluble receptor in serum and cerebrospinal fluid after cerebral trauma. Neuroreport 10:409–412
115. Helmy A, Carpenter KL, Menon DK, Pickard JD, Hutchinson PJ (2011) The cytokine response to human traumatic brain injury: temporal profiles and evidence for cerebral parenchymal production. J Cereb Blood Flow Metab 31:658–670
116. Kossmann T, Stahel PF, Lenzlinger PM, Redl H, Dubs RW, Trentz O, Schlag G, Morganti-Kossmann MC (1997) Interleukin-8 released into the cerebrospinal fluid after brain injury is associated with blood–brain barrier dysfunction and nerve growth factor production. J Cereb Blood Flow Metab 17:280–289
117. Kigerl KA, Lai W, Rivest S, Hart RP, Satoskar AR, Popovich PG (2007) Toll-like receptor (TLR)-2 and TLR-4 regulate inflammation, gliosis, and myelin sparing after spinal cord injury. J Neurochem 102:37–50
118. Valles SL, Blanco AM, Pascual M, Guerri C (2004) Chronic ethanol treatment enhances inflammatory mediators and cell death in the brain and in astrocytes. Brain Pathol 14:365–371
119. Alfonso-Loeches S, Pascual M, Gomez-Pinedo U, Pascual-Lucas M, Renau-Piqueras J, Guerri C (2012) Toll-like receptor 4 participates in the myelin disruptions associated with chronic alcohol abuse. Glia 60:948–964
120. Blanco AM, Valles SL, Pascual M, Guerri C (2005) Involvement of TLR4/type I IL-1 receptor signaling in the induction of inflammatory mediators and cell death induced by ethanol in cultured astrocytes. J Immunol 175:6893–6899
121. Fernandez-Lizarbe S, Pascual M, Gascon MS, Blanco A, Guerri C (2008) Lipid rafts regulate ethanol-induced activation of TLR4 signaling in murine macrophages. Mol Immunol 45:2007–2016
122. Hua F, Ma J, Ha T, Kelley J, Williams DL, Kao RL, Kalbfleisch JH, Browder IW, Li C (2008) Preconditioning with a TLR2 specific ligand increases resistance to cerebral ischemia/reperfusion injury. J Neuroimmunol 199:75–82
123. Leung PY, Stevens SL, Packard AE, Lessov NS, Yang T, Conrad VK, van den Dungen NN, Simon RP, Stenzel-Poore MP (2012) Toll-like receptor 7 preconditioning induces robust neuroprotection against stroke by a novel type I interferon-mediated mechanism. Stroke 43:1383–1389
124. Marsh B, Stevens SL, Packard AE, Gopalan B, Hunter B, Leung PY, Harrington CA, Stenzel-Poore MP (2009) Systemic lipopolysaccharide protects the brain from ischemic injury by reprogramming the response of the brain to stroke: a critical role for IRF3. J Neurosci 29:9839–9849

125. Packard AE, Hedges JC, Bahjat FR, Stevens SL, Conlin MJ, Salazar AM, Stenzel-Poore MP (2012) Poly-IC preconditioning protects against cerebral and renal ischemia-reperfusion injury. J Cereb Blood Flow Metab 32:242–247
126. Pradillo JM, Fernandez-Lopez D, Garcia-Yebenes I, Sobrado M, Hurtado O, Moro MA, Lizasoain I (2009) Toll-like receptor 4 is involved in neuroprotection afforded by ischemic preconditioning. J Neurochem 109:287–294
127. Stevens SL, Ciesielski TM, Marsh BJ, Yang T, Homen DS, Boule JL, Lessov NS, Simon RP, Stenzel-Poore MP (2008) Toll-like receptor 9: a new target of ischemic preconditioning in the brain. J Cereb Blood Flow Metab 28:1040–1047
128. Tasaki K, Ruetzler CA, Ohtsuki T, Martin D, Nawashiro H, Hallenbeck JM (1997) Lipopolysaccharide pre-treatment induces resistance against subsequent focal cerebral ischemic damage in spontaneously hypertensive rats. Brain Res 748:267–270
129. Vartanian KB, Stenzel-Poore MP (2010) Toll-like receptor tolerance as a mechanism for neuroprotection. Transl Stroke Res 1:252–260
130. Hickey EJ, You X, Kaimaktchiev V, Stenzel-Poore M, Ungerleider RM (2007) Lipopolysaccharide preconditioning induces robust protection against brain injury resulting from deep hypothermic circulatory arrest. J Thorac Cardiovasc Surg 133:1588–1596
131. Rosenzweig HL, Minami M, Lessov NS, Coste SC, Stevens SL, Henshall DC, Meller R, Simon RP, Stenzel-Poore MP (2007) Endotoxin preconditioning protects against the cytotoxic effects of TNFalpha after stroke: a novel role for TNFalpha in LPS-ischemic tolerance. J Cereb Blood Flow Metab 27:1663–1674
132. Bahjat FR, Williams-Karnesky RL, Kohama SG, West GA, Doyle KP, Spector MD, Hobbs TR, Stenzel-Poore MP (2011) Proof of concept: pharmacological preconditioning with a toll-like receptor agonist protects against cerebrovascular injury in a primate model of stroke. J Cereb Blood Flow Metab 31:1229–1242
133. Vartanian KB, Stevens SL, Marsh BJ, Williams-Karnesky R, Lessov NS, Stenzel-Poore MP (2011) LPS preconditioning redirects TLR signaling following stroke: TRIF-IRF3 plays a seminal role in mediating tolerance to ischemic injury. J Neuroinflammation 8:140
134. Virca GD, Kim SY, Glaser KB, Ulevitch RJ (1989) Lipopolysaccharide induced hyporesponsiveness to its own action in RAW 264.7 cells. J Biol Chem 264:21951–21956
135. Stevens SL, Leung PY, Vartanian KB, Gopalan B, Yang T, Simon RP, Stenzel-Poore MP (2011) Multiple preconditioning paradigms converge on interferon regulatory factor-dependent signaling to promote tolerance to ischemic brain injury. J Neurosci 31:8456–8463
136. Marsh BJ, Stevens SL, Hunter B, Stenzel-Poore MP (2009) Inflammation and the emerging role of the toll-like receptor system in acute brain ischemia. Stroke 40:S34–7
137. Pappas DJ, Gabatto PA, Oksenberg D, Khankhanian P, Baranzini SE, Gan L, Oksenberg JR (2012) Transcriptional expression patterns triggered by chemically distinct neuroprotective molecules. Neuroscience 226:10–20
138. Longhi L, Gesuete R, Perego C, Ortolano F, Sacchi N, Villa P, Stocchetti N, De Simoni MG (2011) Long-lasting protection in brain trauma by endotoxin preconditioning. J Cereb Blood Flow Metab 31:1919–1929
139. Kwissa M, Nakaya HI, Oluoch H, Pulendran B (2012) Distinct TLR adjuvants differentially stimulate systemic and local innate immune responses in nonhuman primates. Blood 119:2044–2055
140. Rosenzweig HL, Lessov NS, Henshall DC, Minami M, Simon RP, Stenzel-Poore MP (2004) Endotoxin preconditioning prevents the cellular inflammatory response during ischemic neuroprotection in mice. Stroke 35:2576–2581
141. Engblom D, Ek M, Saha S, Ericsson-Dahlstrand A, Jakobsson PJ, Blomqvist A (2002) Prostaglandins as inflammatory messengers across the blood–brain barrier. J Mol Med (Berl) 80:5–15
142. Packard AE, Leung PY, Vartanian KB, Stevens SL, Bahjat FR, Stenzel-Poore MP (2012) TLR9 bone marrow chimeric mice define a role for cerebral TNF in neuroprotection induced by CpG preconditioning. J Cereb Blood Flow Metab 32:2193–200

143. Sandoval KE, Witt KA (2008) Blood–brain barrier tight junction permeability and ischemic stroke. Neurobiol Dis 32:200–219
144. Gesuete R, Packard AE, Vartanian KB, Conrad VK, Stevens SL, Bahjat FR, Yang T, Stenzel-Poore MP (2012) Poly-ICLC preconditioning protects the blood–brain barrier against ischemic injury in vitro through type I interferon signaling. J Neurochem 123(Suppl 2):75–85
145. Weinstock-Guttman B, Ramanathan M, Zivadinov R (2008) Interferon-beta treatment for relapsing multiple sclerosis. Expert Opin Biol Ther 8:1435–1447
146. Roger VL, Go AS, Lloyd-Jones DM et al (2012) Heart disease and stroke statistics–2012 update: a report from the American Heart Association. Circulation 125:e2–e220
147. Altay T, Kang HI, Woo HH, Masaryk TJ, Rasmussen PA, Fiorella DJ, Moskowitz SI (2011) Thromboembolic events associated with endovascular treatment of cerebral aneurysms. J Neurointerv Surg 3:147–150
148. Barber PA, Hach S, Tippett LJ, Ross L, Merry AF, Milsom P (2008) Cerebral ischemic lesions on diffusion-weighted imaging are associated with neurocognitive decline after cardiac surgery. Stroke 39:1427–1433
149. Biondi A, Oppenheim C, Vivas E, Casasco A, Lalam T, Sourour N, Jean LL, Dormont D, Marsault C (2000) Cerebral aneurysms treated by Guglielmi detachable coils: evaluation with diffusion-weighted MR imaging. AJNR Am J Neuroradiol 21:957–963
150. Eggebrecht H, Schmermund A, Voigtlander T, Kahlert P, Erbel R, Mehta RH (2012) Risk of stroke after transcatheter aortic valve implantation (TAVI): a meta-analysis of 10,037 published patients. EuroIntervention 8:129–138
151. Gutierrez IZ, Barone DL, Makula PA, Currier C (1987) The risk of perioperative stroke in patients with asymptomatic carotid bruits undergoing peripheral vascular surgery. Am Surg 53:487–489
152. McKhann GM, Grega MA, Borowicz LM Jr, Baumgartner WA, Selnes OA (2006) Stroke and encephalopathy after cardiac surgery: an update. Stroke 37:562–571
153. Nosan DK, Gomez CR, Maves MD (1993) Perioperative stroke in patients undergoing head and neck surgery. Ann Otol Rhinol Laryngol 102:717–723
154. Faul M, Xu L, Wald MM, Coronado V (2010) Traumatic brain injury in the United States: Emergency Department Visits, Hospitalizations and Deaths, 2002–2006. Centers for Disease Control and Prevention, National Center for Injury Prevention and Control, Atlanta, Georgia
155. Brea D, Sobrino T, Rodriguez-Yanez M, Ramos-Cabrer P, Agulla J, Rodriguez-Gonzalez R, Campos F, Blanco M, Castillo J (2011) Toll-like receptors 7 and 8 expression is associated with poor outcome and greater inflammatory response in acute ischemic stroke. Clin Immunol 139:193–198
156. Park HJ, Kim SK, Yun DH, Kim DH, Chon J, Kim JW, Chung JH (2012) Association of toll-like receptor 2 polymorphisms with National Institute of Health Stroke Scale scores of ischemic stroke patients. J Mol Neurosci 46:536–540
157. Lin YC, Chang YM, Yu JM, Yen JH, Chang JG, Hu CJ (2005) Toll-like receptor 4 gene C119A but not Asp299Gly polymorphism is associated with ischemic stroke among ethnic Chinese in Taiwan. Atherosclerosis 180:305–309
158. Vacchelli E, Galluzzi L, Eggermont A, Fridman WH, Galon J, Sautes-Fridman C, Tartour E, Zitvogel L, Kroemer G (2012) Trial watch: FDA-approved toll-like receptor agonists for cancer therapy. Oncoimmunology 1:894–907

Chapter 8
Ion Transporters in Microglial Function: New Therapeutic Targets for Neuroinflammation in Ischemic Stroke?

Hui Yuan, Yejie Shi, and Dandan Sun

Abstract Microglia are the macrophage immune cells in the CNS and monitor extracellular microenvironment in healthy brains. They can be rapidly activated under pathological conditions and move to a lesion site following chemotactic gradients and unfold their phagocytotic activities to clear tissue debris, damaged cells, or microbes. A growing body of studies illustrated the importance of ion transporters in regulating activation and migration of microglia and peripheral immune cells in cerebral ischemic conditions. This chapter summarized roles of Na^+/H^+ exchanger, Na^+/Ca^{2+} exchanger, and K^+/Cl^- cotransporters in regulation of pH_i, Ca^{2+}-spiking events, cell volume, and membrane signal molecule expression during microglia/peripheral immune cell migration, adhesion, and activation. In light of the detrimental effects of excessive pro-inflammatory response on ischemic brain injury, targeting ion transporters may be a new therapeutic strategy to minimize neuroinflammatory reactions after ischemic stroke.

Abbreviations

NHE	Na^+/H^+ exchanger
NCX	Na^+/Ca^{2+} exchange
MCAO	Middle cerebral artery occlusion
NOX	NADPH oxidase
OGD	Oxygen and glucose deprivation
pH_i	Intracellular pH

H. Yuan • Y. Shi • D. Sun (✉)
Department of Neurology, University of Pittsburgh, Pittsburgh, PA, USA
e-mail: sund@upmc.edu

J. Chen et al. (eds.), *Immunological Mechanisms and Therapies in Brain Injuries and Stroke*, Springer Series in Translational Stroke Research, DOI 10.1007/978-1-4614-8915-3_8,

ROS	Reactive oxygen species
$[Ca^{2+}]_i$	Intracellular Ca^{2+}
$[Na^+]_i$	Intracellular Na^+
KCC	K^+–Cl^- cotransporters
IL	Interleukin
TNF-α	Tumor necrosis factor alpha
ATP	Adenosine triphosphate

Introduction

Microglia are resident macrophages ubiquitously distributed throughout the CNS and serve as neurological sensors. In developing brains, microglia with ameboid morphology phagocytose cellular debris resulting from spontaneous neurodegeneration or nerve fiber remodeling [1, 2]. In the adult brain, microglial cells exhibit a ramified morphology and monitor their extracellular space and cellular neighborhood and are ready to transform to the executive states of the activated microglia. They can be rapidly activated under many pathological conditions, including stroke, brain trauma, and neurodegenerative diseases [3–5]. Microglia can move to a lesion site following chemotactic gradients and unfold their phagocytotic activities to clear tissue debris, damaged cells, or microbes [6].

Recent studies indicate that microglia are highly plastic cells that can assume diverse phenotypes and engage different functional programs in response to specific microenvironmental signals. Interleukin (IL)-4 and IL-10 can induce an "alternatively activated" M2 phenotype of microglia, these microglia are healthier cells with enhanced phagocytic activity and reduced production of inflammatory mediators [7]. In contrast, the "classically activated" M1 microglia typically release destructive pro-inflammatory mediators including reactive oxygen species (ROS), nitrogen species (NO), cytokines (TNF-α and IL-1β), and growth factors [8, 9]. M1-state like microglia represent reduced phagocytosis and increased secretion of pro-inflammatory mediators. Release of chemoattractive factors can recruit and guide immune cell population migrating into the CNS through disrupted blood–brain barrier [6].

The roles of microglia/macrophages in ischemic brain injury remain to be defined. Microglia respond dynamically to ischemic injury: at 1–5 days after ischemic injury, microglia experience an early healthy M2 phenotype and then transition to a sick M1 phenotype up to 14 days after ischemic injury [7]. The early recruitment of M2 microglia after ischemic stroke can represent an endogenous effort to clean ischemic tissue and restrict brain damage [7, 10]. Therefore, microglia can shift from the M2 to M1 phenotypes in ischemic brains.

The failure of ionic homeostasis is a hallmark of ischemic brain damage [11]. Ion exchangers contribute to ionic dysregulation after cerebral ischemia [12]. Emerging new evidence suggests that ion transporters play a role in microglial activation and migration. In this chapter, we summarized recent studies of the ion transporters in microglia, neutrophils, macrophages, and monocytes. A better understanding of the function and regulation of ion transporters in neuroinflammation following cerebral ischemia will provide insight into developing more effective neuroprotective therapy for stroke.

The Na^+/H^+ Exchanger

Expression and Function of NHE-1 in Microglia

NHEs are a family of membrane transport proteins which catalyze the secondary active electroneutral exchange of one Na^+ for one H^+. To date, nine NHE isoforms (NHE1-9) have been cloned in mammalian tissues, and NHE-1 is the most abundantly expressed isoform in brains including microglia [13]. The isoform NHE-1 is found in the plasma membrane of most mammalian cells and normally described as the housekeeping isoform [14]. Other isoforms have a more restricted tissue distribution and appear to regulate more specialized functions: NHE-2 and NHE-3 are expressed predominantly in the kidney and gastrointestinal tract, while NHE-5 is expressed mainly in the brain [15]. NHE-6 and NHE-7 are exclusively localized in intracellular organelles such as mitochondrial and trans-Golgi; these isoforms are expressed in tissues with high metabolism rates such as heart, brain, and skeletal muscle [16, 17]. NHE-8 and NHE-9 are also found to distribute in kidney, stomach, and intestine [18].

Of all the isoforms, NHE-1 is the most extensively characterized member. It resides exclusively on the cell surface but is also present in discrete microdomains of the plasma membrane in different types of cells [13]. NHE-1 exerts two fundamental functions. First, it serves as the principal alkalinizing mechanism in many cell types against the damaging effects of excess intracellular acidification. Together with bicarbonate transporting systems, NHE-1 plays a crucial role in maintaining cytoplasmic acid–base balance. Second, it provides a major resource for Na^+ influx, coupled to Cl^- and H_2O uptake, which is required to restore cell volume to steady-state levels following cell shrinkage induced by external osmolality [19]. The cell-type-specific localization of NHE-1 in distinct subdomains of the plasma membrane also suggests that this exchanger may play more specialized roles in cell function. NHE-1 expression may be a significant factor in regulating cell morphology, adhesion, and migration. It has been reported that NHE-1 plays an important role in remodeling the cortical actin cytoskeleton and cell shape of fibroblasts through its association with the cytoskeletal-associated proteins ezrin, radixin, and moesin (ERMs) [20]. Both the cation translocation and anchorage to the cytoskeleton are required for remodeling focal adhesions at the front and trailing edges of the cell necessary for guided movement [21].

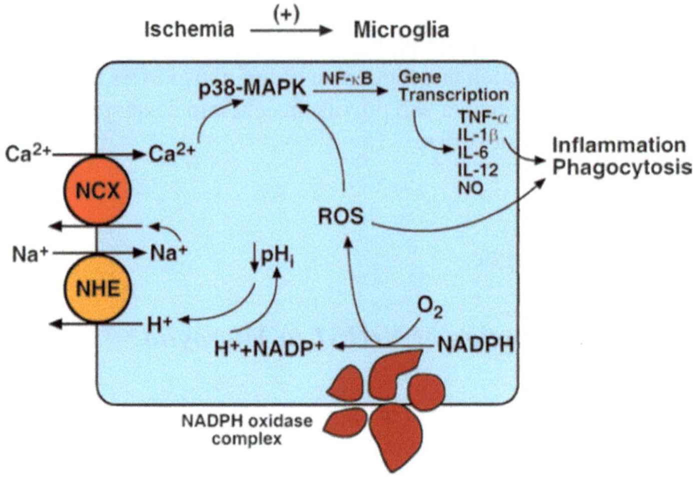

Fig. 8.1 Role of NHE-1 in microglial pro-inflammatory "respiratory burst" and cytokine production. NHE-1 protein is expressed in microglia and its activity is important for microglial function by extruding H^+ and regulation of pH_i during the respiratory burst. Thus, NHE-1 is important in the production of ROS and cytokines and in inflammatory responses in ischemic brains [12]

NHE-1 in Microglial Activation and Neuroinflammation

NHE-1 is the most studied NHE isoform in microglia. One important component for microglial activation and function is the NADPH oxidase (NOX). NOX catalyzes the reduction reaction of molecular oxygen to superoxide anion using NADPH as an electron donor and is the major source of ROS production in microglia [8, 22]. In this process, H^+ accumulates inside microglia as a by-product, causing depolarization and cytoplasmic acidification [18]. NOX is sensitive to intracellular pH and has an optimal pH_i of 7.2 [23]. Intracellular acidosis may impair NOX function. Thus, NHE-1 is required to maintain an optimal pH_i and sustain microglial respiratory burst. It has been shown that microglial activation by several stimuli depends on NHE-1-mediated H^+ homeostasis [24]. Inhibition of NHE-1 with HOE 642 impaired pH_i regulation in microglia under basal conditions. HOE 642 also reduced the production of superoxide anion as well as pro-inflammatory cytokines IL-6, IL-1β, and TNF-α induced by LPS or in vitro ischemia [24]. In a recent study using the mouse middle cerebral artery occlusion (MCAO) model, activation of microglia is significantly reduced with NHE-1 inhibition by HOE642 or transgenic global knockdown of NHE-1 protein [25]. Meanwhile, blocking NHE-1 activity either pharmacologically or by a transgenic knockdown also significantly decreases pro-inflammatory cytokine formation in ischemic brains [25]. Since microglia activation in ischemic brains lasts up to 7–14 days, blocking NHE-1 in microglia may present a new therapeutic target in treating the prolonged inflammatory responses in ischemic brains (Fig. 8.1).

NHE-1 in Microglial Migration

Microglial migration plays an important role in physiological processes, including immune inflammatory response and injury healing after ischemic stroke. Under stroke or neurodegenerative pathological conditions, the microglia with the amoeboid form migrate to the site of injury. Extracellular ATP and ADP released from ischemic and traumatic brain tissues can stimulate microglial migration [26]. Injury to the CNS tissues also releases other chemotactic factors including bradykinin [27], which stimulate microglial migration to the site of injury.

Maintaining alkaline pH_i by NHE-1 is important in microglial migration. pH_i serves as a regulator of cell polarization and migration [28]. However, how the local pH_i environment regulates cell movement is not fully understood. One possible mechanism is through effects of pH_i on regulation of the actin binding protein cofilin activity. Activation of cofilin promotes accumulation of actin at leading edge and is required for cell motility [29, 30]. Cofilin activity is inhibited by pH-dependent phosphoinositide binding at more acidic pH_i [31]. In contrast, more alkaline pH_i environment increases activity of cofilin to promote actin filament dynamics [32]. NHE-1 may regulate microglial migration in part by providing an optimal local alkaline pH_i environment. NHE-1 protein was colocalized with cytoskeletal protein ezrin in lamellipodia of microglia and maintained its more alkaline pH_i [33]. In response to chemoattractant bradykinin (BK), microglia exhibited stimulated migration by the increase in lamellipodial area and protrusion rate, but reduction of lamellipodial persistence time [33]. Interestingly, blocking NHE-1 activity with its potent inhibitor HOE 642 not only acidified microglia, abolished the BK-triggered dynamic changes of lamellipodia, but also reduced microglial motility and microchemotaxis in response to BK [33]. In addition, BK-mediated NHE-1 activation resulted in intracellular Na^+ loading as well as intracellular Ca^{2+} elevation mediated by stimulating a reverse mode operation of Na^+/Ca^{2+} exchange (NCX_{rev}).

Taken together, these studies showed that NHE-1 protein is abundantly expressed in microglial lamellipodia and maintains alkaline pH_i in response to BK stimulation. NHE-1 protein interacts with ERM protein and provides anchoring for actin cytoskeleton in microglia. Meanwhile, NHE-1 and NCX_{rev} play a concerted role in BK-induced microglial migration via Na^+ and Ca^{2+} signaling. Since one of the earliest microglial responses to brain ischemia is its migration to the site of injury, the role of NHE-1 in microglia migration may affect outcome of ischemic stroke injury (Fig. 8.2).

Na^+/Ca^{2+} Exchanger

Expression and Functions of NCX1 in Microglia

The sodium–calcium exchanger (NCX) belongs to the superfamily of Ca^{2+}/cation antiporter membrane proteins. NCX is a high-capacity and low-affinity ionic transporter that exchanges three Na^+ ions for one Ca^{2+} ion [34]. When intracellular Ca^{2+}

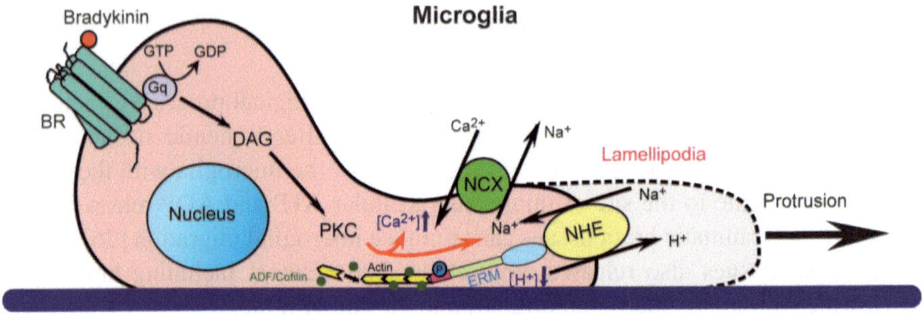

Fig. 8.2 Role of NHE-1 in chemoattractant-induced microglia migration. NHE-1 interacts with ERM proteins and functions as an anchoring point for actin filament, contributing to membrane protrusion and microglial movement. Chemoattractant BK, produced in injured brain tissues, can induce microglial migration. BK stimulates microglial NHE-1 activity to maintain an alkaline pH_i in lamellipodia, which facilitates the pH_i-sensitive actin-binding proteins actin depolymerizing factor (ADF)/cofilin function during microglial movement [33]

concentration rises in the cell, which requires it to be returned to the resting levels, the NCX mechanism couples the uphill Ca^{2+} extrusion with the Na^+ ions entry driven by the Na^+ electrochemical gradient. This forward mode of operation (Ca^{2+} efflux) in NCX, together with Ca^{2+}-ATPase, keeps the 10^4-fold difference in Ca^{2+} across the cell membrane [34]. NCX function depends on the cross-membrane of Na^+ and Ca^{2+} gradients and membrane potential, when $[Na^+]_i$ rises or the cell membrane depolarizes, NCX reverses its mode of operation and extrudes Na^+ in exchange of Ca^{2+} ions [35, 36], a reverse operation of NCX (NCX_{rev}). Therefore, NCX is a major player in regulation of intracellular Ca^{2+} and Na^+ under physiological and pathophysiological conditions.

Three NCX genes, NCX1, NCX2, and NCX3, have been identified and cloned. NCX1 is broadly expressed in the heart, kidney, and brain, whereas NCX2 and NCX3 are exclusively expressed in the brain and skeletal muscle [36, 37]. In vitro studies revealed that among the three different NCX genes, NCX1 is the most highly expressed in microglia [38, 39]. Matsuda et al. reported that exposure of cultured microglia to interferon-γ or nitric oxide enhanced NCX1 transcript levels [40].

NCX in Microglia Activation, Migration, and Neuroinflammation in Cell Cultures

In vitro study performed on the BV2 microglial cell line shows that these cells in anoxic conditions displayed a significant increase of NCX currents and of NCX1 protein expression, but not of NCX2 and NCX3 [41]. The enhanced NCX activity may lead to an increase in $[Ca^{2+}]_i$ after oxygen–glucose deprivation (OGD), which might be attributed to NCX1 in the reverse mode of operation. NCX1 silencing fully prevented the OGD-induced intracellular Ca^{2+} rise [41]. Consistent with this finding, increased NCX1 expression and activation was detected in microglia in other pathological conditions such as interferon-γ or NO exposures [42]. Our recent study showed transgenic knockdown of NCX1 in microglia impairs BK-induced Ca^{2+} rise, chemotaxis, and migration [33].

In Vivo Study of NCX in Microglia Activation, Migration, and Neuroinflammation Following Brain Ischemia

It was reported that NCX1 expression and activity were upregulated in microglia in response to the ischemic injury in MCAO model of mice [41]. One day after permanent MCAO, NCX1 protein expression was detected only in some microglial cells located in the infarct core, but 3–7 days after permanent MCAO, NCX1 expression progressively increased in the Iba1-positive microglia invading the infarct core [41]. In these cells, NCX1 expression was limited to the round phagocytic phenotype, which represents the final stages of microglia activation [43, 44]. Microglial cells isolated from the lesion core region 7 days after permanent MCAO and in cultures show increased NCX1 protein expression [41]. Furthermore, NCX activity in the reverse mode of operation was significantly increased in these cells. Collectively, the upregulation of NCX1 expression and activity in microglial cells after cerebral ischemia suggests NCX1 may play a role in modulating microglial functions in the post-ischemic brain.

K^+–Cl^- Cotransporter and Na^+/HCO_3^- Cotransporter

K^+–Cl^- Cotransporter

K^+–Cl^- cotransporters were characterized by whole cell recording and RT-PCR in microglia [45]. The repertoire of chloride-conducting pathways in murine primary microglial cells includes the K^+–Cl^- cotransporters (KCC1, KCC2, KCC3, and KCC4) as well as swelling-activated chloride channels. K^+–Cl^- cotransporters induce K^+ and Cl^- efflux when swelling-activated chloride channels are activated.

The remarkable feature of microglial cells to respond to distinct changes in their microenvironment is the formation of vibrant lamellipodia, either at the tip of their processes to seal injured capillaries or with a prominent lamellipodium in the case of amoeboid phenotypes to migrate to the site of injury. The polar distribution of KCC transporters and chloride influx promotes microglial lamellipodium formation [45]. Cell death by injuries or neurodegenerative diseases liberates intracellular K^+. Following ischemia, the extracellular K^+ concentration increases from the range of 2.5–3.5 mM to 50–80 mM [46]. Extracellular K^+ increase reverses KCC flux direction and evokes cell swelling of microglia, which may promote microglial lamellipodia formation [45]. In microglial cells, activation of Cl^- influx and concomitant local volume changes may explain the reason why these cells do not swell homogenously in response to osmotic changes in their microenvironment [45].

Na^+–HCO_3^- Cotransporter

The presence of operational Na^+/HCO_3^- cotransporter and Na^+-dependent Cl^-/HCO_3^- exchanger were found in the cultured microglia [47]. Expression of ion transporters Cl^-/HCO_3^- exchangers at the protrusion site of microglia are reported to contribute to the extension of the actin projection in lamellipodium by mediating salt and osmotically obliged water uptake [6]. A migrating cell shows a regulatory volume decrease at its rear end and a regulatory volume increase at its front end [48]. This model has been substantiated by identification of the asymmetric distribution of transporters and ion channels in the polarized migrating cell. In the lamellipodium, Na^+/HCO_3^- cotransporter has been identified [48]. Schwab and coworkers postulate that ion channels and cotransporters, like the Na^+/HCO_3^- cotransporter, are distributed in a polarized way [49] which contributes to cell migration.

Ion Transporters in Neutrophils, Macrophages, and Monocytes

The innate microglia are the predominant immune cells in the brain for the first days after brain ischemia. Microglia were activated and proliferated as early as 1 day after ischemia, followed by an influx of peripheral immune cells such as neutrophils and macrophages into the ischemic brain parenchyma 2 days after ischemia, peaking at day 4 [50]. Macrophage infiltration after ischemia is a later event in most experimental paradigms, starting to infiltrate on day 4 and peaking at day 10 after ischemia [50]. Although the resident microglia were almost exclusively responsible for phagocytosis of dead and dying neurons during the first 3 days after ischemia [51], the delayed peripheral immune cells infiltration may also contribute to the outcome of ischemic brain injury [50]. Ion transporters such as NHE-1 and NCX1 may play a role in activation and function of these peripheral immune cells during their infiltrating and phagocytosis process.

NHE-1 in Neutrophils

Neutrophils are inflammatory leukocyte cells, which are the first of the circulating leukocytes to arrive at the site of injury [52]. Neutrophils kill microorganisms by ingesting them during phagocytosis. Neutrophils rapidly produce superoxide and other ROS under certain specific circumstances, known as the respiratory burst [53, 54]. NHE-1 activity is reported to be stimulated during the respiratory burst in response to inflammation or invading substance [54]. NHE-1 could regulate intracellular pH of neutrophils during the respiratory bursts by extrusion of H^+ following activation of NOX [55]. NHE-1 activity is also involved in neutrophil movement to locations of lesions and infections [56]. The exact mechanisms of NHE-1 in migration of neutrophil cells are unknown. One theory is that NHE-1 controls the electroneutral exchange of sodium for hydrogen, and the change of pH_i and the increase of intracellular cations affect the cell movement [56]. Removal of extracellular sodium impairs neutrophil movement, suggesting the important role of NHE-1 in ion translocation and cell movement function [56]. Moreover, migration of neutrophil cells was significantly reduced in cells that lacked the NHE-1 function [56].

Furthermore, intracellular acidification and NHE inhibition cause an extensive loss of surface L-selectin, which is very likely to reduce neutrophil rolling adhesion and possibly inhibit L-selectin associated signaling processes. This would lead to a decrease in firm adhesion and transmigration, especially during reperfusion period after ischemia injury. Such a mechanism would explain the reduced tissue damage after hypoxia reperfusion because fewer neutrophils would adhere to the endothelium, migrate into the tissue, and cause damage. A reduction in neutrophil adhesion accounts at least in part for the beneficial effects of NHE inhibition on ischemia reperfusion cardiac injury, and suggest the possible utility of therapeutic strategies directed at reducing L-selectin adhesion to reduce neutrophil invasion of tissues during an ischemic event, and prevent tissue injury [57]. On the other hand, NHE-1 causes an increase of osmolytes and water that result in cell swelling and extension of pseudopods [56]. Taken together, NHE-1 plays an important role in neutrophil function, which is involved in H^+ extrusion during the respiratory burst and cell migration.

NCX in Macrophages and Monocytes

Studies in murine macrophages have suggested the presence of NCX in immune cells [58]. NCX activities have also been reported in human lymphocytes and neutrophils investigated using pharmacological agents [59, 60]. In these cells, NCX may contribute either to the extrusion of Ca^{2+} after Ca^{2+}-spiking events or to the elevation in intracellular Ca^{2+} during cell activation which could be important for these cell after they migrate to the brain ischemic damaged area to clearance the dead neurons and synapse.

The mononuclear phagocyte cells are a widely distributed cellular network that has a primary role in innate immunity, tissue inflammation, and remodeling [61–63]. Mononuclear phagocyte cells derive from $CD34^+$ bone marrow progenitor, then differentiating into monocytes [64]. Peripheral monocytes further differentiate into macrophages as they migrate into damaged or injured tissues [65]. Both monocytes and macrophages rely on Ca^{2+} signaling to start their activation programs leading to the release of pro-inflammatory cytokines [66, 67]. Recent study showed that in human macrophages and blood monocytes constitutively express NCX1 and NCX3 [68]. NCX is present in both monocytes and macrophages, indicating that this exchanger is expressed early in the development of mononuclear phagocyte cells, and lasts up to the final steps of cell maturation. NCX may thus have an important role in regulating functions in circulating momocytes and differentiated tissue macrophages. Expression of NCX1 and NCX3 enables the macrophages and monocytes to respond in the classical model of activation. The role of NCX in macrophages and monocytes might be particularly important since NCX3, and partially NCX1, are not dependent on ATP for their function [69]. This means that NCX can contribute to maintaining intracellular Ca^{2+} homeostasis also in pathophysiological conditions such as tissue ischemia or anoxia in which ATP levels are reduced [69]. Under hypoxic conditions, NCX might help preserve the viability and functions of macrophages by limiting Ca^{2+} overload and maintaining intracellular Na^+ levels. On the other hand, activation of NCX promoted mRNA expression and protein release of TNF-α, which is one of the main pro-inflammatory cytokines produced by macrophages, silencing NCX1 in human macrophage significantly reduces the release of TNF-α. Ca^{2+} is involved in the production of TNF-α and other cytokines in macrophages [70] and many stimuli that activate cytokine production also generate Ca^{2+} signals [71–73]. These study shows that human monocytes and macrophages constitutively express functionally active forms of the NCX1 and NCX3. Understanding the role of NCX exchangers in regulating monocyte/macrophage pro-inflammatory functions may reveal novel targets to subsequent inflammatory and immune responses after ischemic stroke.

Conclusion

In summary, new studies show that ion transporters, such as NHE-1 and NCX, are constitutively expressed in microglia and peripheral immune cells including monocytes, neutrophils, and macrophage. These ion transporters are important for microglia and peripheral immune cell function by regulation of pH_i, Ca^{2+}-spiking events, cell volume, and membrane signal molecule expression during microglia/peripheral immune cell activation, adhesion, and migration. Especially, they play a key role in formation of inflammatory responses in ischemic brains. In light of the detrimental effects of excessive pro-inflammatory response on ischemic brain injury, targeting ion transporters may be a new therapeutic strategy to minimize neuroinflammatory reactions after ischemic stroke.

Acknowledgment This work was supported by NIH grants R01NS 48216 and R01NS 38118 (D. Sun).

References

1. Ling EA, Wong WC (1993) The origin and nature of ramified and ameboid microglia: a historical review and current concepts. Glia 7:9–18
2. Moore S, Thanos S (1996) The concept of microglia in relation to central nervous system disease and regeneration. Prog Neurobiol 48:441–460
3. Graeber MB, Streit WJ (2010) Microglia: biology and pathology. Acta Neuropathol 119(1):89–105
4. Jin R, Yang G, Li G (2010) Inflammatory mechanisms in ischemic stroke: role of inflammatory cells. J Leukoc Biol 87(5):779–89
5. Tambuyzer BR, Ponsaerts P, Nouwen EJ (2009) Microglia: gatekeepers of central nervous system immunology. J Leukoc Biol 85:352–370
6. Kettenmann H, Hanisch UK, Noda M, Verkhratsky A (2011) Physiology of microglia. Physiol Rev 91:461–553
7. Hu X, Li P, Guo Y, Wang H, Leak RK, Chen S, Gao Y, Chen J (2012) Microglia/macrophage polarization dynamics reveal novel mechanism of injury expansion after focal cerebral ischemia. Stroke 43(11):3063–70
8. Harrigan TJ, Abdullaev IF, Jourd'heuil D, Mongin AA (2008) Activation of microglia with zymosan promotes excitatory amino acid release via volume-regulated anion channels: the role of NADPH oxidases. J Neurochem 106:2449–2462
9. Yenari MA, Kauppinen TM, Swanson RA (2010) Microglial activation in stroke: therapeutic targets. Neurotherapeutics 7(4):378–91
10. Lalancette-Hebert M, Gowing G, Simard A, Weng YC, Kriz J (2007) Selective ablation of proliferating microglial cells exacerbates ischemic injury in the brain. J Neurosci 27(10):2596–605
11. Mongin AA (2007) Disruption of ionic and cell volume homeostasis in cerebral ischemia: the perfect storm. Pathophysiology 14(3–4):183–93
12. Shi Y, Kim D, Caldwell M, Sun D (2013) The role of Na^+/H^+ exchanger isoform 1 in inflammatory responses: maintaining H^+ homeostasis of immune cells. Adv Exp Med Biol 961:411–8
13. Orlowski J, Grinstein S (2004) Diversity of the mammalian sodium/proton exchanger SLC9 gene family. Pflugers Arch 447:549–565
14. Fliegel L (2005) The Na^+/H^+ exchanger isoform 1. Int J Biochem Cell Biol 37:33–37
15. Zachos NC, Tse M, Donowitz M (2005) Molecular physiology of intestinal Na^+/H^+ exchange. Annu Rev Physiol 67:411–443
16. Brett CL, Wei Y, Donowitz M, Rao R (2002) Human Na^+/H^+ exchanger isoform 6 is found in recycling endosomes of cells, not in mitochondria. Am J Physiol Cell Physiol 282: C1031–C1041
17. Numata M, Orlowski J (2001) Molecular cloning and characterization of a novel (Na^+, K^+)/H^+ exchanger localized to the trans-Golgi network. J Biol Chem 276:17387–17394
18. De Vito P (2006) The sodium/hydrogen exchanger: a possible mediator of immunity. Cell Immunol 240:69–85
19. Rotin D, Grinstein S (1989) Impaired cell volume regulation in Na^+–H^+ exchange-deficient mutants. Am J Physiol 257:C1158–C1165
20. Denker SP, Huang DC, Orlowski J, Furthmayr H, Barber DL (2000) Direct binding of the Na-H exchanger NHE1 to ERM proteins regulates the cortical cytoskeleton and cell shape independently of H+ translocation. Mol Cell 6:1425–1436
21. Denker SP, Barber DL (2002) Cell migration requires both ion translocation and cytoskeletal anchoring by the Na-H exchanger NHE1. J Cell Biol 159(6):1087–96
22. Bedard K, Krause KH (2007) The NOX family of ROS-generating NADPH oxidases: physiology and pathophysiology. Physiol Rev 87(1):245–313

23. Henderson LM, Chappell JB, Jones OT (1988) Internal pH changes associated with the activity of NADPH oxidase of human neutrophils. Further evidence for the presence of an H^+ conducting channel. Biochem J 251:563–567
24. Liu Y, Kintner DB, Chanana V, Algharabli J, Chen X, Gao Y, Chen J, Ferrazzano P, Olson JK, Sun D (2010) Activation of microglia depends on Na^+/H^+ exchange-mediated H^+ homeostasis. J Neurosci 30:15210–15220
25. Shi Y, Chanana V, Watters JJ, Ferrazzano P, Sun D (2011) Role of sodium/hydrogen exchanger isoform 1 in microglial activation and proinflammatory responses in ischemic brains. J Neurochem 119:124–135
26. Honda S, Sasaki Y, Ohsawa K, Imai Y, Nakamura Y, Inoue K, Kohsaka S (2001) Extracellular ATP or ADP induce chemotaxis of cultured microglia through Gi/o-coupled P2Y receptors. J Neurosci 21:1975–1982
27. Ifuku M, Farber K, Okuno Y, Yamakawa Y, Miyamoto T, Nolte C, Merrino VF, Kita S, Iwamoto T, Komuro I, Wang B, Cheung G, Ishikawa E, Ooboshi H, Bader M, Wada K, Kettenmann H, Noda M (2007) Bradykinin-induced microglial migration mediated by B1-bradykinin receptors depends on Ca^{2+} influx via reverse-mode activity of the Na^+/Ca^{2+} exchanger. J Neurosci 27:13065–13073
28. Stock C, Schwab A (2009) Protons make tumor cells move like clockwork. Pflugers Arch 458:981–992
29. Carlier MF, Laurent V, Santolini J, Melki R, Didry D, Xia GX, Hong Y, Chua NH, Pantaloni D (1997) Actin depolymerizing factor (ADF/cofilin) enhances the rate of filament turnover: implication in actin-based motility. J Cell Biol 136:1307–1322
30. Ghosh M, Song X, Mouneimne G, Sidani M, Lawrence DS, Condeelis JS (2004) Cofilin promotes actin polymerization and defines the direction of cell motility. Science 304:743–746
31. Frantz C, Barreiro G, Dominguez L, Chen X, Eddy R, Condeelis J, Kelly MJ, Jacobson MP, Barber DL (2008) Cofilin is a pH sensor for actin free barbed end formation: role of phosphoinositide binding. J Cell Biol 183:865–879
32. Yonezawa N, Nishida E, Sakai H (1985) pH control of actin polymerization by cofilin. J Biol Chem 260:14410–14412
33. Shi Y, Yuan H, Kim D, Chanana V, Baba A, Matsuda T, Cengiz P, Ferrazzano P, Sun D (2013) Stimulation of Na(+)/H(+) exchanger isoform 1 promotes microglial migration. PLoS One 8(8):e74201
34. Annunziato L, Pignataro G, Di Renzo GF (2004) Pharmacology of brain Na^+/Ca^{2+} exchanger: from molecular biology to therapeutic perspectives. Pharmacol Rev 56:633–654
35. Philipson KD, Nicoll DA (2000) Sodium–calcium exchange: a molecular perspective. Annu Rev Physiol 62:111–133
36. Philipson KD, Nicoll DA, Ottolia M, Quednau BD, Reuter H, John S, Qiu Z (2002) The Na^+/Ca^{2+} exchange molecule: an overview. Ann N Y Acad Sci 976:1–10
37. Nicoll DA, Longoni S, Philipson KD (1990) Molecular cloning and functional expression of the cardiac sarcolemmal Na^+–Ca^{2+} exchanger. Science 250:562–565
38. Newell EW, Stanley EF, Schlichter LC (2007) Reversed Na^+/Ca^{2+} exchange contributes to Ca^{2+} influx and respiratory burst in microglia. Channels (Austin) 1:366–376
39. Quednau BD, Nicoll DA, Philipson KD (1997) Tissue specificity and alternative splicing of the Na^+/Ca^{2+} exchanger isoforms NCX1, NCX2, and NCX3 in rat. Am J Physiol 272: C1250–C1261
40. Nagano T, Kawasaki Y, Baba A, Takemura M, Matsuda T (2004) Up-regulation of Na^+–Ca^{2+} exchange activity by interferon-gamma in cultured rat microglia. J Neurochem 90:784–791
41. Boscia F, Gala R, Pannaccione A, Secondo A, Scorziello A, Di Renzo G, Annunziato L (2009) NCX1 expression and functional activity increase in microglia invading the infarct core. Stroke 40:3608–3617
42. Matsuda T, Nagano T, Takemura M, Baba A (2006) Topics on the Na^+/Ca^{2+} exchanger: responses of Na^+/Ca^{2+} exchanger to interferon-gamma and nitric oxide in cultured microglia. J Pharmacol Sci 102:22–26

43. Hanisch UK, Kettenmann H (2007) Microglia: active sensor and versatile effector cells in the normal and pathologic brain. Nat Neurosci 10:1387–1394
44. Ito D, Tanaka K, Suzuki S, Dembo T, Fukuuchi Y (2001) Enhanced expression of Iba1, ionized calcium-binding adapter molecule 1, after transient focal cerebral ischemia in rat brain. Stroke 32:1208–1215
45. Zierler S, Frei E, Grissmer S, Kerschbaum HH (2008) Chloride influx provokes lamellipodium formation in microglial cells. Cell Physiol Biochem 21:55–62
46. Leis JA, Bekar LK, Walz W (2005) Potassium homeostasis in the ischemic brain. Glia 50:407–416
47. Faff L, Ohlemeyer C, Kettenmann H (1996) Intracellular pH regulation in cultured microglial cells from mouse brain. J Neurosci Res 46:294–304
48. Schwab A (2001) Function and spatial distribution of ion channels and transporters in cell migration. Am J Physiol Renal Physiol 280:F739–F747
49. Schwab A (2001) Ion channels and transporters on the move. News Physiol Sci 16:29–33
50. Weinstein JR, Koerner IP, Moller T (2010) Microglia in ischemic brain injury. Future Neurol 5:227–246
51. Schilling M, Besselmann M, Muller M, Strecker JK, Ringelstein EB, Kiefer R (2005) Predominant phagocytic activity of resident microglia over hematogenous macrophages following transient focal cerebral ischemia: an investigation using green fluorescent protein transgenic bone marrow chimeric mice. Exp Neurol 196:290–297
52. Bernardes-Silva M, Anthony DC, Issekutz AC, Perry VH (2001) Recruitment of neutrophils across the blood–brain barrier: the role of E- and P-selectins. J Cereb Blood Flow Metab 21:1115–1124
53. Martins PS, Kallas EG, Neto MC, Dalboni MA, Blecher S, Salomao R (2003) Upregulation of reactive oxygen species generation and phagocytosis, and increased apoptosis in human neutrophils during severe sepsis and septic shock. Shock 20:208–212
54. Robinson JM (2008) Reactive oxygen species in phagocytic leukocytes. Histochem Cell Biol 130:281–297
55. Fukushima T, Waddell TK, Grinstein S, Goss GG, Orlowski J, Downey GP (1996) Na^+/H^+ exchange activity during phagocytosis in human neutrophils: role of Fcgamma receptors and tyrosine kinases. J Cell Biol 132:1037–1052
56. Hayashi H, Aharonovitz O, Alexander RT, Touret N, Furuya W, Orlowski J, Grinstein S (2008) Na^+/H^+ exchange and pH regulation in the control of neutrophil chemokinesis and chemotaxis. Am J Physiol Cell Physiol 294:C526–C534
57. Kaba NK, Schultz J, Law FY, Lefort CT, Martel-Gallegos G, Kim M, Waugh RE, Arreola J, Knauf PA (2008) Inhibition of Na^+/H^+ exchanger enhances low pH-induced L-selectin shedding and beta2-integrin surface expression in human neutrophils. Am J Physiol Cell Physiol 295:C1454–C1463
58. Donnadieu E, Trautmann A (1993) Is there a Na^+/Ca^{2+} exchanger in macrophages and in lymphocytes? Pflugers Arch 424:448–455
59. Balasubramanyam M, Rohowsky-Kochan C, Reeves JP, Gardner JP (1994) Na^+/Ca^{2+} exchange-mediated calcium entry in human lymphocytes. J Clin Invest 94:2002–2008
60. Tintinger GR, Anderson R (2004) Counteracting effects of NADPH oxidase and the Na^+/Ca^{2+} exchanger on membrane repolarisation and store-operated uptake of Ca^{2+} by chemoattractant-activated human neutrophils. Biochem Pharmacol 67:2263–2271
61. Gordon S (2007) The macrophage: past, present and future. Eur J Immunol 37(Suppl 1):S9–17
62. Hume DA, Ross IL, Himes SR, Sasmono RT, Wells CA, Ravasi T (2002) The mononuclear phagocyte system revisited. J Leukoc Biol 72:621–627
63. Mantovani A, Sica A, Sozzani S, Allavena P, Vecchi A, Locati M (2004) The chemokine system in diverse forms of macrophage activation and polarization. Trends Immunol 25:677–686
64. Fogg DK, Sibon C, Miled C, Jung S, Aucouturier P, Littman DR, Cumano A, Geissmann F (2006) A clonogenic bone marrow progenitor specific for macrophages and dendritic cells. Science 311:83–87

65. Gordon S, Taylor PR (2005) Monocyte and macrophage heterogeneity. Nat Rev Immunol 5:953–964
66. Cathcart MK (2004) Regulation of superoxide anion production by NADPH oxidase in monocytes/macrophages: contributions to atherosclerosis. Arterioscler Thromb Vasc Biol 24:23–28
67. Tapper H (1996) The secretion of preformed granules by macrophages and neutrophils. J Leukoc Biol 59:613–622
68. Staiano RI, Granata F, Secondo A, Petraroli A, Loffredo S, Frattini A, Annunziato L, Marone G, Triggiani M (2009) Expression and function of Na^+/Ca^{2+} exchangers 1 and 3 in human macrophages and monocytes. Eur J Immunol 39:1405–1418
69. Secondo A, Staiano RI, Scorziello A, Sirabella R, Boscia F, Adornetto A, Valsecchi V, Molinaro P, Canzoniero LM, Di RG, Annunziato L (2007) BHK cells transfected with NCX3 are more resistant to hypoxia followed by reoxygenation than those transfected with NCX1 and NCX2: Possible relationship with mitochondrial membrane potential. Cell Calcium 42:521–535
70. Watanabe N, Suzuki J, Kobayashi Y (1996) Role of calcium in tumor necrosis factor-alpha production by activated macrophages. J Biochem 120:1190–1195
71. Liu H, Zhang H, Forman HJ (2007) Silica induces macrophage cytokines through phosphatidylcholine-specific phospholipase C with hydrogen peroxide. Am J Respir Cell Mol Biol 36:594–599
72. Mayne M, Holden CP, Nath A, Geiger JD (2000) Release of calcium from inositol 1,4,5-trisphosphate receptor-regulated stores by HIV-1 Tat regulates TNF-alpha production in human macrophages. J Immunol 164:6538–6542
73. Zhou X, Yang W, Li J (2006) Ca^{2+}– and protein kinase C-dependent signaling pathway for nuclear factor-kappaB activation, inducible nitric-oxide synthase expression, and tumor necrosis factor-alpha production in lipopolysaccharide-stimulated rat peritoneal macrophages. J Biol Chem 281:31337–31347

Chapter 9
Inflammation After Acute Brain Injuries Affects the Developing Brain Differently than the Adult Brain

David Fernández-López and Zinaida S. Vexler

Abstract Emerging data show that the mechanisms of injury, including neuroinflammation, differ greatly in the immature brain and adult brain. This chapter will discuss the maturation-dependent contribution of glial and peripheral immune cells, differences in the blood–brain barrier structure and function in relation to injury, as well as the effect of immaturity on the function of several receptors on microglia/macrophages related to innate and acquired immune responses. We will discuss how age-related differential inflammatory and vascular responses to injury would impact development of treatments for newborns affected by stroke.

Introduction

It has become apparent over the last decade that inflammation is not as one-dimensional or necessarily detrimental after brain injury, including stroke, as was traditionally thought. There is also ample evidence that the mechanisms of injury differ greatly in the immature brain and adult brain. This chapter will focus on most recent experimental observations regarding the differential aspects of the neuroinflammatory response after acute brain injury in the postnatal brain as opposed to the adult brain, findings obtained after our previous review on this topic [1]. We will discuss the maturation-dependent contribution of peripheral immune cells, differences in the blood–brain barrier (BBB) structure and function and initiation of neuroinflammation after injury, the intrinsic age-related differences in the resident microglial

D. Fernández-López
University of California San Francisco School of Medicine,
San Francisco, CA, USA

Z.S. Vexler (✉)
Department of Neurology, University of California, San Francisco, CA, USA
e-mail: Zena.Vexler@ucsf.edu

J. Chen et al. (eds.), *Immunological Mechanisms and Therapies in Brain Injuries and Stroke*, Springer Series in Translational Stroke Research, DOI 10.1007/978-1-4614-8915-3_9,

phenotypes, and the consequences of these effects for acute injury. We will then discuss the effect of immaturity on the function of several receptors on microglia/macrophages related to innate and acquired immune responses, toll-like receptors (TLRs), and the scavenger receptor CD36. Finally, we will discuss impact of these recent discoveries on the differential vascular responses to injury in immature and adult brain for development of treatments for newborns affected by stroke.

Brain Maturation and Mechanisms of Brain Susceptibility to Injury

The developmental stage of the brain at the onset of injury is a decisive factor that determines the patterns of brain damage, including cell-type-specific susceptibility, which in turn affects initiation, progression, and final outcome of injury in individual brain regions. For example, in preterm newborns asphyxia- or hypoxia–ischemia (H–I)-induced brain injury generally coincides with a time window of high susceptibility of oligodendrocyte progenitors (OLPs) to excitotoxicity, oxidative stress, and inflammation [2–4] and adversely interferes with normal differentiation of this cell population into mature, myelinating oligodendrocytes [5], thus predisposing the brain to periventricular white matter injury, defective myelination of white matter tracts, and long-term injury. Subplate neurons, a cell type that exists transiently during human fetal brain development, are also prone to H-I, contributing to long-term injury. We will not discuss in detail these aspects, as several recent reviews comprehensively discussed the unique features of fetal brain neuroinflammation and injury in humans and in corresponding injury models in postnatal day 1–3 (P1–P3) rodents [5–8].

The patterns of ischemic injury in full-term newborns are different from those in preterm newborns [9]; injury is no longer diffuse and is mostly manifested focally in gray matter regions, most commonly in the striatum, thalamus, and cortical areas. Studies in animal models of H-I and focal arterial stroke following transient middle cerebral artery occlusion (MCAO) in P7-P10 rodents showed the predominance of neuronal cell death which occurs via the apoptotic–necrotic continuum [10–12].

Although the targeted cell populations and regions affected by cerebral hypoxia and ischemia are different in preterm and term newborns, the “triangle” of the pathophysiological mechanisms that ultimately cause cell death—the excitotoxic, oxidant, and inflammatory components—are similar (Fig. 9.1). Excitotoxicity, triggered by accumulation of glutamate and other excitotoxic molecules in extracellular spaces following neuronal membrane depolarization, glutamate efflux, and failure of its uptake mechanisms, is a common initial damaging process caused by asphyxia, H-I, or focal arterial stroke [13, 14]. The neonatal brain is more excitable and prone to oxidative stress than the adult brain due to higher levels of glutamate receptor expression [15, 16] and a different composition of individual NMDA receptor subunits [17] and interaction with downstream signaling cascades [18]. Several other distinct characteristics of the immature brain, such as high oxygen

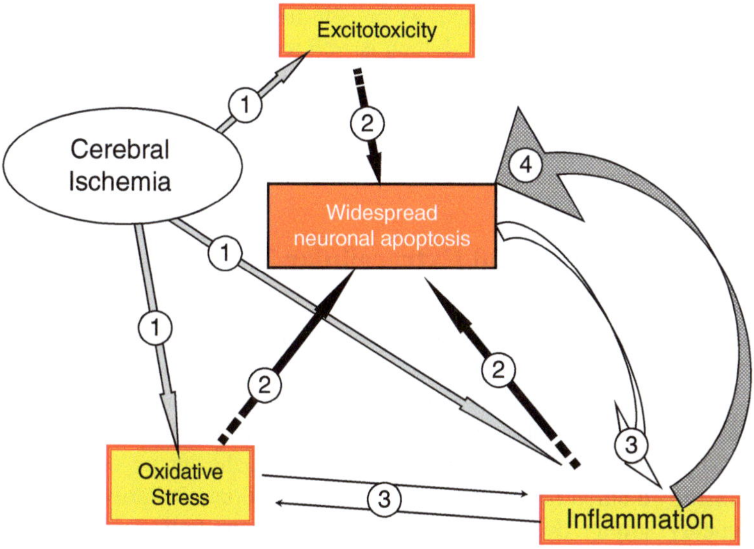

Fig. 9.1 "Triangle" of pathophysiological mechanisms of ischemic brain injury in the immature brain. Cerebral ischemia triggers glutamate excitotoxicity, leads to the accumulation of free radicals, and promotes the activation of inflammatory responses in the affected brain tissue (shown as 1). Each of these primary mechanisms is sufficient to induce cell death, but when they occur simultaneously in the brain widespread cell death occurs (shown as 2). Damage of cell components caused by oxidative stress is one of the signals that induces inflammation. Inflammatory cells are major sources of free radicals and at the same time responders to free radicals, which in turn propagate their own activation. Incomplete removal of apoptotic cells activates inflammation and oxidative stress (shown as 3) and initiates additional apoptotic/necrotic signaling cascades in the neighboring cells (shown as 4). Thus, injury evolution is determined by the presence of these three intrinsically interconnected mechanisms and is further modulated by feedback mechanisms among these cells

consumption, higher iron levels, and low expression of endogenous antioxidant enzymes contribute to higher vulnerability to the deleterious actions of free radicals [19, 20]. However, overexpression of antioxidant enzymes does not necessarily protect the neonatal brain, as was shown for SOD1 overexpressor mouse pups, and is dependent on particular targeted free radical pathways [21, 22] and cell types [23]. Inflammation, the third major contributor to neonatal brain injury, can induce injury by itself as well as enhance excitotoxicity and oxidative stress through the release of cytokines, free radicals, and other toxic products or trigger release of excitotoxic molecules, including glutamate (Fig. 9.1).

Another important distinction between postnatal and adult brain is the major difference in cell death mechanisms in response to brain injury. It has been described more than a decade ago that physiological apoptosis is still prominent in normal brain at term and that the relative contribution of apoptosis to the overall cell death declines with age in injured brain [24–26]. Amplification of necrosis due to failure to remove dying cells was then described [27, 28], leading to a unique hybrid cell death, which was called the "continuum," an intermediate mechanism of cell death that exhibits features of both necrosis and apoptosis (reviewed in [29, 30]).

While the central role of inflammatory cells in insufficient clearance of apoptotic cells and subsequent injury exacerbation is well recognized, whether limited clearance of degenerating cells is due to a lack of recognition of surface markers in dying cells by brain phagocytes or due to ineffective interactions between such cells is not well defined.

Blood–Brain Barrier and Extracellular Matrix

One factor that may contribute to the differential inflammatory response to stroke in adults and neonates is the BBB. BBB disruption after adult stroke is an important contributing factor to injury, but until recently information on BBB function after stroke during the early postnatal period has been scant. Earlier studies have demonstrated the dynamic nature of the BBB function during the first postnatal weeks and, in contrast to common belief, they showed that functional integrity of the BBB to inflammatory stimuli does not increase linearly from birth to adulthood [31]. Instead, based on assessment of permeability to HRP, the BBB was more disrupted in 3-week-old rats than in 2-h-old rats or in young adult rats following intrastriatal injection of the inflammatory cytokine IL-1b [31]. These observations determine the existence of "windows" of susceptibility of the BBB to inflammatory cytokines, although the underlying cellular and molecular mechanisms are still poorly understood.

Using transient MCAO models in P7 rats and in adult rats, we recently discovered that in contrast to the markedly increased leakage of plasma proteins (albumin) and intravascular tracers of approximately 500 Da, 3 kDa, and 70 kDa into injured adult brain 24 h after reperfusion, leakage of these molecules remained low in the injured neonatal brain within 24 h [32]. Although this finding might seem paradoxical, it is known that endothelial tight junctions are functional early during development and that the BBB is integrant at birth in most brain regions [33, 34]. Moreover, several specific components of the BBB known to contribute to vascular impermeability are expressed in higher levels in neonates than in adults (Table 9.1). For example, both gene and protein expression of the basement membrane proteins collagen IV and laminin was much higher in endothelial cells in uninjured neonatal brain cortex than in the adult cortex [32]. Considering that collagen IV and laminin have direct effects on proliferation, migration, and maturation of endothelial cells, higher expression of these proteins may be central to the active vascular outgrowth in the brain that continues during the first three postnatal weeks [35, 36]. It could also affect endogenous mechanisms that preserve BBB integrity but delay and limit angiogenesis [37]. The expression of the tight junction protein occludin was also higher in neonates than in adults, whereas expression of other tight junction proteins, claudin-5 and ZO-1, was better preserved after acute stroke in neonates [32]. Our observation that the gene expression of the proteolytic enzyme MMP-9 and the adhesion molecule E-selectin is lower in endothelial cells isolated from injured neonatal brains than in adult brains also suggests that some mechanisms of interactions

Table 9.1 Gene expression in endothelial cells in adult and neonatal rats subjected to 3 h of MCAO followed by 24 h of reperfusion

Gene name	Neonatal Contra	Ipsi	Ipsi/Contra	Adult Contra	Ipsi	Ipsi/Contra
Collagen IV alpha1	15,622	19,599	1.3	3,278	13,571	4.1[a]
Collagen IV alpha2	15,643	16,910	1.1	1,478	6,966	4.7[a]
Laminin alpha5	1,079	1,413	1.3	656	1,618	2.5[a]
Occludin	9,600	7,132	0.7	8,047	3,034	0.4[a]
Claudin-5	24,730	23,434	0.9	20,181	17,964	0.9
ZO-1	13,473	12,329	0.9	12,887	10,971	0.9
MMP-9	187	88	0.5	42	2,631	63.2[a]
E-selectin	22	1,245	57.6[b]	21	4,414	214.3[b]
VEGFR-2	14,626	12,974	0.9	4,151	10,916	2.6[a]
Angiopoietin-2	9,569	6,066	0.6	2,761	14,568	5.3[a]

Abbreviations: *Ipsi* Ipsilateral; the gene expression is determined in injured tissue. *Contra* contralateral; the gene expression is determined in the matching tissue in contralateral hemisphere. Ipsi/Contra—fold increase in injured vs. contralateral hemisphere. Notice differences in values between Contra hemisphere in the two age groups

[a]Statistical significance (Ipsi/Contra) only this age

[b]Statistical significance (Ipsi/Contra) in both ages

between circulating cells, the extracellular matrix, and endothelial cells are maturation dependent. This result is consistent with low recruitment of circulating immune cells that is discussed later [32]. Together, these observations support the notion of a higher intrinsic resistance of the neonatal BBB to focal transient ischemia than in the adult.

However, the temporo-spatial particulars of cerebral blood flow regulation and the extent of recirculation affect BBB permeability. Focal stroke induced in P7 rats by permanent MCAO combined with transient CCA ligation was associated with BBB disruption and neutrophil accumulation [38]. In another neonatal brain injury model, H-I, increased BBB permeability was demonstrated from 3 to 72 h [39–41]. In the same studies, the injurious effects of H-I were reduced in neonatal MMP-9 knockout mice [39] and abrogated after administration of the broad-spectrum MMP inhibitor GM6001 [41], once again highlighting the important role of MMPs on the degradation of BBB components. Further animal studies should better delineate the dynamics of BBB function and susceptibility to injury during postnatal brain development.

Infiltration of Peripheral Leukocytes and Injury

Under normal conditions, leukocyte entry into the CNS is low and is limited to the vasculature surveillance functions of these cells, but after cerebral ischemia neutrophils rapidly and often transiently infiltrate ischemic tissue in the adult, with the extent and timing of their appearance in the tissue dependent on injury severity [42]. Neutrophils exacerbate reperfusion injury after transient ischemic insult in the adult by at least several mechanisms. They prime the endothelium and contribute to reducing CBF, causing "no-reflow" phenomenon [43], release free radicals and proteolytic enzymes, and stimulate cytokine release from neighboring cells [44]. Neutrophils also possess specific mechanisms that disturb the BBB [45, 46]. Treatments that either induce neutropenia, prevent leukocyte vascular adhesion, and extravasation into the brain parenchyma or inhibit proteolytic enzymes in leukocytes, such as elastase, cathepsin G, or MMP-9, are neuroprotective [46–50]. Peripheral and loosely adherent neutrophils can also induce injury, but the relative roles of transmigrated and intravascular neutrophils in mediating damage are not fully understood. The majority of the genes induced in the blood of human patients with acute stroke are expressed by neutrophils [51]. Neutrophils accumulate within hours after stroke, and neurological outcome is worse and infarct is larger in patients with marked neutrophil accumulation [52].

In contrast, in neonatal brain, infiltration of neutrophils is substantially lower [53] and occurs later [32, 54] in response to H-I and stroke. Indeed, it has been shown that at 24 h after neonatal stroke, neutrophils accumulate in the brain post-capillary venules without transmigrating across the BBB [32] despite the presence of a concentration gradient of the cytokine-induced neutrophil chemoattractant 1 (CINC-1) between the brain and the blood [55]. Interestingly, reduction of CINC-1 concentration in peripheral blood early after injury, by intravenous administration of a neutralizing anti-CINC-1 antibody, led to neutrophil transmigration into injured neonatal brain and, interestingly, increased infiltration, recapitulating the "adult phenotype" (i.e., an increased BBB leakage and worsened injury) [32]. It is yet not clear whether the inability of neutrophils to transmigrate into injured neonatal brain is related to a lower capacity of these cells than in adult brain to respond to their specific chemoattractants, due to inadequate endothelial cell activation after injury, or due to loss of receptors on neutrophils that recognize inducible antigens on activated endothelium. The data on increased transmigration of neutrophils in response to reduction of systemic CINC-1 [32] suggest that neutrophil–endothelial interactions in post-ischemic immature brain differ from that in post-ischemic adult brain.

Monocytes, a leukocyte subpopulation that typically infiltrates injured adult brain later than neutrophils, is also thought to injure the BBB and damage the ischemic tissue. Although macrophages are present in injured neonatal brain in large numbers, especially over time after H-I [56], there is increasing evidence showing that the extent and the dynamics of monocyte infiltration in response to injury are different in the neonatal than in adult brain [57]. In a transient MCAO model in P7 rats, we showed that at a time when circulating monocytes actively infiltrate adult brain after transient MCAO, 24 h after reperfusion, most brain macrophages are comprised of resident microglia ($CD45^{low/int}/CD11b^{+}$ cells), and only a small fraction of the macrophage population are infiltrating monocytes (i.e., $CD45^{high}/CD11b^{+}$ cells) [58]. The contribution of peripheral monocytes to the macrophage brain population increases by 48 h after reperfusion [58], indicating that the invasion by peripheral monocytes after stroke is delayed in the neonatal brain.

The Microglial Phenotypes and Injury

Microglia—endogenous brain macrophages—are the main cell type that provides immuno-surveillance in the brain, but these cells can be toxic [59]. Historically, the concept of a "cytotoxic" role of microglia in stroke came from data showing that the increased number of the macrophages is associated with more severe injury whereas neuroprotective agents decrease both the macrophage numbers and injury severity. As resident brain macrophages, microglia have been typically described as injurious. However, more recent data suggest that they may act as a "double-edged sword" in acute stroke: injurious [60, 61] or beneficial [62–64]. Three aspects of the microglia/macrophage contribution to injury after stroke are frequently overlooked: 1) the origin of these cells, resident microglia vs. peripheral monocytes, 2) the heterogeneity of the microglia/macrophage population, and 3) a switch in the microglial/macrophage phenotype from injurious to beneficial that can occur in response to a changed brain microenvironment. Yet, the origin of macrophages may critically affect their functions and contribution to injury. For example, TNFα is protective after stroke when it is produced in activated microglia, but injurious when it is generated by monocytes [65]. Heterogeneity of the microglial pool [66–68] and microenvironment [69, 70] critically affect an array of functions in these cells. As an example, selective ablation of proliferating ($Mac2^{+}$) microglia (in transgenic mice expressing a mutant thymidine kinase form of herpes simplex virus driven by myeloid-specific CD11b promoter ($CD11bTK^{mt30}$) by ganciclovir treatment) markedly altered the temporal dynamics of inflammatory cytokine expression and significantly increased infarct size and the number of apoptotic neurons in adult brain following MCAO [70].

A classification of macrophage activation—classic activation (M1), alternative activation (M2A), and acquired deactivation (M2B)—has been developed based on their potential responses to infectious conditions [71]. M1/M2A/M2B macrophages were then shown to play distinct roles under sterile inflammation conditions, including stroke and neurodegenerative diseases, and additional M2 markers were described [72]. Microglia also produce mediators that can harm initially but enhance the repair through remodeling of the extracellular matrix [73] during the chronic recovery phase after cerebral ischemia. Therefore, it is not surprising that differing and even opposite effects of microglia have been reported in individual studies of repair after stroke—either adverse [60, 61] or beneficial, by facilitating angiogenesis [62] and supporting neurogenesis [63, 64].

It is yet to be learned whether microglial cells have similar functional phenotypes in the immature and adult brain after injury. In fact, many laboratories [31, 74], including ours [55, 75, 76], have shown that the inflammatory response, the mechanisms of oxidative injury, the modes of microglial activation, and the status of the neurovascular unit are vastly different in neonatal and adult injury. Furthermore, although a marked presence of macrophages 1–4 weeks after H-I has been firmly established, little is known about the relative contribution of microglia and invading differentiated monocytes to brain injury and repair. Intriguingly, several anti-inflammatory drugs known to be protective against adult stroke by reducing macrophage accumulation led to protection after neonatal stroke without directly affecting the inflammatory mechanisms associated with microglia activation [77–80]. The distinction between the local and peripheral components of the brain inflammatory response, especially the contribution of resident microglia and infiltrating monocytes, has been challenging due to the scarcity of reliable differential cell markers for both cell populations. Recently, development of novel experimental techniques aimed at selective pharmacological depletion of microglial cells without affecting peripheral bone marrow-derived monocytes has provided unique tools for the study of the particular role of microglia in the ischemic neonatal brain. Depletion of microglia by intracerebral injection of liposomes containing encapsulated clodronate, a toxin that selectively induces apoptosis in resident microglia without affecting other cell types in the brain [81], led to increased infarct volume and further accumulation of several pro-inflammatory cytokines and chemokines subacutellly after focal arterial stroke in P7 rats [81]. Importantly, in the absence of microglia the production of cytokines and chemokines typically secreted by activated microglia was compensated for by other cell types in the brain, including astrocytes, neurons, and endothelial cells [81]. These data indicated that at least a subpopulation of microglial cells mediates beneficial effects. At the same time, microglia can harm by limiting the removal of neurons dying via caspase-3-dependent apoptosis and inducing secondary inflammation and enhanced necrosis. The lack of microglia did not significantly increase the number of cells with cleaved caspase-3 following acute transient MCAO in P7 rats [23]. These results raised the possibility that the phagocytotic microglial pool might either be low or the recognition mechanisms between dying neurons and the phagocytotic microglia are not sufficiently activated early after injury.

Another potentially important but poorly understood aspect of microglial effects after stroke is the role of these cells in vascular integrity. In adults, microglia have been shown to rapidly respond to localized small lesions induced in brain vessels in vivo by extending their processes towards the injured site in order to "shield" the vessel wall and prevent the leakage of plasma components into the brain parenchyma [82]. The active response of microglia to the degradation of BBB components and vascular leakage has been shown in an animal model of EAE [83], in which accumulation of microglia around leaky vessels correlated with the progression of injury. Our preliminary data [76] also show that after neonatal stroke depletion of microglia leads to increased BBB permeability and vascular degeneration, supporting the notion of a role for microglia in preventing vascular damage after neonatal stroke. It remains to be elucidated whether microglia contribute to the maintenance of vascular integrity by direct cell–cell contact, by paracrine release of vasoprotective factors, or by a combination of both mechanisms.

The Scavenger Receptor CD36 and Oxidative and Inflammatory Injury

There are numerous classes of scavenger receptors with organ-specific and cell type-specific physiological and pathophysiological functions. CD36 is a scavenger class B receptor involved in the modulation of multiple processes, including neuroinflammation, production of free radicals, phagocytosis of apoptotic cells, lipid metabolism, and injury-induced angiogenesis [84]. CD36 is expressed by several cell types, including monocytes, microglia, dendritic cells, and endothelial cells [85]. It is activated by multiple ligands, including phospholipids, advanced glycation end products, TSP-1, and OxLDL, or acts via cooperation with other receptors like vitronectin and TLRs. There is an extensive body of data showing that CD36 is injurious in adult brain, for example, mediating Aβ toxicity and oxLDL-atherosclerosis [86–88]. The effects of CD36, however, are context dependent, as it facilitates hematoma resolution after intracerebral hemorrhage in adult mice by mediating PPARgamma-dependent phagocytosis of red blood cells by microglia [89], leading to reduced neurological deficits [90]. After adult focal stroke, CD36 expression is upregulated in microglia and in peripheral monocytes and is later induced in scar-forming astrocytes in the peri-ischemic brain regions [91]. Adult CD36 knockout mice (CD36ko mice) showed smaller infarcts [91], less free radical formation [91, 92], less NF-kB activation [93], and attenuated accumulation of pro-inflammatory cytokines [94], indicating that CD36 activation contributes to brain injury by multiple mechanisms. Inhibition of the production of CD36 ligands by the antioxidant peptide SS31 also resulted in protection [92].

Compared to adult stroke, the effects of genetic deletion of CD36 differed significantly after neonatal stroke. The absence of CD36 did not improve, but worsened short-term outcome after neonatal stroke. While the volume of "tissue at risk" during MCAO, as defined by diffusion-weighted imaging, was similar in wild-type and CD36ko mice, recovery of the initially affected brain tissue was greatly diminished in CD36ko mice compared to wild-type mice, resulting in larger injury at 24 h after brain reperfusion [95].

Given that the presence of apoptotic neurons is significantly higher in the injured neonatal brain than in the adult brain, the reduced removal of apoptotic cells by macrophage-like cells due to the lack of a scavenger receptor would have deleterious consequences. Consistent with this notion, injury exacerbation was associated with the increased presence of cleaved caspase-3 and diminished engulfment of apoptotic neurons in neonatal CD36ko mice after neonatal stroke [95]. Another likely mechanism responsible for opposite effects in adult and neonates is accumulation of free radicals in microglia/macrophages. Superoxide accumulation after neonatal stroke was low in microglia/macrophages in the injured brain regions within 24 h, a time frame of accumulation of this free radical in microglia/macrophages in injured adult brain [91] (Fig. 9.2). A marked superoxide accumulation was observed in injured vessels after neonatal stroke while no such effect was reported after adult stroke [91]. It is unclear if age differences in free radical production in microglia/macrophages are directly related to distinct effects of the genetic deletion of CD36 on Nf-kB activation after stroke—a significant reduction in the nuclear translocation of NF-kB in the adult but unchanged activation in the neonate—but the production of macrophage chemoattractant chemokines, which were decreased in adult CD36ko mice after adult stroke, remained unaffected in neonatal stroke [95] (Figure 9.2). The distinct age-related effects of CD36 in injury may also depend on downstream intracellular signaling in multiple cell types [95] as well as possible differences in cooperation of CD36 with other receptors, such as TLRs.

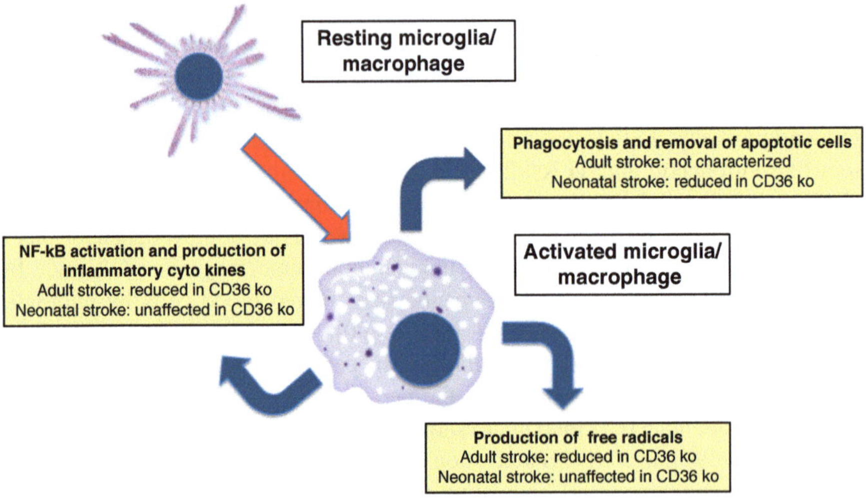

Fig. 9.2 Differential effects of CD36 genetic deletion after adult and neonatal stroke

Toll-Like Receptors and Innate Immune Responses to Neonatal Brain Injury

TLRs are a well-characterized family of proteins involved in innate immune responses to exogenous pathogens that also can be activated by endogenous damage-associated molecular patterns (DAMPs) released from cells affected by local alterations of the tissue homeostasis. In adults, TLR2 and TLR4 expression is induced in activated microglia following stroke [96, 97]. While TLR4 seems to be purely injurious in both experimental [98–100] and human stroke [101], the role of TLR2 is more complex and is not well understood. TLR2 elicits context- and tissue-dependent effects—injurious [102, 103] or beneficial [104, 105]—based on the type(s) of heterodimers that it forms with TLR1 and TLR4 [102]. While TLR downstream signaling pathways are relatively well described, the mechanisms facilitating ligand recognition by TLRs remain poorly understood. Recent results showed that the assembly of the TLR complexes in response to β-amyloid or oxLDL is not ligation of the TLRs but the initial CD36-mediated ligand recognition, which signals to Src kinases to induce TLR heterodimerization [106]. The formation of CD36–TLR4–TLR6 complex mediates injury in the setting of adult stroke [102, 106].

The situation is very different following H-I in neonates than after adult stroke: there is an upregulation of several TLRs, including TRL1, TLR2, and TLR8 [107], but, in contrast with the observations in adults, the expression of TLR4 remains unchanged. The effect of TLR4 on progression of H-I injury is unclear, since deletion of its intracellular effector protein, Myd88, did not result in protection [108]. In contrast, TLR2 deletion led to reduced H-I injury in neonate rats [107]. TLR1 expression was upregulated in neurons in the injured hemisphere following neonatal

H-I, but genetic deletion of this receptor did not affect injury outcome, suggesting a minor involvement of TLR1 in injury evolution after neonatal H-I [107]. Even without stroke, activation of brain TLR2 and TLR4 by exogenous ligands adversely interfered with normal postnatal brain development [109]. Selective TLR2 and TLR4 agonists administered from P3 to P11 (a time window during which oligodendrocyte progenitors are selectively vulnerable) impaired normal postnatal myelination and reduced the number of hippocampal neurons [109]. Maternal exposure to LPS (endotoxin) also has negative consequences on the offspring, including decreased myelination and increased reactive astrogliosis [110, 111]. Moreover, exposure of pregnant mice to LPS induced changes in the gene expression of factors involved in key cell processes for brain development, including cell stress, migration, and cell death and altered certain behaviors through young adulthood [112].

The systemic activation of TLRs also results in the induction of inflammatory responses in the neonatal brain. Peripheral administration of serial doses of LPS in rats during the first postnatal week led to a transient increase in BBB permeability in the white matter regions [113]. Strikingly, adult animals that received LPS systemically during the first postnatal week also showed increased BBB permeability, indicating that activation of TLR4 early after birth could induce permanent alterations in BBB function [113]. The mechanisms by which peripheral LPS can induce brain inflammation are not very clear, although several hypotheses have been proposed [114]. LPS does not seem to cross the intact BBB [115], but activation of TLR4 in circulating immune cells may induce release of inflammatory mediators that can affect the neurovascular endothelium and cause alterations of the BBB. Also, LPS can bind to the TLR4 present on endothelial cells and activate synthesis and release of inflammatory cytokines and prostaglandins to the brain parenchyma, where they can thereby act on neural cells and propagate inflammation [116, 117]. Finally, cytokines released from circulating cells activated by LPS can be translocated across the disrupted BBB and reach brain parenchymal cells [118, 119].

While data on the involvement of TLR in neonatal H-I and focal stroke are accumulating, the underlying immaturity-related differences in the mechanisms remain mostly unknown and may be related to differences in ligand levels and the ways in which CD36 forms heterodimers with multiple TLRs. It is also possible that overactivation of TLR4 receptors early postnatal reprograms brain responses later in life.

Remarks on Inflammation and Long-term Injury and Repair after Stroke

Although we limited the discussion of inflammation to its role in acute stroke, inflammation is an important modulator of repair. Repair is very complex [120, 121]. Several studies have demonstrated that post-ischemic angiogenesis enhanced pharmacologically [122, 123] or by cell-based therapy [124, 125] promotes functional recovery, while suppression of angiogenesis by anti-inflammatory strategies, like a MMP inhibitor [73], or by disruption of SDF1 or Ang1/Tie2 signaling [126] worsens functional recovery. Neurogenesis itself is regulated by non-neuronal cells [127], including microglia. Microglia can damage newly born neurons or support them through production of anti-inflammatory factors [63, 64], such as IGF1 and IL-10. LPS-activated microglia block neurogenesis whereas microglia activated by IL-4/T-helper cells induce neurogenesis [67]. Microglia contribute to the maintenance of hippocampal neurogenesis and spatial learning in the adult by enriched environment [68]. Data are also emerging that the effect of neuroprogenitor cells to repair injured brain may be due to reshaping brain microenvironment rather than due to engrafted cells themselves.

In the neonatal brain, extensive proliferation of neural stem/progenitor cells in the SVZ has been described in response to H-I and stroke [128–131], but relatively few newly born neurons survive and mature. Surprisingly, angiogenesis remains relatively low for at least a week after stroke in the neonate [37], likely adversely affecting neurogenesis. Future studies would define the effects of brain microenvironment, hostile or supportive, in neurogenesis and survival of newly generated neurons, information necessary to facilitate neurogenesis and synaptic plasticity.

References

1. Vexler ZS, Yenari MA (2009) Does inflammation after stroke affect the developing brain differently than adult brain? Dev Neurosci 31(5):378–93
2. Back SA, Luo NL, Borenstein NS, Levine JM, Volpe JJ, Kinney HC (2001) Late oligodendrocyte progenitors coincide with the developmental window of vulnerability for human perinatal white matter injury. J Neurosci 21(4):1302–12
3. Kaur C, Ling EA (2009) Periventricular white matter damage in the hypoxic neonatal brain: role of microglial cells. Prog Neurobiol 87(4):264–80
4. Oka A, Belliveau MJ, Rosenberg PA, Volpe JJ (1993) Vulnerability of oligodendroglia to glutamate: pharmacology, mechanisms, and prevention. J Neurosci 13(4):1441–53
5. Dahlstrom A, McRae A, Polinsky R, Nee L, Sadasivan B, Ling EA (1994) Alzheimer's disease cerebrospinal fluid antibodies display selectivity for microglia. Investigations with cell cultures and human cortical biopsies. Mol Neurobiol 9(1–3):41–54
6. Buser JR, Maire J, Riddle A, Gong X, Nguyen T, Nelson K et al (2012) Arrested preoligodendrocyte maturation contributes to myelination failure in premature infants. Ann Neurol 71(1):93–109
7. Kinney HC, Haynes RL, Xu G, Andiman SE, Folkerth RD, Sleeper LA et al (2012) Neuron deficit in the white matter and subplate in periventricular leukomalacia. Ann Neurol 71(3):397–406
8. McQuillen PS, Ferriero DM (2004) Selective vulnerability in the developing central nervous system. Pediatr Neurol 30(4):227–35
9. Kirton A, Armstrong-Wells J, Chang T, Deveber G, Rivkin MJ, Hernandez M et al (2011) Symptomatic neonatal arterial ischemic stroke: the International Pediatric Stroke Study. Pediatrics 128(6):e1402–10
10. Manabat C, Han BH, Wendland M, Derugin N, Fox CK, Choi J et al (2003) Reperfusion differentially induces caspase-3 activation in ischemic core and penumbra after stroke in immature brain. Stroke 34(1):207–13
11. Northington FJ, Graham EM, Martin LJ (2005) Apoptosis in perinatal hypoxic-ischemic brain injury: how important is it and should it be inhibited? Brain Res Brain Res Rev 50(2):244–57

12. Renolleau S, Fau S, Goyenvalle C, Joly LM, Chauvier D, Jacotot E et al (2007) Specific caspase inhibitor Q-VD-OPh prevents neonatal stroke in P7 rat: a role for gender. J Neurochem 100(4):1062–71
13. Gonzalez FF, Ferriero DM (2009) Neuroprotection in the newborn infant. Clin Perinatol 36(4):859–80
14. McLean C, Ferriero D (2004) Mechanisms of hypoxic-ischemic injury in the term infant. Semin Perinatol 28(6):425–32
15. Jensen FE (2002) The role of glutamate receptor maturation in perinatal seizures and brain injury. Int J Dev Neurosci 20(3–5):339–47
16. Johnston MV (2005) Excitotoxicity in perinatal brain injury. Brain Pathol 15(3):234–40
17. Knox R, Zhao C, Miguel-Perez D, Wang S, Yuan J, Ferriero D et al (2012) Enhanced NMDA receptor tyrosine phosphorylation and increased brain injury following neonatal hypoxia-ischemia in mice with neuronal Fyn overexpression. Neurobiol Dis 51:113–9
18. Jiang X, Mu D, Sheldon RA, Glidden DV, Ferriero DM (2003) Neonatal hypoxia-ischemia differentially upregulates MAGUKs and associated proteins in PSD-93-deficient mouse brain. Stroke 34(12):2958–63
19. Ferriero DM (2001) Oxidant mechanisms in neonatal hypoxia-ischemia. Dev Neurosci 23(3):198–202
20. Siddappa AJ, Rao RB, Wobken JD, Leibold EA, Connor JR, Georgieff MK (2002) Developmental changes in the expression of iron regulatory proteins and iron transport proteins in the perinatal rat brain. J Neurosci Res 68(6):761–75
21. Sheldon RA, Aminoff A, Lee CL, Christen S, Ferriero DM (2007) Hypoxic preconditioning reverses protection after neonatal hypoxia-ischemia in glutathione peroxidase transgenic murine brain. Pediatr Res 61(6):666–70
22. Sheldon RA, Jiang X, Francisco C, Christen S, Vexler ZS, Tauber MG et al (2004) Manipulation of antioxidant pathways in neonatal murine brain. Pediatr Res 56(4):656–62
23. Faustino J, Wang X, Jonhson C, Klibanov A, Derugin N, Wendland M et al (2011) Microglial cells contribute to endogenous brain defenses after acute neonatal focal stroke. J Neurosci 31(36):12992–3001
24. Hu BR, Liu CL, Ouyang Y, Blomgren K, Siesjo BK (2000) Involvement of caspase-3 in cell death after hypoxia-ischemia declines during brain maturation. J Cereb Blood Flow Metab 20(9):1294–300
25. Han BH, Xu D, Choi J, Han Y, Xanthoudakis S, Roy S et al (2002) Selective, reversible caspase-3 inhibitor is neuroprotective and reveals distinct pathways of cell death following neonatal hypoxic- ischemic brain injury. J Biol Chem 277(33):30128–36
26. Zhu C, Wang X, Xu F, Bahr BA, Shibata M, Uchiyama Y et al (2005) The influence of age on apoptotic and other mechanisms of cell death after cerebral hypoxia-ischemia. Cell Death Differ 12(2):162–76
27. West T, Atzeva M, Holtzman DM (2006) Caspase-3 deficiency during development increases vulnerability to hypoxic-ischemic injury through caspase-3-independent pathways. Neurobiol Dis 22(3):523–37
28. Carloni S, Carnevali A, Cimino M, Balduini W (2007) Extended role of necrotic cell death after hypoxia-ischemia-induced neurodegeneration in the neonatal rat. Neurobiol Dis 27(3):354–61
29. Blomgren K, Leist M, Groc L (2007) Pathological apoptosis in the developing brain. Apoptosis 12(5):993–1010
30. Northington FJ, Zelaya ME, O'Riordan DP, Blomgren K, Flock DL, Hagberg H et al (2007) Failure to complete apoptosis following neonatal hypoxia-ischemia manifests as "continuum" phenotype of cell death and occurs with multiple manifestations of mitochondrial dysfunction in rodent forebrain. Neuroscience 149(4):822–33
31. Anthony DC, Bolton SJ, Fearn S, Perry VH (1997) Age-related effects of interleukin-1 beta on polymorphonuclear neutrophil-dependent increases in blood–brain barrier permeability in rats. Brain 120(Pt 3):435–44

32. Fernandez-Lopez D, Faustino J, Daneman R, Zhou L, Lee SY, Derugin N et al (2012) Blood–brain barrier permeability is increased after acute adult stroke but not neonatal stroke in the rat. J Neurosci 32(28):9588–600
33. Kniesel U, Risau W, Wolburg H (1996) Development of blood–brain barrier tight junctions in the rat cortex. Brain Res Dev Brain Res 96(1–2):229–40
34. Engelhardt B (2003) Development of the blood–brain barrier. Cell Tissue Res 314(1):119–29
35. Iwai M, Cao G, Yin W, Stetler RA, Liu J, Chen J (2007) Erythropoietin promotes neuronal replacement through revascularization and neurogenesis after neonatal hypoxia/ischemia in rats. Stroke 38(10):2795–803
36. Robertson PL, Du Bois M, Bowman PD, Goldstein GW (1985) Angiogenesis in developing rat brain: an in vivo and in vitro study. Brain Res 355(2):219–23
37. Fernandez-Lopez D, Faustino J, Derugin N, Vexler ZS (2012) Acute and chronic vascular responses to experimental focal arterial stroke in the neonate rat. Transl Stroke Res 4(2):179–188
38. Benjelloun N, Renolleau S, Represa A, Ben-Ari Y, Charriaut-Marlangue C (1999) Inflammatory responses in the cerebral cortex after ischemia in the P7 neonatal Rat. Stroke 30(9):1916–23, discussion 23–4
39. Svedin P, Hagberg H, Savman K, Zhu C, Mallard C (2007) Matrix metalloproteinase-9 gene knock-out protects the immature brain after cerebral hypoxia-ischemia. J Neurosci 27(7):1511–8
40. Muramatsu K, Fukuda A, Togari H, Wada Y, Nishino H (1997) Vulnerability to cerebral hypoxic-ischemic insult in neonatal but not in adult rats is in parallel with disruption of the blood–brain barrier. Stroke 28(11):2281–8
41. Chen W, Hartman R, Ayer R, Marcantonio S, Kamper J, Tang J et al (2009) Matrix metalloproteinases inhibition provides neuroprotection against hypoxia-ischemia in the developing brain. J Neurochem 111(3):726–36
42. Emerich DF, Dean RL 3rd, Bartus RT (2002) The role of leukocytes following cerebral ischemia: pathogenic variable or bystander reaction to emerging infarct? Exp Neurol 173(1):168–81
43. Mori E, del Zoppo GJ, Chambers JD, Copeland BR, Arfors KE (1992) Inhibition of polymorphonuclear leukocyte adherence suppresses no-reflow after focal cerebral ischemia in baboons. Stroke 23(5):712–8
44. Amantea D, Nappi G, Bernardi G, Bagetta G, Corasaniti MT (2009) Post-ischemic brain damage: pathophysiology and role of inflammatory mediators. FEBS J 276(1):13–26
45. Gautam N, Olofsson AM, Herwald H, Iversen LF, Lundgren-Akerlund E, Hedqvist P et al (2001) Heparin-binding protein (HBP/CAP37): a missing link in neutrophil-evoked alteration of vascular permeability. Nat Med 7(10):1123–7
46. Afshar-Kharghan V, Thiagarajan P (2006) Leukocyte adhesion and thrombosis. Curr Opin Hematol 13(1):34–9
47. Garcia JH, Liu KF, Yoshida Y, Lian J, Chen S, del Zoppo GJ (1994) Influx of leukocytes and platelets in an evolving brain infarct (Wistar rat). Am J Pathol 144(1):188–99
48. Matsuo Y, Kihara T, Ikeda M, Ninomiya M, Onodera H, Kogure K (1995) Role of neutrophils in radical production during ischemia and reperfusion of the rat brain: effect of neutrophil depletion on extracellular ascorbyl radical formation. J Cereb Blood Flow Metab 15(6): 941–7
49. Gidday JM, Gasche YG, Copin JC, Shah AR, Perez RS, Shapiro SD et al (2005) Leukocyte-derived matrix metalloproteinase-9 mediates blood–brain barrier breakdown and is proinflammatory after transient focal cerebral ischemia. Am J Physiol Heart Circ Physiol 289(2):H558–68
50. Tonai T, Shiba K, Taketani Y, Ohmoto Y, Murata K, Muraguchi M et al (2001) A neutrophil elastase inhibitor (ONO-5046) reduces neurologic damage after spinal cord injury in rats. J Neurochem 78(5):1064–72
51. Tang Y, Xu H, Du X, Lit L, Walker W, Lu A et al (2006) Gene expression in blood changes rapidly in neutrophils and monocytes after ischemic stroke in humans: a microarray study. J Cereb Blood Flow Metab 26:1089–102

52. Akopov SE, Simonian NA, Grigorian GS (1996) Dynamics of polymorphonuclear leukocyte accumulation in acute cerebral infarction and their correlation with brain tissue damage. Stroke 27(10):1739–43
53. Bona E, Andersson AL, Blomgren K, Gilland E, Puka-Sundvall M, Gustafson K et al (1999) Chemokine and inflammatory cell response to hypoxia-ischemia in immature rats. Pediatr Res 45(4 Pt 1):500–9
54. Hudome S, Palmer C, Roberts RL, Mauger D, Housman C, Towfighi J (1997) The role of neutrophils in the production of hypoxic-ischemic brain injury in the neonatal rat. Pediatr Res 41(5):607–16
55. Denker S, Ji S, Lee SY, Dingman A, Derugin N, Wendland M et al (2007) Macrophages are comprised of resident brain microglia not infiltrating peripheral monocytes acutely after neonatal stroke. J Neurochem 100(4):893–904
56. McRae A, Gilland E, Bona E, Hagberg H (1995) Microglia activation after neonatal hypoxic-ischemia. Brain Res Dev Brain Res 84(2):245–52
57. Dingman A, Lee SY, Derugin N, Wendland MF, Vexler ZS (2006) Aminoguanidine inhibits caspase-3 and calpain activation without affecting microglial activation following neonatal transient ischemia. J Neurochem 96:1467–79
58. Denker SP, Ji S, Dingman A, Lee SY, Derugin N, Wendland MF et al (2007) Macrophages are comprised of resident brain microglia not infiltrating peripheral monocytes acutely after neonatal stroke. J Neurochem 100(4):893–904
59. Kreutzberg GW (1996) Microglia: a sensor for pathological events in the CNS. Trends Neurosci 19(8):312–8
60. Monje ML, Toda H, Palmer TD (2003) Inflammatory blockade restores adult hippocampal neurogenesis. Science 302(5651):1760–5
61. Iosif RE, Ekdahl CT, Ahlenius H, Pronk CJ, Bonde S, Kokaia Z et al (2006) Tumor necrosis factor receptor 1 is a negative regulator of progenitor proliferation in adult hippocampal neurogenesis. J Neurosci 26(38):9703–12
62. Fantin A, Vieira JM, Gestri G, Denti L, Schwarz Q, Prykhozhij S et al (2010) Tissue macrophages act as cellular chaperones for vascular anastomosis downstream of VEGF-mediated endothelial tip cell induction. Blood 116(5):829–40
63. Watanabe H, Abe H, Takeuchi S, Tanaka R (2000) Protective effect of microglial conditioning medium on neuronal damage induced by glutamate. Neurosci Lett 289(1):53–6
64. Lu YZ, Lin CH, Cheng FC, Hsueh CM (2005) Molecular mechanisms responsible for microglia-derived protection of Sprague–Dawley rat brain cells during in vitro ischemia. Neurosci Lett 373(2):159–64
65. Lambertsen KL, Clausen BH, Babcock AA, Gregersen R, Fenger C, Nielsen HH et al (2009) Microglia protect neurons against ischemia by synthesis of tumor necrosis factor. J Neurosci 29(5):1319–30
66. Schwartz M, Butovsky O, Bruck W, Hanisch UK (2006) Microglial phenotype: is the commitment reversible? Trends Neurosci 29(2):68–74
67. Butovsky O, Ziv Y, Schwartz A, Landa G, Talpalar AE, Pluchino S et al (2006) Microglia activated by IL-4 or IFN-gamma differentially induce neurogenesis and oligodendrogenesis from adult stem/progenitor cells. Mol Cell Neurosci 31(1):149–60
68. Ziv Y, Ron N, Butovsky O, Landa G, Sudai E, Greenberg N et al (2006) Immune cells contribute to the maintenance of neurogenesis and spatial learning abilities in adulthood. Nat Neurosci 9(2):268–75
69. Petersen MA, Dailey ME (2004) Diverse microglial motility behaviors during clearance of dead cells in hippocampal slices. Glia 46(2):195–206
70. Lalancette-Hebert M, Gowing G, Simard A, Weng YC, Kriz J (2007) Selective ablation of proliferating microglial cells exacerbates ischemic injury in the brain. J Neurosci 27(10): 2596–605
71. Gordon S (2003) Alternative activation of macrophages. Nat Rev Immunol 3(1):23–35
72. Martinez FO, Helming L, Gordon S (2009) Alternative activation of macrophages: an immunologic functional perspective. Annu Rev Immunol 27:451–83

73. Zhao BQ, Wang S, Kim HY, Storrie H, Rosen BR, Mooney DJ et al (2006) Role of matrix metalloproteinases in delayed cortical responses after stroke. Nat Med 12(4):441–5
74. Fullerton HJ, Ditelberg JS, Chen SF, Sarco DP, Chan PH, Epstein CJ et al (1998) Copper/zinc superoxide dismutase transgenic brain accumulates hydrogen peroxide after perinatal hypoxia ischemia. Ann Neurol 44(3):357–64
75. Faustino J, Liu B, Lee SY, Derugin N, Wendland MF, Vexler ZS, (eds) (2009) Blockade of endogenous cytokine-induced neutrophil chemoattractant protein 1 exacerbates injury after neonatal stroke. Stroke meeting
76. Fernandez Lopez D, Faustino J, Daneman R, Zhou L, Lee SY, Derugin N et al (2012) Blood–brain barrier permeability is increased after acute adult stroke but not neonatal stroke. J Neurosci 32(28):9588–600
77. Dingman A, Lee SY, Derugin N, Wendland MF, Vexler ZS (2006) Aminoguanidine inhibits caspase-3 and calpain activation without affecting microglial activation following neonatal transient cerebral ischemia. J Neurochem 96(5):1467–79
78. Fox C, Dingman A, Derugin N, Wendland MF, Manabat C, Ji S et al (2005) Minocycline confers early but transient protection in the immature brain following focal cerebral ischemia-reperfusion. J Cereb Blood Flow Metab 25(9):1138–49
79. Tikka T, Fiebich BL, Goldsteins G, Keinanen R, Koistinaho J (2001) Minocycline, a tetracycline derivative, is neuroprotective against excitotoxicity by inhibiting activation and proliferation of microglia. J Neurosci 21(8):2580–8
80. van den Tweel ER, van Bel F, Kavelaars A, Peeters-Scholte CM, Haumann J, Nijboer CH et al (2005) Long-term neuroprotection with 2-iminobiotin, an inhibitor of neuronal and inducible nitric oxide synthase, after cerebral hypoxia-ischemia in neonatal rats. J Cereb Blood Flow Metab 25(1):67–74
81. Faustino JV, Wang X, Johnson CE, Klibanov A, Derugin N, Wendland MF et al (2011) Microglial cells contribute to endogenous brain defenses after acute neonatal focal stroke. J Neurosci 31(36):12992–3001
82. Nimmerjahn A, Kirchhoff F, Helmchen F (2005) Resting microglial cells are highly dynamic surveillants of brain parenchyma in vivo. Science 308(5726):1314–8
83. Davalos D, Ryu J, Merlini M, Baeten K, Lemoan N, Murray SG, et al. (2012) In vivo imaging of blood–brain barrier disruption in the course of neuroinflammatory disease. Soc Neurosci
84. Silverstein RL, Febbraio M (2009) CD36, a scavenger receptor involved in immunity, metabolism, angiogenesis, and behavior. Sci Signal 2(72):re3
85. Primo L, Ferrandi C, Roca C, Marchio S, di Blasio L, Alessio M et al (2005) Identification of CD36 molecular features required for its in vitro angiostatic activity. Faseb J 19(12):1713–5
86. Park YM, Febbraio M, Silverstein RL (2009) CD36 modulates migration of mouse and human macrophages in response to oxidized LDL and may contribute to macrophage trapping in the arterial intima. J Clin Invest 119(1):136–45
87. Rahaman SO, Lennon DJ, Febbraio M, Podrez EA, Hazen SL, Silverstein RL (2006) A CD36-dependent signaling cascade is necessary for macrophage foam cell formation. Cell Metab 4(3):211–21
88. Silverstein RL, Li W, Park YM, Rahaman SO (2010) Mechanisms of cell signaling by the scavenger receptor CD36: implications in atherosclerosis and thrombosis. Trans Am Clin Climatol Assoc 121:206–20
89. Zhao X, Sun G, Zhang J, Strong R, Song W, Gonzales N et al (2007) Hematoma resolution as a target for intracerebral hemorrhage treatment: role for peroxisome proliferator-activated receptor gamma in microglia/macrophages. Ann Neurol 61(4):352–62
90. Zhao X, Grotta J, Gonzales N, Aronowski J (2009) Hematoma resolution as a therapeutic target: the role of microglia/macrophages. Stroke 40(3 Suppl):S92–4
91. Cho S, Park EM, Febbraio M, Anrather J, Park L, Racchumi G et al (2005) The class B scavenger receptor CD36 mediates free radical production and tissue injury in cerebral ischemia. J Neurosci 25(10):2504–12

92. Cho S, Szeto HH, Kim E, Kim H, Tolhurst AT, Pinto JT (2007) A novel cell-permeable antioxidant peptide, SS31, attenuates ischemic brain injury by down-regulating CD36. J Biol Chem 282(7):4634–42
93. Kunz A, Abe T, Hochrainer K, Shimamura M, Anrather J, Racchumi G et al (2008) Nuclear factor-kappaB activation and postischemic inflammation are suppressed in CD36-null mice after middle cerebral artery occlusion. J Neurosci 28(7):1649–58
94. Kim E, Tolhurst AT, Qin LY, Chen XY, Febbraio M, Cho S (2008) CD36/fatty acid translocase, an inflammatory mediator, is involved in hyperlipidemia-induced exacerbation in ischemic brain injury. J Neurosci 28(18):4661–70
95. Woo MS, Wang X, Faustino J, Derugin N, Wendland MF, Zhou P, et al. (2012) Genetic deletion of CD36 enhances injury after acute neonatal stroke. Ann Neurol 72(6):961–70
96. Caso JR, Pradillo JM, Hurtado O, Lorenzo P, Moro MA, Lizasoain I (2007) Toll-like receptor 4 is involved in brain damage and inflammation after experimental stroke. Circulation 115(12):1599–608
97. Lehnardt S, Lehmann S, Kaul D, Tschimmel K, Hoffmann O, Cho S et al (2007) Toll-like receptor 2 mediates CNS injury in focal cerebral ischemia. J Neuroimmunol 190(1–2):28–33
98. Saito S, Matsuura M, Tominaga K, Kirikae T, Nakano M (2000) Important role of membrane-associated CD14 in the induction of IFN-beta and subsequent nitric oxide production by murine macrophages in response to bacterial lipopolysaccharide. Eur J Biochem 267(1):37–45
99. Lehnardt S, Massillon L, Follett P, Jensen FE, Ratan R, Rosenberg PA et al (2003) Activation of innate immunity in the CNS triggers neurodegeneration through a Toll-like receptor 4-dependent pathway. Proc Natl Acad Sci U S A 100(14):8514–9
100. Kilic U, Kilic E, Matter CM, Bassetti CL, Hermann DM (2008) TLR-4 deficiency protects against focal cerebral ischemia and axotomy-induced neurodegeneration. Neurobiol Dis 31(1):33–40
101. Beschorner R, Schluesener HJ, Gozalan F, Meyermann R, Schwab JM (2002) Infiltrating CD14+ monocytes and expression of CD14 by activated parenchymal microglia/macrophages contribute to the pool of CD14+ cells in ischemic brain lesions. J Neuroimmunol 126(1–2):107–15
102. Abe T, Shimamura M, Jackman K, Kurinami H, Anrather J, Zhou P et al (2010) Key role of CD36 in Toll-like receptor 2 signaling in cerebral ischemia. Stroke 41(5):898–904
103. Ziegler G, Harhausen D, Schepers C, Hoffmann O, Rohr C, Prinz V et al (2007) TLR2 has a detrimental role in mouse transient focal cerebral ischemia. Biochem Biophys Res Commun 359(3):574–9
104. Rolls A, Shechter R, London A, Ziv Y, Ronen A, Levy R et al (2007) Toll-like receptors modulate adult hippocampal neurogenesis. Nat Cell Biol 9(9):1081–8
105. Hua F, Ma J, Ha T, Kelley JL, Kao RL, Schweitzer JB et al (2009) Differential roles of TLR2 and TLR4 in acute focal cerebral ischemia/reperfusion injury in mice. Brain Res 1262:100–8
106. Stewart CR, Stuart LM, Wilkinson K, van Gils JM, Deng J, Halle A et al (2010) CD36 ligands promote sterile inflammation through assembly of a Toll-like receptor 4 and 6 heterodimer. Nat Immunol 11(2):155–61
107. Stridh L, Smith PL, Naylor AS, Wang X, Mallard C (2011) Regulation of Toll-like receptor 1 and −2 in neonatal mice brains after hypoxia-ischemia. J Neuroinflammation 8:45
108. Wang X, Stridh L, Li W, Dean J, Elmgren A, Gan L et al (2009) Lipopolysaccharide sensitizes neonatal hypoxic-ischemic brain injury in a MyD88-dependent manner. J Immunol 183(11):7471–7
109. Du X, Fleiss B, Li H, D'Angelo B, Sun Y, Zhu C et al (2011) Systemic stimulation of TLR2 impairs neonatal mouse brain development. PLoS One 6(5):e19583
110. Bell MJ, Hallenbeck JM (2002) Effects of intrauterine inflammation on developing rat brain. J Neurosci Res 70(4):570–9
111. Rousset CI, Chalon S, Cantagrel S, Bodard S, Andres C, Gressens P et al (2006) Maternal exposure to LPS induces hypomyelination in the internal capsule and programmed cell death in the deep gray matter in newborn rats. Pediatr Res 59(3):428–33

112. Oskvig DB, Elkahloun AG, Johnson KR, Phillips TM, Herkenham M (2012) Maternal immune activation by LPS selectively alters specific gene expression profiles of interneuron migration and oxidative stress in the fetus without triggering a fetal immune response. Brain Behav Immun 26(4):623–34
113. Stolp HB, Johansson PA, Habgood MD, Dziegielewska KM, Saunders NR, Ek CJ (2011) Effects of neonatal systemic inflammation on blood–brain barrier permeability and behaviour in juvenile and adult rats. Cardiovasc Psychiatry Neurol 2011:469046
114. Mallard C (2012) Innate immune regulation by toll-like receptors in the brain. ISRN Neurol 2012:701950
115. Singh AK, Jiang Y (2004) How does peripheral lipopolysaccharide induce gene expression in the brain of rats? Toxicology 201(1–3):197–207
116. Quan N, Whiteside M, Kim L, Herkenham M (1997) Induction of inhibitory factor kappaBalpha mRNA in the central nervous system after peripheral lipopolysaccharide administration: an in situ hybridization histochemistry study in the rat. Proc Natl Acad Sci U S A 94(20):10985–90
117. Verma S, Nakaoke R, Dohgu S, Banks WA (2006) Release of cytokines by brain endothelial cells: a polarized response to lipopolysaccharide. Brain Behav Immun 20(5):449–55
118. Banks WA, Kastin AJ, Durham DA (1989) Bidirectional transport of interleukin-1 alpha across the blood–brain barrier. Brain Res Bull 23(6):433–7
119. Banks WA, Kastin AJ, Gutierrez EG (1994) Penetration of interleukin-6 across the murine blood–brain barrier. Neurosci Lett 179(1–2):53–6
120. Zhang RL, Zhang ZG, Chopp M (2005) Neurogenesis in the adult ischemic brain: generation, migration, survival, and restorative therapy. Neuroscientist 11(5):408–16
121. Beck H, Plate KH (2009) Angiogenesis after cerebral ischemia. Acta Neuropathol 117(5):481–96
122. Wang L, Zhang Z, Wang Y, Zhang R, Chopp M (2004) Treatment of stroke with erythropoietin enhances neurogenesis and angiogenesis and improves neurological function in rats. Stroke 35(7):1732–7
123. Shimamura M, Sato N, Sata M, Kurinami H, Takeuchi D, Wakayama K et al (2007) Delayed postischemic treatment with fluvastatin improved cognitive impairment after stroke in rats. Stroke 38(12):3251–8
124. Xiong Y, Mahmood A, Chopp M (2010) Angiogenesis, neurogenesis and brain recovery of function following injury. Curr Opin Investig Drugs 11(3):298–308
125. Li L, Jiang Q, Zhang L, Ding G, Gang Zhang Z, Li Q et al (2007) Angiogenesis and improved cerebral blood flow in the ischemic boundary area detected by MRI after administration of sildenafil to rats with embolic stroke. Brain Res 1132(1):185–92
126. Ohab JJ, Fleming S, Blesch A, Carmichael ST (2006) A neurovascular niche for neurogenesis after stroke. J Neurosci 26(50):13007–16
127. Battista D, Ferrari CC, Gage FH, Pitossi FJ (2006) Neurogenic niche modulation by activated microglia: transforming growth factor beta increases neurogenesis in the adult dentate gyrus. Eur J Neurosci 23(1):83–93
128. Plane JM, Liu R, Wang TW, Silverstein FS, Parent JM (2004) Neonatal hypoxic-ischemic injury increases forebrain subventricular zone neurogenesis in the mouse. Neurobiol Dis 16(3):585–95
129. Ong J, Plane JM, Parent JM, Silverstein FS (2005) Hypoxic-ischemic injury stimulates subventricular zone proliferation and neurogenesis in the neonatal rat. Pediatr Res 58(3):600–6
130. Yang Z, Covey MV, Bitel CL, Ni L, Jonakait GM, Levison SW (2007) Sustained neocortical neurogenesis after neonatal hypoxic/ischemic injury. Ann Neurol 61(3):199–208
131. Gonzalez FF, McQuillen P, Mu D, Chang Y, Wendland M, Vexler Z et al (2007) Erythropoietin enhances long-term neuroprotection and neurogenesis in neonatal stroke. Dev Neurosci 29(4–5):321–30

Chapter 10
Systemic Immune Responses after Experimental Stroke

Halina Offner and Patricia D. Hurn

Abstract Stroke-induced pathology involves biphasic induction of an early massive peripheral immune cell activation followed by delayed, progressive splenic apoptosis and consequent immunosuppression. In this chapter, we review key immunological aspects of stroke induction and immunosuppression in the context of an emerging concept called "brain–spleen injury cycling." Moreover, we discuss the contribution to stroke severity of immune cells, including T and B cells, regulatory IL-10 secreting B cells and the programmed death receptor/ligand inhibitory pathway. Finally, we propose two novel therapeutic approaches that may limit stroke pathology and ameliorate downstream immunosuppression.

Introduction

It is now increasingly clear that human stroke creates not just a single organ insult, but a complex interaction between two great physiological systems: the CNS and the peripheral immune system. Until recently, the events behind how stroke induces pathology in distant immune organs (e.g., spleen and thymus) has been relatively unstudied. Furthermore, the significance of, and mechanisms underlying, cerebral ischemia-induced immune dysfunction remain poorly understood. However, using animal and cell models, we have observed that systemic immunopathology evolves in tandem with the maturing central cerebral infarct. This chapter summarizes evidence from our work and that of others, characterizing the systemic immune

H. Offner (✉)
Neurology and Anesthesiology and Perioperative Medicine,
Oregon Health and Science University, Portland, OR, USA
e-mail: offnerva@ohsu.edu

P.D. Hurn
The University of Texas System, Austin, TX, USA

J. Chen et al. (eds.), *Immunological Mechanisms and Therapies in Brain Injuries and Stroke*, Springer Series in Translational Stroke Research, DOI 10.1007/978-1-4614-8915-3_10,

response after experimental stroke, the cell players, and their interactions with the injured brain. Several key concepts are discussed. The first is that post-ischemic immunopathology is biphasic in nature. Within hours of brain events, intra-splenic immune cell activation and enormous local cytokine elaboration occur, followed by a delayed, progressive splenic apoptosis and consequent immunosuppression. As activated immunocytes are released from the spleen, they are transported across a dysfunctional blood–brain barrier, fully primed to contribute to cerebral inflammation. We call this phenomenon "brain–spleen injury cycling," a second key concept of the chapter. Next, a variety of immunocytes are discussed as part of brain–spleen cycling, all of which impact outcomes for the injured brain. Lastly, the implications of brain–spleen injury cycling for future immunotherapy are presented.

Significance of Stroke-Induced Immunopathology

Stroke and infection are strongly intertwined. Evolving data suggests that chronic infection may precede acute ischemic stroke and alter stroke risk, presumably through inflammatory processes that lead to progression/destabilization of atherosclerotic plaques. Acute infection can also represent a risk factor for cerebral infarction, and stroke severity is an important predictor of subsequent infection [1, 2]. However, most striking is the vulnerability of patients to infection after a cerebral ischemic event, a vulnerability that arises in part through altered immune function. Animals employed in models of cerebral ischemia and human stroke survivors demonstrate an acute phase response within hours of the ischemic insult, chiefly increased white blood cell count and blood-borne markers of inflammation. In individuals with large strokes, immunodeficiency follows within days of the acute phase response, typically exhibiting lymphopenia and elevated plasma levels of IL-10 and IL-6 [3, 4]. The presence of immune system dysfunction is most ubiquitously recognized in these patients by the enhanced susceptibility to respiratory and urinary tract infections. Such infections account for significant morbidity and mortality in stroke patients. Therefore, it is imperative to understand the causes and mechanisms of post-stroke immunopathology, the subject of this chapter.

Early Work: Characterizing Stroke-Induced Immune System Pathology in Mice

While post-ischemic inflammation within brain has been well studied in models of experimental focal stroke, systemic inflammatory responses have been poorly characterized. To initiate a study of this problem, we quantified changes in cell numbers in the spleen, and mRNA and protein levels for cytokines, chemokines, and chemokine receptors (CCR) in brain, spinal cord, peripheral lymphoid organs (spleen, lymph node, blood, and cultured mononuclear cells from these sources), and blood plasma 6 and 22 h after reversible middle cerebral artery occlusion (MCAO) or sham MCAO in male C57BL/6 mice [5, 6]. Infarction at 22 h was present in all animals, and damage was consistent with previous work in this model. In the post-ischemic brain, there were striking differences in cytokines, chemokines, and chemokine receptor levels. Cortex and striatum ipsilateral to the occlusion demonstrated pronounced increases in expression of inflammatory cytokines (TNF-α, IL-1β, and IL-6) and chemokines (RANTES, IP-10, and MIP-2), as well as noninflammatory factors (TGF-β1, IL-10, and IL-13), but not IFN-γ or *FoxP3* relative to equivalent regions in sham MCAO-treated brains. In most instances, there was already substantial basal expression of chemokine receptors, and MCAO further enhanced expression of only CCR3 and CCR8. After 22 h of reperfusion, tissue ipsilateral to the occlusion showed a similar pattern, but generally lower levels of expression of cytokines and chemokines, with the exception of IL-6 and MIP-2, which were notably increased. However, more widespread changes in expression of chemokine receptors were evident at 22 h, including fivefold additional increase in message of CCR1 and CCR2, and a 40-fold increase in message of CCR5, and lower but significant changes in CCR3, CCR7, and CCR8. These measurements were consistent with what has become well known about post-ischemic cerebral inflammation.

The novel aspect lies in the connection between the ischemic lesion in brain and evolving inflammatory changes in distant peripheral immune cell populations. Mononuclear cells were isolated from various lymphoid organs 6 and 22 h after MCAO or sham MCAO, and cytokines were assessed by CBA and ELISA in supernatants of cultures stimulated for an additional 24 h with plate-bound anti-CD3/CD28 antibodies. The most striking and consistent changes induced in the MCAO mice versus sham-MCAO mice were observed in the spleen. At both the 6 and 22 h time points, activated spleen cells from stroke-injured mice secreted significantly enhanced levels of the inflammatory factors TNF-α, IFN-γ, IL-6, MCP-1, and IL-2 (Fig. 10.1), with increased secretion of the anti-inflammatory factor, IL-10 only at the 22 h time point. Moreover, unstimulated spleen tissue from stroke mice had increased expression of message for MIP-2, CCR2, CCR7, and CCR8 at the 6 h time point, and MIP-2, IP-10, CCR1, and CCR2 at the 22 h time point (not shown). Similar increases in secretion of TNF-α, IL-6, IL-2, and IFN-γ (LN only) were observed only at the 22 h time point in activated lymph node and blood mononuclear cells (Fig. 10.1).

These data demonstrated for the first time that focal cerebral ischemia results in dynamic and widespread activation of inflammatory cytokines, chemokines, and chemokine receptors in the peripheral immune system. A major finding was the rapid and widespread increase in production of inflammatory factors (TNF-α, IL-6, IL-2, MCP-1, and MIP-2) by basal and activated splenocytes that occurred as early as 6 h after stroke, with similar changes occurring later in the spleen as well as in lymph nodes and blood. While there were many similarities in the pattern of expression of splenic versus brain cytokines and chemokines, there were also striking differences. In particular, splenic T cells activated with anti-CD3/CD28 antibodies produced a significant increase in IFN-γ (not observed in brain), but absent levels of IL-1β (observed only in brain). Others have evaluated the effects of MCAO on

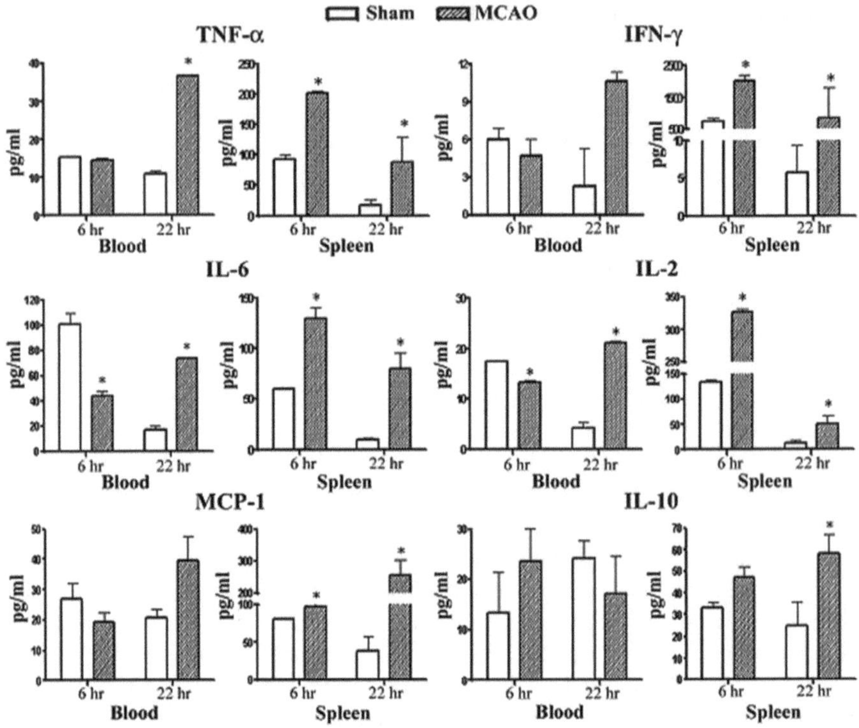

Fig. 10.1 Effects of stroke on cytokines secreted from stimulated splenocytes and blood cells. Spleens and blood were collected 6 and 22 h after vascular occlusion and immune cells were stimulated for 48 h with plate-bound anti-CD3/CD28 antibodies. Supernatants were evaluated for levels of secreted factors, including TNF-α, IFN-γ, IL-6, MCP-1, IL-2, and IL-10. *Asterisk* indicates a significant difference in expression in stroke mice versus sham-treated mice (Previously published, *Journal of Cerebral Blood Flow and Metabolism*, 2006;26:654)

lymphoid tissue, demonstrating extensive loss of lymphocytes in spleen and thymus, a shift from T helper cell (Th1) to Th2 cytokine production and increased lymphocyte apoptosis by 12 h of reperfusion [7, 8]. We observed enormous splenic T-cell cytokine/chemokine production that was readily measured during early reperfusion (minimum 6 h). These observations are significant in that they may play a role in the adaptive immune response to stroke. There is a likelihood that once the brain no longer possesses a functional blood–brain barrier, then exposure of normally cloistered brain structural elements occurs, initiating autoimmune-like processes and cell-mediated immune defenses in the periphery. However, in the absence of previously activated memory T cells, naïve autoimmune responses to newly released brain antigens would require days, rather than hours, to manifest.

Given the distance of the splenic T cells from the evolving infarction and the low level of infiltrating mononuclear cells present in the injured brain [9], it is unlikely that the splenic inflammatory cells measured at 6 h represent emigrating cells from the brain. It is more likely that the early phase of splenic inflammation and the spread of activated lymphocytes to lymph nodes and blood results from sympathetic neural stimulation initiated in response to brain injury.

Splenic Activation Is Followed by Destruction and Suppressed Immune Function

An important early study demonstrated reduced numbers of immune cells and increased percentages of TUNEL$^+$ B cells, T cells, and NK cells in blood, spleen, and thymus in mice after focal cerebral ischemia [10]. The reduced cell numbers accounted for decreased production of IFN-γ, resulting in increased mortality caused by bacteremia and pneumonia. Further work by this group showed that immune cells in peripheral lymphoid organs decrease, accompanied by decreased secretion of TNF-α and IFN-γ and increased susceptibility to spontaneous bacterial infections [8]. We evaluated how MCAO affected spleen cell numbers, morphology, and function through 4 days of recovery [5, 6] and so characterized a second phase of post-stroke immunopathology. The total number of mononuclear cells per spleen fell by 6 h post-MCAO and continued to fall by 22 h post-occlusion (Table 10.1). By 96 h, spleen cell numbers were drastically decreased, as compared to sham-MCAO-treated mice or mice naïve to injury. Furthermore, splenocyte proliferation in response to mitogen stimulation (ConA) was suppressed at 22 h after MCAO. To determine if the reduction in spleen cell numbers and responsiveness was due to cell death, splenocytes were harvested from animals at 22 h post-MCAO and evaluated for the presence of apoptotic markers Annexin V and TUNEL. Annexin V staining was increased in MCAO versus sham-MCAO B lymphocytes and CD4$^+$ T lymphocytes, and a moderate increase in TUNEL$^+$ splenocytes was detected in situ 22 h after MCAO [5, 6].

Furthermore, both spleens and thymi were grossly reduced in volume by 96 h after MCAO [5, 6]. Hematoxylin and Eosin stained spleens from MCAO mice showed a gross loss of tissue in both the white and red pulp, loss of lymphoid isles, and clear apoptosis morphologically, while spleens from sham-MCAO animals

Table 10.1 Spleen Cell Yield (×10^6) from sham and MCAO B6 male mice

Time/status	Sham MCAO	MCAO
6 h	50±14 (4)*	46±4 (3)*
22 h	35±5 (15)*	16±3 (16)**, *
96 h	80±10 (2)	6±3 (4)**, *
Naïve mice: 91±8 (4)		

Reprinted from *Neuroscience*, 158(3), Offner H, Vandenbark AA, Hurn PD. Effect of experimental stroke on peripheral immunity: CNS ischemia induces profound immunosuppression, 1098–1111, 2009, with permission from Elsevier

*$p<0.001$ versus Naïve mice (number of mice)

**$p<0.002$ versus sham-MCAO mice

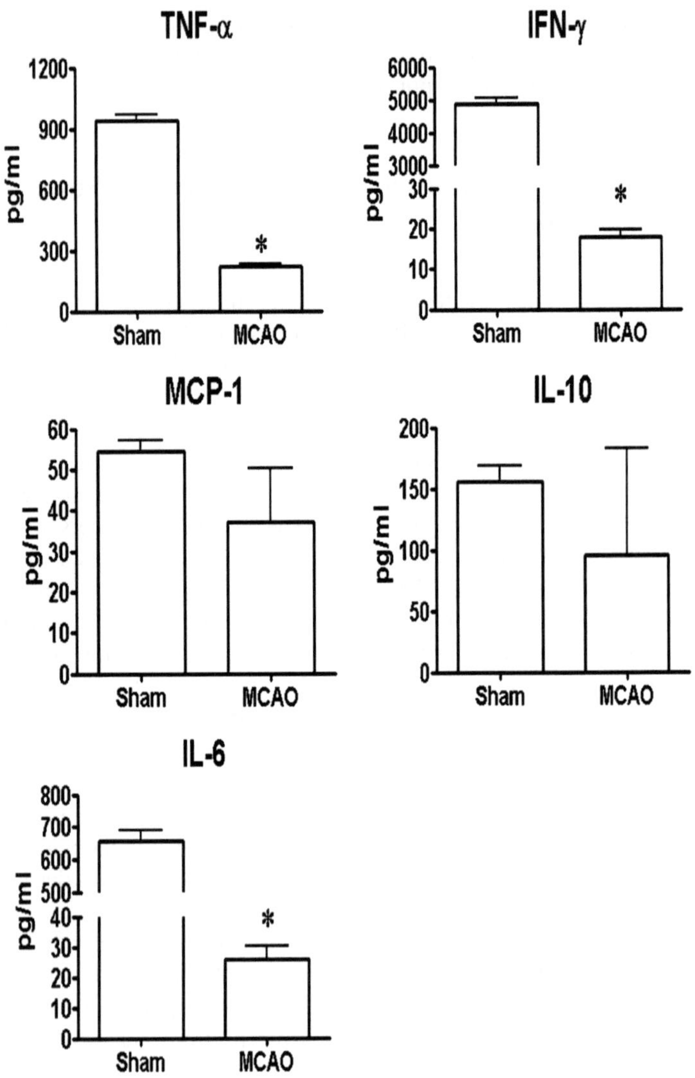

Fig. 10.2 Effects of stroke on cytokines secreted from stimulated splenocytes. Spleens were collected 96 h after vascular occlusion and immune cells were stimulated for 48 h with plate-bound anti-CD3/CD28 antibodies. Supernatants were evaluated for levels of secreted factors, including TNF-α, IFN-γ, IL-6, MCP-1, and IL-10. *Asterisk* indicates a significant difference in expression in MCAO versus sham-MCAO mice (Previously published, *The Journal of Immunology*, 176:6523–6531, 2006. Copyright 2006. The American Association of Immunologists, Inc.)

appeared grossly normal. Total spleen size and cell number (Table 10.1) and T-cell response to both ConA and anti-CD3 mAb were all strongly decreased [5, 6]. The 96 h MCAO spleens had strongly reduced secretion and message levels for inflammatory cytokines, TNF-α, IFN-γ, and IL-6 (Figs. 10.2 and 10.3). These changes were reflected by a striking increase in TUNEL$^+$ cells and in propidium iodide positive

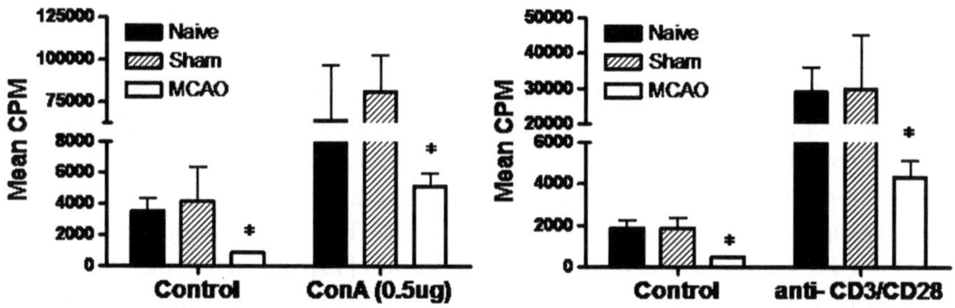

Fig. 10.3 Strongly reduced T-cell proliferation 96 h after MCAO. Splenocytes were obtained from naïve, sham MCAO, and MCAO mice 96 h after stroke and evaluated for proliferation responses 3 days after stimulation with 0.5 μg ConA, anti-CD3 mAb or medium. *Asterisk* indicates significant reduction in response compared to naïve or sham-MCAO mice (Previously published, *The Journal of Immunology*, 176:6523–6531, 2006. Copyright 2006. The American Association of Immunologists, Inc.)

(PI^+) splenocytes. Curiously, despite near total loss of splenocytes after MCAO, there were no changes in the percentage composition of T or B lymphocytes, or macrophages. However, splenocytes from 96 h MCAO mice had an increase in *FoxP3* mRNA, and a threefold increase in $CD4^+FoxP3^+$ regulatory T lymphocytes (Tregs) as compared to naïve or sham-MCAO mice. In blood, there was a twofold decrease in white cell counts/ml, but a dramatic increase in the percentage of $CD11b^+VLA\text{-}4^-$ macrophages in MCAO (66 %) versus sham MCAO (9 %). This suggests that by 96 h, the VLA-4 *positive* T and B cells have likely migrated to the brain or other tissues, causing an increase in circulating macrophages. These results strongly indicate a general loss of B and T cells and macrophage/dendritic cells from the spleen, while Treg cells were relatively enriched, and appearance of macrophages/DC is readily detectable in the blood. We interpreted the cause of suppressed responsiveness to anti-CD3 and ConA by splenic T cells to be linked to the combination of reduced T-cell numbers and increased Treg activity.

Accordingly, cell death represented only one of the processes that contributed to post-MCAO splenic atrophy and loss of T-cell responsiveness to mitogen stimulation. The relatively selective reduction in B cells in spleen and blood, the pronounced increase in splenic $CD4^+$ Treg cells, and the increased presence of circulating monocytes/macrophages that all occur by 96 h after MCAO are also important, and in the following ways. First, B cells constitute about 60 % of splenic and blood mononuclear cells in mice. By 96 h after stroke, the percentage of B cells was reduced by about half to about 30 % of the remaining splenic and blood mononuclear cells. When translated to total cells, this represents a reduction from 45 million to only 2.5 million B cells per spleen, and a >80 % reduction in the number of B cells/ml of blood. This degree of B-cell loss just 4 days after stroke would undoubtedly compromise the ability of the humoral immune system to provide protection against invading microorganisms.

Second, $CD4^+CD25^+$ Tregs are considered to be "master regulators" of the immune system. Initial descriptions indicated that Treg cells expressed very high levels of the IL-2 receptor, CD25, and consequently these cells are often referred to as $CD4^+CD25^+$ or $CD4^+CD25^{bright}$. In normal mice, Tregs limit inflammation and inhibit autoimmune diseases [11–13]. The forkhead/winged helix transcription factor gene, *FoxP3*, is strongly linked to the regulatory function of $CD4^+CD25^+$ Tregs [14–16] and has become a useful intracellular marker for their identification. Although a normal complement of Tregs specific for self tissue determinants may maintain self-tolerance [17], it is now appreciated that an overabundance of Tregs may impede immunosurveillance against autologous tumor cells [18] and may suppress the ability of $CD4^+CD25^-$ effector T cells to eliminate parasites [19]. Taken together, these findings document the importance of the $CD4^+CD25^+$ Tregs subpopulation in regulating autoreactive as well as protective T-effector cells in vivo.

Third, the pronounced increase in the percentage of $CD11b^+$ macrophages/monocytes in blood may also be important. These circulating cells are clearly viable and do not express the VLA-4 trafficking marker that would otherwise permit these cells to infiltrate into the tissues, including the brain. Certain subtypes of macrophages and dendritic cells (that are also $CD11b^+$) can potentiate activation of Treg cells and reduce the activation of T-effector cells [20].

The underlying process that results in widespread immunosuppression and concurrent systemic infections after stroke induction in animals or humans is not well understood. However, it is conceivable that sympathetic signaling to the spleen and thymus after MCAO hyperstimulates the activation of these immune organs, the local elaboration of massive amounts of cytokines/chemokines, and the precipitation of further dangerous immunopathological sequelae. These sequelae include splenic apoptosis, selective splenocyte depletion (i.e., a relative overabundance of Tregs could inhibit protective immune cells, including $CD4^+$ and $CD8^+$ T cells, B cells, and natural killer cells), and consequent abnormal immune function.

Brain Antigens May Impact Post-stroke Immune Activation

A key assumption in studies of post-stroke immune activation is that the cellular adaptive immune response must be triggered through an encounter with brain-derived antigens, either in soluble form or as presented by macrophages or dendritic cells. Under ordinary circumstances, the CNS is uniquely "immune privileged" in that brain is isolated from the immune system by a functional blood–brain barrier. Following injury, antigen-presenting cells have long been thought to leave the brain and to be transported by the blood or cerebrospinal fluid to lymphoid tissue, including the cervical lymph nodes. Consistent with this maxim is the observation that brain-derived antigen immunoreactivities, to the NMDA receptor subunit NR-2A and to myelin, are present in palatine tonsils and cervical lymph nodes of patients after acute stroke [21]. This finding suggests that transfer of neural antigen to lymphoid tissue is an important mechanism of immune system activation and control.

Leakage of brain autoantigens such as myelin basic protein or myelin oligodendrocyte glycoprotein (MOG) to the periphery results in a variety of consequences for the immune system and for the injured brain. We have evaluated the functional consequences of this autoimmune response in murine experimental stroke, using an adoptive transfer process and MOG reactive cells [22]. Transferring MOG-reactive splenocytes which secrete toxic Th1 cytokines (e.g., IFN-γ and TNF-α) into severe combined immunodeficient (SCID) mice produced significant exacerbation of infarct volume and neurological deficits by 96 h post-MCAO. Furthermore, animals receiving MOG-stimulated splenocytes exhibited a higher percentage of immune cells in the ischemic hemisphere than did control mice. These data demonstrate clearly that MOG-stimulated splenocytes traffic into injured brain in a selective manner and emphasize the role of CNS antigen in activation of peripheral immune cells.

Brain–Spleen–Brain Injury Cycling Hypothesis

This previous work has directed our working hypothesis that evolving cerebral ischemic injury elicits a cycle of injury from brain-to-spleen-to-brain (Fig. 10.4). We suggest that selected cell sources and mechanisms of deleterious brain–spleen–brain pathobiology contribute in different ways to CNS outcomes and immune capacity. As shown in Fig. 10.4, the evolving brain injury "signals" for splenic activation which leads to apoptosis and drastic loss of immune cells. The activated spleen also releases cells into the blood, followed by trafficking across the microvasculature replete with inflammatory display of adhesion molecules and chemokines. Cells released from the spleen and engaged in trafficking into brain may also be temporally specific, different in intensity, and controlled by different mechanisms as discussed below.

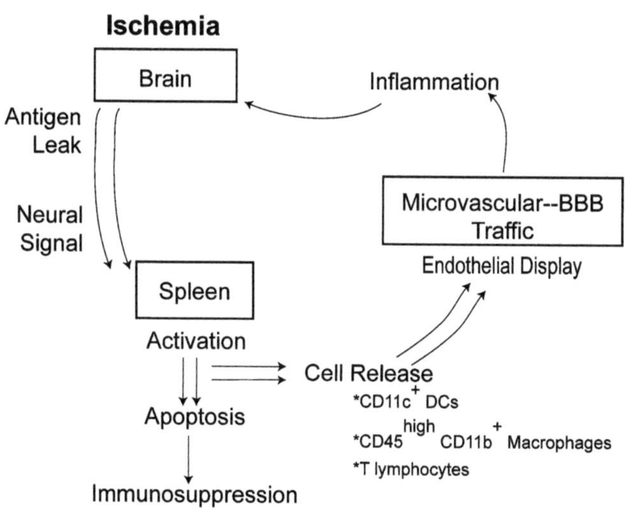

Fig. 10.4 Concept of Brain–spleen–brain injury cycling

The Adaptive Immune System Contributes to Stroke Lesion Development

We hypothesized that activated peripheral T and B lymphocytes would home to injured brain and alter the trajectory of the evolving infarct. To test this directly, we employed C.B-17$^{scid/scid}$ mice that lack T and B lymphocytes with a consequent loss of all immune functions that require these cells. The immunodeficiency in SCID mice results from a mutation on chromosome 16, which causes deficient recombinase activity required for immunoglobulin and T-cell receptor gene rearrangement [23]. Despite the lack of T and B cells, SCID mice have normal functioning macrophages, dendritic cells, and Natural killer cells (but not NK-T cells that have a rearranged T-cell receptor (TCR)), as well as neutrophils. We compared early (22 h) histological outcomes in this strain versus the WT background strain, C57BL/6, as a means of determining the contribution of T and B cells to focal ischemia, while minimizing the potential vulnerability of SCID mice to postoperative immune challenges that would confound stroke outcome. Intra-ischemic physiological parameters, anesthesia requirement and residual cortical perfusion as assessed by LDF were not different in SCID versus WT mice, suggesting that the insult was comparable.

We found that both cortical and total hemispheric infarct volumes were strikingly reduced at 22 h after MCAO by ~40 % in male SCID ($n=10$) versus C57BL/6 control ($n=9$) mice ($p<0.01$) [24]. Striatal infarction was not altered in the SCID mice, suggesting that the core of the evolving infarction was not protected by the lack of T and B cells. One explanation may be that vascular collateralization is more pronounced in cortex, offering an effective route for T-cell entry into brain during reperfusion. In accordance with our previous study [5, 6], WT mice had increased message levels in post-ischemic brain for TNF-α, IL-1β, IL-6, IL-10, IP-10, CCR1, CCR2, CCR3, and CCR5 as a relative expression of MCAO versus sham WT brains. In contrast, expression levels for most lymphocyte-derived cytokines, chemokines, and chemokine receptors in SCID mouse brains were greatly reduced in the ischemic hemisphere, particularly IL-10, CCR2, and CCR3. Only mRNA expression for IL-1β was significantly elevated by ischemia in SCID brains, likely the product of resident or infiltrating macrophages or CNS parenchymal cells rather than T or B cells.

As discussed above, focal cerebral ischemia in immunologically intact C57BL/6 mice resulted in a marked reduction in spleen cell numbers when evaluated at 22 h post-ischemia (16 ± 3 million in MCAO versus 35 ± 5 million in sham versus 91 ± 8 million in Naïve mice [5, 6]). Naïve SCID mice had an even greater reduction in splenocyte numbers (2.7 ± 1.4 million cells) as would be expected in the absence of T and B cells. Sham-treated SCID mice had only 1.8 ± 0.9 million cells per spleen, and MCAO treatment further reduced the cell number to 1.2 ± 0.6 million per spleen ($p<0.05$). It is noteworthy that the remaining cell population in spleen after MCAO in both WT and SCID mice included a striking increase in CD11b^{+}VLA-4^{-} macrophages. In contrast with WT C57BL/6 mice in which MCAO (22 h) induced marked increases in splenic mRNA expression for IFN-γ, TNF-α, IL-6, MIP-2, IP-10, CCR1, and CCR2, MCAO in SCID mice induced increases only in splenic IFN-γ and MIP-2 message and decreases in IP-10 and CCR5 message. The increased expression of IFN-γ and MIP-2 in SCID splenocytes after MCAO may derive from the enriched population of $CD11b^{+}VLA\text{-}4^{-}$ macrophages or other residual splenocyte populations that are mainly natural killer cells, monocytes, macrophages, and dendritic cells.

This study demonstrated three important findings. First, loss of T and B lymphocytes through a genetic mutation in SCID mice resulted in significant improvement in early ischemic histological damage, thus indicating that lymphocytes are strongly involved in the size of the evolving infarct, and are a significant source of selected inflammatory mediators in brain. The target region is cortex, presumably in penumbral areas rather than in the core of the infarct. Second, when T and B cells are absent, post-ischemic induction of inflammatory mediators in brain was largely suppressed within the window of our observations. Only IL-1β was elevated in ischemic SCID brain relative to sham-operated mice. Third, post-stroke loss of splenocytes was blunted but not completely ablated in SCID mice. Further, post-stroke expression of intra-splenic cytokines/chemokines was also blocked with the exception of IFN-γ and MIP-2. Our data are consistent with a previous report that lymphocyte-deficient $Rag1^{-/-}$ mice sustain smaller infarct volumes and improved neurological deficits after MCAO [25]. Importantly, $CD4^{+}$ and $CD8^{+}$ T lymphocytes contributed largely to post-ischemic intravascular inflammatory and prothrombotic responses in cerebral venules.

The early (24 h) appearance of T-cell infiltration into brain after MCAO [26, 27] may indicate that recruitment of activated cells is antigen-nonspecific, perhaps generated by sympathetic signaling from brain to the periphery as discussed above. Alternatively, leakage of brain antigens across a compromised blood–brain barrier could initiate a peripheral immune response. Activated T cells have the capacity to infiltrate brain and could contribute to expansion of the ischemic penumbra, an area that already contains infiltrating neutrophils after 24 h. The functionality of this response has received much interest because of reports that tolerance to brain antigens can be induced with beneficial effects on stroke severity [28]. The context of lymphocyte activation is likely important, as pro-inflammatory $CD4^{+}CD28^{-}$ lymphocyte subsets in blood are well recognized in clinical ischemic stroke, and rising $CD4^{+}CD28^{-}$ counts are associated with increased risk of stroke recurrence and death [29]. In summary, our study quantified the relative contribution of T and B lymphocytes to production of inflammatory mediators in the context of a developing infarct and emphasized that spleen-derived immunocytes are a potential target for therapeutic intervention.

B-Cell Deficiency Exacerbates Stroke Outcomes and Alters Cerebral Inflammatory Cell Invasion

As discussed above, MCAO triggers early signaling from the ischemic brain to spleen, resulting in a massive production of inflammatory factors and transmigration of splenocytes to the circulation and brain. Whereas inflammatory cells from the periphery have now been shown to contribute to CNS damage and cell death, other regulatory immune cells can reduce inflammation and limit damage within the brain. A major conundrum in the immunology of stroke is how to enhance the early immunoregulation that limits CNS inflammation while preventing excessive systemic suppression. To do this in a strategic manner requires a full understanding of the involved inflammatory and regulatory immune pathways. To this end, we evaluated the ability of regulatory B cells from the peripheral immune system to exert immunosuppressive effects, diminish stroke lesion size, and protect from neurological damage [30]. For this comparison, we induced MCAO in B-cell-deficient $\mu MT^{-/-}$ mice versus WT C57BL/6 mice.

Our studies demonstrated conclusively that the B-cell-deficient mice sustained significantly larger total hemispheric infarcts at 48 h ($p < 0.05$) and a worsened functional outcome at 24 h ($p = 0.02$) and 48 h ($p = 0.002$) after reperfusion, indicating that the presence of functional B cells limited stroke severity. Leukocytes are major effectors of inflammatory damage after experimental brain ischemia [31, 32]. To determine if the lack of B cells altered leukocyte composition in brain after MCAO, numbers of infiltrating $Gr1^+$ neutrophils, $CD3^+$ T cells, $CD11b^+CD45^{low}$ microglia, and $CD11b^+CD45^{high}$ macrophages were evaluated by flow cytometry. After 48 h reperfusion, accumulation of all of these leukocyte subtypes was significantly greater in the affected hemisphere of MCAO-treated $\mu MT^{-/-}$ mice as compared to MCAO-treated WT mice. Lack of B cells in $\mu MT^{-/-}$ mice further permitted significant increases in the absolute number of IFN-γ- and TNF-α-secreting $CD3^+$ T cells and MHC class II^+ and TNF-α-secreting microglia and macrophages in the lesioned ipsilateral hemisphere of MCAO mice at 48 h reperfusion. In addition to the cell types mentioned above, we detected ~7,000 $CD19^+$ B cells per hemisphere in both naïve and sham-treated WT mouse brains and modest but significant increases to ~10,000 B cells in the non-ischemic hemisphere and 12,000 B cells in the ischemic hemisphere of WT mice 48 h after MCAO.

Regulatory Effects of IL-10-Secreting B Cells on MCAO

To specifically implicate B cells as a key protective cell type, $CD19^+$ B cells were obtained and enriched to 99 % purity by negative selection from splenocytes of transgenic green fluorescent (GFP^+) mice, and 50 million GFP^+CD19^+ B cells were injected i.p. into $\mu MT^{-/-}$ B-cell-deficient mice 1 day prior to MCAO. The B-cell-deficient animals that received adoptively transferred GFP^+CD19^+ B cells had reduced infarct volumes ($p > 0.05$) after 48 h reperfusion compared to no cell transfer (PBS) controls (Fig. 10.5, top panel), as well as a lower mortality rate. Consistent with smaller infarction size, neurological outcome scores were also improved in B-cell-restored $\mu MT^{-/-}$ mice with stroke after 48 h reperfusion compared to no cell (PBS) transferred control mice. These findings clearly demonstrate that WT $CD19^+$ B cells can restore improved ischemic outcomes in B-cell-deficient $\mu MT^{-/-}$ mice.

Because of the significant B-cell-dependent activity in limiting stroke infarct size and functional outcome demonstrated above, we hypothesized that the protective

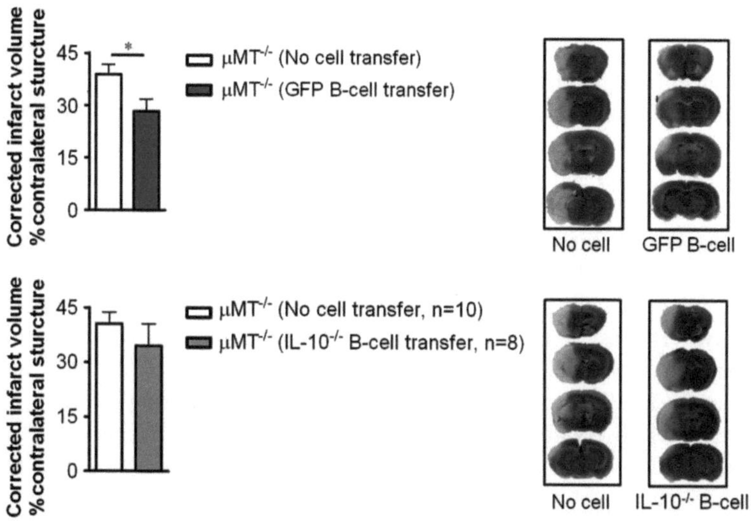

Fig. 10.5 Transfer of IL-10$^{-/-}$ CD19^{+} B cells does not alter infarct volume or improve behavioral outcome of μMT$^{-/-}$ micecompared to transferred GFP B cells that reduced infarct size (*top*), transfer of 50 million CD19^{+} B cells from IL-10$^{-/-}$ mice had no effect on infarct volume of total hemisphere (Mean ± SEM) at 48 h reperfusion after 60 min MCAO. Statistical analysis was performed using the Student's *t*-test. There was no significant difference of infarct volumes between no cell (PBS) transferred (n=9) and IL-10$^{-/-}$ B-cell (n=8)-transferred μMT$^{-/-}$ mice. (*Right*) Representative TTC-stained cerebral sections of the MCAO modeled to analyze infarct volume. Localization of the ischemic lesion did not differ between no cell and IL-10$^{-/-}$ B-cell-transferred μMT$^{-/-}$ mice. Transfer of IL-10$^{-/-}$ B cells did not significantly improve functional outcome after 24 or 48 h reperfusion (not shown). The statistical analysis was performed using the Mann–Whitney *U*-test (Previously published, *The Journal of Neuroscience*, 2011;31(23):5886–8563)

actions of CD19^{+} B cells might be linked to IL-10 production, a major regulatory cytokine known to be produced by both B cells and T cells. Thus, intracellular staining of IL-10 was carried out in CD19^{+} B cells and CD3^{+} T cells harvested from immune organs after MCAO and stimulated ex vivo with LPS, PMA, and ionomycin. We found an increased percentage of IL-10-secreting CD19^{+} (CD3 negative) B cells in blood but not in spleen or lymph nodes in WT mice but not μMT$^{-/-}$ mice after 60 MCAO and 48 h reperfusion. These data demonstrated enhanced availability of B cells with potential to limit ischemic and neurological outcomes after MCAO through secretion of IL-10.

To specifically address the mechanism of B cells as the protective cell type producing IL-10, highly enriched populations of B cells were transferred from IL-10$^{-/-}$ donors to B-cell-deficient recipient mice prior to MCAO. As shown in Fig. 10.5 (bottom panel), the B-cell-deficient animals that received adoptively transferred IL-10$^{-/-}$ B cells did not exhibit significantly reduced infarct volumes after 60 min MCAO followed by 48 h reperfusion compared to no cell transfer (PBS) controls. Moreover, there were no differences in mortality rates or neurological outcome scores after 24 or 48 h reperfusion between PBS and IL-10$^{-/-}$ B-cell transfer groups.

Taken together, these data clearly demonstrate that WT $CD19^+$ B cells can restore improved ischemic outcomes that were shown to be lacking in B-cell-deficient $\mu MT^{-/-}$ mice through the secretion of IL-10.

To further evaluate possible regulatory effects of B cells on T-cell cytokine production during MCAO, inflammatory factors were quantified in blood and spleens after 60 min MCAO and 48 h reperfusion in WT mice, WT B-cell-restored $\mu MT^{-/-}$ mice, and IL-$10^{-/-}$ B-cell-restored $\mu MT^{-/-}$ mice. We found that the percentages of both IFN-γ and TNF-α-secreting $CD3^+$ T cells were significantly increased in blood and spleen from B-cell-deficient versus WT mice, with a reduction to WT levels of these peripheral T cells in $\mu MT^{-/-}$ mice after restoration with WT B cells, but not with IL-$10^{-/-}$ B cells. Thus, WT B cells with the potential for IL-10 secretion limited both inflammatory cytokine production of peripheral T cells and infiltration of inflammatory T cells into the MCAO-affected hemisphere during MCAO.

These novel observations demonstrated the previously unrecognized activity of WT B cells to limit infarct volume and functional neurological deficits as well as to inhibit activation and recruitment of inflammatory T cells, macrophages, and microglia into the growing CNS infarct after experimental stroke in mice. Regulatory activities were not only significantly decreased in MCAO-treated B-cell-deficient $\mu MT^{-/-}$ mice but also were fully restored after passive transfer of WT B cells, thus implicating unequivocally the protective activity of regulatory B cells. These regulatory functions were associated with increased percentages of IL-10-secreting $CD19^+$ B cells in blood, but not IL-10-secreting T cells, including Treg cells that have received much previous attention as possible immune regulators in stroke [33], suggesting that the Treg protective effects are likely mediated through alternative mechanisms.

A key function of B cells is their secretion of IL-10, an anti-inflammatory cytokine that has been studied extensively in stroke [34]. IL-10-deficient mice developed larger infarcts after permanent focal ischemia [35], whereas exogenous administration of IL-10 by multiple routes reduced infarct volumes after MCAO (discussed in [30]). In the clinic, early worsening of stroke was associated with lower IL-10 plasma levels in patients with subcortical infarcts or lacunar stroke, but not in patients with cortical lesions [36]. Conversely, excessive levels of IL-10 may predispose to increased infections [37]. Taken together, these findings suggest that local secretion of IL-10 by circulating or CNS-infiltrating B cells may be preferable to systemic delivery. Recent studies have identified a subpopulation of $CD1d^{high}CD5^+CD19^+$ "regulatory B cells," and we also investigated changes in these $CD1d^{high}CD5^+$ Breg cells in stroke. We observed that the IL-10 secreting population was not restricted to either the $CD1d^{high}$ or the $CD5^+$ population post-stroke. We thus concluded that IL-10 secretion is more specific than the $CD1d^{high}CD5^+19^+$ B-cell subset markers for identification of Breg cells in stroke.

In conclusion, our study provides new insights into the endogenous inflammatory response after acute brain ischemia. Specifically, we have described a previously unknown role for B cells as cerebroprotective immunomodulators after stroke, a function that affects diverse cytokine-dependent and cellular inflammatory targets through the anti-inflammatory effects of IL-10.

Regulatory Effects of the Programmed Death (PD)-1/PD-Ligand Co-inhibitory Pathway in MCAO Involves PD-L$^+$ B Cells That Inhibit PD-1$^+$ Inflammatory Macrophages, Microglia, and T Cells

Our earlier studies demonstrating profound loss of immunocytes in the spleen and thymus after MCAO, accompanied by dramatic upregulation of apoptotic markers [5, 6], suggested stroke-induced aberrations in the PD-1/PD-L co-inhibitory pathway. PD-1 (CD279) is an Ig-superfamily member containing an immunoreceptor tyrosine-based inhibitory motif [38] and an immunoreceptor tyrosine-based switch motif that are inducibly expressed by activated T cells, B cells, natural killer cells, monocytes, and some dendritic cell subsets [39, 40]. Binding of PD-1 to either of two ligands, PD-L1 or PD-L2 with overlapping expression patterns, induces inhibitory signals that control induction and maintenance of peripheral T-cell tolerance and immune homeostasis [41, 42]. To assess the role of PD-1 in stroke development, infarct volume, neurological outcome, and infiltration of inflammatory cells into brain were evaluated in cohorts of PD-1-deficient versus C57BL/6 WT mice treated with 60 min of focal cerebral ischemia and 96 h of reperfusion. Loss of PD-1 resulted in significantly larger hemispheric infarct volumes in cortex ($p=0.01$), striatum ($p=0.01$), and total hemisphere ($p=0.0001$) relative to WT mice (Fig. 10.6). In a further cohort, there was significant recovery of neurological scores in WT mice that did not occur in PD-1KO mice. These data clearly implicate the role of

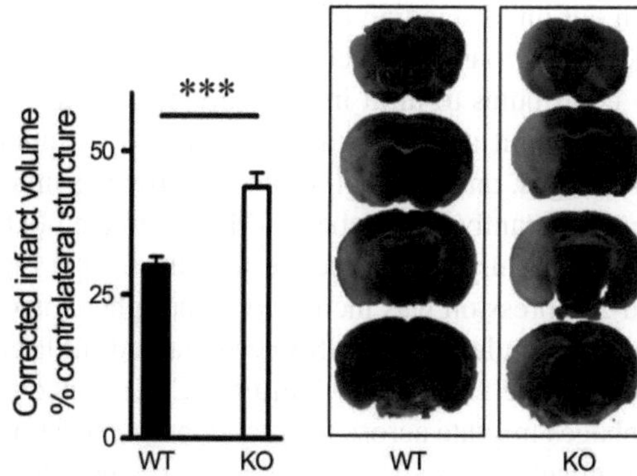

Fig. 10.6 Deficiency of PD-1 exacerbates ischemic infarct volume and worsens behavioral recovery after MCAO. (*Left*) Infarct volumes in the hemisphere, corrected for the presence of edema, as significantly increased (***$p=0.0001$) in PD-1 KO ($n=9$) versus WT ($n=9$) mice after 60 min MCAO and 96 h reperfusion. Values represent mean ± SEM. (*Right*) Representative 2,3,5-triphenyltetrazolium chloride-stained cerebral sections of the MCAO modeled to analyze infarct volume. Neurological dysfunction scores (mean ± SEM) after reperfusion were significantly worsened in PD-1 KO ($n=25$) versus WT ($n=26$) mice (not shown). *KO* knockout, *WT* wild-type (Previously published, *Stroke*, 2011;42:2578–2583)

PD-1 in limiting functional and histological damage after MCAO in both cortex and striatum. This range of effects mediated through PD-1 is broader than with regulatory IL-10-secreting B cells that affect infarct size in only the cortex, suggesting that PD-1 protective mechanisms may affect brain parenchymal cells in addition to infiltrating immunocytes.

As discussed above, leukocytes are major effectors of inflammatory damage after experimental brain ischemia. To determine if the loss of PD-1 altered leukocyte composition, cytokine production and activation in brain after MCAO, numbers of infiltrating $CD3^+$ T cells, $Gr1^+$ neutrophils, $CD11b^+CD45^{low}$ microglia, and $CD11b^+CD45^{high}$ macrophages were evaluated by flow cytometry. After 96 h reperfusion, accumulation of all of these leukocyte subtypes was significantly greater and there were three- to five-fold increases in the absolute numbers of IFN-γ- and TNF-α-secreting $CD3^+$ T cells in the ischemic versus non-ischemic hemisphere of MCAO-treated PD-1-KO mice as compared to MCAO-treated WT mice. Moreover, the percentages of $CD3^+$ T cells secreting TNF-α and IFN-γ were significantly increased in both blood and spleen. PD-1 expression was strongly upregulated on both microglia and macrophages within the ischemic CNS lesion after MCAO in WT mice and obviously could not be expressed similarly in PD-1-KO mice. The effect of the loss of PD-1 was to permit significant two- to three-fold increases in the absolute numbers of MHC class II^+ and TNF-α-secreting microglia and macrophages in the ischemic hemisphere after activation ex vivo. Of note, there was approximately two- to three-fold more infiltrating macrophages present in the ischemic hemisphere than T cells or microglia after MCAO. These results clearly demonstrate enhanced infiltration of inflammatory cells into the affected CNS in PD-1-deficient mice after MCAO.

This study demonstrated two important and novel findings which have potentially high impact in our understanding of immunological mechanisms of ischemic brain injury. First, a previously unrecognized activity of the PD-1/PD-L co-inhibitory pathway contributes to limit infarct volume and functional neurological deficits, as well as to inhibit activation and recruitment of inflammatory T cells, granulocytes, macrophages, and microglia into the growing infarct. These regulatory activities were significantly diminished in MCAO-treated PD-1-deficient mice, thus implicating unequivocally the protective activity of this regulatory pathway. Second, PD-L1 and 2 expression was increased on peripheral and CNS B cells and PD-1 expression was upregulated on CNS microglia and infiltrating macrophages within the lesioned brain hemisphere 96 h after MCAO. Since peripheral B cells, T cells, and macrophages migrate across the blood–brain barrier to contribute to the ischemic injury, our novel results suggest a previously undescribed regulatory circuit in which $PD\text{-}L1/2^+$, IL-10-secreting B cells may directly inhibit T cells and regulate activation and release of neurotoxic factors by $PD\text{-}1^+$ microglia and macrophages. These putative interactions are illustrated in Fig. 10.7. These findings implicate the PD-1/PD-L immunoregulatory pathway as a novel target for protection from CNS damage in experimental stroke.

The cytoplasmic domain of PD-1 contains an immunoreceptor tyrosine-based switch motif (ITSM) sequence, and it was later demonstrated that the tyrosine within the ITSM motif is essential for binding the protein tyrosine phosphatases

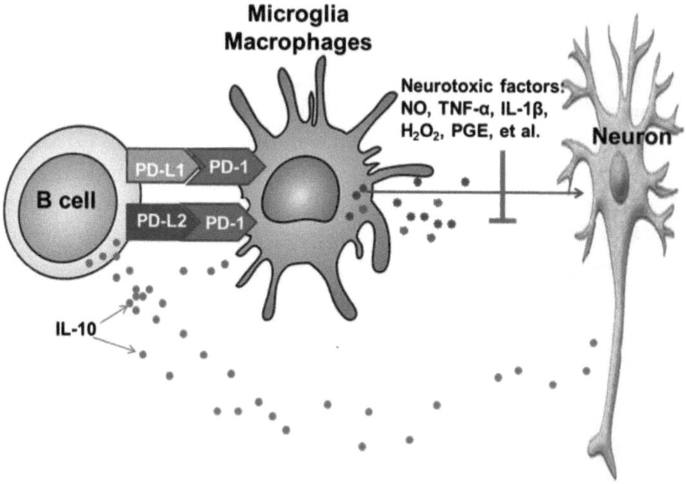

Fig. 10.7 B-cell regulation of microglial activation may occur both through release of IL-10 and the PD-1/PD-L co-inhibitory pathway. MCAO causes increased secretion of IL-10 and enhanced expression of PD-L1 and PD-L2 by peripheral B cells. When these B cells are attracted to the growing infarct, they cross the blood–brain barrier and inhibit activation (Previously published, *Stroke*, 2011;42:2578–2583)

SHP-1 and SHP-2 [42] that mediate inhibitory PD-1 function. In T cells, inhibition of activation requires co-ligation of the TCR [43]. In macrophages (and presumably microglia), PD-1 can be upregulated and cell activation inhibited by an interferon-sensitive response element (ISRE) [44], Toll-like receptor (TLR)-2, TLR3 and TLR4, and other agents [45]. This is of particular importance in stroke, where microglia activation and release of CNS neurotoxic factors is known to occur through ligation of TLR2 and TLR4 [46–48], possibly by heat-shock protein 60 released from CNS cells undergoing necrotic or apoptotic cell death [49]. Thus, ligation of PD-1 expressed on macrophages and microglia by PD-L expressed by regulatory B cells could result in the inhibition of a key neurodestructive process in stroke.

Potential New Treatment Options for Stroke Include Estrogen (E2), G1, and Recombinant T-Cell Receptor Ligands

Reduced risk and severity of stroke in adult females is thought to depend on normal endogenous levels of estrogen (E2+17β-estradiol), a well-known neuroprotectant and immunomodulator currently in clinical trials in multiple sclerosis. In male mice, experimental stroke induces immunosuppression of the peripheral immune system, characterized by a reduction in spleen size and cell numbers and decreased cytokine and chemokine expression. However, stroke-induced immunosuppression has not been well studied in female mice. To test the hypothesis that estradiol deficiency exacerbates immunosuppression after focal stroke in females, we evaluated the effect of MCAO on infarct size and peripheral and central nervous system (CNS) immune responses in ovariectomized female mice with or without add-back of sustained, controlled levels of (1) 17-β-estradiol (E2) administered by subcutaneous implant or (2) the putative membrane estrogen receptor agonist, G1. Both E2- and G1-replacement 1 week prior to MCAO decreased infarct volume in brain cortex, striatum, and total hemisphere and partially restored splenocyte numbers [50]. Moreover, E2-replacement increased splenocyte proliferation in response to stimulation with anti-CD3/CD28 antibodies and normalized aberrant mRNA expression for cytokines, chemokines, and chemokine receptors, and percentage of $CD4^+CD25^+FoxP3^+$ Treg cells observed in E2-deficient animals. These beneficial changes in peripheral immunity after E2 and G1 replacement were accompanied by a profound reduction in expression of the chemokine, MIP-2, and a 40-fold increased expression of CCR7 in the lesioned brain hemisphere of E2-treated mice. These results demonstrate that both E2 and G1 replacement in ovariectomized female mice prior to MCAO reduced stroke volumes and ameliorated stroke-induced peripheral immunosuppression. As was observed with the PD-1/PD-L co-inhibitory pathway, treatment with E2 or G1 limited stroke volume in cortex, striatum, and total hemispheres, indicating protective effects on both CNS parenchymal cells and infiltrating immunocytes. It is of potential importance that E2-mediated protection versus Experimental Autoimmune Encephalomyelitis (EAE) and its modulation of IL-17 requires B cells [51] and expression of estrogen receptor (ER)-α, the membrane ER, GPR30 [52, 53], and PD-1 [52–54]. It now appears that this regulatory circuit involves E2 activation of PD-L on B cells through ER-α and GPR30, resulting in upregulation of PD-1 on $CD4^+FoxP3^+$ Treg cells [55] that block pathogenic T cells and possibly other inflammatory immunocytes. Involvement of the same elements suggests that a similar regulatory circuit may be at play in stroke as well.

Recombinant T-cell Receptor Ligands (*RTLs*) represent a second promising therapeutic approach for treatment of experimental stroke. RTLs are comprised of a single exon containing covalently linked β1 and α1 domains of MCH class II molecules that may also have linked antigenic peptides (Fig. 10.8) [56]. These molecules not only regulate cognate inflammatory T cells but also bind to and downregulate expression of CD74 (the Class II invariant chain) that serves as the primary receptor for a key pathogenic cytokine called Macrophage Migration Inhibitory Factor (MIF). RTLs block MIF effects on monocytes, macrophages, and possibly microglia, resulting in loss of chemotactic infiltration into tissues (e.g., the CNS during EAE), reduced survival of APC and decreased T-cell activation and cytokine release [57]. RTLs can reverse ongoing clinical signs of EAE [58, 59] and other autoimmune diseases and has been tested successfully in a Phase I clinical trial for multiple sclerosis [60, 61]. As discussed above, MCAO induces a biphasic effect on the immune response that involves early activation of peripheral leukocytes followed by severe immunodepression and atrophy of spleen and thymus. In tandem, the developing infarct is exacerbated by influx of numerous inflammatory cell types, including T and B lymphocytes. These features of stroke prompted our use of RTLs for therapy of MCAO. We tested the hypothesis that RTL would

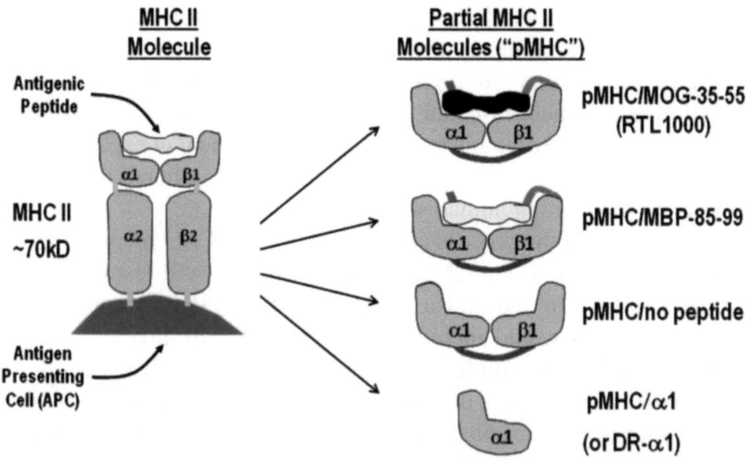

Fig. 10.8 A family of MHC class-II-derived RTL constructs that might be effective in the treatment of MCAO and potentially human stroke. RTL constructs are comprised of an MHC class II β1 domain covalently linked to an MHC class II α1 domain linked (or not) to an antigenic peptide, thus providing a complementary shape to a cognate T-cell receptor. These partial MHC II constructs (pMHC) can be produced using any class II combination that restricts immunodominant antigenic peptides. The figure shows schematically RTLs made from DRB1*1501 MHC class II, including RTL1000 (pDR2/hMOG-35-55), pDR2/MBP-85-99 (a different neuroantigen-specific RTL), pDR2/no peptide, and DR-α1 (no DR-β1 domain)

improve ischemic outcome in brain without exacerbating defects in peripheral immune system function.

Four daily doses of RTL were administered subcutaneously to C57BL/6 mice at onset of reperfusion after MCAO, and lesion size and cellular composition were assessed in brain, and cell numbers were assessed in spleen and thymus. Treatment with RTL551 (I-A^b molecule linked to MOG-35-55 peptide) reduced cortical and total stroke lesion size by ~50 %, inhibited the accumulation of inflammatory cells, particularly macrophages/activated microglial cells and dendritic cells, and mitigated splenic atrophy [62]. Treatment with RTL1000 (HLA-DR2 moiety linked to human MOG-35-55 peptide) similarly reduced the stroke lesion size in HLA-DR2 transgenic mice. Subsequent studies [63] demonstrated that RTL551 treatment produced a significant reduction in cortex, striatum, and hemisphere infarct volumes as well as improved sensorimotor outcome when administered on four consecutive days beginning 3 h after MCAO. A third set of experiments demonstrated that RTL551 given 4 h after MCAO could reduce infarct size in both cortex and striatum at 24 h and in cortex at 96 h after MCAO and inhibited the accumulation of inflammatory cells in brain at both time points [64]. At 24 h post-MCAO, RTL551 reduced the frequency of the activation marker, CD44, on T cells in blood and in the ischemic hemisphere. Moreover, RTL551 reduced expression of the chemokine receptors, CCR5 in lymph nodes and spleen, and CCR7 in the blood and lymph nodes. This treatment regime represents a more clinically relevant time point that would be compatible with the sole current stroke treatment, thrombolytic recombinant tissue plasminogen activator (rtPA), and we are now in the process of determining if these two agents are compatible with each other to allow simultaneous treatment of stroke subjects within 4 h of infarction. Our results are the first to demonstrate successful treatment of experimental stroke using immunomodulatory RTL constructs administered after ischemia, suggesting therapeutic potential in human stroke.

We currently are in the process of determining if MCAO involves MIF attraction of immunocytes into the CNS, in which case we will test other RTL constructs that also bind to and downregulate CD74, the MIF receptor (Fig. 10.8). Such RTL constructs would include pMHC/MBP-85-99 directed at a different neuroantigen, pMHC/no peptide that has no bound neuroantigen peptide, and the DR-α1 domain that represents the current minimal CD74 binding moiety. Should DR-α1 or a subsumed determinant that can still bind to and downregulate CD74 retain the ability to treat MCAO, this monomorphic DR-α1 sequence would ultimately be the best choice for treatment of stroke subjects due to its universal tolerance among humans. That is, the DR-α1 construct could be injected into any patient without the need for tissue typing would otherwise be needed to match HLA-DR types between stroke recipients and polymorphic DR-β chains present in other two-domain RTL constructs.

Acknowledgments The authors wish to thank Ms. Sherry Cass and Ms. Zefora Alderman for assistance in preparing the manuscript. This work is supported by NIH Grant NS10733275. This material is based upon work supported in part by the Department of Veterans Affairs, Veterans Health Administration, Office of Research and Development, Biomedical Laboratory Research and Development. The contents do not represent the views of the Department of Veterans Affairs or the United States Government.

Conflict of Interest Dr. Offner and OHSU have a significant financial interest in Artielle ImmunoTherapeutics, Inc., a company that may have a commercial interest in the results of this research and technology. This potential conflict of interest has been reviewed and managed by the OHSU and VAMC Conflict of Interest in Research Committees.

References

1. Emsley HCA, Hopkins SJ (2008) Acute ischaemic stroke and infection: recent and emerging concepts. Lancet Neurol 7:341–53
2. Engel O, Meisel A (2010) Models of infection before and after stroke: investigating new targets. Infect Disord Drug Targets 10:98–104
3. Emsley HCA, Smith CJ, Gavin CM, Georgiou RF, Vail A, Barberan EM, Hallenbeck JM, del Zoppo GJ, Rothwell NJ, Tyrrell PJ, Hopkins SJ (2003) An early and sustained peripheral inflammatory response in acute ischaemic stroke: relationships with infection and atherosclerosis. J Neuroimmunol 139:93–101
4. Smith CJ, Emsley HCA, Gavin CM, Georgiou RF, Vail A, Barberan EM, del Zoppo GJ, Hallenbeck JM, Rothwell NJ, Hopkins SJ, Tyrrell PJ (2004) Peak plasma interleukin-6 and

other peripheral markers of inflammation in the first week of ischemic stroke correlate with brain infarct volume, stroke severity and long-term outcome. BMC Neurol 4:2
5. Offner H, Subramanian S, Parker SM, Afentoulis ME, Vandenbark AA, Hurn PD (2006) Experimental stroke induces massive, rapid activation of the peripheral immune system. J Cereb Blood Flow Metab 26:654–65
6. Offner H, Subramanian S, Parker SM, Wang C, Afentoulis ME, Lewis A, Vandenbark AA, Hurn PD (2006) Splenic atrophy in experimental stroke is accompanied by increased regulatory T cells and circulating macrophages. J Immunol 176:6523–31
7. Gendron A, Teitelbaum J, Cossette C, Nuara S, Dumont M, Geadah D, du Souch P, Kouassi E (2002) Temporal effects of left versus right middle cerebral artery occlusion on spleen lymphocyte subsets and mitogenic response in Wistar rats. Brain Res 955:85–97
8. Prass K, Meisel C, Hoflich C, Braun J, Halle E, Wolf T, Ruscher K, Victorov IV, Priller J, Dirnagl U, Volk HD, Meisel A (2003) Stroke-induced immunodeficiency promotes spontaneous bacterial infections and is mediated by sympathetic activation reversal by poststroke T helper cell type 1-like immunostimulation. J Exp Med 198:725–36
9. Stevens SL, Bao J, Hollis J, Lessov NS, Clark WM, Stenzel-Poore MP (2002) The use of flow cytometry to evaluate temporal changes in inflammatory cells following focal cerebral ischemia in mice. Brain Res 932:110–9
10. Prass K, Ruscher K, Karsch M, Isaev N, Megow D, Priller J, Scharff A, Dirnagl U, Meisel A (2002) Desferrioxamine induces delayed tolerance against cerebral ischemia in vivo and in vitro. J Cereb Blood Flow Metab 22:520–5
11. Maloy KJ, Powrie F (2001) Regulatory T cells in the control of immune pathology. Nat Immunol 2:816–22
12. Roncarolo MG, Levings MK (2000) The role of different subsets of T regulatory cells in controlling autoimmunity. Curr Opin Immunol 12:676–83
13. Shevach EM (2000) Regulatory T cells in autoimmunity. Annu Rev Immunol 18:423–49
14. Fontenot JD, Gavin MA, Rudensky AY (2003) Foxp3 programs the development and function of CD4+CD25+ regulatory T cells. Nat Immunol 4:330–6
15. Hori S, Nomura T, Sakaguchi S (2003) Control of regulatory T cell development by the transcription factor Foxp3. Science 299:1057–61
16. Khattri R, Cox T, Yasayko S, Ramsdell F (2003) An essential role for Scurfin in CD4+CD25+ T regulatory cells. Nat Immunol 4:337–42
17. Sakaguchi S, Sakaguchi N (2005) Regulatory T cells in immunologic self-tolerance and autoimmune disease. Int Rev Immunol 24:211–26
18. Shimizu J, Yamazaki S, Sakaguchi S (1999) Induction of tumor immunity by removing CD25+CD4+ T cells: a common basis between tumor immunity and autoimmunity. J Immunol 163:5211–8
19. Belkaid Y, Piccirillo CA, Mendez S, Shevach EM, Sachs DL (2002) CD4+CD25+ regulatory T cells control Leishmania major persistence and immunity. Nature 420:502–7
20. Polanczyk MJ, Hopke C, Vandenbark AA, Offner H (2006) Estrogen-mediated immunomodulation involves reduced activation of effector T cells, potentiation of Treg cells, and enhanced expression of the PD-1 costimulatory pathway. J Neurosci Res 84:370–8
21. Planas AM, Gómez-Choco M, Urra X, Gorina R, Caballero M, Chamorro Á (2012) Brain-derived antigens in lymphoid tissue of patients with acute stroke. J Immunol 188:2156–63
22. Ren X, Akiyoshi K, Grafe MR, Vandenbark AA, Hurn PD, Herson PS, Offner H (2012) Myelin specific cells infiltrate MCAO lesions and exacerbate stroke severity. Metab Brain Dis 27:7–15
23. Jones RE, Bourdette DN, Whitham RH, Offner H, Vandenbark AA (1993) Induction of experimental autoimmune encephalomyelitis in severe combined immunodeficient mice reconstituted with allogeneic or xenogeneic hematopoietic cells. J Immunol 150:4620–9
24. Hurn PD, Subramanian S, Parker SM, Afentoulis ME, Kaler LJ, Vandenbark AA, Offner H (2007) T- and B-cell-deficient mice with experimental stroke have reduced lesion size and inflammation. J Cereb Blood Flow Metab 27:1798–805

25. Yilmaz G, Arumugam TV, Strokes KY, Granger DN (2006) Role of T lymphocytes and interferon- in ischemic stroke. Circulation 113:2105–12
26. Jander S, Kraemer M, Schroeter M, Witte OW, Stoll G (1995) Lymphocytic infiltration and expression of intercelluar adhesion molecule-1 in photochemically induced ischemia of the rat cortex. J Cereb Blood Flow Metab 15:42–51
27. Schroeter M, Jander S, Witte OW, Stoll G (1994) Local immune responses in the rat cerebral cortex after middle cerebral artery occlusion. J Neuroimmunol 55:195–203
28. Becker KJ, McCarron RM, Ruetzler C, Laban O, Sternberg E, Flanders KC, Hallenbeck JM (1997) Immunologic tolerance to myelin basic protein decreases stroke size after transient focal cerebral ischemia. Proc Natl Acad Sci USA 94:10873–8
29. Nadareishvili ZG, Li H, Wright V, Maric D, Warach S, Hallenbeck JM, Dambrosia J, Barker JL, Baird AE (2004) Elevated pro-inflammatory CD4+CD28- lymphocytes and stroke recurrence and death. Neurology 63:1446–51
30. Ren X, Akiyoshi K, Dziennis S, Vandenbark AA, Herson PS, Hurn PD, Offner H (2011) Regulatory B cells limit CNS inflammation and neurologic deficits in murine experimental stroke. J Neurosci 31:8556–63
31. Gee JM, Kalil A, Shea C, Becker KJ (2007) Lymphocytes: potential mediators of postischemic injury and neuroprotection. Stroke 38(Suppl 2):783–8
32. Wang Q, Tang SN, Yenari MA (2007) The inflammatory response in stroke. J Neuroimmunol 184:53–68
33. Liesz A, Suri-Payer E, Veltkamp C, Doerr H, Sommer C, Rivest S, Giese T, Veltkamp R (2009) Regulatory T cells are key cerebroprotective immunomodulators in acute experimental stroke. Nat Med 15:192–9
34. Planas AM, Gorina R, Chamorro A (2006) Signalling pathways mediating inflammatory responses in brain ischaemia. Biochem Soc Trans 34:1267–70
35. Grilli M, Barbieri I, Basudev H, Brusa R, Casati C, Lozza G, Ongini E (2000) Interleukin-10 modulates neuronal threshold of vulnerability to ischaemic damage. Eur J Neurosci 12:2265–72
36. Vila N, Castillo J, Dávlos A, Esteve A, Planas AM, Chamorro Á (2003) Levels of anti-inflammatory cytokines and neurological worsening in acute ischemic stroke. Stroke 34:671–5
37. Chamorro Á, Amaro S, Vargas M, Obach V, Cervera A, Torres F, Planas AM (2006) Interleukin 10, monocytes and increased risk of early infection in ischaemic stroke. J Neurol Neurosurg Psychiatry 77:1279–81
38. Ishida Y, Agata Y, Shibahara K, Honjo T (1992) Induced expression of PD-1, a novel member of the immunoglobulin gene superfamily, upon programmed cell death. EMBO J 11:3887–95
39. Keir ME, Francisco LM, Sharpe AH (2007) PD-1 and its ligands in T-cell immunity. Curr Opin Immunol 19:309–14
40. Nishimura H, Minato N, Nakano T, Honjo T (1998) Immunological studies on PD-1 deficient mice: implication of PD-1 as a negative regulator for B cell responses. Int Immunol 10:1563–72
41. Francisco LM, Sage PT, Sharpe AH (2010) The PD-1 pathway in tolerance and autoimmunity. Immunol Rev 236:219–42
42. Okazaki T, Honjo T (2006) The PD-1-PD-L pathway in immunological tolerance. Trends Immunol 27:195–201
43. Chemnitz JM, Parry RV, Nichols KE, June CH, Riley JL (2004) SHP-1 and SHP-2 associate with immunoreceptor tyrosine-based switch motif of programmed death 1 upon primary human T cell stimulation, but only receptor ligation prevents T cell activation. J Immunol 173:945–54
44. Cho HY, Lee SW, Seo SK, Choi IW, Choi I, Lee SW (2008) Interferon-sensitive response element (ISRE) is mainly responsible for IFN-alpha-induced upregulation of programmed death-1 (PD-1) in macrophages. Biochim Biophys Acta 1779:811–9
45. Yao S, Wang S, Zhu Y, Luo L, Zhu G, Flies S, Xu H, Ruff W, Broadwater M, Choi I, Tamada K, Chen L (2009) PD-1 on dendritic cells impedes innate immunity against bacterial infection. Blood 113:5811–8

46. Arslan F, Keogh B, McGuirk P, Parker AE (2010) TLR2 and TLR4 in ischemia reperfusion injury. Mediators Inflamm 2010:704202
47. Caso JR, Pradillo JM, Hurtado O, Lorenzo P, Moro MA, Lizasoain I (2007) Toll-like receptor 4 is involved in brain damage and inflammation after experimental stroke. Circulation 115:1599–608
48. Urra X, Cervera Á, Obach V, Climent N, Planas A, Chamorro Á (2009) Monocytes are major players in the prognosis and risk of infection after acute stroke. Stroke 40:1262–8
49. Lehnardt S, Schott E, Trimbuch T, Laubisch D, Krueger C, Wulczyn G, Nitsch R, Weber JR (2008) A vicious cycle involving release of heat shock protein 60 from injured cells and activation of toll-like receptor 4 mediates neurodegeneration in the CNS. J Neurosci 28:2320–31
50. Zhang B, Subramanian S, Dziennis S, Jia J, Uchida M, Akiyoshi K, Migliati E, Lewis AD, Vandenbark AA, Offner H, Hurn PD (2010) Estradiol and G1 reduce infarct size and improve immunosuppression after experimental stroke. J Immunol 184:4087–94
51. Bodhankar S, Wang C, Vandenbark AA, Offner H (2011) Estrogen-induced protection against experimental autoimmune encephalomyelitis is abrogated in the absence of B cells. Eur J Immunol 41:1165–75
52. Wang C, Dehghani B, Li Y, Kaler LJ, Proctor T, Vandenbark AA, Offner H (2009) Membrane estrogen receptor regulates experimental autoimmune encephalomyelitis through upregulation of programmed death 1. J Immunol 182:3294–303
53. Wang C, Dehghani B, Li Y, Kaler LJ, Vandenbark AA, Offner H (2009) Oestrogen modulates experimental autoimmune encephalomyelitis and interleukin-17 production via programmed death 1. Immunology 126:329–35
54. Polanczyk MJ, Hopke C, Vandenbark AA, Offner H (2007) Treg suppressive activity involves estrogen-dependent expression of programmed death-1 (PD-1). Int Immunol 19:337–43
55. Bodhankar S, Vandenbark AA, Offner H (2012) Estrogen treatment of experimental autoimmune encephalomyelitis requires E2 receptor positive B-cells that unregulate PD-1 on CD4+FoxP3+ Treg cells. Immunology 137(4):282–93
56. Burrows GG, Chang JW, Bächinger H-P, Bourdette DN, Offner H, Vandenbark AA (1999) Design, engineering and production of functional single-chain T cell receptor ligands. Protein Eng 12:771–8
57. Vandenbark AA, Meza-Romero R, Benedek G, Andrew S, Huan J, Chou YK, Buenafe AC, R D, Reiter Y, Mooney JL, Offner H, Burrows GG (2013) A novel regulatory pathway for autoimmune disease: Binding of partial MHC class II constructs to monocytes reduces CD74 expression and induces both specific and bystander T-cell tolerance. J Autoimmunity 40:96–110
58. Sinha S, Subramanian S, Proctor TM, Kaler LJ, Grafe M, Dahan R, Huan J, Vandenbark AA, Burrows GG, Offner H (2007) A promising therapeutic approach for multiple sclerosis: recombinant T-cell receptor ligands modulate experimental autoimmune encephalomyelitis by reducing interleukin-17 production and inhibiting migration of encephalitogenic cells into the CNS. J Neurosci 27:12531–9
59. Vandenbark AA, Rich C, Mooney J, Zamora A, Wang C, Huan J, Fugger L, Offner H, Jones R, Burrows GG (2003) Recombinant TCR ligand induces tolerance to myelin oligodendrocyte glycoprotein 35–55 peptide and reverses clinical and histological signs of chronic experimental autoimmune encephalomyelitis in HLA-DR2 transgenic mice. J Immunol 171:127–33
60. Offner H, Sinha S, Burrows GG, Ferro AJ, Vandenbark AA (2011) RTL therapy for multiple sclerosis: a phase I clinical study. J Neuroimmunol 231:7–14
61. Yadav V, Bourdette DN, Bowen JD, Lynch SG, Mattson D, Preiningerova J, Bever CT Jr, Simon J, Goldstein A, Burrows GG, Offner H, Ferro AJ, Vandenbark AA (2012) Recombinant T-cell receptor ligand (RTL) for treatment of multiple sclerosis: a double blind, placebo-controlled, Phase 1. Dose-escalation study. Autoimmune Dis 2012:954739
62. Subramanian S, Zhang B, Kosaka Y, Burrows GG, Grafe MR, Vandenbark AA, Hurn PD, Offner H (2009) Recombinant T cell receptor ligand treats experimental stroke. Stroke 40:2539–45

63. Akiyoshi K, Dziennis S, Palmateer J, Ren X, Vandenbark AA, Offner H, Herson PS, Hurn PD (2011) Recombinant T Cell receptor ligands improve outcome after experimental cerebral ischemia. Transl Stroke Res 2:404–10
64. Dziennis S, Mader S, Akiyoshi K, Ren X, Ayala P, Burrows GG, Vandenbark AA, Herson PS, Hurn PD, Offner HA (2011) Therapy with recombinant T-cell receptor ligand reduces infarct size and infiltrating inflammatory cells in brain after middle cerebral artery occlusion in mice. Metab Brain Dis 26:123–33

Chapter 11
Infectious Burden and Risk of Stroke

Jorge M. Luna and Mitchell S.V. Elkind

Abstract Atherosclerosis and prothrombotic vascular states are complex pathologies of the tightly coupled immune and cardiovascular systems that, when dysregulated, synergistically act to elevate risk of acute cerebral ischemic events. Infectious exposures are atypical cardiovascular risk factors requiring methodological and paradigmatic departures from traditional risk factor investigation approaches. Several parameters of pathogen activity have been mechanistically mapped to immune and vascular processes that induce and aggravate atherosclerosis and can upregulate physiological states that are directly linked to triggering ischemic events. Elevated specific antibody titers in serum samples have been exploited to create objective measures of historical exposures to pathogens. This chapter reviews the following domains supporting Infectious Burden (IB) [or synonymously pathogen burden (PB)] and its association with stroke risk: (1) vascular, immune, and infectious dynamic frameworks for research; (2) hypothesized pathogen mechanisms influencing stroke risk, and important individual characteristics that may biologically interact with infection; (3) molecular exposure measurement tools and single pathogen

J.M. Luna, M.P.H (✉)
Department of Epidemiology, Mailman School of Public Health, New York, NY, USA

Department of Neurology, Columbia University College of Physicians and Surgeons, New York, NY, USA

Department of Industrial and Operations Research, School of Engineering and Applied Sciences, Columbia University, New York, NY, USA
e-mail: jl2708@mail.cumc.columbia.edu

M.S.V. Elkind, M.D., M.S.
Department of Epidemiology, Mailman School of Public Health, New York, NY, USA

Department of Neurology, Columbia University College of Physicians and Surgeons, New York, NY, USA

Gertrude H. Sergievsky Center, Columbia University, New York, NY, USA

J. Chen et al. (eds.), *Immunological Mechanisms and Therapies in Brain Injuries and Stroke*, Springer Series in Translational Stroke Research, DOI 10.1007/978-1-4614-8915-3_11,

association with stroke; (4) statistical measurement approaches to IB and selected evidence that evaluates IB associates with stroke; and (5) the current state in public health and clinical intervention strategies.

Introduction

Atherosclerosis and prothrombotic vascular states are complex pathologies of the tightly coupled immune and cardiovascular systems that, when dysregulated, synergistically act to elevate risk of acute cerebral ischemic events. Stroke continues to exert a large societal burden, regardless of the metric used to assess burden— mortality, morbidity, QALYs, economic productivity, and economic resources [1]. Approximately 795,000 persons suffer strokes annually in the USA (88 % are of ischemic stroke subtypes), with 137,000 resulting in death [2]. Of the remaining stroke cases, 10 % will require long-term care facilities, and 40 % will have long-term, moderate-to-severe impairment. In 2008, the economic burden of stroke was estimated to be USD $18.8 billion in direct costs and USD $15.5 billion in indirect costs, with costs projected to increase [3]. In global contexts, understanding alternate prevention targets for intervening on cerebrovascular disease may align with broader development goals and efforts that seek to address the dual burden of non-communicable and infectious diseases [4].

Recently, chronic and acute infections have been implicated as exposures that impact systemic and local immune system functioning and are capable of triggering inflammatory cascades that increase the initiation and progression of atherosclerosis; the destabilization of atherosclerotic plaque; and the creation pro-coagulant, prothrombotic states [5–7]. The notion of infectious agents playing an active role in the development of vascular lesions and increasing risk of thrombus formation to impact the risk of stroke dates back as early as 1911, when Frothingham wrote of atherosclerosis: "…it is not known whether these lesions were due to the actual presence of the influenza bacillus or to concentrated toxic action followed by thrombus formation and organization…" [8]. No single pathogen is believed to fulfill Koch's postulates of sufficient cause of stroke, and we know from experimental hyperlipidemic animal studies that atherosclerosis can develop even in microbe-free environments [9]. However, the impact of acute infection, the high frequency of recurrent acute infections, and exposure to chronic multiple infections over the life course are all active area of investigation to assess the relationship between "infectious burden" and cerebrovascular ischemic injuries. The causal role of pathogens and atherosclerotic plaque remains controversial, and some have even referred to these pathogens as "innocent bystanders" that demonstrate a propinquity to site of acute inflammation and are not causally related to disease severity [10].

This chapter reviews the following domains supporting Infectious Burden (IB) [or synonymously pathogen burden (PB)] and its association with stroke risk: (1) vascular, immune, and infectious dynamic frameworks for research; (2) hypothesized pathogen mechanisms influencing stroke risk, and important individual characteristics that may biologically interact with infection; (3) molecular exposure measurement tools and single pathogen association with stroke; (4) statistical measurement approaches to IB and selected evidence that evaluates IB associates with stroke; and (5) the current state in public health and clinical intervention strategies.

Dynamics of Infection and Cardiovascular Disease

Infectious exposures are atypical cardiovascular risk factors requiring methodological and paradigmatic departures from traditional risk factor investigation approaches. In dissimilarity to vascular risk factors such as hypertension, infectious agents are contagious exposures that can be transmitted between individual subjects in clinical and population studies. Infections, when treated as exposures, can give rise to endogenous effects, emergent properties, and dynamic feedback loops, such as in the case of herd immunity in vaccine effectiveness trials [11]. It should be noted in this context that several noninfectious but contagious risk factors, such as obesity and smoking, are already under investigation in cardiovascular research through use of social network models, notably by Christakis et al. [12, 13] (see Table 11.1).

Pathogen Mechanisms Affecting Cerebrovascular Risk

Several parameters of pathogen activity have been mechanistically mapped to immune and vascular processes that induce and aggravate atherosclerosis and can upregulate physiological states that are directly linked to triggering ischemic events. Several mechanisms identified in molecular, cell culture, and animals models of research can affect cerebrovascular risk, including blood coagulopathies and fibrinolysis [14–21]; inflammatory cytokine profiles [22–25]; heightened adhesion and chemokine molecules expression [26–28]; impaired lipid metabolism [29, 30]; hemodynamic stress [31]; repeated exposure to acute phase response proteins, metalloproteases, and plasminogen activator inhibitors that weaken fibrous cap stability [32–41]; enhanced smooth muscle cell growth [30, 42, 43]; self-antigen responses resulting from pathogen epitope cross-reactivity [44–50]; and infection-related immune senescence linked to chronic low-grade inflammation [51–54].

Host Factor Determinants of Vascular Response to Infectious Stimuli

Host factors commonly studied in cardiovascular disease research may play important roles in governing progression of both IB and atherosclerosis, due to biologically confounding and interacting effects on immune cells and inflammatory processes.

Table 11.1 Dynamics of infection and cardiovascular disease

Transmission Dynamics	
Evidence	"Infection burden" pathogens may be able to circulate differentially through social networks and can establish endemic status in larger social network topologies. Several different *C. pneumoniae* atherosclerotic plaque isolates have been identified in investigations of antibiotic susceptibility profiles [178]. Antigenic molecular analysis of small sample studies confirm multiple serovar of *C. pneumoniae* have been confirmed to be in existence [179]. If there are differences in circulating strains, with regard to virulence or pathogenicity, then it may be possible that regional studies may observe different association with vascular outcomes due to interstrain differences. For example, *C. pneumoniae* pathogen has been implicated in carotid atherosclerosis in numerous US and German studies, yet studies in Israel, Japan, and Australia have not shown associations [10]. *C. pneumoniae* has been characterized as endemic in certain regions such as Western nations and Japan, but the demographic distribution is not consistent across regions, with Japan experiencing a higher prevalence among younger age groups [180]. Within the US longitudinal studies of HPY have shown racial differences of seroconversion rates at age 1–3 years (13 % black vs 4 % white) and also at ages 18–23 years (43 % black vs. 8 % white) [181]. Variable spread through population may be consistent with antigenic or molecular variants, but alternate explanations, such as variability by host genetic determinants, or socioeconomic factors, have also been postulated.
Modeling implications	Spatiotemporal patterns of disease prevalence are manifestation of complex realities affecting disease transmission, including spatial heterogeneity of host composition and dynamic human interaction networks. Commonly applied generalized linear models, such as logistic and Cox stochastic regression models, may be ill suited to capture arising nonlinear dynamics and feedback. Parameter estimation technique behind these data models are heavily impacted by violations to the so-called *stable unit treatment value assumptions,* and naturally arising feedback mechanisms [182]. For example, herd immunity would cause an overestimation in the effect of a vaccine in trials prevent infections, as it would not account for disrupted transmission chains. Infectious disease epidemiology has dealt with these dynamics by employing various data and simulation models to capture nonlinearity of effects and the intersubject interaction mechanisms through social network graph models [13] and agent-based models [183].
Inflammatory dynamics	
Evidence	Evidence suggests that functioning of T cells, monocytes, and other cells of the immune system is primal to atherogenesis, erosion, and rupture [184]. Our understanding of atherosclerosis is no longer characterized by parallel disease progression of vascular plaque formation and immune cell activity, instead we view the immune systems and vascular functioning as tightly coupled, interacting systems that give rise to systemic disease [185]. From an immune system perspective, atherosclerosis is marked by the shifting of immune cells profile of arterial tissue, beginning from a low density of dendritic cells and macrophages in healthy tissues towards a shifting to a high distributions of macrophages, dendritic cells, foam cells, lymphocytes in intima region of plaque lesion and high degree of infiltration by adventitial leukocytes in smooth muscle regions, with homology to tertiary lymphoid tissues [186]. Macrophages and mast cells actively secrete metalloproteases that can enzymatically degrade extracellular plaque matrix, causing expansion of plaque in positive (adventitial) or negative (lumen) remodeling, potentially weakening fibrous cap stability [187]. The presence of new immune cells in vessel tissue, shift of expression of cytokine and chemokine proteins, and ability to override anti-inflammatory, autoregulatory feedback processes alters localized vessel functioning, with the severity and trajectory of vascular damage modified by genetic factors, behavioral and environmental exposures over the course of decades.

Modeling implications	Integrating immune system dynamics and atherosclerosis progression not only forces a reconceptualization IB but can also enable the development of useful predictive models of atherosclerosis progression and strategies for intervention. Useful methodological approaches in development include complex adaptive network models to capture nonlinear immune system dynamics [188]; qualitative loop analysis models to understand dynamic casual structure for design of important confounder measurements [189]; or system dynamic models to simulate deployment of interventional trials [190].
Life-course dynamics	
Evidence	Cumulative life-course models readily provide insight into multidecade exposure that lead to vessel wall damage in key phases of initiation, progression, and atherothrombosis [191]. In vascular research, life-course approaches have prompted a rethink of prevalent beliefs that atherosclerosis progression is continuous and linear, as longitudinal repeated measures studies suggests that atherosclerosis undergoes acute periods of growth, and remodeling [192]. Life-course epidemiology has also helped reassess timing of infection, duration of infection, and chronic disease outcomes. Novel findings included measles exposures in utero and development of Crohn's disease [193], human papillomavirus exposure, and throat and neck cancer [194] or even relationships between childhood infection exposures and allergy development later in life [195]. Longitudinal studies, currently underway, seek to understand the natural history of the blood microbiome and with repeated measures of the tracking changes in detection of bacterial ribosomal DNA gene sequences and relating changes to incident of cardiovascular (245). The immune system similarly undergoes aging associated changes across the lifespan, including harmful transition to senesced states that may result from chronic pathogen exposures [53].
Modeling implications	Estimating causal effects in the context of time-varying exposures or attributing causality of infection exposure at specific time point, where previous history of the same infection may act as confounder, require more complex modeling tools. Time-varying Cox models are one approach to correlate changing exposures to changing outcomes over time; however, fully parameterized Cox models require assumption about hazard functions over time [196]. G-estimation methods and inverse probability weighting allows investigators to partition effect of an exposure at a given time point, adjusting for past states of exposure [197]. Alternatively, dynamical models as used in pharmacokinetic simulation, based on sets of differential equations, may capture immune, cytokine, and plaque dynamics over time [198].

Evidence of complex pathogen involvement with cardiovascular risk factors is mounting, and existing studies have found correlations with *depressive symptoms* [55–64], *stress* [65, 66], *genotypic variants of human immune system* [67–79], *health seeking behaviors* [116, 117], *and social vulnerability and low socioeconomic position* [80–86].

Measurement of Infection: Molecular Tools

Elevated specific antibody titers in serum samples have been exploited to create objective measures of historical exposures to pathogens. Antibody-based serologic tests are designed to be sensitive and specific for clinical diagnostics in acute infection cases. The utility of these tests with regard to identifying chronic infections has not been extensively validated [87]. A study that comparatively assessed community-acquired *C. pneumoniae* molecular testing specificity, sensitivity, and positive predictive values from a range of molecular detection modalities—PCR methods, tissue cultures, complement fixation test, microimmunofluorescence (MIF), and immunosorbent assay (ELISA) serology tests—recommended MIF as most efficient to determining *C. pneumoniae* prevalence in a community [88]. Infectious Disease Society of America guidelines have called for widespread etiological identification of pathogens involved in community-acquired pneumonia, with specific testing guidance calls for use of MIF and ELISA assays to differentiate acute primary infection, as compared to reinfection, with MIF considered the gold standard serology test [89, 90]. These practices have increased epidemiologic understanding of the prevalence of pathogens associated with vascular risk [91]. Many published studies have also used enzyme immunoassay (EIA) to assess history of pathogen exposures [92]. These definitions from infectious disease research have been adopted by cardiovascular researchers seeking to measure histories of pathogen infections and investigate associations with stroke events.

Acute primary infection is characterized by an increase in immunoglobulin M (IgM) and a subsequent rise of immunoglobulin G (IgG) in convalescent samples (IgM levels dissipate in 2–3 months after infection); in contrast, reinfection is marked by increases in IgG or immunoglobulin A, without elevated IgM levels; and chronic infection is indicated by persistent IgG and IgA elevated titers [93]. There has been limited use of repeated serum sampling in CVD research to establish acute, chronic, and colonization assessments of pathogen infection. Biomarker specimens and testing are subject to general measurement error from common study procedures that include circadian timing and type of biological specimen collection; storage and manipulation; laboratory variation, error, and batching; and within individual changes over time [94]. Additionally, unmeasured individual parameters at the time of sampling, such as prevalent infections, autoimmune conditions, medication status, or even fasting requirements, may induce bias if associated with exposure status [95].

Measurement of Burden: Statistical Tools

Epstein et al. noted that increasing levels of multiple coinfections align with biological plausibility and dose–response mechanisms believed to impact vascular disease [96]. The notion of IB encapsulates pro-inflammatory factors, such as time dynamics in the length of infectious exposure, extent of immune response provoked by multiple coinfections, measurement of inflammatory biomarkers, and frequency and severity of recurrent infections over time. We define IB using a life-course epidemiology lens to be the cumulative life-course exposure to infectious agents eliciting strong inflammatory responses that are most likely to lead to advanced atherosclerosis and be at the greatest risk for vascular events.

Several serology-based statistical approaches to quantifying IB have been proposed, but the lack of standardized or validated approaches remains a major impediment to cross-study comparisons and generalized causal inference. Pathogen selection for inclusion into "burden" measures has often been arbitrary, based on convenience samples of available serology measures, and simple aggregation into burden measure may not always be sensitive to possible pathogen subtype clustering, such as viral vs. bacterial burden, or mechanistic subtypes, such as endothelial reactivity burden, plaque progression burden, or plaque destabilization burden. In the majority of studies that assessed IB, seropositivity has been assessed as a dichotomous variable (positive or negative) for infection and each seropositive value was included in burden measures with untested assumptions of equal weight in the ability to impact vascular risk. The most common serology IB measure is constructed as a *sum of any seropositivity* with varying parameters of (1) MIF, ELISA, and EIA testing types, (2) variable number of pathogens, ranging from 3 to 8 pathogens, and (3) differential usage of IgG, IgA, and IgM antibodies to establish positivity according to commercial manufacturer cutoffs. The construct validity of measuring sum of serology as a surrogate measure of persistent infection remains controversial [97].

Alternative statistical summary measures have been utilized that may minimize measurement error of chronic infection. Aiello et al. utilized the continuous antibody concentration measures available from MIF testing to create upper quantile concentrations of serum IgG as cutoffs to indicate chronic inflammation [98]. Elkind et al. formulated an individual pathogen stroke-associated IgG weight combined into an index measure of IB, allowing for an empirically derived index value based on long-term stroke risk [99]. Aging research into the "allostatic load" hypothesis has adopted several alternate algorithms to cluster or group measurements from multiple domains, using methods like recursive partitioning, k-means cluster analysis, and even genetic programming-based symbolic regression algorithms [100].

Other examples of IB measurements are derived from clinical history measures. Dental pathogen studies utilize measures that can include sum of dental caries, along with direct plaque bacterial testing that correlate with "burden" [101]. Administrative databases—mined for histories of acute infection clinical encounters—have provided time-varying exposure measurements that serve as surrogate measures of disease chronicity and have been associated with vascular outcomes [102].

Chronic Infectious Burden and Non-stroke Associations: Socioeconomic Position, Cytokines, Insulin Resistance, CAD, PAD, and Non-stroke CVD Endpoint

IB has been correlated with many factors associated with cardiovascular risk. IB measured as sum of serology with equal weighting was found to be associated with *socioeconomic position*, with findings showing that IB was associated with greater BMI, waist/hip ratio, blood pressure, and incidence of diabetes in the Whitehall II study [103] and IB was found to be associated with low educational attainment in the MESA study (this study used variable IB measure that incorporated sum of serology with >median titer value) [98]. IB was not linked with cytokine markers CRP, IL6, or fibrinogen levels [104]. IB was found to explain 42 % of variance of insulin resistance, with an association that dissipated after adjustment for CRP [105]. IB was associated with reduced percent change in coronary blood flow ($p<0.01$), reduced ability to decrease vascular resistance ($p<0.02$) [106], and reduced brachial flow-mediated vasodilation ($p<0.05$) [107], but not in all studies [108]. IB was also associated with PAD in women [109]. Among cardiac outcomes, some studies have found association between IB and CAD [96, 110, 111], but not in all CAD studies [112], as well as associations between IB and MI [113, 114] and IB and cardiovascular mortality [115].

Selected Studies of Chronic and Recurrent IB and Carotid Atherosclerosis

Cross-sectional investigations of infectious burden have found association with presence of carotid plaque using weighted indexes [116], but associations were not detected using equal weighting sum of serology scores [10, 117]. Prospective studies of incident carotid plaque found associations with sum of IB (serology and equal weight) [118], and studies of progression of carotid plaque found association with IB (sum of serology and highest quartile) [119]. It may be that unweighted sum of serologies may be associated with incident carotid plaque (infection to initiate lesions), but elevated titers in highest quantiles may be associated with progression (chronic inflammation). Weighted index measures may be more sensitive and detect associations even in cross-sectional studies. We highlight examples of chronic IB and recurrent IB studies that detected association with carotid plaque.

Chronic Infectious Burden and Carotid Atherosclerosis

In a cross-sectional analysis of the Northern Manhattan study (NOMAS)—a population-based prospective stroke risk factor study—the authors evaluated the association between maximal carotid plaque thickness (MCPT) and weighting approach to create an index of IB. The IB measure was comprised of two bacterial and three viral pathogens (ELISA IgG assays were conducted for *C. pneumoniae*, *H. pylori*, CMV, HSV1 and 2) and the relationship with stroke was assessed in a multiracial population with heterogeneous levels of SEP. Individual pathogen weights were estimated from Cox model regression associations with incident stroke events. An IB index was then created as sum of weights for each positive serology for an individual and used as an independent predictor of stroke risk (n=891; mean age=67.2) [116]. Carotid atherosclerotic plaque thickness was estimated from off-line assessments of high-resolution B-mode ultrasonography images, as the maximum of measured MCPT bilaterally, in both common and internal carotid arteries. The IB index (IBI) was moderately associated with increased carotid plaque (adjusted change in MCPT, mm per SD=0.09, 95 % CI, 0.02–0.15) among a subset of the NOMAS cohort. Adjusting for CRP and leukocyte count did not modify estimates. A secondary analysis reviewing cross-sectional associations of IBI with high-risk plaque subtypes, such as unstable irregular plaque, found increasing IBI to be linked with irregular plaque (adjusted OR per SD IBI=1.76, 95 % CI, 1.1–2.8). Positive cross-sectional associations of the novel weighted burden measurement and two measures of carotid atherosclerosis support the hypothesis that prevalent IBI may work through atherogenic processes to increase atherosclerosis, and in particular rupture-prone irregular plaque formations. Irregular plaque is thought to be linked to unstable plaque formations that can lead to ischemic or procoagulant states. The NOMAS study samples from an urban, largely Hispanic population may have greater variability of socioeconomic position and ethnicity that are postulated to be upstream causal factors of IB, enabling the detection of the IB and atherosclerosis association.

Recurrent Infectious Burden and Carotid Atherosclerosis

An analysis of the Bruneck cohort by Kiechl et al. (n=826; mean age=54.1; mean follow-up=5 years) added further domains of IB termed "chronic infection" status through use of clinical history investigation that included recurrent chronic respiratory, urinary tract (recurrent lower tract and pyelonephritis), dental pathogen, and other infectious conditions (chronic pancreatitis, diverticulitis, periodontitis, recurrent bacterial skin infections, and diabetic foot ulcers) [50]. The association of chronic infection status and the development of carotid plaques were estimated by logistic regression analysis, after adjusting for clinical risk factor and social status. Chronic infection was associated with an increase of relative odds of incident carotid atherosclerosis among those individuals without plaque at baseline (adjusted OR=4.10, 95 % CI, 2.37–7.10). Analysis for interaction with CRP levels showed elevated relative odds of incident atherosclerosis among those with >1 mg/dL hsCRP levels (p<0.001). Identifying and staging individuals as free of carotid atherosclerosis at baseline allow this prospective study to provide support for an atherogenic role of recurrent respiratory infection through use of clinical databases.

Chronic Infectious Burden and Composite Cardiovascular Endpoints, Stroke

This section reviews available literature evaluating IB and composite cardiovascular endpoints that include stroke or stroke events directly. The review includes case–control, cohort, and nested-cohort analytic designs with various methods of constructing IB. A concise summary of this section is presented in Table 11.2.

Case–Control Studies

A case–control study (*n* cases/controls = 91/86) evaluated the relationship between prevalent antibodies for three atypical respiratory pathogens (positive IgG or IgM serology for *C. pneumoniae*, *M. pneumoniae,* and *L. pneumoniae* according to manufacturer thresholds) with imaging-confirmed total stroke or TIA cases, and concurrently admitted non-cardiovascular disease patients [120]. This small study found that individuals with a burden of all three pathogens as measured by IgG serologies were at increased risk of all stroke or TIA (adjusted OR Score 3 vs. Score 1), but the categorical modeling of the variable (with all four levels) was not overall significant in the model ($p = 0.15$). Notably, this study utilized samples that excluded infective conditions for control, but no screen was applied to case introducing possible biases; the reference group ($n = 21$) and high-risk group ($n = 10$) were very small in size inducing large confidence intervals and did not provide variable estimates for first stroke event or recurrent stroke although status was available.

Another case–control study of non-cardiogenic stroke patients (*n* case/controls = 59/53) evaluated the impact of IB in younger individuals aged <65. The authors imposed age as a selection criterion to increase the likelihood that stroke would be driven by inflammatory factors, and not traditional risk factors [71]. IB (created by summing seropositive ELISA for IgG HSV1 and 2, IgM IgG CMV, IgG EBV, IgG HHV-6, and IgG IgA *C. pneumoniae*, whose quantitative OD values were in the highest tertile) and IL-8 and CD14 promoter polymorphisms were assessed for their relationship to stroke. No association between IB or immune polymorphisms and stroke risk was found, but this study sample may have been underpowered.

Cohort Studies

In 1999, a cohort study explored multiple coinfections and risk of combined cardiovascular outcomes, including stroke events [121]. Ridker et al. selected 122 case-patients who had experienced an incident vascular event (MI, stroke, cardiovascular death, or coronary revascularization), and 244 controls, from the underlying

Women's Health Study cohort, matched on age and smoking status. This study used multivariate conditional logistic regression to estimate the rate ratio (RR) for future cardiovascular events and the number of pathogens detected in serum samples (based on seropositivity for MIF IgG CPNEU; ELISA IgG HPY, HSV and CMV), but found no association between seropositivity for 4 coinfections (adjusted RR=1.2, 95 % CI 0.6–2.7) and vascular events. This analysis was adjusted for traditional cardiovascular risk factors, overcame possible survival biases of case–control studies and potential confounding by socioeconomic position due to selection from a uniform pool of female health professionals. A similar analysis by Ridker et al. based on the Physician Health Study cohort did not aggregate impacts of multiple coinfections [122].

A 2002 analysis, among patients admitted for diagnostic heart catheterization, classified study population according to stage of atherosclerosis (no atherosclerosis, limited disease, or advance disease) at baseline and evaluated both progression of atherosclerosis (based on sonography imaging of both peripheral arteries and carotid arteries) with logistic models and time to cardiovascular mortality using Cox proportional hazard models [123]. This analysis used an 8-item IB construct (seropositivity for ELISA IgG measurements for HSV-1, HSV-2, CMV, *H. influenza*, *H. pylori*, and EBV; and EIA IgA measurements for *M. pneumoniae* and *C. pneumoniae*) [119]. Models were adjusted for clinical risk factors and baseline stage of atherosclerosis, although adjustments for indicators of socioeconomic status were not included. Individuals in the highest total IB group (adjusted OR=2.45, 95 % CI, 1.18–5.10) and bacterial IB group (adjusted OR=2.12, 95 % CI, 1.32–3.41) were at increased risk of atherosclerosis progression, when compared to the referent group. When analysis was stratified by baseline severity of atherosclerotic disease, those with advanced disease and 6–8 total infections experienced a higher risk of cardiovascular mortality (adjusted HR=10.82, 95 % CI 2.00–58.54), relative to those with limited disease and only 0–3 total pathogen burden.

Findings from the Framingham Heart Study in an analysis from 2002 ($n=1187$; mean age=69; mean follow-up=10 years) showed no association between chronic coinfection, as evidenced through serology markers, and time to incident cardiovascular events (combined MI, coronary artery death, and atherothrombotic stroke) [124]. A three IB measure was created, with positive serology established by serum ELISA IgG HPY, *C. pneumoniae* CMV. Seropositivity for one (adjusted HR=0.88, 95 % CI, 0.50–1.56), two (adjusted HR=0.78, 95 % CI, 0.45–1.37), or three pathogens (adjusted HR=0.77, 95 % CI, 0.44–1.35) was not associated with time to vascular endpoints, after adjustment for clinical risk factors. Additionally, Haider et al. demonstrated that variable cutoffs for infection positivity did not meaningfully impact associations with cardiovascular disease, but they did not specify whether the IB measure used normal or high threshold cutoffs. This study found no association between IB and composite vascular outcomes, but did not report stroke-specific hazard ratios or a pathogen-weighting scheme. The homogenous composition of the Framingham population may enhance internal validity, but it may limit detection of IB associations with stroke. IB and stroke mechanistic linkages may require variability in complex social and biological processes such as socioeconomic position,

Table 11.2 Select studies review association between chronic, recurrent, and acute infectious burden and composite cardiovascular endpoints and stroke

Reference	Author	Year	Location	Design	Follow-up	Sample size [Sample age]	Infect burden
Chronic infectious burden, composite cardiovascular endpoints and stroke							
[120]	Ngeh et al.	2005	UK	Case–control (Hosp. stroke patients)	N/A	Case $n=91$ [Median 80 years] Control $n=86$ [Median 82 years]	Sum of score, Equal weight
[71]	Kis et al	2007	Hungary	Case–control (Young stroke patients)	N/A	Case $n=59$ [Mean 52.8 years] Control $n=53$ [Median 50.4 years]	Sum of score, Upper tertile
[121]	Ridker et al.	1999	USA	Nested Case–control (Woman Health Study)	3 years	Case $n=122$ [Mean 59.3 years] Control $n=244$ [Mean 59.3 years]	Sum of score, Equal weight
[119]	Espinola-Klein et al.	2002	Germany	Prospective (Heart catheterization patients)	3.2 years	$n=572$ [Mean 62.8 years]	Sum of score, Equal weight
[124]	Haider et al.	2002	USA	Prospective (Framingham)	10 years	$n=1,187$ [Mean age$=69$]	Sum of score, Equal weight

Pathogens	Outcome	Findings	Adjustment Variables		
			Diabetes	Immune	SEP
ELISA IgG IgM LPNEU, MPNEU, CPNEU	(i) CT confirmed stroke/TIA	(i) Infectious burden 3 vs. 0: Adj. RR = 6.67, 95 % CI, 1.22–37.04	YES	NO	NO
ELISA IgG HSV1&2, IgM IgG CMV, IgG EBV, IgG HHV-6, & IgG IgA CPNEU	(i) Non-standardized, non-cardiogenic stroke	(i) Infectious burden categorical: Chi-sqr $p = 0.657$	NO	NO	NO
ELISA IgG HPY,HSV, CMV MIF IgG CPNEU	(i) CVD Composite: MI, stroke, cardiovascular death, or coronary revascularization	(i) Infectious burden 4 vs. 0–1: Adj. RR = 1.2, 95 % CI, 0.6–2.7	YES	NO	NO
ELISA IgG HSV1&2, CMV, H. influenza, HPY, EBV; EIA IgA MPNEU, CPNEU	(i) Atheroprogression to stenosis: Units in increase in 1 SD per pathogen (ii) CVD Death: MI, stroke, cardiovascular death, or coronary revascularization	(i) Infectious burden 6–8 vs. 0–3: Adj. OR = 2.45, 95 % CI, 1.18–5.10 (i) Bacterial Burden 3–4 vs. 0–1: Adj. OR=2.12, 95 % CI, 1.32–3.41 (i) Viral Burden 3–4 vs 0–1: Adj. OR=0.99, 95 % CI, 0.58–1.69 (ii) Infectious burden 6–8 vs. 0–3: Adj. OR = 2.87, 95 % CI, 1.21–9.65 (ii) Infectious burden 6–8 advanced disease vs. 0–3 limited disease: Adj. HR = 10.82, 95 % CI 2.00–58.54	YES	CRP	NO
ELISA IgG HPY, CPNEU, CMV	(i) CVD Composite: MI, coronary artery death, and atherothrombotic stroke	(i) Infectious burden 3 vs. 0: Adj. HR = 0.77, 95 % CI, 0.44–1.35	YES	NO	NO

(continued)

Table 11.2 (continued)

Reference	Author	Year	Location	Design	Follow-up	Sample size [Sample age]	Infect burden
[125]	Smieja	2003	Canada	Prospective (High-risk cohort)	4.5 years	$n=3{,}618$ [Mean age = 65.4]	Sum of score, Equal weight
[126]	Corrado et al.	2006	Italy	Prospective (High-risk cohort)	5 years	$n=668$ [Mean age = 63.2]	Sum of score, Equal weight
[105]	Dai et al.	2007	Taiwan	Prospective (High-risk cohort)	3 years	$n=568$ [Mean age = 62.5]	Sum of score, Equal weight
[116]	Elkind et al.	2010	US	Prospective (NOMAS)	Median 7.6 years	$n=1{,}625$ [Mean age = 69.5]	Index, Weighted Sum of score, Equal weight
Recurrent infectious burden, composite cardiovascular endpoints and stroke							
[127]	Grau et al.	2009	Germany	Case–control (Hosp. stroke Patients)	N/A	Case $n=370$ [Mean age = 60.7] Control $n=370$ [Mean age = 60.6)	Chronic bronchitis Frequent FLU-like illness
Acute infectious burden, composite cardiovascular endpoints and stroke							

Pathogens	Outcome	Findings	Adjustment Variables		
			Diabetes	Immune	SEP
MIF IgG CPNEU EIA IgG HPY, CMV, and HAV	(i) CVD Composite: MI, Stroke, CVD death, revascularization (ii) Stroke	(i) Infectious burden 4 vs. 0–1: Adj. HR=1.41, 95 % CI, 1.02–1.96 (ii) Infectious burden 4 vs. 0–1 Adj. HR=0.75, 95 % CI, 0.37–1.50	YES	NO	NO
ELISA IgG antibodies against CPNEU, HPY, CMV, HSV1&2, HAV, HBV, HCV	(i) CVD Composite:TIA, Stroke, AMI, PAD, CVD death (ii) Severity of Carotid Plaque (3-Level)	(i) Infectious burden continuous: Adj. HR=10.9, 95 % CI, 6.2–19.5 (ii) Infectious burden continuous: Chi-Sqr p=<0.0001	YES	NO	NO
ELISA IgG CPNEU, HPY, CMV, HSV1&2, HAV, HBV, HCV	(i) CVD Composite: MI, Stroke, CVD death	(i) Infectious burden continuous: Adj. HR=1.13, 95 % CI, 0.85–1.51)	YES Met Synd	CRP	NO
ELISA IgG CPNEU , *H. pylori*, CMV, HSV1&2	(i) Risk of Total Stroke	(i) Infectious Burden Index per SD: Adj. HR=1.40, 95 % CI, 1.03–1.91	YES	CRP	YES
	(i) Total Stroke Outcome (ii) Ischemic Stroke Subtype (iii) Large artery atherosclerosis Stroke Subtype (iv) Stroke or TIA	(i) ≥3Month per year chronic bronchitis: Adj. OR=2.63, 95 % CI, 1.17–5.94 (ii) >3 Month per year chronic bronchitis: Adj. 2.45, 95 % CI, 1.10–5.46 (iii) >3 Month per year chronic bronchitis: Adj. OR=4.77, 95 % CI, 1.18–19.3) (iv) >2 Flu-like illness per year: Adj. OR=3.54, 95 % CI, 1.52–8.27	YES	NO	NO

(continued)

Table 11.2 (continued)

Reference	Author	Year	Location	Design	Follow-up	Sample size [Sample age]	Infect burden
[128]	Elkind et al.	2011	US	Case-crossover Prospective	Hazard period 14–90 days	Case-Crossover $n=669$ [Mean age = 74] Prospective $n=5,888$ [mean age = 72.8]	Hospitalization for infection

Pathogens	Outcome	Findings	Adjustment Variables		
			Diabetes	Immune	SEP
ICD9 code for Respiratory, Assorted, Urinary, Skin, Bacteremia, and Osteomyelitis	(i) Ischemic Stroke	(i) Case-crossover 14-days: OR=8.0, 95 % CI, 1.6–77.3 (i) Case-crossover 30-days: OR=7.3, 95 % CI, 1.9–40.9 (i) Case-crossover 90-days: OR=3.4, 95 % CI, 1.8–6.5 (i) Survival analysis 14-days: Adj. HR=3.9, 95 % CI, 1.9–7.9) (i) Survival analysis 30-days: Adj. HR=2.4, 95 % CI, 1.3–4.4 (i) Survival analysis 90-days: Adj. HR=2.4, 95 % CI, 1.6–3.4).	N/A	NO	NO

stress, race-ethnicity, and pathogen strain exposures. Homogenous populations like Framingham may not detect IB associations because sampling may lack variability in causal antecedents of IB to stroke pathways. The analogy for the Framingham results would be akin to trying to detect an association between HA1C with stroke using homogenously nondiabetic populations—there would be no detectable association.

Secondary analysis of the Heart Outcome Prevention Study (HOPE) Canadian trial evaluated an IB measure using summation of four positive serologies (MIF IgG for *C. pneumoniae* and EIA IgG for HPY, CMV, and HAV). This study population was recruited from individuals previously experiencing high-risk conditions such as CAD, stroke, PAD, or diabetes [125]. Individuals with the highest pathogen score, relative to the lowest score group, were at increased risk for time to incident MI, stroke, or CV mortality (adjusted HR = 1.41, 95 % CI, 1.02–1.96), with adjustments for clinical risk factor but not socioeconomic indicators. Total pathogen score was associated with MI (adjusted HR Score 4 vs. Score 0–1 = 1.57, 95 % CI, 1.08–2.30), but not stroke (adjusted HR Score 4 vs. Score 0–1 = 0.75, 95 % CI, 0.37–1.50), relative to lowest score group. The magnitude of effect for total pathogen proved to be a stronger predictor than any single serological marker.

An Italian prospective study created three categories of carotid plaque severity based on intima-media thickness using color duplex Doppler scanning and measured a 4-pathogen IB score (sum of positive ELISA IgG serologies for *C. pneumoniae* HPY, cytotoxic *H. pylori*, CMV) at baseline. The cohort was prospectively followed for 5 years and found individuals positive for all four pathogen antibodies were at increased risk of combined clinical events (TIA, stroke, MIA, or vascular death) in logistic models (adjusted OR = 10.9, 95 % CI, 6.2–19.5) [126]. Cross-sectional analysis from the study indicated IB was associated with baseline severity of carotid atherosclerosis, as measured by IMT thickness ($p < 0.0001$).

A Taiwanese study of coronary angiography patients compared summary score measures of metabolic syndrome and IB and the relationship with time to major adverse cardiovascular events (including cardiovascular mortality, MI, and stroke) defined by standard criteria and imaging. The IB score was comprised of the sum of seven pathogens (seropositivity for ELISA IgG antibodies against *C. pneumoniae* HPY, CMV, HSV1 and 2, HAV, HBV, and HCV). IB score, modeled as a continuous variable, was not found to be associated with a composite CVD outcome (adjusted HR = 1.13, 95 %CI, 0.85–1.51) in models that adjusted for metabolic syndrome and CRP. The authors claimed that metabolic syndrome was more prominent in the model, but it may be possible that inflammation as measured by CRP and individual elements of the metabolic syndrome, like diabetes, may be mediators of the causal pathway of IB. Adjustment for mediator and exposure in the same model diminished the observed effect of IB. Estimates of IB without CRP were not presented.

A prospective analysis of the NOMAS cohort data evaluated the association between ischemic stroke and IB index. The IB index created as sum of weights for each positive serology for an individual was used as an independent predictor of stroke risk. IB index was found to be associated with an increased risk of ischemic stroke (adjusted HR = 1.40, 95 % CI, 1.03–1.91) in fully adjusted models [99].

Key factors may have enabled detection of IBI effect on stroke, such as weighted measures to capture the impact of the "pathogen burden" construct without a prior assumption of association strength, allowing for bacterial pathogens and viral pathogen to contribute differently to overall risk. The NOMAS study offered long-term periods of observation, with minimal loss to follow-up, perhaps permitting sufficient time to discern an impact of baseline IB on atherosclerosis progression and plaque rupture Similar to the Framingham results, the sum of positive pathogen serologies in NOMAS was not associated with vascular endpoints (not shown in paper), but a reinterpretation of IB with empirically derived weights did establish an association. The constituency of NOMAS—sampled from the general population and with greater variability of determinants of socioeconomic position, such as race/ethnicity—may also play a role in detection of an association. This analysis also detected effect of heterogeneity for the IB–stroke association by diabetes status, with elevated risk among diabetics (adjusted HR per SD HR = 1.63, 95 % CI, 1.16–2.29). This interaction is suggestive that diabetes status may influence how IB may promote stroke events.

The set of available analysis in the arena of infectious burden and stroke risk provides a compelling but incomplete analysis of causal relationships. Many important questions remain unanswered by existing observational study designs. Key limitations include but are not limited to *measurement issues* (optimal measurement of chronic infection: upper quantile, weighted index, and combining clinical histories), *selecting counterfactuals* (appropriateness of reference groups: small size and comparability of individuals who have no burden), *modeling considerations* (completeness and suitability of adjustment variables: SEP, healthy behaviors, immune factors, genetic, as well as limitation of models to capture dynamics), *incomplete causal models* (issues include using confounders and mediators in same models, limited attempted to falsify proposed mechanisms linking burden to stroke risk, and clarity in designating clear atherosclerotic and thrombosis outcomes: initiation, progression, or rupture), *pathogen heterogeneity and natural histories* (unclear effects of strain pathogenicity and limited understanding of aberrant microbiome behavior), and finally *time dynamics* (issues include timing of exposure, effect of immune senescence, and staging of atherosclerosis at the time of study observation).

Recurrent Infectious Burden and Stroke

The high recurrence of respiratory infections can be designated an alternate form of IB. A German study of stroke patients (defined by TOAST criteria) and randomly selected controls matched on age, sex, and area of residence measured burden as self-reported frequency of chronic bronchitis or recurrence of flu-like symptoms [127]. Chronic bronchitis was assessed by questionnaire and recurrent flu-like symptoms were assessed by inquiring for the frequency of respiratory infection in the past 5 years. The study found that chronic bronchitis ≥3 months per year was associated with total stroke (adjusted OR = 2.63, 95 % CI, 1.17–5.94) and found ischemic stroke subtypes (adjusted 2.45, 95 % CI, 1.10–5.46) and large artery atherosclerosis subtypes (adjusted OR=4.77, 95 % CI, 1.18–19.3) associated with recurrent respiratory infections in multivariate conditional logistic regression models, adjusted for school education. Additionally, recurrent exposure to flu-like illness (nonspecific measure of respiratory infection) was associated with stroke and TIA (adjusted OR=3.54, 95 % CI, 1.52–8.27). This study found the association between chronic bronchitis and stroke was modified by education levels, with those in the low-level education having a twofold increase in pathogen burden–stroke association, compared to those with higher educational attainment. Additionally, the study found that health seeking behavior during infection was associated with stroke events, and healthy behaviors may be an important covariate for adjustment. This study did not validate self-reports with medical records and may be subject to differential recall of exposure by case subjects.

Acute Infectious Burden and Stroke

Acute infection has been implicated in elevated short-term stroke risk. Predictors of short-term stroke risk, operating in the range of days and weeks, operate under different dynamic mechanisms, as compared to traditional vascular risk factors, whose processes are in the scale of decades [129]. Study designs that can evaluate short-term temporal associations between infection and stroke have used various time-series, time-varying Cox models, and case-crossover analytic designs.

A study from Toschke et al. utilized a 3H-algorithm time-series algorithm to analyze 2,874 incident strokes from South London Stroke Registry (SLSR) with a capture area of 271,817 inhabitants [130]. The analysis was able to detect increases in ischemic stroke incidence with 2 weeks of peak influenza epidemic curves and increases in hemorrhagic stroke at 4-week post-peak influenza periods. This study is suggestive of increased hazard periods in the populations during influenza epidemics can persist 2–4-week post-peak transmission periods.

Individual cohort data also provides temporal evidence of acute hazard experiences immediately following exposure to acute infections and suggests a role for pathogens as acute precipitants of cerebral infarctions. A dual analysis of case-crossover ($n=669$; mean age=74; hazard periods=14–90 days) and survival analysis ($n=5,888$; mean age=72.8; hazard period=14–90 days) approaches using collected data from the Cardiovascular Health Study found strong associations between short-term hazard periods and incident ischemic strokes. Case-crossover analysis utilizes an earlier exposure period of an individual as his own control, to minimize possible confounding of exposure and outcomes associations, under the assumption of major changes in the individual between exposure periods and is modeled using conditional logistic models [131]. Time-dependent exposure in survival analysis allows investigators to specify functional form of predictors over time (i.e., short-term exposure immediately after infection, no exposure post an a priori cutoff period) and also adjust for possible residual confounding factors, such as age, that may remain in case-crossover analysis [102]. The CHS case-crossover analysis found a strong association of increased risk for ischemic stroke in periods immediately following infection at 14 days (OR=8.0, 95 % CI, 1.6–77.3), 30 days (OR=7.3, 95 % CI, 1.9–40.9), and 90 days (OR=3.4, 95 % CI, 1.8–6.5), with risk declining as time progressed. Risk of ischemic stroke in time intervals after acute hospitalization for infection, as assessed by Cox models, was increased at 14 days (adjusted HR=3.9, 95 % CI, 1.9–7.9), 30 days (adjusted HR=2.4, 95 % CI, 1.3–4.4), and 90 days (adjusted HR=2.4, 95 % CI, 1.6–3.4). Associations were found to be inversely modified by IMT, suggesting that thicker, older plaque may represent more stable plaque formation with decreased ischemic risk, as compared to new or intermediate plaque. Small limitations, such as small exposure sizes, use of administrative data for exposure assessment, or possible carryover effects of infection-driven plaque remodeling, would likely not impact large effect size, dose–response associations.

Risk Mitigation Strategies and Interventions

The data presented in this chapter is suggestive that chronic exposure to infectious agents, through asymptomatic, persistent infections or clinically acute manifestations, may elevate cerebrovascular risk. Further etiological and epidemiological investigations are warranted to develop more refined models of causal effect of joint infections and vascular disease, and in particular studies that can compare and contrast measurement of burden. Current available data does generate tempting intervention targets and suggest possible clinical strategies to affect cerebrovascular risk.

Current Clinical and Population Health Approaches

A series of randomized placebo-controlled trials of macrolide antibiotics were the first large-scale attempts to intervene on the hypothesized infectious pathways that influence coronary artery and myocardial infarction risk [132]. Antibiotic trials that evaluated stroke outcomes data found no preventative effects on stroke among patients with stable CAD in the ACADEMIC trial (nonfatal stroke outcome, $p=0.31$) and the ACES trial (composites outcome risk reduction =–0.13 %, 95 % CI, −0.73 to 0.26) or among patients with acute coronary syndrome within the CLARIFY (stroke events HR=0.69, 95 % CI, 0.9–5.1), the ANTIBIO [stroke events during inpatient period ($p=0.060$) or stroke events during 12-month follow-up ($p=0.634$) with effect estimates not available], and PROVE-IT trials (% Risk Change= +0.02, with p-values or confidence intervals not available) [133–136]. The results of these trials corroborated with large observational studies of claims databases ($n=354{,}358$) that found no association between antibiotic use and myocardial infarction [137]. The trials and recent meta-analysis surrounding these trials show a lack of efficacy of macrolide antibiotics to reduce coronary and cerebrovascular events [132]. More recent reevaluations of these antibiotic trials suggest the null-findings of the trial do not falsify a causal role for pathogen burden. The trials have been criticized for several shortcomings, including not addressing the role of pathogens in progression versus plaque stability; failure to target viral pathogens; and the possibility of decreased effectiveness of antibiotics to impact bacteria in protective latent phase as observed in biofilm formations [138]. Recent studies have suggested previously unknown serum variability in complement C4 deficiencies may modify antibiotic effectiveness, with demonstrable risk reduction in randomized trials of clarithromycin among complement C4 deficient individuals [139].

An alternate risk reduction strategy has focused on influenza. The influenza vaccine has proven an effective strategy at reducing overall risk of MI and cardiovascular death, as documented in some studies case–control studies [140, 141] and cohort studies [142], but many other studies found no protective effect among older high-risk adults [143–145]. Recent large-scale cohort studies conducted in China ($n=36{,}636$) reported a large efficacy associated with joint inoculation of 23-valent pneumococcal and influenza vaccination on reducing stroke events (adjusted HR, 0.67; 95 % CI, 0.54–0.83) [146]. Recent meta-analysis of existing of four randomized clinical trials from Netherlands [147], Argentina [148], Poland [149], and Thailand [150] found the flu vaccine provided an approximate halving of risk for MACE events (OR=0.52; $p=0.0002$) and has prompted recommendation for both MI and stroke prevention purposes in high-risk individuals. The evidence prompted the American Heart Association and American College of Cardiology to issue a scientific advisory, recommending seasonal influenza vaccines for individuals with coronary artery disease and severe atherosclerosis [151]. However, mismatch of influenza vaccine against prevalent strains due to antigenic drift or emergence of new strain through antigenic shift can limit the effectiveness of influenza vaccines on any given year [152]. Many researchers have questioned the efficacy of influenza vaccines when administered to elderly populations with senesced immune systems [153].

Novel Applications: Retooling Pharmacotherapies

Recent developments in the use of statins have suggested their anti-inflammatory and positive immunomodulatory impacts may have broader clinical applications than lipid regulation alone [154]. Enthusiasm with regard to expanded use of statins stems from studies that observed protective benefits during acute infection. Reviews of bacterial sepsis studies found several observational studies detected protective effects against mortality among statins users [155]. A multistate evaluation of patients hospitalized with laboratory-confirmed influenza in the USA found statin users had a 50 % reduction in mortality (adjusted OR= 0.59, 95 % CI, 0.38–0.92) after adjusting for age; race; cardiovascular, lung, and renal disease; influenza vaccination status; and antiviral administration [156]. Use of statins has been suggested as a possible prophylaxis in the event of a pandemic [157], and secondary analysis of the JUPITER trial found that incident pneumonia was reduced in the statin arm [158]. It is possible that statin administration during acute phases of infection may reduce atherothrombotic risk and reduce ischemic cerebrovascular event risk in patients. Similar investigations of antiplatelet therapies for cardiovascular risk reduction postinfection may warrant investigation, although effects of antiviral drugs agents such as oseltamivir have been shown to be inhibited by clopidogrel [159]. Antivirals such as oseltamivir have been found to reduce stroke/TIA risk in large claims databases and may represent another strategy for stroke–risk reduction that is currently implementable [160]. Finally, recent studies suggest influenza infection may complicate acute stroke treatments, such as tissue plasminogen activator [161] and may alter post-stroke survival [162].

Downstream, pleiotropic, and pro-inflammatory cytokines have been identified as drug targets for anti-inflammatory treatment trials using human monoclonal antibody therapies directed at high vascular risk individuals [163]. Canakinumab, a biopharmaceutical drug inhibiting interleukin-1β directly, has been shown to significantly reduce hs-CRP ($p<0.02$), fibrinogen ($p<0.001$), and IL6 levels ($p<0.008$), with adverse events very similar to the placebo group [164]. An alternate strategy under evaluation is low-dose methotrexate, an inexpensive treatment for rheumatoid arthritis, with efficacy to reduce homocysteine levels, TNFα, IL-6, and CRP [165, 166]. Trials of these two inflammatory targets intend to link changes in surrogate outcome with reduction in cardiovascular events, and evaluation of possible interaction with high IB may elucidate possible roles of infectious agent and inflammation. Upstream immune cells targets are also trending areas of research, but are further from clinical application. Recent findings demonstrate that infusion of CD8+ Treg cells, involved in dampening immune response postinfection, can downregulate cytokine production and stimulation of new T cells to reduce symptoms of autoimmune conditions [167]. Research into drug compounds to expand naturally occurring populations of CD8+ Treg cells may allow for future top-down immune modulatory approaches.

Novel Applications: Optimizing Clinical Practice

Associations between hospitalizations for infectious syndromes and prospective stroke risk are increasingly studied, but inpatient strategies to mitigate stroke risk are not. Leveraging biomedical informatics databases may be an opportunity to optimize care of high-risk cardiovascular patients by increasing clinical cognition of elevated risk [168]. Risk stratification for cardiovascular events, based on the prospective Framingham Risk Scores [169] or similar predictive measures, for patients hospitalized for infection may be one way to identify high-risk vascular patients and target them for prophylaxis, such as high-dose statins. The positive predictive value of Framingham Risk Scores in the acute infection setting has not been validated, and combination with inpatient severity indexes that include mechanical ventilation, septic shock, acute respiratory distress syndrome, and CURB-65 scores (a mortality prediction tool that incorporates measures of confusion onset, urea concentrations, respiratory rates, blood pressure and >65 years age) warrant further investigation [170, 171]. New risk detection algorithms can be incorporated into decision support prompts for high-risk vascular inpatients at critical care junctions, such as during care provider order entry (CPOE) and may be another point of intervention to optimize screening and possible therapeutic intervention [172]. Ensuring continuity of medication for comorbid cardiovascular risk factors through use of medical reconciliation strategies can serve as platform for reducing stroke risk [173]. Presently, medication discrepancies remain prevalent even in elite medical institutions (unintentional medication discrepancies at 1.4 per patient) [174] and cardiovascular drugs are also the most frequent drug class involved in medical reconciliation errors at discharge [175]. Discontinuation of statins has been implicated in elevated MI risk [176] and increased mortality following stroke [177]; and ensuring medication adherence during hospitalization and post-discharge may reduce discontinuation-associated morbidity and mortality. The prospective impact of discharge medical reconciliation errors on long-term vascular outcomes remains understudied. Finally, patient–physician encounters for infectious exposure represent another opportunity for cardiovascular risk factor screening and opportunity to provide care, but current scope of specialist practices remains increasing stove-piped. Surveys of infectious disease-certified physicians that evaluate influenza patients report less comfort in prescribing medications for common comorbid conditions than general medicine physicians [178]. Broader sensitivity training of infectious disease specialists to update treatment decisions, in light of cardiovascular comorbid conditions, may create opportunities to provide care.

Conclusion

In summary, emerging molecular, clinical, and epidemiological evidence evaluating infectious burden constructs provide suggestive, but not irrefutable, evidence linking (1) multiple, chronic coinfection, (2) acute infections, and (3) recurrent infection to carotid plaque and stroke risk. This chapter has focused on clinical measurements tools, as well as prevalent study designs, as important precursors for valid causal inference. Minimizing measurement error of the infection burden construct and identification of gold standards for measuring chronic infection is a requisite for any serious attempts to falsify the IB hypothesis. Future developments in the field may feature trials that investigate influence IB by mitigating inflammatory and prothrombotic downstream pathways, or by evaluating large payer datasets with precise tracking of lab testing and outpatient clinical visits, or by further efforts to link IB with in vivo imaging of plaque progression or unstable plaque. In the interim, lesson learned from investigating IB suggests there are ample opportunities to optimize clinical care practices to reduce risk of cerebrovascular events.

References

1. Alwan A, World Health Organization (2011) Global status report on noncommunicable diseases 2010. World Health Organization, Geneva, Switzerland
2. White H, Boden-Albala B, Wang C, Elkind MSV, Rundek T, Wright CB et al (2005) Ischemic stroke subtype incidence among whites, blacks, and hispanics: the Northern Manhattan Study. Circulation 111(10):1327–31
3. Roger VL, Go AS, Lloyd-Jones DM, Benjamin EJ, Berry JD, Borden WB et al (2012) Heart disease and stroke statistics–2012 update: a report from the American Heart Association. Circulation 125(1):e2–220
4. Bygbjerg IC (2012) Double burden of noncommunicable and infectious diseases in developing countries. Science 337(6101):1499–501
5. Elkind MS (2010) Infectious burden: a new risk factor and treatment target for atherosclerosis. Infect Disord Drug Targets 10(2):84–90
6. Corrales-Medina VF, Madjid M, Musher DM (2010) Role of acute infection in triggering acute coronary syndromes. Lancet Infect Dis 10(2):83–92
7. Wang CM, Kaltenboeck B (2010) Exacerbation of chronic inflammatory diseases by infectious agents: fact or fiction? World J Diabetes 1(2):27–35
8. Frothingham C (1911) The relation between acute infectious diseases and arterial lesions. Arch Intern Med 7:153–62
9. Tedgui A, Mallat Z (2006) Cytokines in atherosclerosis: pathogenic and regulatory pathways. Physiol Rev 86(2):515–81
10. Hagiwara N, Toyoda K, Inoue T, Shimada H, Ibayashi S, Iida M et al (2007) Lack of association between infectious burden and carotid atherosclerosis in Japanese patients. J Stroke Cerebrovasc Dis 16(4):145–52
11. Rashid H, Khandaker G, Booy R (2012) Vaccination and herd immunity: what more do we know? Curr Opin Infect Dis 25(3):243–9. doi:10.1097/QCO.0b013e328352f727
12. Cohen-Cole E, Fletcher JM (2008) Is obesity contagious? Social networks vs. environmental factors in the obesity epidemic. J Health Econ 27(5):1382–7
13. Christakis NA, Fowler JH (2008) The collective dynamics of smoking in a large social network. N Engl J Med 358(21):2249–58
14. Visser MR, Tracy PB, Vercellotti GM, Goodman JL, White JG, Jacob HS (1988) Enhanced thrombin generation and platelet binding on herpes simplex virus-infected endothelium. Proc Natl Acad Sci U S A 85(21):8227–30
15. Etingin OR, Silverstein RL, Friedman HM, Hajjar DP (1990) Viral activation of the coagulation cascade: molecular interactions at the surface of infected endothelial cells. Cell 61(4):657–62
16. Visser MR, Vercellotti GM (1993) Herpes simplex virus and atherosclerosis. Eur Heart J 14(Suppl K):39–42
17. WuDunn D, Spear PG (1989) Initial interaction of herpes simplex virus with cells is binding to heparan sulfate. J Virol 63(1):52–8
18. van Dam-Mieras MC, Muller AD, van Hinsbergh VW, Mullers WJ, Bomans PH, Bruggeman CA (1992) The procoagulant response of cytomegalovirus infected endothelial cells. Thromb Haemost 68(3):364–70
19. Dechend R, Maass M, Gieffers J, Dietz R, Scheidereit C, Leutz A et al (1999) Chlamydia pneumoniae infection of vascular smooth muscle and endothelial cells activates NF-κB and induces tissue factor and PAI-1 expression : a potential link to accelerated arteriosclerosis. Circulation 100(13):1369–73
20. Kreutz RP, Bliden KP, Tantry US, Gurbel PA (2005) Viral respiratory tract infections increase platelet reactivity and activation: an explanation for the higher rates of myocardial infarction and stroke during viral illness. J Thromb Haemost 3(9):2108–9
21. Kreutz RP, Tantry US, Bliden KP, Gurbel PA (2007) Inflammatory changes during the 'common cold' are associated with platelet activation and increased reactivity of platelets to agonists. Blood Coagul Fibrinolysis 18(8):713–8. doi:10.1097/MBC.0b013e328201c77e

22. Loppnow H, Buerke M, Werdan K, Rose-John S (2011) Contribution of vascular cell-derived cytokines to innate and inflammatory pathways in atherogenesis. J Cell Mol Med 15(3):484–500
23. Dudding L, Haskill S, Clark BD, Auron PE, Sporn S, Huang ES (1989) Cytomegalovirus infection stimulates expression of monocyte-associated mediator genes. J Immunol 143(10):3343–52
24. Liao W (1996) Endotoxin: possible roles in initiation and development of atherosclerosis. J Lab Clin Med 128(5):452–60
25. Lin TM, Chen CH, Wu HL, Wang CH, Chen YL, Eng HL (2008) The association of C (−260)-->T polymorphism in CD14 promoter and Chlamydia pneumoniae infection in ischemic stroke patients. Am J Clin Pathol 130(4):595–601
26. Ager A (1993) Lymphocyte-vascular endothelial cell interactions in the immune response. Clin Exp Immunol 93(Suppl 1):5–6
27. Sedmak DD, Knight DA, Vook NC, Waldman JW (1994) Divergent patterns of ELAM-1, ICAM-1, and VCAM-1 expression on cytomegalovirus-infected endothelial cells. Transplantation 58(12):1379–85
28. Etingin OR, Silverstein RL, Hajjar DP (1991) Identification of a monocyte receptor on herpesvirus-infected endothelial cells. Proc Natl Acad Sci 88(16):7200–3
29. Hajjar DP, Fabricant CG, Minick CR, Fabricant J (1986) Virus-induced atherosclerosis. Herpesvirus infection alters aortic cholesterol metabolism and accumulation. Am J Pathol 122(1):62–70
30. Nachtigal M, Legrand A, Greenspan P, Nachtigal SA, Nagpal ML (1990) Immortalization of rabbit vascular smooth muscle cells after transfection with a fragment of the BglII N region of herpes simplex virus type 2 DNA. Intervirology 31(2–4):166–74
31. Katritsis DG, Pantos J, Efstathopoulos E (2007) Hemodynamic factors and atheromatic plaque rupture in the coronary arteries: from vulnerable plaque to vulnerable coronary segment. Coron Artery Dis 18(3):229–37
32. Fleet JC, Clinton SK, Salomon RN, Loppnow H, Libby P (1992) Atherogenic diets enhance endotoxin-stimulated interleukin-1 and tumor necrosis factor gene expression in rabbit aortae. J Nutr 122(2):294–305
33. van der Wal AC, Piek JJ, de Boer OJ, Koch KT, Teeling P, van der Loos CM et al (1998) Recent activation of the plaque immune response in coronary lesions underlying acute coronary syndromes. Heart 80(1):14–8
34. Loftus IM, Naylor AR, Goodall S, Crowther M, Jones L, Bell PRF et al (2000) Increased matrix metalloproteinase-9 activity in unstable carotid plaques: a potential role in acute plaque disruption. Stroke 31(1):40–7
35. Sugamata R, Dobashi H, Nagao T, Yamamoto K-i, Nakajima N, Sato Y et al (2012) Contribution of neutrophil-derived myeloperoxidase in the early phase of fulminant acute respiratory distress syndrome induced by influenza virus infection. Microbiol Immunol 56(3):171–82
36. Schultz DR, Arnold PI (1990) Properties of four acute phase proteins: C-reactive protein, serum amyloid A protein, alpha 1-acid glycoprotein, and fibrinogen. Semin Arthritis Rheum 20(3):129–47
37. Woodhouse PR, Meade TW, Khaw K-T (1997) Plasminogen activator inhibitor-1, the acute phase response and vitamin C. Atherosclerosis 133(1):71–6
38. van der Wal AC, Becker AE, van der Loos CM, Das PK (1994) Site of intimal rupture or erosion of thrombosed coronary atherosclerotic plaques is characterized by an inflammatory process irrespective of the dominant plaque morphology. Circulation 89(1):36–44
39. Hansson GK, Robertson A-KL, Söderberg-Nauclér C (2006) Inflammation and atherosclerosis. Ann Rev Pathol: Mechanisms of Disease 1(1):297–329
40. Kloner RA, Poole WK, Perritt RL (1999) When throughout the year is coronary death most likely to occur? A 12-year population-based analysis of more than 220,000 cases. Circulation 100(15):1630–4
41. Williams ES, Zipes DP (2003) The winter's tale—and toll. Am Heart J 146(6):941–3

42. Speir E, Modali R, Huang ES, Leon MB, Shawl F, Finkel T et al (1994) Potential role of human cytomegalovirus and p53 interaction in coronary restenosis. Science 265(5170):391–4
43. Kovacs A, Weber ML, Burns LJ, Jacob HS, Vercellotti GM (1996) Cytoplasmic sequestration of p53 in cytomegalovirus-infected human endothelial cells. Am J Pathol 149(5):1531–9
44. Petri M, Perez-Gutthann S, Spence D, Hochberg MC (1992) Risk factors for coronary artery disease in patients with systemic lupus erythematosus. Am J Med 93(5):513–9
45. Rosner S, Ginzler EM, Diamond HS, Weiner M, Schlesinger M, Fries JF et al (1982) A multicenter study of outcome in systemic lupus erythematosus. II. Causes of death. Arthritis Rheum 25(6):612–7
46. Medina G, Casaos D, Jara LJ, Vera-Lastra O, Fuentes M, Barile L et al (2003) Increased carotid artery intima-media thickness may be associated with stroke in primary antiphospholipid syndrome. Ann Rheum Dis 62(7):607–10
47. Kaplan MJ, McCune WJ (2003) New evidence for vascular disease in patients with early rheumatoid arthritis. Lancet 361(9363):1068–9
48. Alard J-E, Dueymes M, Youinou P, Jamin C (2007) Modulation of endothelial cell damages by anti-Hsp60 autoantibodies in systemic autoimmune diseases. Autoimmun Rev 6(7):438–43
49. Mayr M, Metzler B, Kiechl S, Willeit J, Schett G, Xu Q et al (1999) Endothelial cytotoxicity mediated by serum antibodies to heat shock proteins of Escherichia coli and Chlamydia pneumoniae: immune reactions to heat shock proteins as a possible link between infection and atherosclerosis. Circulation 99(12):1560–6
50. Kiechl S, Egger G, Mayr M, Wiedermann CJ, Bonora E, Oberhollenzer F et al (2001) Chronic infections and the risk of carotid atherosclerosis: prospective results from a large population study. Circulation 103(8):1064–70
51. Johnson SA, Cambier JC (2004) Ageing, autoimmunity and arthritis: senescence of the B cell compartment—implications for humoral immunity. Arthritis Res Ther 6(4):131–9
52. High KP, Akbar AN, Nikolich-Zugich J (2012) Translational research in immune senescence: assessing the relevance of current models. Semin Immunol 24(5):373–82
53. Almanzar G, Schwaiger S, Jenewein B, Keller M, Herndler-Brandstetter D, Wurzner R et al (2005) Long-term cytomegalovirus infection leads to significant changes in the composition of the CD8+ T-cell repertoire, which may be the basis for an imbalance in the cytokine production profile in elderly persons. J Virol 79(6):3675–83
54. Pawelec G, Akbar A, Beverley P, Caruso C, Derhovanessian E, Fulop T et al (2010) Immunosenescence and cytomegalovirus: where do we stand after a decade? Immun Ageing 7(1):13
55. Maes M (1995) Evidence for an immune response in major depression: a review and hypothesis. Prog Neuropsychopharmacol Biol Psychiatry 19(1):11–38
56. Whipple MO, Lewis TT, Sutton-Tyrrell K, Matthews KA, Barinas-Mitchell E, Powell LH et al (2009) Hopelessness, depressive symptoms, and carotid atherosclerosis in women: The Study of Women's Health Across the Nation (SWAN) Heart Study. Stroke 40(10):3166–72
57. Edmondson D, Newman JD, Whang W, Davidson KW (2013) Emotional triggers in myocardial infarction: do they matter? Eur Heart J 34(4):300–6
58. Miller GE, Freedland KE, Duntley S, Carney RM (2005) Relation of depressive symptoms to C-reactive protein and pathogen burden (cytomegalovirus, herpes simplex virus, Epstein-Barr virus) in patients with earlier acute coronary syndromes. Am J Cardiol 95(3):317–21
59. Banerjee C, Moon YP, Paik MC, Rundek T, Mora-McLaughlin C, Vieira JR et al (2012) Duration of diabetes and risk of ischemic stroke: the Northern Manhattan Study. Stroke 43:1212–7
60. Sullivan T, Ben Amor Y (2012) The co-management of tuberculosis and diabetes: challenges and opportunities in the developing world. PLoS Med 9(7):e1001269
61. Ishiguro T, Takayanagi N, Yamaguchi S, Yamakawa H, Nakamoto K, Takaku Y et al (2013) Etiology and factors contributing to the severity and mortality of community-acquired pneumonia. Intern Med 52(3):317–24
62. Rizzo A, Paolillo R, Iafusco D, Prisco F, Romano CC (2012) Chlamydia pneumoniae infection in adolescents with type 1 diabetes mellitus. J Med Microbiol 61(Pt 11):1584–90

63. Jeon CY, Haan MN, Cheng C, Clayton ER, Mayeda ER, Miller JW et al (2012) Helicobacter pylori infection is associated with an increased rate of diabetes. Diabetes Care 35(3):520–5
64. Oldstone MB, Nerenberg M, Southern P, Price J, Lewicki H (1991) Virus infection triggers insulin-dependent diabetes mellitus in a transgenic model: role of anti-self (virus) immune response. Cell 65(2):319–31
65. Glaser R, Kiecolt-Glaser J (2005) How stress damages immune system and health. Discov Med 5(26):165–9
66. Glaser R, Kiecolt-Glaser JK, Speicher CE, Holliday JE (1985) Stress, loneliness, and changes in herpesvirus latency. J Behav Med 8(3):249–60
67. Kostulas N, Kivisakk P, Huang Y, Matusevicius D, Kostulas V, Link H (1998) Ischemic stroke is associated with a systemic increase of blood mononuclear cells expressing interleukin-8 mRNA. Stroke 29(2):462–6
68. Grau AJ, Reis A, Buggle F, Al-Khalaf A, Werle E, Valois N et al (2001) Monocyte function and plasma levels of interleukin-8 in acute ischemic stroke. J Neurol Sci 192(1–2):41–7
69. Risley P, Jerrard-Dunne P, Sitzer M, Buehler A, von Kegler S, Markus HS (2003) Promoter polymorphism in the endotoxin receptor (CD14) is associated with increased carotid atherosclerosis only in smokers: the Carotid Atherosclerosis Progression Study (CAPS). Stroke 34(3):600–4
70. Park MH, Min JY, Koh SB, Kim BJ, Park MK, Park KW et al (2006) Helicobacter pylori infection and the CD14 C(−260)T gene polymorphism in ischemic stroke. Thromb Res 118(6):671–7
71. Kis Z, Sas K, Gyulai Z, Treso B, Petrovay F, Kapusinszky B et al (2007) Chronic infections and genetic factors in the development of ischemic stroke. New Microbiol 30(3):213–20
72. Tjernstrom F, Hellmer G, Nived O, Truedsson L, Sturfelt G (1999) Synergetic effect between interleukin-1 receptor antagonist allele (IL1RN*2) and MHC class II (DR17, DQ2) in determining susceptibility to systemic lupus erythematosus. Lupus 8(2):103–8
73. Schrijver HM, Crusius JB, Uitdehaag BM, Garcia Gonzalez MA, Kostense PJ, Polman CH et al (1999) Association of interleukin-1beta and interleukin-1 receptor antagonist genes with disease severity in MS. Neurology 52(3):595–9
74. Bioque G, Crusius JB, Koutroubakis I, Bouma G, Kostense PJ, Meuwissen SG et al (1995) Allelic polymorphism in IL-1 beta and IL-1 receptor antagonist (IL-1Ra) genes in inflammatory bowel disease. Clin Exp Immunol 102(2):379–83
75. Gretarsdottir S, Thorleifsson G, Reynisdottir ST, Manolescu A, Jonsdottir S, Jonsdottir T et al (2003) The gene encoding phosphodiesterase 4D confers risk of ischemic stroke. Nat Genet 35(2):131–8
76. Ameziane N, Beillat T, Verpillat P, Chollet-Martin S, Aumont MC, Seknadji P et al (2003) Association of the Toll-like receptor 4 gene Asp299Gly polymorphism with acute coronary events. Arterioscler Thromb Vasc Biol 23(12):e61–4
77. Arroyo-Espliguero R, Avanzas P, Jeffery S, Kaski JC (2004) CD14 and toll-like receptor 4: a link between infection and acute coronary events? Heart 90(9):983–8
78. Hernandez-Pacheco G, Aguilar-Garcia J, Flores-Dominguez C, Rodriguez-Perez JM, Perez-Hernandez N, Alvarez-Leon E et al (2003) MHC class II alleles in Mexican patients with rheumatic heart disease. Int J Cardiol 92(1):49–54
79. Romo N, Fito M, Guma M, Sala J, Garcia C, Ramos R et al (2011) Association of atherosclerosis with expression of the LILRB1 receptor by human NK and T-cells supports the infectious burden hypothesis. Arterioscler Thromb Vasc Biol 31(10):2314–21
80. Dowd JB, Zajacova A, Aiello A (2009) Early origins of health disparities: burden of infection, health, and socioeconomic status in US children. Soc Sci Med 68(4):699–707
81. Baltrus PT, Lynch JW, Everson-Rose S, Raghunathan TE, Kaplan GA (2005) Race/ethnicity, life-course socioeconomic position, and body weight trajectories over 34 years: The Alameda County Study. Am J Public Health 95(9):1595–601
82. Mensah GA, Mokdad AH, Ford ES, Greenlund KJ, Croft JB (2005) State of disparities in cardiovascular health in the United States. Circulation 111(10):1233–41

83. Brunner E, Marmot M, Canner R, Beksinska M, Davey Smith G, O'Brien J (1996) Childhood social circumstances and psychosocial and behavioural factors as determinants of plasma fibrinogen. Lancet 347(9007):1008–13
84. Galobardes B, Smith GD, Lynch JW (2006) Systematic review of the influence of childhood socioeconomic circumstances on risk for cardiovascular disease in adulthood. Ann Epidemiol 16(2):91–104
85. Zajacova A, Dowd JB, Aiello AE (2009) Socioeconomic and race/ethnic patterns in persistent infection burden among U.S. adults. J Gerontol A Biol Sci Med Sci 64(2):272–9
86. Moayyedi P, Axon AT, Feltbower R, Duffett S, Crocombe W, Braunholtz D et al (2002) Relation of adult lifestyle and socioeconomic factors to the prevalence of Helicobacter pylori infection. Int J Epidemiol 31(3):624–31
87. Karunakaran KP, Blanchard JF, Raudonikiene A, Shen C, Murdin AD, Brunham RC (2002) Molecular detection and seroepidemiology of the Chlamydia pneumoniae bacteriophage (PhiCpn1). J Clin Microbiol 40(11):4010–4
88. Verkooyen RP, Willemse D, Hiep-van Casteren SC, Joulandan SA, Snijder RJ, van den Bosch JM et al (1998) Evaluation of PCR, culture, and serology for diagnosis of Chlamydia pneumoniae respiratory infections. J Clin Microbiol 36(8):2301–7
89. Cosentini R, Blasi F, Raccanelli R, Rossi S, Arosio C, Tarsia P et al (1996) Severe community-acquired pneumonia: a possible role for Chlamydia pneumoniae. Respiration 63(2):61–5
90. Kauppinen MT, Saikku P, Kujala P, Herva E, Syrjala H (1996) Clinical picture of community-acquired Chlamydia pneumoniae pneumonia requiring hospital treatment: a comparison between chlamydial and pneumococcal pneumonia. Thorax 51(2):185–9
91. Mandell LA, Wunderink RG, Anzueto A, Bartlett JG, Campbell GD, Dean NC et al (2007) Infectious Diseases Society of America/American Thoracic Society Consensus Guidelines on the Management of Community-Acquired Pneumonia in Adults. Clinical Infectious Diseases 44(Suppl 2):S27–72
92. Lequin RM (2005) Enzyme immunoassay (EIA)/enzyme-linked immunosorbent assay (ELISA). Clin Chem 51(12):2415–8
93. Blasi F, Tarsia P, Arosio C, Fagetti L, Allegra L (1998) Epidemiology of Chlamydia pneumoniae. Clin Microbiol Infect 4(Suppl 4):S1–6
94. Tworoger S, Hankinson S (2006) Use of biomarkers in epidemiologic studies: minimizing the influence of measurement error in the study design and analysis. Cancer Causes Control 17(7):889–99
95. Aiello AE, Kaplan GA (2009) Socioeconomic position and inflammatory and immune biomarkers of cardiovascular disease: applications to the panel study of income dynamics. Biodemography Soc Biol 55(2):178–205
96. Epstein SE, Zhou YF, Zhu J (1999) Infection and atherosclerosis: emerging mechanistic paradigms. Circulation 100(4):e20–8
97. Dowell SF, Peeling RW, Boman J, Carlone GM, Fields BS, Guarner J et al (2001) Standardizing Chlamydia pneumoniae assays: recommendations from the Centers for Disease Control and Prevention (USA) and the Laboratory Centre for Disease Control (Canada). Clin Infect Dis 33(4):492–503
98. Aiello AE, Diez-Roux A, Noone AM, Ranjit N, Cushman M, Tsai MY et al (2009) Socioeconomic and psychosocial gradients in cardiovascular pathogen burden and immune response: the multi-ethnic study of atherosclerosis. Brain Behav Immun 23(5):663–71
99. Elkind Mv RPMYP et al (2010) Infectious burden and risk of stroke: the Northern Manhattan Study. Arch Neurol 67(1):33–8
100. Juster RP, McEwen BS, Lupien SJ (2010) Allostatic load biomarkers of chronic stress and impact on health and cognition. Neurosci Biobehav Rev 35(1):2–16
101. Desvarieux M, Demmer RT, Rundek T, Boden-Albala B, Jacobs DR, Sacco RL et al (2005) Periodontal microbiota and carotid intima-media thickness: the Oral Infections and Vascular Disease Epidemiology Study (INVEST). Circulation 111(5):576–82
102. Fisher LD, Lin DY (1999) Time-dependent covariates in the COX proportional-hazards regression model. Annu Rev Public Health 20(1):145–57

103. Steptoe A, Shamaei-Tousi A, Gylfe A, Henderson B, Bergstrom S, Marmot M (2007) Socioeconomic status, pathogen burden and cardiovascular disease risk. Heart 93(12):1567–70
104. Nazmi A, Diez-Roux AV, Jenny NS, Tsai MY, Szklo M, Aiello AE (2010) The influence of persistent pathogens on circulating levels of inflammatory markers: a cross-sectional analysis from the Multi-Ethnic Study of Atherosclerosis. BMC Public Health 10:706
105. Dai DF, Lin JW, Kao JH, Hsu CN, Chiang FT, Lin JL et al (2007) The effects of metabolic syndrome versus infectious burden on inflammation, severity of coronary atherosclerosis, and major adverse cardiovascular events. J Clin Endocrinol Metab 92(7):2532–7
106. Prasad A, Zhu J, Halcox JP, Waclawiw MA, Epstein SE, Quyyumi AA (2002) Predisposition to atherosclerosis by infections: role of endothelial dysfunction. Circulation 106(2):184–90
107. Marchesi S, Lupattelli G, Sensini A, Lombardini R, Brozzetti M, Roscini AR et al (2007) Racial difference in endothelial function: role of the infective burden. Atherosclerosis 191(1):227–34
108. Khairy P, Rinfret S, Tardif J-C, Marchand R, Shapiro S, Brophy J et al (2003) Absence of association between infectious agents and endothelial function in healthy young men. Circulation 107(15):1966–71
109. Bloemenkamp DGM, Mali WPTM, Tanis BC, Rosendaal FR, van den Bosch MAAJ, Kemmeren JM et al (2002) Chlamydia pneumoniae, Helicobacter pylori and cytomegalovirus infections and the risk of peripheral arterial disease in young women. Atherosclerosis 163(1):149–56
110. Mattila KJ (1989) Viral and bacterial infections in patients with acute myocardial infarction. J Intern Med 225(5):293–6
111. Georges JL, Rupprecht HJ, Blankenberg S, Poirier O, Bickel C, Hafner G et al (2003) Impact of pathogen burden in patients with coronary artery disease in relation to systemic inflammation and variation in genes encoding cytokines. Am J Cardiol 92(5):515–21
112. Rothenbacher D, Brenner H, Hoffmeister A, Mertens T, Persson K, Koenig W (2003) Relationship between infectious burden, systemic inflammatory response, and risk of stable coronary artery disease: role of confounding and reference group. Atherosclerosis 170(2):339–45
113. Zhu J, Nieto FJ, Horne BD, Anderson JL, Muhlestein JB, Epstein SE (2001) Prospective study of pathogen burden and risk of myocardial infarction or death. Circulation 103(1):45–51
114. Mattila K, Nieminen MS, Valtonen VV, Rasi VP, Kesaniemi VP, Syrjala SL et al (1989) Association between dental health and acute myocardial infarction. BMJ 298:779–82
115. Rupprecht HJ, Blankenberg S, Bickel C, Rippin G, Hafner G, Prellwitz W et al (2001) Impact of viral and bacterial infectious burden on long-term prognosis in patients with coronary artery disease. Circulation 104(1):25–31
116. Elkind MS, Ramakrishnan P, Moon YP, Boden-Albala B, Liu KM, Spitalnik SL et al (2010) Infectious burden and risk of stroke: the Northern Manhattan Study. Arch Neurol 67(1):33–8
117. Szklo M, Ding J, Tsai MY, Cushman M, Polak JF, Lima J et al (2009) Individual pathogens, pathogen burden and markers of subclinical atherosclerosis: the Multi-Ethnic Study of Atherosclerosis. J Cardiovasc Med 10(10):747–51. doi:10.2459/JCM.0b013e32832cacab
118. Mayr M, Kiechl S, Willeit J, Wick G, Xu Q (2000) Infections, immunity, and atherosclerosis: associations of antibodies to Chlamydia pneumoniae, Helicobacter pylori, and cytomegalovirus with immune reactions to heat-shock protein 60 and carotid or femoral atherosclerosis. Circulation 102(8):833–9
119. Espinola-Klein C, Rupprecht HJ, Blankenberg S, Bickel C, Kopp H, Victor A et al (2002) Impact of infectious burden on progression of carotid atherosclerosis. Stroke 33(11):2581–6
120. Ngeh J, Goodbourn C (2005) Chlamydia pneumoniae, Mycoplasma pneumoniae, and Legionella pneumophila in elderly patients with stroke (C-PEPS, M-PEPS, L-PEPS): a case–control study on the infectious burden of atypical respiratory pathogens in elderly patients with acute cerebrovascular disease. Stroke 36(2):259–65

121. Ridker PM, Hennekens CH, Buring JE, Kundsin R, Shih J (1999) Baseline IgG antibody titers to Chlamydia pneumoniae, Helicobacter pylori, herpes simplex virus, and cytomegalovirus and the risk for cardiovascular disease in women. Ann Intern Med 131(8):573–7
122. Ridker PM, Hennekens CH, Stampfer MJ, Wang F (1998) Prospective Study of Herpes Simplex Virus, Cytomegalovirus, and the risk of future Myocardial infarction and stroke. Circulation 98(25):2796–9
123. Espinola-Klein C, Rupprecht HJ, Blankenberg S, Bickel C, Kopp H, Victor A et al (2002) Impact of infectious burden on progression of carotid atherosclerosis. Stroke 33(11):2581–6
124. Haider AW, Wilson PW, Larson MG, Evans JC, Michelson EL, Wolf PA et al (2002) The association of seropositivity to Helicobacter pylori, Chlamydia pneumoniae, and cytomegalovirus with risk of cardiovascular disease: a prospective study. J Am Coll Cardiol 40(8):1408–13
125. Smieja M, Gnarpe J, Lonn E, Gnarpe H, Olsson G, Yi Q et al (2003) Multiple infections and subsequent cardiovascular events in the Heart Outcomes Prevention Evaluation (HOPE) Study. Circulation 107(2):251–7
126. Corrado E, Rizzo M, Tantillo R, Muratori I, Bonura F, Vitale G et al (2006) Markers of inflammation and infection influence the outcome of patients with baseline asymptomatic carotid lesions: a 5-year follow-up study. Stroke 37(2):482–6
127. Grau AJ, Preusch MR, Palm F, Lichy C, Becher H, Buggle F (2009) Association of symptoms of chronic bronchitis and frequent flu-like illnesses with stroke. Stroke 40(10):3206–10
128. Elkind MSV, Carty CL, O'Meara EL, Lumley T, Lefkowitz D, Kronmal RA et al (2011) Hospitalization for infection and risk of acute ischemic stroke: the cardiovascular health study. Stroke 42:1851-6. doi:10.1161/STROKEAHA.110.608588
129. Elkind MS (2007) Why now? Moving from stroke risk factors to stroke triggers. Curr Opin Neurol 20(1):51–7
130. Toschke AM, Heuschmann PU, Wood O, Wolfe CDA (2009) Temporal relationship between influenza infections and subsequent first-ever stroke incidence. Age Ageing 38(1):100–3
131. Maclure M (1991) The case-crossover design: a method for studying transient effects on the risk of acute events. Am J Epidemiol 133(2):144–53
132. Andraws R, Berger JS, Brown DL (2005) Effects of antibiotic therapy on outcomes of patients with coronary artery disease: a meta-analysis of randomized controlled trials. JAMA 293(21):2641–7
133. Muhlestein JB, Anderson JL, Carlquist JF, Salunkhe K, Horne BD, Pearson RR et al (2000) Randomized secondary prevention trial of azithromycin in patients with coronary artery disease: primary clinical results of the ACADEMIC study. Circulation 102(15):1755–60
134. Grayston JT, Kronmal RA, Jackson LA, Parisi AF, Muhlestein JB, Cohen JD et al (2005) Azithromycin for the secondary prevention of coronary events. N Engl J Med 352(16):1637–45
135. Sinisalo J, Mattila K, Valtonen V, Anttonen O, Juvonen J, Melin J et al (2002) Effect of 3 months of antimicrobial treatment with clarithromycin in acute non–Q-wave coronary syndrome. Circulation 105(13):1555–60
136. Cannon CP, Braunwald E, McCabe CH, Grayston JT, Muhlestein B, Giugliano RP et al (2005) Antibiotic treatment of Chlamydia pneumoniae after acute coronary syndrome. N Engl J Med 352(16):1646–54
137. Luchsinger JA, Pablos-Mendez A, Knirsch C, Rabinowitz D, Shea S (2002) Relation of antibiotic use to risk of myocardial infarction in the general population. Am J Cardiol 89(1):18–21
138. Kozarov E (2012) Bacterial invasion of vascular cell types: vascular infectology and atherogenesis. Future Cardiol 8(1):123–38
139. Paakkanen R, Palikhe A, Seppänen M, Nieminen MS, Vauhkonen H, Saikku P et al (2009) Beneficial effect of clarithromycin in patients with acute coronary syndrome and complement C4 deficiencies. Scand Cardiovasc J 43(6):395–401
140. Siscovick DS, Raghunathan TE, Lin D, Weinmann S, Arbogast P, Lemaitre RN et al (2000) Influenza vaccination and the risk of primary cardiac arrest. Am J Epidemiol 152(7):674–7
141. Naghavi M, Barlas Z, Siadaty S, Naguib S, Madjid M, Casscells W (2000) Association of

influenza vaccination and reduced risk of recurrent myocardial infarction. Circulation 102(25):3039–45
142. Wang CS, Wang ST, Lai CT, Lin LJ, Chou P (2007) Impact of influenza vaccination on major cause-specific mortality. Vaccine 25(7):1196–203
143. Heffelfinger JD, Heckbert SR, Psaty BM, Weiss NS, Thompson WW, Bridges CB et al (2006) Influenza vaccination and risk of incident myocardial infarction. Hum Vaccin 2(4):161–6
144. Jackson LA, Yu O, Heckbert SR, Psaty BM, Malais D, Barlow WE et al (2002) Influenza vaccination is not associated with a reduction in the risk of recurrent coronary events. Am J Epidemiol 156(7):634–40
145. Armstrong BG, Mangtani P, Fletcher A, Kovats S, McMichael A, Pattenden S et al (2004) Effect of influenza vaccination on excess deaths occurring during periods of high circulation of influenza: cohort study in elderly people. BMJ 329(7467):660
146. Hung IFN, Leung AYM, Chu DWS, Leung D, Cheung T, Chan C-K et al (2010) Prevention of acute myocardial infarction and stroke among elderly persons by dual pneumococcal and influenza vaccination: a prospective cohort study. Clin Infect Dis 51(9):1007–16
147. Hak E, van Loon S, Buskens E, van Essen GA, de Bakker D, Tacken MAJB et al (2003) Design of the Dutch prevention of influenza, surveillance and management (PRISMA) study. Vaccine 21(15):1719–24
148. Gurfinkel EP, Leon de la Fuente R, Mendiz O, Mautner B (2004) Flu vaccination in acute coronary syndromes and planned percutaneous coronary interventions (FLUVACS) study: one-year follow-up. Eur Heart J 25(1):25–31
149. Ciszewski A, Bilinska ZT, Brydak LB, Kepka C, Kruk M, Romanowska M et al (2008) Influenza vaccination in secondary prevention from coronary ischaemic events in coronary artery disease: FLUCAD study. Eur Heart J 29(11):1350–8
150. Phrommintikul A, Kuanprasert S, Wongcharoen W, Kanjanavanit R, Chaiwarith R, Sukonthasarn A (2011) Influenza vaccination reduces cardiovascular events in patients with acute coronary syndrome. Eur Heart J 32(14):1730–5
151. Davis MM, Taubert K, Benin AL, Brown DW, Mensah GA, Baddour LM et al (2006) Influenza vaccination as secondary prevention for cardiovascular disease: A Science Advisory From the American Heart Association/American College of Cardiology: Endorsed by the American Association of Cardiovascular and Pulmonary Rehabilitation, the American Association of Critical Care Nurses, the American Association of Heart Failure Nurses, the American Diabetes Association, the Association of Black Cardiologists, Inc., the Heart Failure Society of America, and the Preventive Cardiovascular Nurses Association.: The American Academy of Nurse Practitioners supports the recommendations of this scientific advisory.: This science advisory is consistent with the recommendations of the Centers for Disease Control and Prevention and the Advisory Committee on Immunization Practices. Circulation 114(14):1549–53
152. Schaffner W (2006) The impact of 'mismatched strains' on influenza vaccinations. US Infectious Disease:28–30
153. Simonsen L, Taylor RJ, Viboud C, Miller MA, Jackson LA (2007) Mortality benefits of influenza vaccination in elderly people: an ongoing controversy. Lancet Infect Dis 7(10):658–66
154. Almuti K, Rimawi R, Spevack D, Ostfeld RJ (2006) Effects of statins beyond lipid lowering: potential for clinical benefits. Int J Cardiol 109(1):7–15
155. Kopterides P, Falagas ME (2009) Statins for sepsis: a critical and updated review. Clin Microbiol Infect 15(4):325–34
156. Vandermeer ML, Thomas AR, Kamimoto L, Reingold A, Gershman K, Meek J et al (2012) Association between use of statins and mortality among patients hospitalized with laboratory-confirmed influenza virus infections: a multistate study. J Infect Dis 205(1):13–9
157. Fedson DS (2006) Pandemic influenza: a potential role for statins in treatment and prophylaxis. Clin Infect Dis 43(2):199–205
158. Novack V, MacFadyen J, Malhotra A, Almog Y, Glynn RJ, Ridker PM (2012) The effect of rosuvastatin on incident pneumonia: results from the JUPITER trial. CMAJ 184(7):E367–72
159. Shi D, Yang J, Yang D, LeCluyse EL, Black C, You L et al (2006) Anti-influenza prodrug

oseltamivir is activated by carboxylesterase human carboxylesterase 1, and the activation is inhibited by antiplatelet agent clopidogrel. J Pharmacol Exp Therap 319(3):1477–84
160. Madjid M, Curkendall S, Blumentals WA (2009) The influence of oseltamivir treatment on the risk of stroke after influenza infection. Cardiology 113(2):98–107
161. Muhammad S, Haasbach E, Kotchourko M, Strigli A, Krenz A, Ridder DA et al (2011) Influenza virus infection aggravates stroke outcome. Stroke 42(3):783–91
162. Grabska K, Gromadzka G, Czlonkowska A (2011) Infections and ischemic stroke outcome. Neurol Res Int 2011:691348
163. Verma S, Gupta M, Ridker PM (2012) Therapeutic targeting of inflammation in atherosclerosis: we are getting closer. Can J Cardiol 28(6):619–22
164. Ridker PM, Howard CP, Walter V, Everett B, Libby P, Hensen J et al (2012) Effects of interleukin-1beta inhibition with canakinumab on hemoglobin A1c, lipids, C-reactive protein, interleukin-6, and fibrinogen: a phase IIb randomized, placebo-controlled trial. Circulation 126(23):2739–48
165. Roman MJ, Salmon JE (2007) Cardiovascular manifestations of rheumatologic diseases. Circulation 116(20):2346–55
166. Ridker PM (2009) Testing the inflammatory hypothesis of atherothrombosis: scientific rationale for the cardiovascular inflammation reduction trial (CIRT). J Thromb Haemost 7:332–9
167. Leavenworth JW, Tang X, Kim H-J, Wang X, Cantor H (2013) Amelioration of arthritis through mobilization of peptide-specific CD8+ regulatory T cells. J Clin Invest
168. Patel VL, Currie LM (2005) Clinical cognition and biomedical informatics: issues of patient safety. Int J Med Inform 74(11–12):869–85
169. Eichler K, Puhan MA, Steurer J, Bachmann LM (2007) Prediction of first coronary events with the Framingham score: a systematic review. Am Heart J 153(5):722–31, 731.e1–8
170. Capelastegui A, Quintana JM, Bilbao A, España PP, Garin O, Alonso J et al (2012) Score to identify the severity of adult patients with influenza A (H1N1) 2009 virus infection at hospital admission. Eur J Clin Microbiol Infect Dis 31(10):2693–701
171. Mandell LA, Wunderink RG, Anzueto A, Bartlett JG, Campbell GD, Dean NC et al (2007) Infectious Diseases Society of America/American Thoracic Society consensus guidelines on the management of community-acquired pneumonia in adults. Clin Infect Dis 44(Suppl 2):S27–72
172. Miller RA, Waitman LR, Chen S, Rosenbloom ST (2005) The anatomy of decision support during inpatient care provider order entry (CPOE): empirical observations from a decade of CPOE experience at Vanderbilt. J Biomed Inform 38(6):469–85
173. Greenwald JL, Halasyamani LK, Greene J, LaCivita C, Stucky E, Benjamin B et al (2010) Making inpatient medication reconciliation patient centered, clinically relevant, and implementable: a consensus statement on key principles and necessary first steps. Jt Comm J Qual Patient Saf 36(11):504–13, 481
174. Pippins JR, Gandhi TK, Hamann C, Ndumele CD, Labonville SA, Diedrichsen EK et al (2008) Classifying and predicting errors of inpatient medication reconciliation. J Gen Intern Med 23(9):1414–22
175. Unroe KT, Pfeiffenberger T, Riegelhaupt S, Jastrzembski J, Lokhnygina Y, Colón-Emeric C (2010) Inpatient medication reconciliation at admission and discharge: a retrospective cohort study of age and other risk factors for medication discrepancies. Am J Geriatr Pharmacother 8(2):115–26
176. Spencer FA, Fonarow GC, Frederick PD, Wright RS, Every N, Goldberg RJ et al (2004) Early withdrawal of statin therapy in patients with non-ST-segment elevation myocardial infarction: national registry of myocardial infarction. Arch Intern Med 164(19):2162–8
177. Colivicchi F, Bassi A, Santini M, Caltagirone C (2007) Discontinuation of statin therapy and clinical outcome after ischemic stroke. Stroke 38(10):2652–7
178. Fultz SL, Goulet JL, Weissman S, Rimland D, Leaf D, Gibert C et al (2005) Differences between infectious diseases—certified physicians and general medicine—certified physicians in the level of comfort with providing primary care to patients. Clin Infect Dis 41(5):738–43

179. Gieffers J, Solbach W, Maass M (1998) In vitro susceptibilities of Chlamydia pneumoniae strains recovered from atherosclerotic coronary arteries. Antimicrob Agents Chemother 42(10):2762–4
180. Jantos CA, Heck S, Roggendorf R, Sen-Gupta M, Hegemann JH (1997) Antigenic and molecular analyses of different Chlamydia pneumoniae strains. J Clin Microbiol 35(3):620–3
181. Kanamoto Y, Ouchi K, Mizui M, Ushio M, Usui T (1991) Prevalence of antibody to Chlamydia pneumoniae TWAR in Japan. J Clin Microbiol 29(4):816–8
182. Malaty HM, El-Kasabany A, Graham DY, Miller CC, Reddy SG, Srinivasan SR et al (2002) Age at acquisition of Helicobacter pylori infection: a follow-up study from infancy to adulthood. Lancet 359(9310):931–5
183. Schwartz S, Gatto NM, Campbell UB (2012) Extending the sufficient component cause model to describe the stable unit treatment value assumption (SUTVA). Epidemiol Perspect Innov 9:3
184. Erasto P, Hoti F, Auranen K (2012) Modeling transmission of multitype infectious agents: application to carriage of Streptococcus pneumoniae. Stat Med 31(14):1450–63
185. Hansson GK, Hermansson A (2011) The immune system in atherosclerosis. Nat Immunol 12(3):204–12
186. Libby P, Ridker PM, Maseri A (2002) Inflammation and atherosclerosis. Circulation 105(9):1135–43
187. Galkina E, Ley K (2009) Immune and inflammatory mechanisms of atherosclerosis*. Annu Rev Immunol 27(1):165–97
188. Tangco RV (2009) Public health implications of the inflammatory concept of atherosclerosis. Acta Med Philipp 43:70–7
189. Narang V, Decraene J, Wong S-Y, Aiswarya B, Wasem A, Leong S et al (2012) Systems immunology: a survey of modeling formalisms, applications and simulation tools. Immunol Res 53(1–3):251–65
190. Mohtashemi M, Higgs B, Levins R (2006) Infection and atherosclerosis: is there an association? Complex Systems 16:259–76
191. Fusaro VA, Patil P, Chi C-L, Contant CF, Tonellato PJ (2013) A systems approach to designing effective clinical trials using simulations. Circulation 127:517–26
192. Ben-Shlomo Y, Kuh D (2002) A life course approach to chronic disease epidemiology: conceptual models, empirical challenges and interdisciplinary perspectives. Int J Epidemiol 31(2):285–93
193. Bruschke AVG, Kramer JR Jr, Bal ET, Haque IU, Detrano RC, Goormastic M (1989) The dynamics of progression of coronary atherosclerosis studied in 168 medically treated patients who underwent coronary arteriography three times. Am Heart J 117(2):296–305
194. Ekbom A, Daszak P, Kraaz W, Wakefield AJ (1996) Crohn's disease after in-utero measles virus exposure. Lancet 348(9026):515–7
195. Liang C, Marsit CJ, McClean MD, Nelson HH, Christensen BC, Haddad RI et al (2012) Biomarkers of HPV in head and neck squamous cell carcinoma. Cancer Res 72(19):5004–13
196. Yazdanbakhsh M, Kremsner PG, van Ree R (2002) Allergy, parasites, and the hygiene hypothesis. Science 296(5567):490–4
197. Fisher LD, Lin DY (1999) Time-dependent covariates in the Cox proportional-hazards regression model. Annu Rev Public Health 20:145–57
198. Witteman JCM, D'Agostino RB, Stijnen T, Kannel WB, Cobb JC, de Ridder MAJ et al (1998) G-estimation of causal effects: isolated systolic hypertension and cardiovascular death in the Framingham Heart Study. Am J Epidemiol 148(4):390–401
199. Lee J, Hwang Y, Kang W, Seong SJ, Lim M-S, Lee HW et al (2011) Population pharmacokinetic/pharmacodynamic modeling of clopidogrel in Korean healthy volunteers and stroke patients. J Clin Pharmacol 52:985–95

Chapter 12
Inflammatory Biomarkers in Patients with Acute Brain Injuries

Glen C. Jickling

Abstract Acute brain injury as a result of stroke and traumatic brain injury are leading causes of disability and mortality. Methods to improve patient diagnosis and prognosis for these common conditions are needed. Molecular biomarkers are one method that has been evaluated to diagnose acute brain injury, determine its cause, and predict outcomes and response to therapy. Markers that have been identified include a variety of proteins, nucleic acids, and lipids that relate to the pathophysiology of acute brain injury. Inflammation plays an important role in acute brain injury and several molecules involved in inflammation have been identified as biomarkers. However, biomarkers for acute brain injury is a developing field that requires additional study is to identify markers for use in clinical practice. These studies will include evaluating a larger number of candidate markers using proteomic, genomic, metabolomic, and lipidomic approaches. Additionally, novel markers such as microRNA and the use of panels that integrate multiple markers may also prove to be valuable tools in acute brain injury. In this chapter, we provide a summary of identified inflammatory biomarkers in acute brain injury, and how these biomarkers could add to patient care.

Introduction

Acute brain injury due to stroke and traumatic brain injury (TBI) are leading causes of morbidity and mortality [1]. To reduce the impact of these common disorders, blood-based biomarkers have been evaluated as tools to improve patient diagnosis and prognosis. As such they may help direct patients as quickly as possible to the care they require to reduce disability and improve outcomes. The pathophysiology

G.C. Jickling (✉)
Department of Neurology, University of California, Davis, CA, USA
e-mail: gcjickling@ucdavis.edu

J. Chen et al. (eds.), *Immunological Mechanisms and Therapies in Brain Injuries and Stroke*, Springer Series in Translational Stroke Research, DOI 10.1007/978-1-4614-8915-3_12,

of brain injury has guided the development of biomarkers in stroke and brain trauma, including markers of brain tissue damage, markers of immune response and inflammation, markers of endothelial dysfunction, and markers of hemostasis. Evaluated markers have included a variety of proteins, nucleic acids, and lipids which have been studied both as individual markers and in biomarker panels.

Biomarkers are important tools to aid in the diagnosis and prognosis of patients with neurological disease. A clinical diagnosis is often imperfect, with variable agreement on final diagnoses. Improving diagnostic methods can reduce healthcare costs by directing patients to appropriate therapy earlier, and thus potentially preventing irreversible neurological deficits. Patients who otherwise would be misdiagnosed can be identified and obtain treatment. Improving prognostic abilities is also important, as it has implications to patients and families making decisions regarding care. Prognosis can also alter treatment, either by identifying patients most likely to benefit or patients who may experience harmful side effects.

Identifying biomarkers of acute brain injury that can meet the rigorous requirements of a clinical biomarker has been challenging. Indeed such a clinical biomarker must not only be sufficiently sensitive and specific but also cost-effective, feasible to use in a clinical setting, provide timely results, maintain discriminatory ability between different laboratories, and importantly address a relevant clinical need. Though progress has been made, further investigation is required.

Inflammation plays an important role in acute brain injury and several molecules involved in inflammation have been identified as biomarkers for the diagnosis and prognosis of acute brain injury. Acute injury of brain tissue elicits both an innate and adaptive immune response. Following injury a number of molecules termed damage-associated molecular pattern molecules (DAMPs) are release and lead to the activation of cerebral microglia. Injury also results in the release of a number of cytokines and chemokines that lead to the activation of circulating leukocytes. Cerebral endothelium also express adhesion molecules in response to injury that promote leukocyte adhesion and migration across the blood–brain barrier into brain tissue. Damaged cerebral tissue also exposes a number of new antigens to the peripheral immune system, eliciting an adaptive immune response. Furthermore, in ischemic stroke, inflammation also plays an important role in atherosclerosis, vascular disease, hemostasis, and vascular risk factors important to disease pathogenesis. In this chapter, we provide a focused review of inflammatory biomarkers in acute ischemic stroke and acute traumatic brain injury. For each condition, we highlight inflammatory biomarkers that may improve diagnosis or prognosis and how these biomarkers could add to patient care.

Diagnostic Biomarkers in Ischemic Stroke

Biomarkers in blood for the diagnosis of ischemic stroke have long been sought; however, to date none have been implemented in clinical practice. A biomarker to identify cerebral infarction is similar in concept to the use of troponin to identify

Table 12.1 Biomarkers in blood to distinguish ischemic stroke from controls and other neurological disorders

Biomarker	Description of biomarker
S100B [12, 167]	Calcium-binding protein from glial cells
GFAP [15, 16, 168, 169]	Intermediate filament protein, astrocyte marker
NSE [167]	Neuronal glycolytic enzyme
NMDA-R-Ab [5, 6, 11]	Antibody to the NMDA-receptor
MBP [168, 170]	Myelin sheath protein
C-reactive protein [32, 171]	Acute phase protein
VCAM-1 [12]	Adhesion molecule
MMP-9 [12, 13, 50]	Proteolytic enzyme
ApoC-I ApoC-III [17]	Lipoproteins
D-dimer [27, 28, 172]	Fibrin degradation product
von Willebrand factor [12, 13]	Glycoprotein
PARK7 [173]	RNA-binding protein
NDKA [173]	Nucleoside kinase
BDNF [13]	Growth factor
LDL-oxidized [174]	Oxidized lipoprotein
Malondialdehyde [175]	Oxidized lipid product
H-FABP [176]	Lipid-binding protein
RNA Profile [19–22, 177, 178]	Nucleic acid

ApoC apolipoprotein C, *BDNF* brain-derived neurotrophic factor, *CRP* C-reactive protein, *GFAP* glial fibrillary acid protein, *H-FABP* heat fatty acid binding protein, *LDL* low-density lipoprotein, *MBP* myelin basic protein, *NMDA-R-Ab* *N*-methyl D-aspartate receptor antibody, *NSE* neuron-specific enolase, *PARK7* Parkinson's disease 7, *RNA* ribonucleic acid, *VCAM-1* vascular cell adhesion protein 1

myocardial infarction. Troponin is a useful marker in a patient with chest pain to distinguish cardiac ischemia from other common causes. Likewise, in a patient presenting with acute focal neurological deficits, a marker of focal brain ischemia could distinguish ischemic stroke from other common causes of neurological dysfunction such as hemorrhagic stroke, traumatic brain injury, migraine, seizure, brain neoplasm, and other neurological events that mimic ischemic stroke. This could aid in the triage of stroke patients and implementation of additional diagnostic tests and therapies specific to ischemic stroke. Such a marker would likely be complementary to neurovascular imaging, just as troponin is complementary to electrocardiogram and cardiac imaging in myocardial ischemia.

Several biomarkers have been identified for diagnosis of ischemic stroke and have been well reviewed by others [2–4]. A summary of these diagnostic markers is shown in Table 12.1. Of interest are antibodies to the glutamate NMDA-receptor (NMDA-R, NR2A/NR2B subunits) that have been shown to distinguish ischemic stroke from controls at 3 h with 97 % sensitivity and 98 % specificity [5, 6]. Though NMDA-R antibodies have also been associated with hypertension, atherosclerosis, prior stroke, epilepsy, systemic lupus erythematosus, and encephalitis, it is possible that specific immunogenic epitopes to the NMDA-R may be present in stroke and thus permit its identification [7–10]. More recently a NMDA-receptor peptide in blood has been shown to distinguish ischemic stroke from controls, stroke mimics, and patients with vascular risk factors with 92 % sensitivity and 96 % specificity [11]. These interesting results require further study in larger cohorts.

Though most individual biomarkers identified to date have not had sufficient sensitivity or specificity to identify ischemic stroke on their own, panels of biomarkers may show promise. A panel of four biomarkers (S100B, vWF, MMP9, and VCAM) was able to separate ischemic stroke from controls with 90 % sensitivity and specificity [12]. Likewise, a panel of five biomarkers [S100B, vWF, MMP9, B-type neurotrophic growth factor (BNGF), and MCP-1] was able to distinguish ischemic stroke from healthy controls with 92 % sensitivity and 93 % specificity [13].

Biomarkers that distinguish ischemic stroke from other neurological disorders is of greater clinical utility. Several disorders can be challenging to distinguish from ischemic stroke, including hemorrhagic stroke, seizure, migraine, and brain neoplasm [14]. Separating ischemic stroke from stroke mimics has generally been more difficult because many of the diseases that mimic stroke can also influence markers studied. Distinguishing hemorrhagic stroke from ischemic stroke is of great importance, particularly in the acute setting where thrombolysis can benefit ischemic stroke but worsen hemorrhagic stroke. In this regard, the protein GFAP has shown promise [15, 16] as have apolipoprotein CI and apolipoprotein CIII [17]. Few inflammatory markers have been able to distinguish ischemic from hemorrhagic stroke. When MMP-9 was included in a panel of four biomarkers (S100B, MMP9, D-dimer, and BDNF), ischemic stroke could be distinguished from stroke mimics including hemorrhagic stroke with 85 % sensitivity but only 34 % specificity [18].

Studies of RNA expressed in circulating leukocytes have been carried out to distinguish ischemic stroke from controls and patients with vascular risk factors. The use of RNA as a diagnostic marker is an emerging filed that is supported by its clinical use in the diagnosis of breast cancer (Mammaprint and Oncotype), coronary artery disease (CardioDx), and infectious diseases (SARS, HIV, Hepatitis C, and HPV). RNA could be evaluated in the form of a microarray, as in the case of Mammaprint, or using a PCR-based assay. A microarray study of RNA isolated from peripheral blood mononuclear cells identified 190 genes differentially expressed in ischemic stroke compared to controls. A 22 gene panel of the 190 genes could separate ischemic stroke from controls with 78 % sensitivity and 80 % specificity [19]. A subsequent study of whole blood RNA identified 1,335 genes expressed in acute ischemic stroke compared to controls [20]. An 18-gene panel of the 1,335 genes could distinguish ischemic stroke from controls in all patients at 24 h. This 18-gene panel has more recently been shown to distinguish ischemic stroke ($n=70$, 199 samples) from controls with 93.5 % sensitivity and 89.5 % specificity [21]. A panel of 9 genes (5 of which are in the 18 gene panel) has also been shown to distinguish stroke from controls [22]. Thus, a multigene approach shows promise as a method to identify acute ischemic stroke, though further larger studies comparing to stroke mimics are required.

Whole proteome studies of ischemic stroke have also begun to be performed. As with whole genome RNA studies, the technology to perform whole proteome analysis continues to develop [23, 24]. In microdialysates of 6 patients with ischemic stroke, 53 proteins including a number of inflammatory markers were found to be associated with cerebral infarction [25]. In a separate study of 6 ischemic strokes, 132 protein spots were found to be different in ischemic stroke compared to controls, 39 of which were evaluated by mass spectrometry [26]. It is likely that larger whole proteome studies will be conducted as the technology improves, and potentially identify sensitive and specific markers to diagnose ischemic stroke.

Diagnostic Biomarkers to Determine Cause of Stroke

Identifying the cause of ischemic stroke is a critical step to preventing future stroke, as cause guides treatment. Patients with a stroke due to cardiac disease such as atrial fibrillation benefit from anticoagulation, whereas patients with a large vessel stroke due to carotid stenosis benefit from revascularization either with endarterectomy or stenting. However, the cause of stroke remains unknown or cryptogenic in about 35 % of ischemic strokes despite extensive investigation. Biomarkers to predict a cause of stroke may have applications to identify cause in cryptogenic stroke. The three major causes of stroke where blood biomarkers have been evaluated are cardioembolic, large vessel atherosclerotic, and small vessel lacunar. These are reviewed below and summarized in Table 12.2.

Cardioembolic stroke comprise a number of cardiac disorders that are known to be high risk for ischemic stroke. Several biomarkers show potential to distinguish cardioembolic from non-cardioembolic ischemic stroke. Brain Natriuretic Peptide (BNP) and D-dimer have shown some ability to identify cardioembolic stroke, though sensitivity and specificity may not sufficient for clinical use [27–31]. An RNA profile of 37 genes mostly associated with immune response has been shown to separate cardioembolic stroke due to atrial fibrillation from non-atrial fibrillation causes with >90 % sensitivity and specificity.

Table 12.2 Biomarkers to predict cause of ischemic stroke

Biomarker	Cause of stroke	Description of biomarker
IL-6 [179]	Cardioembolic, lacunar	Inflammatory cytokine
TNF-α [179]	Cardioembolic, lacunar	Inflammatory cytokine
ICAM-1 [38, 180, 181]	Lacunar, large vessel	Adhesion molecule
BNP [27–31]	Cardioembolic	Vasoactive peptide hormone
Fibrinogen [32, 182]	Large vessel	Glycoprotein
D-dimer [28, 29, 183]	Cardioembolic, large vessel	Fibrin degradation product
Von Willebrand factor [184]	Cardioembolic	Glycoprotein
C-reactive protein [32, 36]	Cardioembolic, large vessel, lacunar	Acute phase protein
Thrombomodulin [38]	Lacunar	Thrombin cofactor
RNA Profiles [34, 35, 39]	Cardioembolic, large vessel, lacunar	Nucleic Acid

BNP brain natriuretic peptide, *ICAM-1* intracellular adhesion molecule 1, *IL-6* interleukin-6, *RNA* ribonucleic acid, *TNF-α* tumor necrosis factor alpha

Large vessel atherosclerosis is considered to be the cause of stroke in patients with either occlusive or stenotic (≥50 % diameter reduction or <50 % diameter reduction with plaque ulceration or thrombosis) arterial disease of presumed atherosclerosis origin occurring in a clinically relevant extracranial or intracranial artery. In patients with large vessel disease that also have a second cause of stroke, it remains difficult to determine whether large vessel atherosclerosis is the responsible cause of stroke. Additionally, in stroke patients with atherosclerotic plaque causing mild stenosis (<50 %) without ulceration or thrombosis, it remains uncertain whether large vessel disease is the cause of stroke. In such patients, biomarkers indicating if large vessel disease is the likely cause of stroke may have utility. In this regard, the inflammatory marker C-reactive protein has been combined with D-dimer, D-dimer/fibrinogen ratio, and erythrocyte sedimentation to separate large vessel from cardioembolic stroke [32]. A profile of RNA differentially expressed in circulating leukocytes has also been shown to distinguish cardioembolic from large vessel ischemic stroke [33, 34]. In a study of 194 samples from 76 acute ischemic strokes, a 40-gene panel was able to distinguish cardioembolic from large vessel ischemic stroke with >95 % sensitivity and specificity at each of 3, 5, and 24 h after stroke onset [34]. The identified genes relate to differences in inflammation between cardioembolic and large vessel stroke. When these panels were applied to patients with cryptogenic stroke, 17 % were predicted to be large vessel and 41 % to be cardioembolic stroke [34, 35]. Of the cryptogenic strokes predicted to be cardioembolic, 27 % were predicted to have atrial fibrillation.

Small vessel lacunar stroke is considered in patients with a brain infarct <15–20 mm in diameter occurring in regions of the penetrating cerebral arteries and associated with a lacunar syndrome. Frequently, however, infarcts larger than 20 mm in diameter occur in the same regions, and it remains unclear whether these are due to small vessel lacunar disease or other pathophysiology. A biomarker may provide supporting evidence that such a stroke is due to small vessel disease. Only a few biomarkers have been associated with small vessel ischemic strokes. CRP has been shown to be lower in small vessel lacunar stroke compared to large vessel stroke ($n=116$) [36]. Lacunar stroke has also been shown to have higher levels of thrombomodulin, ICAM-1, tissue factor, and homocysteine compared to controls. However, it remains unclear whether these markers are different compared to other stroke subtypes [37, 38]. RNA expressed in circulating leukocytes has been shown to differ between lacunar and non-lacunar strokes [39]. Identified genes correspond to differences in inflammation and endothelial function in lacunar stroke. White matter hyperintensities (WMH) on MRI are may also be a manifestation of cerebral small vessel disease. WMH have been associated with a higher plasma level of intercellular adhesion molecule 1 (ICAM-1) [40]. Additionally, a profile of differentially expressed RNA in blood has also been associated with WMH, and many of the genes were associated with inflammation [41]. Cerebral microbleeds are also considered a manifestation of cerebral small vessel disease. Several inflammatory markers have been shown to be increased in patients with microbleeds, including hsCRP, IL-6, and IL-18 [42].

Prognostic Biomarkers in Ischemic Stroke

Predicting an outcome based on a blood sample acquired at time of stroke presentation could have practical value in the clinical setting. Patients at increased risk for a specific outcome could have therapy altered or an intervention performed in an attempt to shift this risk. A number of stroke outcomes have been evaluated to determine whether a blood biomarker may have utility to stratify patients and identify target groups for therapy.

Biomarkers to Predict Poor Stroke Outcome

Identifying stroke patients likely to have a poor outcome may help in patient management. Those likely to have a poor outcome might benefit from more aggressive early therapy, or rehabilitation. Levels of IL-6 have been associated with worse outcomes in stroke. In 250 ischemic strokes, IL6 levels were increased in the 14 patients who died at 1 year [43]. In another study, poor stroke outcome was also associated with IL-6 (OR 2.4, 95 % CI 1.4–4.2) and Ln NT pro-BNP (OR 2.2, 95 % CI 1.2–4.0) [44]. However, neither IL-6 nor Ln NT pro-BNP were able to improve upon outcome prediction achieved by the NIHSS score and age alone (c-statistic 0.84). Furthermore, IL-6 may be a marker of systemic comorbid disease, and thus may not specifically relate to ischemic stroke outcome.

Infarct volume is often used as a surrogate measure of outcome, with larger infarcts being associated with worse outcomes. Several biomarkers have been associated with infarct volume, including S-100B, MMP, IL-6, TNF-α, ICAM-1, and glutamate. A larger infarct volume would be expected to increase release of brain-specific markers into the circulation. This is supported by studies that show levels of S-100B [45, 46] and NSE [46–48], tau [47], and glutamate [49], each correlates with final infarct volume [45, 48]. A larger infarct volume might also lead to a larger inflammatory response to ischemic brain tissue. This is supported by studies showing inflammatory markers correlate with infarct volume, including TNF-α [50], IL-6 [51, 52], ICAM-1 [50], MMP-2, and MMP-9 [50, 53].

Biomarkers to Predict Hemorrhagic Transformation

Hemorrhagic transformation (HT) is a major complication in ischemic stroke that is increased when thrombolysis (tissue plasminogen activator, tPA) is used. Preventing HT is important to improve stroke outcomes, as it is associated with both increased morbidity and mortality. Identifying patients at increased risk of HT could identify a patient group where management is modified to reduce the risk of HT. For example, in high-risk patients, the dose of tPA could be altered or an additional therapy such as minocycline could be administered [54, 55]. Several blood biomarkers have been evaluated for their ability to predict HT in ischemic stroke including MMP-9, c-FN, PAI-1, TAFI, and S100B [55–60].

The inflammatory marker MMP-9 has been found to be increased in strokes that develop hemorrhagic transformation, both before and after treatment with t-PA [56, 61–65]. Serum MMP-9 levels ≥140 ng/ml predicted hemorrhagic transformation in ischemic stroke patients with a sensitivity of 87 % and specificity of 90 % [61]. MMP-9 activity may also be promoted by hyperglycemia which also is associated hemorrhagic transformation [66, 67]. MMP-9 levels also correlate with disruption of the blood–brain barrier, a key feature in HT [68].

Biomarkers to Predict Early Neurological Deterioration

Early neurological deterioration (END) is defined as worsening of neurological status from admission to 48–72 h after admission. Identifying patients at risk of END may be useful to initiate therapies early to potentially prevent or reduce deterioration. Several biomarkers have been associated with END including ferritin, TNF-α, ICAM-1, MMP-9, MMP-13, nitric oxide, glutamate, GABA, and S100B [69, 70]. Among the inflammatory biomarkers, predictors of END included a plasma ferritin > 275 ng/ml [71] and plasma IL-6 > 21.5 pg/ml [51]. Both MMP-9 and MMP-13 have been associated with infarct volume expansion in stroke [53]. In lacunar stroke, TNF-α > 14 pg/ml and ICAM-1 > 208 pg/ml have also been shown to correlate with END [72].

Biomarkers to Predict Risk of Future Stroke

Biomarkers have shown some success to identify patients at risk for a first stroke as well as those who are more likely to have a recurrent ischemic stroke. The enzyme lipoprotein-associated phospholipase A2 (Lp-PLA2) is produced by immune cells. Increased levels have been associated with an increase in primary stroke occurrence and risk of recurrent stroke [73–78]. C-reactive protein (CRP) has also been shown to be predictive of stroke risk [75]. Despite the increased risk of stroke predicted by Lp-PLA2 and CRP, the benefit in terms of reducing stroke has yet to be demonstrated. Trials of the Lp-PLA2 inhibitor Darapladib may provide additional insight to the role of Lp-PLA2 in stroke.

Biomarkers to Identify Candidates for Decompressive Hemicraniectomy

Decompressive hemicraniectomy following stroke can improve stroke outcomes in selected patients with large cortical ischemic infarcts when performed early [79].

A biomarker to identify ischemic strokes at risk for malignant cerebral infarction and those likely to benefit from decompressive hemicraniectomy would be useful. MMP-9 when combined with cellular-fibronectin (c-Fn) has been shown to be predictive of malignant cerebral infarction [70]. Admission MMP-9 level >140 ng/ml predicted malignant MCA infarction with 64 % sensitivity and 88 % specificity. The protein S-100B has also been associated with malignant infarction [80].

Biomarkers to Predict Response to Stroke Prevention Therapy

Biomarkers might be useful to guide decisions regarding stroke prevention therapies [81]. Often it is unclear which antiplatelet agent a stroke patient should first be started on. A recent study showed that in stroke patients with a monocyte chemotactic protein-1 (MCP-1) > 217 pg/ml, outcomes were better at 90 days in those treated with aspirin plus extended release dipyridamole (aggrenox), compared to aspirin alone [82]. Though the reasons for this difference remain unclear, biomarkers may have a role to select antiplatelet agent.

Traumatic Brain Injury

Brain trauma can result from direct impact to the head (blunt force trauma and falls), by shearing forces from a rapid change in acceleration/deceleration (motor vehicle accident), a rapid change in pressure (blast exposure), or by penetrating injury by a high-velocity object (bullet and blast shrapnel). Traumatic brain injury (TBI) is a major health issue with over 1.5 million persons experiencing a TBI annually in the USA alone. Of these about 50,000 patients die, and 500,000 are hospitalized [83–85]. In patients with mild TBI, 40–50 % experience persistent neurological problems for months–years following injury. Deficits include impairment in cognition, behavioral function, and depression. Every year, an estimated 90,000 patients with TBI suffer permanent disabilities [86]. A biomarker of TBI would provide a needed indicator of brain tissue damage. Being able to measure changes in tissue injury at a molecular level during the event would add to TBI management and therapeutic development.

In TBI, mechanical damage to neuronal and vascular structures results in a primary brain injury. A secondary injury can subsequently occur as a result of molecular events that further promote cell death. In primary injury prevention is the key, through combined use of education programs, safety devices (seat belts and safety laws) and protective equipment (helmets and armor). Medical intervention may be able to reduce secondary injury by reducing further neural injury, complications, and promoting processes of repair. Inflammation following TBI is an important component of TBI that is involved not only in brain damage but also in repair. Several studies have evaluated inflammatory biomarkers in TBI to aid in the diagnosis, prediction of complications, and prognosis of TBI.

Table 12.3 Biomarkers for the diagnosis of TBI

Biomarker	Description of biomarker
S100B [87–89]	Calcium-binding protein from glial cells
GFAP [90–94]	Intermediate filament protein, astrocyte marker
NSE [95–97]	Neuronal glycolytic enzyme
MBP [98, 99]	Myelin sheath protein
Brain type-fatty acid binding protein [100]	Carrier protein for fatty acid
Alpha II spectrin [101–105]	Neuronal cytoskeletal molecule
Cleaved tau [106–109]	Protein to stabilize microtubules
Phosphorylated neurofilament H [110–112]	Neuronal intermediate filaments
Ubiquitin C-terminal hydrolase [113–116]	Neuronal deubiquitnating enzyme
IL-6 [117–121]	Cytokine
IL-1β [121, 123, 127]	Cytokine
TNF-α [117, 121, 128–133]	Cytokine
IL-8 [121, 134–136]	Chemokine
IL-10 [117, 137, 138]	Cytokine
TGF-β [118]	Cytokine
C-reactive protein [139]	Acute phase protein
ICAM-1 [140]	Adhesion molecule
MCP-1 [141]	Chemokine
F2-isoprostane [142–144]	Lipid peroxidation product
RNA profile [145]	Nucleic acid
miR-16, miR-92a, and miR-765 [147]	Nucleic acid
mir671-5p, mir-US4, mir1285, and mir455-3p [148]	Nucleic acid

GFAP glial fibrillary acid protein, *ICAM-1* intracellular adhesion molecule 1, *IL* interleukin, *miR* microRNA, *MBP* myelin basic protein, *NSE* neuron-specific enolase, *RNA* ribonucleic acid, *TGF* transforming growth factor, *TNF* tumor necrosis factor

Diagnostic Biomarkers for TBI

The diagnosis of TBI relies on history of injury, physical exam, and diagnostic investigations. Neuroimaging is frequently used to identify parenchymal injury, edema, ischemia, and hemorrhage. Electroencephalogram (EEG) can identify changes associated with brain injury and the presence of seizure activity. Biomarkers may have a role to aid in the diagnosis of TBI. For example, a biomarker may aid in distinguishing mild TBI from posttraumatic stress disorder. Biomarkers may also aid in the detection of TBI that may be otherwise missed, and/or enable a more rapid diagnosis of TBI to permit earlier implementation of treatment.

Several molecules released as a result of brain injury have evaluated as biomarkers to identify TBI including S100B [87–89], GFAP [90–94], NSE [95–97], and MBP [98, 99] (Table 12.3). However, these markers have also been associated with ischemic and hemorrhagic stroke, and thus are not specific to TBI. A number of additional markers have also been associated with TBI, including brain-type fatty acid binding protein [100], alpha II spectrin [101–105], cleaved tau [106–109], phosphorylated neurofilament H [110–112], and ubiquitin C-terminal hydrolase [113–116]. Several of these markers have been combined into panels and are currently undergoing validation studies to determine their ability to identify TBI.

A number of inflammatory cytokines may also serve as biomarkers of TBI. Cytokines contribute to the inflammatory response to damaged brain tissue, activating microglial cells and recruiting circulating leukocytes to site of brain injury. Among the most studied are IL-6, IL-1β, and TNF-α. IL-6 is increased in the CSF of TBI patients [117], and this increase is greater than that observed in the serum [118, 119]. Levels of IL-6 in the CSF peak between days 3 and 6 after TBI [120]. IL-6 has also been shown to be elevated in brain tissue obtained from patients with TBI [121]. IL-1β is increased in the CSF and microdialysates of patients with TBI [122–126]. IL-1β is also increased in brain tissue of patients with TBI, and this increase occurs within hours of injury [121, 123, 127]. TNF-α is also increased in the CSF and serum of patients with TBI [117, 128–133], and elevated in brain tissue obtained from TBI patients [121].

Other cytokines that are increased in TBI include IL-8, IL-10, IL-2, and INF-gamma. IL-8 is increased in the CSF [134, 135] and brain tissue of patients with TBI [121, 136]. IL-10 is increased in the CSF of patients with TBI [117, 137, 138]. Levels of TGF-β are also increased in the CSF of TBI, with levels peaking at 24 h and remaining elevated for several weeks [118]. Thus, in patients with TBI, there is extensive inflammation both in the central nervous system and in the peripheral circulation.

As our understanding of the inflammatory response in TBI has increased, studies evaluating inflammatory molecules as biomarkers of TBI have begun to be performed. One study has shown a panel of cytokines measured in CSF can distinguish patients with TBI from controls. Among the cytokines measured were IL-1β, IL-6, IL-12p70, IL-10, IL-8, and MIP-1α [122]. Other inflammatory markers that have been studied include CRP, ICAM-1, MCP-1, and F2-isoprostane. C-reactive protein (CRP) in combination with serum amyloid A has been shown to distinguish TBI from healthy control subjects [139]. However, CRP also increases as a result of injury to other non-brain tissue and thus is not specific to TBI. Intracellular adhesion molecule 1 (ICAM-1) levels are increased in TBI and correlate with degree of BBB disruption [140]. The chemokine monocyte chemoattractant protein-1 (MCP-1) also is increased at least in experimental TBI [141]. F2-isoprostane, which is a marker of lipid peroxidation in inflammatory disease, has also been shown to be increased in the serum and CSF of TBI subjects [142–144]. Thus, a number of inflammatory markers are associated with TBI, and with further validation may serve as biomarkers to identify TBI.

A question that requires further evaluation is whether the inflammatory response to brain injury in TBI differs from the inflammatory response observed in other acute brain injuries such as ischemic stroke. In a small study of whole genome RNA expression in TBI, some differences between immune response in TBI and stroke were identified. The study compared 15 acute TBI subjects to 20 acute ischemic strokes and 15 healthy controls using U133A Affymetrix microarrays [145].

RNA was isolated from peripheral blood mononuclear cells, thus identified genes reflect aspects of the immune response to TBI. TBI subjects compared to controls had a number of genes in common to stroke subjects including MMP9, CD36, CD14, IL13RA1, TLR2, and CD163. This suggests that features of the peripheral immune response to acute brain injury in TBI are similar to that in ischemic stroke. However, there were differences in RNA expression between TBI and stroke, including differential expression of MMP25 and CLC (eosinophil lysophospholipase) [145]. Differences in immune response between cerebral trauma and ischemia are supported by a rodent study demonstrating increased levels of p43/pro-EMAPII (a proinflammatory cytokine) in TBI compared to ischemic stroke [146]. Further study evaluating differences in immune response to brain injury in TBI compared to ischemic stroke and forms of acute brain injury are required.

Preliminary studies have also identified differences in plasma microRNA in TBI. A study of 10 TBI compared to 10 healthy controls suggested differences in several plasma miRNA in TBI including miR-16, miR-92a, and miR-765 [147]. Another study comparing 9 TBI subjects to 9 non-TBI subjects identified different levels of 4 miRNA (mir671-5p, mir-US4, mir1285, and mir455-3p), 13 snoRNA, and 1 scaRNA [148]. The presence of miRNA differentially present in TBI is supported by a rodent study reporting several miRNA in the serum and CSF of rats with TBI including miR-let7i, mir-122, mir200b, and mir-340-5p [149, 150].

Prognostic Biomarkers of TBI

Biomarkers of TBI Outcome

Markers of poor outcome following TBI could add to the information used by physicians, patients, and families making decisions regarding patient care. A biomarker of TBI outcome could also identify patients likely to benefit from a specific therapy or rehabilitation program. Several biomarkers in blood have been associated with outcome in TBI. Increase baseline serum levels of S100B, NSE, GFAP, and tau have all been associated with a poor TBI outcome at 6 months [89, 97, 109, 151, 152].

A number of inflammatory markers have also been associated with outcome in TBI. Peak levels of IL-1α in the CSF have been shown to correlate with TBI outcome at 3 months [153]. IL-1α levels in the CSF have also been shown to correlate with Glasgow outcome score [154]. The severity of TBI also correlates with the ratio of venous to arterial IL-6 levels [155]; IL-6 measured by microdialysis correlates with Glasgow outcome score [156]. Mortality as a result of TBI has been associated with several inflammatory biomarkers measured at baseline. IL-6 in CSF was found to be predictive of survival [156]. In children with TBI, increased IL-10 levels correlated with mortality [137]. Though TNF-α is increase in TBI, it does not appear to correlate with mortality in TBI [128].

Biomarkers of Response to Hypothermia

Hypothermia is frequently induced in patients with TBI as a treatment to minimize brain injury. Hypothermia has widespread effects including modulation of inflammatory response. Indeed in TBI patients treated with hypothermia, IL-6 levels were decreased [157]. Biomarkers may have a role in the use of hypothermia, though further study is required. A biomarker may be able to identify patients likely to benefit from hypothermia, to help determine optimal duration of cooling, or time of rewarming. Of interest, levels of F2-isoprostane have been shown to be higher in males with TBI compared to females [158]. This finding may relate to gender differences that have been observed in response to hypothermia [159].

Biomarkers of Intracranial Pressure

Elevated intracranial pressure can increase morbidity and mortality in TBI [160]. A biomarker of intracranial pressure could be used to identify subjects in greater need of ICP reduction therapy, potentially prior to a rise in ICP. Serum levels of cleaved-tau have been shown to distinguish TBI patients with ICP > 30 mm Hg from those with ICP < 30 mm Hg [109]. Other markers associated with increased ICP include GFAP, S100B [92], retinal binding protein [139]. Of interest, levels of ceruloplasmin and total serum copper measured within the first 24 h of TBI were able to identify subjects whose ICP remained <20 mm Hg for the duration of study with 100 % specificity and 67 % sensitivity [161]. IL-1β has been shown to induce ceruloplasmin expression in TBI and is affected by IL-1R1 receptor [162]. Thus, the immune response to TBI may be a method to monitor ICP in TBI.

Biomarkers to Predict Multiple Organ Failure and Acute Respiratory Distress Syndrome in TBI

TBI frequently occurs in the setting of injury to other body organs. Multiple organ failure (MOF) and acute respiratory distress syndrome (ARDS) are two common causes of death in TBI patients with polytrauma [163, 164]. Both MOF and ARDS have been associated with a systemic inflammatory response. Identifying patients at risk for MOF prior to organ failure, or ARDS prior to respiratory distress, may provide a window to intervene and prevent or reduce the effects of these complications. IL-6 has been shown to correlate with MOF and ARDS [165]. Likewise, serum levels of sTNF-R p55 and sTNF-R- p75 are associated with the development of MOF [165]. A genome-wide study of RNA expressed in peripheral blood monocytes in trauma patients also suggests a proinflammatory response is associated with MOF [166]. In 13 trauma subjects, the 3 who died had overexpressed proinflammatory mRNA patterns. These changes were observed as early as 90 min after trauma.

Conclusions

The translation of a blood-based biomarker in acute brain injury to clinical practice has proven difficult. This challenge may relate in part to the heterogeneity of stroke and traumatic brain injury, as well as to the presence of the blood–brain barrier which restricts release of brain-specific markers into the circulation. However, biomarkers studies have provided insight into the pathophysiology of acute brain injury. The role of the immune system in stroke and TBI is of recognized importance, and human studies of inflammatory markers are often cited as supportive evidence. Though progress in the development of clinical biomarkers for stroke and TBI has been challenging, there are several markers that show promise, such as LpPLA2 and CRP for risk stratification in stroke, and cytokine profile, cleaved-tau, and ubiquitin C-terminal hydrolase in TBI. High-throughput screening of RNA and protein inflammatory markers are being performed and may identify novel markers with sufficient sensitivity and specificity to be of clinical use. With further study, biomarkers will likely will play a larger role in the management of stroke and TBI and continue to provide insight into the pathophysiology of these common disorders.

References

1. Lloyd-Jones D, Adams R, Carnethon M, De Simone G, Ferguson TB, Flegal K et al (2008) Heart disease and stroke statistics–2009 update: a report from the American Heart Association Statistics Committee and Stroke Statistics Subcommittee. Circulation 119(3):e182
2. Foerch C, Montaner J, Furie KL, Ning MM, Lo EH (2009) Invited article: searching for oracles? Blood biomarkers in acute stroke. Neurology 73(5):393–399
3. Jensen MB, Chacon MR, Sattin JA, Levine RL, Vemuganti R (2009) Potential biomarkers for the diagnosis of stroke. Expert Rev Cardiovasc Ther 7(4):389–393
4. Whiteley W, Chong WL, Sengupta A, Sandercock P (2009) Blood markers for the prognosis of ischemic stroke: a systematic review. Stroke 40(5):e380–e389
5. Dambinova SA, Khounteev GA, Izykenova GA, Zavolokov IG, Ilyukhina AY, Skoromets AA (2003) Blood test detecting autoantibodies to N-methyl-D-aspartate neuroreceptors for evaluation of patients with transient ischemic attack and stroke. Clin Chem 49(10):1752–1762
6. Dambinova SA, Khounteev GA, Skoromets AA (2002) Multiple panel of biomarkers for TIA/stroke evaluation. Stroke 33(5):1181–1182
7. Dalmau J, Tuzun E, Wu HY, Masjuan J, Rossi JE, Voloschin A et al (2007) Paraneoplastic anti-N-methyl-D-aspartate receptor encephalitis associated with ovarian teratoma. Ann Neurol 61(1):25–36
8. DeGiorgio LA, Konstantinov KN, Lee SC, Hardin JA, Volpe BT, Diamond B (2001) A subset of lupus anti-DNA antibodies cross-reacts with the NR2 glutamate receptor in systemic lupus erythematosus. Nat Med 7(11):1189–1193
9. Ganor Y, Goldberg-Stern H, Lerman-Sagie T, Teichberg VI, Levite M (2005) Autoimmune epilepsy: distinct subpopulations of epilepsy patients harbor serum autoantibodies to either glutamate/AMPA receptor GluR3, glutamate/NMDA receptor subunit NR2A or double-stranded DNA. Epilepsy Res 65(1–2):11–22
10. Husebye ES, Sthoeger ZM, Dayan M, Zinger H, Elbirt D, Levite M et al (2005) Autoantibodies to a NR2A peptide of the glutamate/NMDA receptor in sera of patients with systemic lupus erythematosus. Ann Rheum Dis 64(8):1210–1213

11. Dambinova SA, Bettermann K, Glynn T, Tews M, Olson D, Weissman JD et al (2012) Diagnostic potential of the NMDA receptor peptide assay for acute ischemic stroke. PLoS One 7(7):e42362
12. Lynch JR, Blessing R, White WD, Grocott HP, Newman MF, Laskowitz DT (2004) Novel diagnostic test for acute stroke. Stroke 35(1):57–63
13. Reynolds MA, Kirchick HJ, Dahlen JR, Anderberg JM, McPherson PH, Nakamura KK et al (2003) Early biomarkers of stroke. Clin Chem 49(10):1733–1739
14. Kim MH, Kang SY, Kim MC, Lee WI (2010) Plasma biomarkers in the diagnosis of acute ischemic stroke. Ann Clin Lab Sci 40(4):336–341
15. Foerch C, Curdt I, Yan B, Dvorak F, Hermans M, Berkefeld J et al (2006) Serum glial fibrillary acidic protein as a biomarker for intracerebral haemorrhage in patients with acute stroke. J Neurol Neurosurg Psychiatry 77(2):181–184
16. Foerch C, Niessner M, Back T, Bauerle M, De Marchis GM, Ferbert A et al (2012) Diagnostic accuracy of plasma glial fibrillary acidic protein for differentiating intracerebral hemorrhage and cerebral ischemia in patients with symptoms of acute stroke. Clin Chem 58(1):237–245
17. Allard L, Lescuyer P, Burgess J, Leung KY, Ward M, Walter N et al (2004) ApoC-I and ApoC-III as potential plasmatic markers to distinguish between ischemic and hemorrhagic stroke. Proteomics 4(8):2242–2251
18. Laskowitz DT, Kasner SE, Saver J, Remmel KS, Jauch EC (2009) Clinical usefulness of a biomarker-based diagnostic test for acute stroke: the Biomarker Rapid Assessment in Ischemic Injury (BRAIN) study. Stroke 40(1):77–85
19. Moore DF, Li H, Jeffries N, Wright V, Cooper RA Jr, Elkahloun A et al (2005) Using peripheral blood mononuclear cells to determine a gene expression profile of acute ischemic stroke: a pilot investigation. Circulation 111(2):212–221
20. Tang Y, Xu H, Du X, Lit L, Walker W, Lu A et al (2006) Gene expression in blood changes rapidly in neutrophils and monocytes after ischemic stroke in humans: a microarray study. J Cereb Blood Flow Metab 26(8):1089–1102
21. Stamova B, Xu H, Jickling G, Bushnell C, Tian Y, Ander BP et al (2010) Gene expression profiling of blood for the prediction of ischemic stroke. Stroke 41(10):2171–2177
22. Barr TL, Conley Y, Ding J, Dillman A, Warach S, Singleton A et al (2010) Genomic biomarkers and cellular pathways of ischemic stroke by RNA gene expression profiling. Neurology 75(11):1009–1014
23. Jin G, Zhou X, Wang H, Wong S (2010) The challenges in blood proteomic biomarker discovery. TP, editor, Springer, New York, NY
24. Sideso E, Papadakis M, Wright C, Handa A, Buchan A, Kessler B et al (2010) Assessing the quality and reproducibility of a proteomic platform for clinical stroke biomarker discovery. Transl Stroke Res 1:304–314
25. Dayon L, Turck N, Garci-Berrocoso T, Walter N, Burkhard PR, Vilalta A et al (2011) Brain extracellular fluid protein changes in acute stroke patients. J Proteome Res 10(3):1043–1051
26. Cuadrado E, Rosell A, Colome N, Hernandez-Guillamon M, Garcia-Berrocoso T, Ribo M et al (2010) The proteome of human brain after ischemic stroke. J Neuropathol Exp Neurol 69(11):1105–1115
27. Ageno W, Finazzi S, Steidl L, Biotti MG, Mera V, Melzi D'Eril G et al (2002) Plasma measurement of D-dimer levels for the early diagnosis of ischemic stroke subtypes. Arch Intern Med 162(22):2589–2593
28. Tombul T, Atbas C, Anlar O (2005) Hemostatic markers and platelet aggregation factors as predictive markers for type of stroke and neurological disability following cerebral infarction. J Clin Neurosci 12(4):429–434
29. Isenegger J, Meier N, Lammle B, Alberio L, Fischer U, Nedeltchev K et al (2010) D-dimers predict stroke subtype when assessed early. Cerebrovasc Dis 29(1):82–86
30. Nakagawa K, Yamaguchi T, Seida M, Yamada S, Imae S, Tanaka Y et al (2005) Plasma concentrations of brain natriuretic peptide in patients with acute ischemic stroke. Cerebrovasc Dis 19(3):157–164

31. Shimizu H, Murakami Y, Inoue S, Ohta Y, Nakamura K, Katoh H et al (2002) High plasma brain natriuretic polypeptide level as a marker of risk for thromboembolism in patients with nonvalvular atrial fibrillation. Stroke 33(4):1005–1010
32. Alvarez-Perez FJ, Castelo-Branco M, Alvarez-Sabin J (2011) Usefulness of measurement of fibrinogen, D-dimer, D-dimer/fibrinogen ratio, C reactive protein and erythrocyte sedimentation rate to assess the pathophysiology and mechanism of ischaemic stroke. J Neurol Neurosurg Psychiatry 82(9):986–992
33. Xu H, Tang Y, Liu DZ, Ran R, Ander BP, Apperson M et al (2008) Gene expression in peripheral blood differs after cardioembolic compared with large-vessel atherosclerotic stroke: biomarkers for the etiology of ischemic stroke. J Cereb Blood Flow Metab 28(7):1320–1328
34. Jickling GC, Xu H, Stamova B, Ander BP, Zhan X, Tian Y et al (2010) Signatures of cardioembolic and large-vessel ischemic stroke. Ann Neurol 68(5):681–692
35. Jickling GC, Stamova B, Ander BP, Zhan X, Liu D, Sison SM et al (2012) Prediction of cardioembolic, arterial, and lacunar causes of cryptogenic stroke by gene expression and infarct location. Stroke 43(8):2036–2041
36. Suwanwela NC, Chutinet A, Phanthumchinda K (2006) Inflammatory markers and conventional atherosclerotic risk factors in acute ischemic stroke: comparative study between vascular disease subtypes. J Med Assoc Thai 89(12):2021–2027
37. Hassan A, Hunt BJ, O'Sullivan M, Bell R, D'Souza R, Jeffery S et al (2004) Homocysteine is a risk factor for cerebral small vessel disease, acting via endothelial dysfunction. Brain 127(Pt 1):212–219
38. Hassan A, Hunt BJ, O'Sullivan M, Parmar K, Bamford JM, Briley D et al (2003) Markers of endothelial dysfunction in lacunar infarction and ischaemic leukoaraiosis. Brain 126(Pt 2):424–432
39. Jickling GC, Stamova B, Ander BP, Zhan X, Tian Y, Liu D et al (2011) Profiles of lacunar and nonlacunar stroke. Ann Neurol 70(3):477–485
40. Markus HS, Hunt B, Palmer K, Enzinger C, Schmidt H, Schmidt R (2005) Markers of endothelial and hemostatic activation and progression of cerebral white matter hyperintensities: longitudinal results of the Austrian Stroke Prevention Study. Stroke 36(7):1410–1414
41. Xu H, Stamova B, Jickling G, Tian Y, Zhan X, Ander BP et al (2010) Distinctive RNA expression profiles in blood associated with white matter hyperintensities in brain. Stroke 41(12):2744–2749
42. Miwa K, Tanaka M, Okazaki S, Furukado S, Sakaguchi M, Kitagawa K (2011) Relations of blood inflammatory marker levels with cerebral microbleeds. Stroke 42(11):3202–3206
43. Shenhar-Tsarfaty S, Ben Assayag E, Bova I, Shopin L, Fried M, Berliner S et al (2010) Interleukin-6 as an early predictor for one-year survival following an ischaemic stroke/transient ischaemic attack. Int J Stroke 5(1):16–20
44. Whiteley W, Rumley A, Sattar N, Wardlaw JM, Lowe GDO, Welsh P et al (2010) POS11 Blood markers and poor outcome after acute cerebrovascular disease: a prospective cohort study. Journal of Neurology, Neurosurgery & Psychiatry 81(11):e70
45. Foerch C, Singer OC, Neumann-Haefelin T, du Mesnil de Rochemont R, Steinmetz H, Sitzer M (2005) Evaluation of serum S100B as a surrogate marker for long-term outcome and infarct volume in acute middle cerebral artery infarction. Arch Neurol 62(7):1130–1134
46. Missler U, Wiesmann M, Friedrich C, Kaps M (1997) S-100 protein and neuron-specific enolase concentrations in blood as indicators of infarction volume and prognosis in acute ischemic stroke. Stroke 28(10):1956–1960
47. Wunderlich MT, Lins H, Skalej M, Wallesch CW, Goertler M (2006) Neuron-specific enolase and tau protein as neurobiochemical markers of neuronal damage are related to early clinical course and long-term outcome in acute ischemic stroke. Clin Neurol Neurosurg 108(6): 558–563
48. Wunderlich MT, Wallesch CW, Goertler M (2004) Release of neurobiochemical markers of brain damage is related to the neurovascular status on admission and the site of arterial occlusion in acute ischemic stroke. J Neurol Sci 227(1):49–53

49. Castillo J, Davalos A, Naveiro J, Noya M (1996) Neuroexcitatory amino acids and their relation to infarct size and neurological deficit in ischemic stroke. Stroke 27(6):1060–1065
50. Sotgiu S, Zanda B, Marchetti B, Fois ML, Arru G, Pes GM et al (2006) Inflammatory biomarkers in blood of patients with acute brain ischemia. Eur J Neurol 13(5):505–513
51. Vila N, Castillo J, Davalos A, Chamorro A (2000) Proinflammatory cytokines and early neurological worsening in ischemic stroke. Stroke 31(10):2325–2329
52. Montaner J, Rovira A, Molina CA, Arenillas JF, Ribo M, Chacon P et al (2003) Plasmatic level of neuroinflammatory markers predict the extent of diffusion-weighted image lesions in hyperacute stroke. J Cereb Blood Flow Metab 23(12):1403–1407
53. Rosell A, Alvarez-Sabin J, Arenillas JF, Rovira A, Delgado P, Fernandez-Cadenas I et al (2005) A matrix metalloproteinase protein array reveals a strong relation between MMP-9 and MMP-13 with diffusion-weighted image lesion increase in human stroke. Stroke 36(7):1415–1420
54. Murata Y, Rosell A, Scannevin RH, Rhodes KJ, Wang X, Lo EH (2008) Extension of the thrombolytic time window with minocycline in experimental stroke. Stroke 39(12):3372–3377
55. Matsukawa N, Yasuhara T, Hara K, Xu L, Maki M, Yu G et al (2009) Therapeutic targets and limits of minocycline neuroprotection in experimental ischemic stroke. BMC Neurosci 10(1):126
56. Montaner J (2009) Blood biomarkers to guide stroke thrombolysis. Front Biosci (Elite Ed) 1:200–208
57. Castellanos M, Leira R, Serena J, Blanco M, Pedraza S, Castillo J et al (2004) Plasma cellular-fibronectin concentration predicts hemorrhagic transformation after thrombolytic therapy in acute ischemic stroke. Stroke 35(7):1671–1676
58. Castellanos M, Sobrino T, Millán M, García M, Arenillas J, Nombela F et al (2007) Serum cellular fibronectin and matrix metalloproteinase-9 as screening biomarkers for the prediction of parenchymal hematoma after thrombolytic therapy in acute ischemic stroke: a multicenter confirmatory study. Stroke 38(6):1855–1859
59. Ribo M, Montaner J, Molina CA, Arenillas JF, Santamarina E, Quintana M et al (2004) Admission fibrinolytic profile is associated with symptomatic hemorrhagic transformation in stroke patients treated with tissue plasminogen activator. Stroke 35(9):2123–2127
60. Foerch C, Wunderlich MT, Dvorak F, Humpich M, Kahles T, Goertler M et al (2007) Elevated serum S100B levels indicate a higher risk of hemorrhagic transformation after thrombolytic therapy in acute stroke. Stroke 38(9):2491–2495
61. Castellanos M, Leira R, Serena J, Pumar JM, Lizasoain I, Castillo J et al (2003) Plasma metalloproteinase-9 concentration predicts hemorrhagic transformation in acute ischemic stroke. Stroke 34(1):40–46
62. Montaner J, Alvarez-Sabin J, Molina C, Angles A, Abilleira S, Arenillas J et al (2001) Matrix metalloproteinase expression after human cardioembolic stroke: temporal profile and relation to neurological impairment. Stroke 32(8):1759–1766
63. Montaner J, Alvarez-Sabin J, Molina CA, Angles A, Abilleira S, Arenillas J et al (2001) Matrix metalloproteinase expression is related to hemorrhagic transformation after cardioembolic stroke. Stroke 32(12):2762–2767
64. Montaner J, Molina CA, Monasterio J, Abilleira S, Arenillas JF, Ribo M et al (2003) Matrix metalloproteinase-9 pretreatment level predicts intracranial hemorrhagic complications after thrombolysis in human stroke. Circulation 107(4):598–603
65. Mun-Bryce S, Rosenberg GA (1998) Matrix metalloproteinases in cerebrovascular disease. J Cereb Blood Flow Metab 18(11):1163–1172
66. Demchuk AM, Morgenstern LB, Krieger DW, Linda Chi T, Hu W, Wein TH et al (1999) Serum glucose level and diabetes predict tissue plasminogen activator-related intracerebral hemorrhage in acute ischemic stroke. Stroke 30(1):34–39
67. Uemura S, Matsushita H, Li W, Glassford AJ, Asagami T, Lee KH et al (2001) Diabetes mellitus enhances vascular matrix metalloproteinase activity: role of oxidative stress. Circ Res 88(12):1291–1298

68. Barr TL, Latour LL, Lee KY, Schaewe TJ, Luby M, Chang GS et al (2010) Blood-brain barrier disruption in humans is independently associated with increased matrix metalloproteinase-9. Stroke 41(3):e123–e128
69. Castellanos M, Sobrino T, Pedraza S, Moldes O, Pumar JM, Silva Y et al (2008) High plasma glutamate concentrations are associated with infarct growth in acute ischemic stroke. Neurology 71(23):1862–1868
70. Serena J, Blanco M, Castellanos M, Silva Y, Vivancos J, Moro MA et al (2005) The prediction of malignant cerebral infarction by molecular brain barrier disruption markers. Stroke 36(9):1921–1926
71. Davalos A, Castillo J, Marrugat J, Fernandez-Real JM, Armengou A, Cacabelos P et al (2000) Body iron stores and early neurologic deterioration in acute cerebral infarction. Neurology 54(8):1568–1574
72. Castellanos M, Castillo J, Garcia MM, Leira R, Serena J, Chamorro A et al (2002) Inflammation-mediated damage in progressing lacunar infarctions: a potential therapeutic target. Stroke 33(4):982–987
73. Elkind MS, Tai W, Coates K, Paik MC, Sacco RL (2009) Lipoprotein-associated phospholipase A2 activity and risk of recurrent stroke. Cerebrovasc Dis 27(1):42–50
74. Gorelick PB (2008) Lipoprotein-associated phospholipase A2 and risk of stroke. Am J Cardiol 101(12A):34F–40F
75. Ballantyne CM, Hoogeveen RC, Bang H, Coresh J, Folsom AR, Chambless LE et al (2005) Lipoprotein-associated phospholipase A2, high-sensitivity C-reactive protein, and risk for incident ischemic stroke in middle-aged men and women in the Atherosclerosis Risk in Communities (ARIC) study. Arch Intern Med 165(21):2479–2484
76. Oei HH, van der Meer IM, Hofman A, Koudstaal PJ, Stijnen T, Breteler MM et al (2005) Lipoprotein-associated phospholipase A2 activity is associated with risk of coronary heart disease and ischemic stroke: the Rotterdam Study. Circulation 111(5):570–575
77. Wassertheil-Smoller S, Hendrix SL, Limacher M, Heiss G, Kooperberg C, Baird A et al (2003) Effect of estrogen plus progestin on stroke in postmenopausal women: the Women's Health Initiative: a randomized trial. JAMA 289(20):2673–2684
78. Thompson A, Gao P, Orfei L, Watson S, Di Angelantonio E, Kaptoge S et al (2010) Lipoprotein-associated phospholipase A(2) and risk of coronary disease, stroke, and mortality: collaborative analysis of 32 prospective studies. Lancet 375(9725):1536–1544
79. Vahedi K, Vicaut E, Mateo J, Kurtz A, Orabi M, Guichard JP et al (2007) Sequential-design, multicenter, randomized, controlled trial of early decompressive craniectomy in malignant middle cerebral artery infarction (DECIMAL Trial). Stroke 38(9):2506–2517
80. Foerch C, Otto B, Singer OC, Neumann-Haefelin T, Yan B, Berkefeld J et al (2004) Serum S100B predicts a malignant course of infarction in patients with acute middle cerebral artery occlusion. Stroke 35(9):2160–2164
81. Maas MB, Furie KL (2009) Molecular biomarkers in stroke diagnosis and prognosis. Biomark Med 3(4):363–383
82. Worthmann H, Dengler R, Schumacher H, Schwartz A, Eisert WG, Lichtinghagen R et al (2012) Monocyte chemotactic protein-1 as a potential biomarker for early anti-thrombotic therapy after ischemic stroke. Int J Mol Sci 13(7):8670–8678
83. Haskins WE, Kobeissy FH, Wolper RA, Ottens AK, Kitlen JW, McClung SH et al (2005) Rapid discovery of putative protein biomarkers of traumatic brain injury by SDS-PAGE-capillary liquid chromatography-tandem mass spectrometry. J Neurotrauma 22(6):629–644
84. Pineda JA, Wang KK, Hayes RL (2004) Biomarkers of proteolytic damage following traumatic brain injury. Brain Pathol 14(2):202–209
85. Smith DH, Uryu K, Saatman KE, Trojanowski JQ, McIntosh TK (2003) Protein accumulation in traumatic brain injury. Neuromolecular Med 4(1–2):59–72
86. Crooks CY, Zumsteg JM, Bell KR (2007) Traumatic brain injury: a review of practice management and recent advances. Phys Med Rehabil Clin N Am 18(4):681–710, vi
87. Savola O, Pyhtinen J, Leino TK, Siitonen S, Niemela O, Hillbom M (2004) Effects of head and extracranial injuries on serum protein S100B levels in trauma patients. J Trauma 56(6):1229–1234, discussion 34

88. Unden J, Astrand R, Waterloo K, Ingebrigtsen T, Bellner J, Reinstrup P et al (2007) Clinical significance of serum S100B levels in neurointensive care. Neurocrit Care 6(2):94–99
89. Vos PE, Lamers KJ, Hendriks JC, van Haaren M, Beems T, Zimmerman C et al (2004) Glial and neuronal proteins in serum predict outcome after severe traumatic brain injury. Neurology 62(8):1303–1310
90. Metting Z, Wilczak N, Rodiger LA, Schaaf JM, van der Naalt J (2012) GFAP and S100B in the acute phase of mild traumatic brain injury. Neurology 78(18):1428–1433
91. Vos PE, Jacobs B, Andriessen TM, Lamers KJ, Borm GF, Beems T et al (2010) GFAP and S100B are biomarkers of traumatic brain injury: an observational cohort study. Neurology 75(20):1786–1793
92. Pelinka LE, Kroepfl A, Leixnering M, Buchinger W, Raabe A, Redl H (2004) GFAP versus S100B in serum after traumatic brain injury: relationship to brain damage and outcome. J Neurotrauma 21(11):1553–1561
93. Missler U, Wiesmann M, Wittmann G, Magerkurth O, Hagenstrom H (1999) Measurement of glial fibrillary acidic protein in human blood: analytical method and preliminary clinical results. Clin Chem 45(1):138–141
94. Nylen K, Ost M, Csajbok LZ, Nilsson I, Blennow K, Nellgard B et al (2006) Increased serum-GFAP in patients with severe traumatic brain injury is related to outcome. J Neurol Sci 240(1–2):85–91
95. Fridriksson T, Kini N, Walsh-Kelly C, Hennes H (2000) Serum neuron-specific enolase as a predictor of intracranial lesions in children with head trauma: a pilot study. Acad Emerg Med 7(7):816–820
96. Herrmann M, Curio N, Jost S, Grubich C, Ebert AD, Fork ML et al (2001) Release of biochemical markers of damage to neuronal and glial brain tissue is associated with short and long term neuropsychological outcome after traumatic brain injury. J Neurol Neurosurg Psychiatry 70(1):95–100
97. Raabe A, Grolms C, Keller M, Dohnert J, Sorge O, Seifert V (1998) Correlation of computed tomography findings and serum brain damage markers following severe head injury. Acta Neurochir (Wien) 140(8):787–791, discussion 91–92
98. Berger RP, Adelson PD, Pierce MC, Dulani T, Cassidy LD, Kochanek PM (2005) Serum neuron-specific enolase, S100B, and myelin basic protein concentrations after inflicted and noninflicted traumatic brain injury in children. J Neurosurg 103(1 Suppl):61–68
99. Mao Q, Chen J, Li N, Li C, Mao B, Wang R et al (1995) The value of serum myelin basic protein in assessment of severity of acute closed head trauma. Hua Xi Yi Ke Da Xue Xue Bao 26(2):135–137
100. Pelsers MM, Hanhoff T, Van der Voort D, Arts B, Peters M, Ponds R et al (2004) Brain- and heart-type fatty acid-binding proteins in the brain: tissue distribution and clinical utility. Clin Chem 50(9):1568–1575
101. Mondello S, Robicsek SA, Gabrielli A, Brophy GM, Papa L, Tepas J et al (2010) AlphaII-spectrin breakdown products (SBDPs): diagnosis and outcome in severe traumatic brain injury patients. J Neurotrauma 27(7):1203–1213
102. Brophy GM, Pineda JA, Papa L, Lewis SB, Valadka AB, Hannay HJ et al (2009) AlphaII-Spectrin breakdown product cerebrospinal fluid exposure metrics suggest differences in cellular injury mechanisms after severe traumatic brain injury. J Neurotrauma 26(4):471–479
103. Pineda JA, Lewis SB, Valadka AB, Papa L, Hannay HJ, Heaton SC et al (2007) Clinical significance of alphaII-spectrin breakdown products in cerebrospinal fluid after severe traumatic brain injury. J Neurotrauma 24(2):354–366
104. Cardali S, Maugeri R (2006) Detection of alphaII-spectrin and breakdown products in humans after severe traumatic brain injury. J Neurosurg Sci 50(2):25–31
105. Ringger NC, O'Steen BE, Brabham JG, Silver X, Pineda J, Wang KK et al (2004) A novel marker for traumatic brain injury: CSF alphaII-spectrin breakdown product levels. J Neurotrauma 21(10):1443–1456
106. Ma M, Lindsell CJ, Rosenberry CM, Shaw GJ, Zemlan FP (2008) Serum cleaved tau does not predict postconcussion syndrome after mild traumatic brain injury. Am J Emerg Med 26(7):763–768

107. Bazarian JJ, Zemlan FP, Mookerjee S, Stigbrand T (2006) Serum S-100B and cleaved-tau are poor predictors of long-term outcome after mild traumatic brain injury. Brain Inj 20(7):759–765
108. Gabbita SP, Scheff SW, Menard RM, Roberts K, Fugaccia I, Zemlan FP (2005) Cleaved-tau: a biomarker of neuronal damage after traumatic brain injury. J Neurotrauma 22(1):83–94
109. Zemlan FP, Jauch EC, Mulchahey JJ, Gabbita SP, Rosenberg WS, Speciale SG et al (2002) C-tau biomarker of neuronal damage in severe brain injured patients: association with elevated intracranial pressure and clinical outcome. Brain Res 947(1):131–139
110. Zurek J, Fedora M (2012) The usefulness of S100B, NSE, GFAP, NF-H, secretagogin and Hsp70 as a predictive biomarker of outcome in children with traumatic brain injury. Acta Neurochir (Wien) 154(1):93–103, discussion
111. Anderson KJ, Scheff SW, Miller KM, Roberts KN, Gilmer LK, Yang C et al (2008) The phosphorylated axonal form of the neurofilament subunit NF-H (pNF-H) as a blood biomarker of traumatic brain injury. J Neurotrauma 25(9):1079–1085
112. Shaw G, Yang C, Ellis R, Anderson K, Parker Mickle J, Scheff S et al (2005) Hyperphosphorylated neurofilament NF-H is a serum biomarker of axonal injury. Biochem Biophys Res Commun 336(4):1268–1277
113. Papa L, Lewis LM, Silvestri S, Falk JL, Giordano P, Brophy GM et al (2012) Serum levels of ubiquitin C-terminal hydrolase distinguish mild traumatic brain injury from trauma controls and are elevated in mild and moderate traumatic brain injury patients with intracranial lesions and neurosurgical intervention. J Trauma Acute Care Surg 72(5):1335–1344
114. Mondello S, Linnet A, Buki A, Robicsek S, Gabrielli A, Tepas J et al (2012) Clinical utility of serum levels of ubiquitin C-terminal hydrolase as a biomarker for severe traumatic brain injury. Neurosurgery 70(3):666–675
115. Brophy GM, Mondello S, Papa L, Robicsek SA, Gabrielli A, Tepas J 3rd et al (2011) Biokinetic analysis of ubiquitin C-terminal hydrolase-L1 (UCH-L1) in severe traumatic brain injury patient biofluids. J Neurotrauma 28(6):861–870
116. Papa L, Akinyi L, Liu MC, Pineda JA, Tepas JJ 3rd, Oli MW et al (2010) Ubiquitin C-terminal hydrolase is a novel biomarker in humans for severe traumatic brain injury. Crit Care Med 38(1):138–144
117. Csuka E, Morganti-Kossmann MC, Lenzlinger PM, Joller H, Trentz O, Kossmann T (1999) IL-10 levels in cerebrospinal fluid and serum of patients with severe traumatic brain injury: relationship to IL-6, TNF-alpha, TGF-beta1 and blood–brain barrier function. J Neuroimmunol 101(2):211–221
118. Morganti-Kossmann MC, Hans VH, Lenzlinger PM, Dubs R, Ludwig E, Trentz O et al (1999) TGF-beta is elevated in the CSF of patients with severe traumatic brain injuries and parallels blood-brain barrier function. J Neurotrauma 16(7):617–628
119. Pleines UE, Morganti-Kossmann MC, Rancan M, Joller H, Trentz O, Kossmann T (2001) S-100 beta reflects the extent of injury and outcome, whereas neuronal specific enolase is a better indicator of neuroinflammation in patients with severe traumatic brain injury. J Neurotrauma 18(5):491–498
120. Kossmann T, Hans V, Imhof HG, Trentz O, Morganti-Kossmann MC (1996) Interleukin-6 released in human cerebrospinal fluid following traumatic brain injury may trigger nerve growth factor production in astrocytes. Brain Res 713(1–2):143–152
121. Frugier T, Morganti-Kossmann MC, O'Reilly D, McLean CA (2010) In situ detection of inflammatory mediators in post mortem human brain tissue after traumatic injury. J Neurotrauma 27(3):497–507
122. Buttram SD, Wisniewski SR, Jackson EK, Adelson PD, Feldman K, Bayir H et al (2007) Multiplex assessment of cytokine and chemokine levels in cerebrospinal fluid following severe pediatric traumatic brain injury: effects of moderate hypothermia. J Neurotrauma 24(11):1707–1717
123. Holmin S, Hojeberg B (2004) In situ detection of intracerebral cytokine expression after human brain contusion. Neurosci Lett 369(2):108–114

124. Hutchinson PJ, O'Connell MT, Rothwell NJ, Hopkins SJ, Nortje J, Carpenter KL et al (2007) Inflammation in human brain injury: intracerebral concentrations of IL-1alpha, IL-1beta, and their endogenous inhibitor IL-1ra. J Neurotrauma 24(10):1545–1557
125. Mellergard P, Aneman O, Sjogren F, Pettersson P, Hillman J (2008) Changes in extracellular concentrations of some cytokines, chemokines, and neurotrophic factors after insertion of intracerebral microdialysis catheters in neurosurgical patients. Neurosurgery 62(1):151–157, discussion 7-8
126. Winter CD, Iannotti F, Pringle AK, Trikkas C, Clough GF, Church MK (2002) A microdialysis method for the recovery of IL-1beta, IL-6 and nerve growth factor from human brain in vivo. J Neurosci Methods 119(1):45–50
127. Clark RS, Kochanek PM, Chen M, Watkins SC, Marion DW, Chen J et al (1999) Increases in Bcl-2 and cleavage of caspase-1 and caspase-3 in human brain after head injury. FASEB J 13(8):813–821
128. Crespo AR, Da Rocha AB, Jotz GP, Schneider RF, Grivicich I, Pinheiro K et al (2007) Increased serum sFas and TNFalpha following isolated severe head injury in males. Brain Inj 21(4):441–447
129. Goodman JC, Robertson CS, Grossman RG, Narayan RK (1990) Elevation of tumor necrosis factor in head injury. J Neuroimmunol 30(2–3):213–217
130. Ross SA, Halliday MI, Campbell GC, Byrnes DP, Rowlands BJ (1994) The presence of tumour necrosis factor in CSF and plasma after severe head injury. Br J Neurosurg 8(4):419–425
131. Is M, Coskun A, Sanus GZ, Tanriverdi T, Kafadar AM, Hanimoglu H et al (2007) High-sensitivity C-reactive protein levels in cerebrospinal fluid and serum in severe head injury: relationship to tumor necrosis factor-alpha and interleukin-6. J Clin Neurosci 14(12):1163–1171
132. Hayakata T, Shiozaki T, Tasaki O, Ikegawa H, Inoue Y, Toshiyuki F et al (2004) Changes in CSF S100B and cytokine concentrations in early-phase severe traumatic brain injury. Shock 22(2):102–107
133. Shiozaki T, Hayakata T, Tasaki O, Hosotubo H, Fuijita K, Mouri T et al (2005) Cerebrospinal fluid concentrations of anti-inflammatory mediators in early-phase severe traumatic brain injury. Shock 23(5):406–410
134. Kossmann T, Stahel PF, Lenzlinger PM, Redl H, Dubs RW, Trentz O et al (1997) Interleukin-8 released into the cerebrospinal fluid after brain injury is associated with blood-brain barrier dysfunction and nerve growth factor production. J Cereb Blood Flow Metab 17(3):280–289
135. Whalen MJ, Carlos TM, Kochanek PM, Wisniewski SR, Bell MJ, Clark RS et al (2000) Interleukin-8 is increased in cerebrospinal fluid of children with severe head injury. Crit Care Med 28(4):929–934
136. Stefini R, Catenacci E, Piva S, Sozzani S, Valerio A, Bergomi R et al (2008) Chemokine detection in the cerebral tissue of patients with posttraumatic brain contusions. J Neurosurg 108(5):958–962
137. Bell MJ, Kochanek PM, Doughty LA, Carcillo JA, Adelson PD, Clark RS et al (1997) Comparison of the interleukin-6 and interleukin-10 response in children after severe traumatic brain injury or septic shock. Acta Neurochir Suppl 70:96–97
138. Bell MJ, Kochanek PM, Doughty LA, Carcillo JA, Adelson PD, Clark RS et al (1997) Interleukin-6 and interleukin-10 in cerebrospinal fluid after severe traumatic brain injury in children. J Neurotrauma 14(7):451–457
139. Hergenroeder G, Redell JB, Moore AN, Dubinsky WP, Funk RT, Crommett J et al (2008) Identification of serum biomarkers in brain-injured adults: potential for predicting elevated intracranial pressure. J Neurotrauma 25(2):79–93
140. Pleines UE, Stover JF, Kossmann T, Trentz O, Morganti-Kossmann MC (1998) Soluble ICAM-1 in CSF coincides with the extent of cerebral damage in patients with severe traumatic brain injury. J Neurotrauma 15(6):399–409
141. Bye N, Habgood MD, Callaway JK, Malakooti N, Potter A, Kossmann T et al (2007) Transient neuroprotection by minocycline following traumatic brain injury is associated with attenuated microglial activation but no changes in cell apoptosis or neutrophil infiltration. Exp Neurol 204(1):220–233

142. Wagner AK, Bayir H, Ren D, Puccio A, Zafonte RD, Kochanek PM (2004) Relationships between cerebrospinal fluid markers of excitotoxicity, ischemia, and oxidative damage after severe TBI: the impact of gender, age, and hypothermia. J Neurotrauma 21(2):125–136
143. Bayir H, Adelson PD, Wisniewski SR, Shore P, Lai Y, Brown D et al (2009) Therapeutic hypothermia preserves antioxidant defenses after severe traumatic brain injury in infants and children. Crit Care Med 37(2):689–695
144. Seifman MA, Adamides AA, Nguyen PN, Vallance SA, Cooper DJ, Kossmann T et al (2008) Endogenous melatonin increases in cerebrospinal fluid of patients after severe traumatic brain injury and correlates with oxidative stress and metabolic disarray. J Cereb Blood Flow Metab 28(4):684–696
145. Grond-Ginsbach C, Hummel M, Wiest T, Horstmann S, Pfleger K, Hergenhahn M et al (2008) Gene expression in human peripheral blood mononuclear cells upon acute ischemic stroke. J Neurol 255(5):723–731
146. Yao C, Williams AJ, Ottens AK, Lu XC, Liu MC, Hayes RL et al (2009) P43/pro-EMAPII: a potential biomarker for discriminating traumatic versus ischemic brain injury. J Neurotrauma 26(8):1295–1305
147. Redell JB, Moore AN, Ward NH 3rd, Hergenroeder GW, Dash PK (2010) Human traumatic brain injury alters plasma microRNA levels. J Neurotrauma 27(12):2147–2156
148. Pasinetti GM, Ho L, Dooley C, Abbi B, Lange G (2012) Select non-coding RNA in blood components provide novel clinically accessible biological surrogates for improved identification of traumatic brain injury in OEF/OIF Veterans. Am J Neurodegener Dis 1(1):88–98
149. Balakathiresan N, Bhomia M, Chandran R, Chavko M, McCarron RM, Maheshwari RK (2012) MicroRNA let-7i is a promising serum biomarker for blast-induced traumatic brain injury. J Neurotrauma 29(7):1379–1387
150. Redell JB, Zhao J, Dash PK (2011) Altered expression of miRNA-21 and its targets in the hippocampus after traumatic brain injury. J Neurosci Res 89(2):212–221
151. Bandyopadhyay S, Hennes H, Gorelick MH, Wells RG, Walsh-Kelly CM (2005) Serum neuron-specific enolase as a predictor of short-term outcome in children with closed traumatic brain injury. Acad Emerg Med 12(8):732–738
152. Liliang PC, Liang CL, Weng HC, Lu K, Wang KW, Chen HJ et al (2010) Tau proteins in serum predict outcome after severe traumatic brain injury. J Surg Res 160(2):302–307
153. Singhal A, Baker AJ, Hare GM, Reinders FX, Schlichter LC, Moulton RJ (2002) Association between cerebrospinal fluid interleukin-6 concentrations and outcome after severe human traumatic brain injury. J Neurotrauma 19(8):929–937
154. Chiaretti A, Genovese O, Aloe L, Antonelli A, Piastra M, Polidori G et al (2005) Interleukin 1beta and interleukin 6 relationship with paediatric head trauma severity and outcome. Childs Nerv Syst 21(3):185–193, discussion 94
155. Minambres E, Cemborain A, Sanchez-Velasco P, Gandarillas M, Diaz-Reganon G, Sanchez-Gonzalez U et al (2003) Correlation between transcranial interleukin-6 gradient and outcome in patients with acute brain injury. Crit Care Med 31(3):933–938
156. Winter CD, Pringle AK, Clough GF, Church MK (2004) Raised parenchymal interleukin-6 levels correlate with improved outcome after traumatic brain injury. Brain 127(Pt 2):315–320
157. Aibiki M, Maekawa S, Nishiyama T, Seki K, Yokono S (1999) Activated cytokine production in patients with accidental hypothermia. Resuscitation 41(3):263–268
158. Bayir H, Marion DW, Puccio AM, Wisniewski SR, Janesko KL, Clark RS et al (2004) Marked gender effect on lipid peroxidation after severe traumatic brain injury in adult patients. J Neurotrauma 21(1):1–8
159. Suzuki T, Bramlett HM, Dietrich WD (2003) The importance of gender on the beneficial effects of posttraumatic hypothermia. Exp Neurol 184(2):1017–1026
160. Treggiari MM, Schutz N, Yanez ND, Romand JA (2007) Role of intracranial pressure values and patterns in predicting outcome in traumatic brain injury: a systematic review. Neurocrit Care 6(2):104–112
161. Dash PK, Redell JB, Hergenroeder G, Zhao J, Clifton GL, Moore A (2010) Serum ceruloplasmin and copper are early biomarkers for traumatic brain injury-associated elevated intracranial pressure. J Neurosci Res 88(8):1719–1726

162. Kuhlow CJ, Krady JK, Basu A, Levison SW (2003) Astrocytic ceruloplasmin expression, which is induced by IL-1beta and by traumatic brain injury, increases in the absence of the IL-1 type 1 receptor. Glia 44(1):76–84
163. Bose D, Tejwani NC (2006) Evolving trends in the care of polytrauma patients. Injury 37(1):20–28
164. Pape HC (2012) Classification of patients with multiple injuries—is the polytrauma patient defined adequately in 2012? Injury 43(2):127–128
165. Maier B, Lefering R, Lehnert M, Laurer HL, Steudel WI, Neugebauer EA et al (2007) Early versus late onset of multiple organ failure is associated with differing patterns of plasma cytokine biomarker expression and outcome after severe trauma. Shock 28(6):668–674
166. Biberthaler P, Bogner V, Baker HV, Lopez MC, Neth P, Kanz KG et al (2005) Genome-wide monocytic mRNA expression in polytrauma patients for identification of clinical outcome. Shock 24(1):11–19
167. Hill MD, Jackowski G, Bayer N, Lawrence M, Jaeschke R (2000) Biochemical markers in acute ischemic stroke. CMAJ 162(8):1139–1140
168. Lamers KJ, Vos P, Verbeek MM, Rosmalen F, van Geel WJ, van Engelen BG (2003) Protein S-100B, neuron-specific enolase (NSE), myelin basic protein (MBP) and glial fibrillary acidic protein (GFAP) in cerebrospinal fluid (CSF) and blood of neurological patients. Brain Res Bull 61(3):261–264
169. Dvorak F, Haberer I, Sitzer M, Foerch C (2009) Characterisation of the diagnostic window of serum glial fibrillary acidic protein for the differentiation of intracerebral haemorrhage and ischaemic stroke. Cerebrovasc Dis 27(1):37–41
170. Jauch EC, Lindsell C, Broderick J, Fagan SC, Tilley BC, Levine SR (2006) Association of serial biochemical markers with acute ischemic stroke: the National Institute of Neurological Disorders and Stroke recombinant tissue plasminogen activator Stroke Study. Stroke 37(10):2508–2513
171. Di Napoli M, Schwaninger M, Cappelli R, Ceccarelli E, Di Gianfilippo G, Donati C et al (2005) Evaluation of C-reactive protein measurement for assessing the risk and prognosis in ischemic stroke: a statement for health care professionals from the CRP Pooling Project members. Stroke 36(6):1316–1329
172. Fon EA, Mackey A, Cote R, Wolfson C, McIlraith DM, Leclerc J et al (1994) Hemostatic markers in acute transient ischemic attacks. Stroke 25(2):282–286
173. Allard L, Burkhard PR, Lescuyer P, Burgess JA, Walter N, Hochstrasser DF et al (2005) PARK7 and nucleoside diphosphate kinase A as plasma markers for the early diagnosis of stroke. Clin Chem 51(11):2043–2051
174. Uno M, Kitazato KT, Nishi K, Itabe H, Nagahiro S (2003) Raised plasma oxidised LDL in acute cerebral infarction. J Neurol Neurosurg Psychiatry 74(3):312–316
175. Cano CP, Bermudez VP, Atencio HE, Medina MT, Anilsa A, Souki A et al (2003) Increased serum malondialdehyde and decreased nitric oxide within 24 hours of thrombotic stroke onset. Am J Ther 10(6):473–476
176. Zimmermann-Ivol CG, Burkhard PR, Le Floch-Rohr J, Allard L, Hochstrasser DF, Sanchez JC (2004) Fatty acid binding protein as a serum marker for the early diagnosis of stroke: a pilot study. Mol Cell Proteomics 3(1):66–72
177. Jickling GC, Zhan X, Stamova B, Ander BP, Tian Y, Liu D et al (2012) Ischemic transient neurological events identified by immune response to cerebral ischemia. Stroke 43(4):1006–1012
178. Zhan X, Jickling GC, Tian Y, Stamova B, Xu H, Ander BP et al (2011) Transient ischemic attacks characterized by RNA profiles in blood. Neurology 77(19):1718–1724
179. Tuttolomondo A, Di Sciacca R, Di Raimondo D, Serio A, D'Aguanno G, La Placa S et al (2009) Plasma levels of inflammatory and thrombotic/fibrinolytic markers in acute ischemic strokes: relationship with TOAST subtype, outcome and infarct site. J Neuroimmunol 215(1–2):84–89
180. Mocco J, Choudhri TF, Mack WJ, Laufer I, Lee J, Kiss S et al (2001) Elevation of soluble intercellular adhesion molecule-1 levels in symptomatic and asymptomatic carotid atherosclerosis. Neurosurgery 48(4):718–721, discussion 21-2

181. DeGraba TJ, Siren AL, Penix L, McCarron RM, Hargraves R, Sood S et al (1998) Increased endothelial expression of intercellular adhesion molecule-1 in symptomatic versus asymptomatic human carotid atherosclerotic plaque. Stroke 29(7):1405–1410
182. Turaj W, Slowik A, Pulyk R, Adamski M, Szczudlik A (2006) Comparison of plasma concentrations of fibrinogen in patients with ischemic stroke due to large vessel disease and small vessel disease. Neurol Neurochir Pol 40(4):297–301
183. Montaner J, Perea-Gainza M, Delgado P, Ribo M, Chacon P, Rosell A et al (2008) Etiologic diagnosis of ischemic stroke subtypes with plasma biomarkers. Stroke 39(8):2280–2287
184. Sato M, Suzuki A, Nagata K, Uchiyama S (2006) Increased von Willebrand factor in acute stroke patients with atrial fibrillation. J Stroke Cerebrovasc Dis 15(1):1–7

Chapter 13
In Vivo Imaging of Neuroinflammation in Acute Brain Injury

Alexander Thiel

Abstract Neuroinflammation is a dynamic process which undergoes significant changes in spatial distribution and intensity within hours and days after an acute brain injury. At present non-invasive in vivo imaging methods like positron emission tomography (PET) offer the only possibility to capture this dynamics longitudinally in the same subject and the entire brain. Amongst the multitude of cellular and non-cellular mechanisms which constitute the complex neuroinflammatory reaction, microglia and macrophages have been the primary targets for developing non-invasive imaging methods. This chapter is an introduction into the basic principles of microglia imaging with PET, its application to ischaemic stroke and traumatic brain injury in both animal models and clinical imaging in patients. Future developments towards magnetic resonance imaging (MRI) of neuroinflammation and imaging of specific enzyme activity in the neuroinflammatory cascade are discussed.

Activated Microglia and Macrophages as Imaging Target

Neuroinflammation is a complex reaction of the brain to acute or chronic injuries comprising many cellular (e.g. macrophages, astrocytes, lymphocytes, etc.) and non-cellular (e.g. cytokines, antibodies, etc.) components and their responses to noxious stimuli. Whether neuroinflammation as such should be regarded as a beneficial physiological response or a pathological process is still unsettled and seems to depend on the context in which the inflammatory reaction takes place, its temporal and spatial relationships to the injury and the mediators involved [1]. Imaging neuroinflammation in vivo and non-invasively is thus of significant scientific and clinical interest but appears challenging because these processes are not linked to a single

A. Thiel (✉)
Department of Neurology and Neurosurgery, McGill University, Montreal, QC, Canada
e-mail: alexander.thiel@mcgill.ca

J. Chen et al. (eds.), *Immunological Mechanisms and Therapies in Brain Injuries and Stroke*, Springer Series in Translational Stroke Research, DOI 10.1007/978-1-4614-8915-3_13,

specific cell type or metabolic or enzymatic reaction but rather constitute a complex cascade of such mechanisms. In addition, the activity of these mechanisms varies in space and time thus rendering identification of a representative surrogate maker as suitable imaging target even more difficult [2].

Morphological imaging methods like computed tomography (CT) or MRI provide information about tissue composition (e.g. water or lipid content), water movement (e.g. diffusion) or permeability of the blood–brain barrier (e.g. contrast agents). Although these tissue properties are altered during the course of an inflammatory process, they are not inflammation specific and have thus traditionally been used to monitor inflammation activity (e.g. contrast enhancement to differentiate acute from chronic demyelinating lesions in multiple sclerosis) rather than to characterize the inflammatory process per se [3]. Similarly, unspecific PET or SPECT parameters are cerebral blood flow (CBF), or cerebral metabolic rate for glucose (CMRGlu) which are in general correlated with inflammation activity without being specific surrogate markers [4].

More specific targets are cellular components of neuroinflammation which, especially in case of an acute injury, are present early after the injury and whose activity level and spatial distribution may allow a more specific quantification of inflammation intensity. Macrophages invading the injured brain from the blood or microglia, the resident macrophages of the brain, constitute such imaging targets [5, 6]. In the normal healthy brain, microglia cells are constantly monitoring the microenvironment and as such are thought to play a role in extracelluar homeostasis. In case of an injury, microglia as well as blood macrophages undergo a transformational change into phagocytes, proliferate, release pro-inflammatory compounds and increase expression of immunomodulatory surface antigens. This transformational change is usually referred to as "activation" and is observed in microglia as well as blood macrophages alike (Fig. 13.1). Since at present it is not possible to distinguish between resident macrophages (microglia) and blood macrophages with in vivo imaging methods [7], the term "activated microglia" (AMG) will be used for both entities.

The TSPO-Receptor System

One of the molecular changes which happen during the transformation of microglia cells in to phagocytes is the expression of surface proteins on the outer mitochondrial membrane (Fig. 13.1). AMG expresses the 18 kDa translocator protein (TSPO) which acts as transport molecule for cholesterol over the mitochondrial membrane and consists of a voltage-gated anion channel and an adenine nucleotide carrier, the mitochondrial transition pore [8]. Outside the brain, it is frequently expressed in steroid synthesizing tissue (e.g. adrenal glands) and has first been discovered in the rat kidney as a high-affinity binding site for diazepam [9] and thus also been named "peripheral benzodiazepine receptor" (PBR). The exact role of TSPO in cells of the CNS remains to be elucidated, it has however been implied in regulation of endogenous steroid synthesis as well as in blocking certain apoptosis pathways. Microglia and macrophages are not the only CNS cells that express TSPO, an upregulation can also be observed in astrocytes [10].

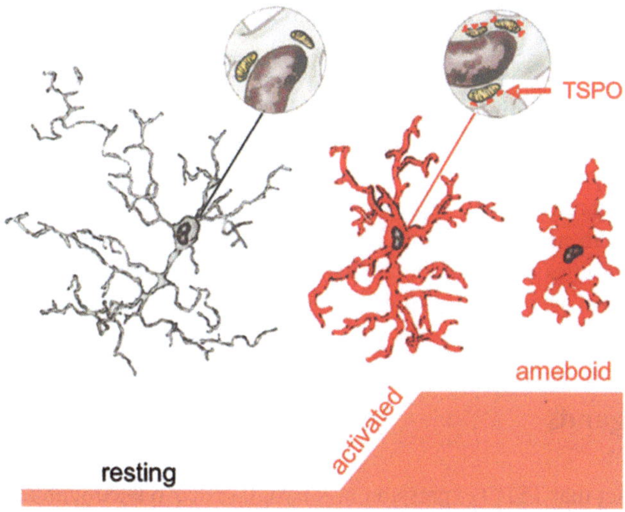

Fig. 13.1 After exposure to noxious stimuli, a resting microglia cell (*left*) starts expressing the 18 kDa translocator at the outer mitochondrial membrane (*middle*) and undergoes morphological transformation into phagocytes (*right*). From Banati, Glia 2002 Nov;40(2):206–17

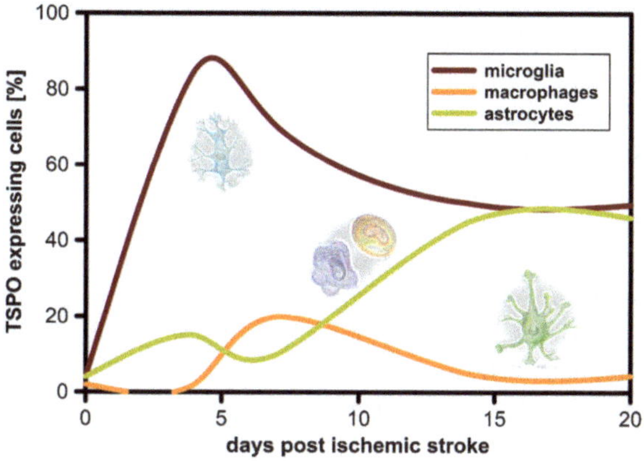

Fig. 13.2 Presence of microglia cells peaks early after an ischaemic injury and stays elevated for several weeks, followed by invasion of macrophages from the blood. TSPO-expressing astrocytes only occur later in the course of the disease

In acute brain injury (e.g. stroke), the time course of TSPO expression in microglia and astrocytes appears to be different with microglia TSPO expression being observed early in the course of an injury and late expression in astrocytes (Fig. 13.2). As far as human pathology is concerned, immunostaining for TSPO was co-localized with microglia rather than astrocytes in a range of different pathologies. The time course of in vivo binding of TSPO ligands also appears to correlate with TSPO expression on microglia and macrophages rather than astrocytes, indicating that either TSPO expression in microglia is higher than TSPO expression in astrocytes or that the ligand affinity is higher for TSPO receptors from microglia than for astrocyte TSPO receptors [11].

A differential interplay between astrocytes and microglia has been postulated based on animal experiments using models for neurodegenerative diseases or specific toxins. Here, irreversible neuronal damage seemed to induce strong TSPO expression in microglia but not in astrocytes, whereas reversible neuronal damage is supposed to cause TSPO expression in astrocytes thus possibly indicating a role of astrocytes in neuroprotection [12, 13]. Further work in this direction and evidence from studies in human brain tissue is however required before firm conclusions as per the role of TSPO expressing astrocytes in human diseases can be drawn.

TSPO Ligands

Given the fact that TSPO expression in microglia is correlated with their activation state, there has been considerable interest in developing ligands for TSPO to quantify this subpopulation of inflammatory cells [14]. At present, TSPO ligands are divided into seven different chemical classes (Table 13.1). Substances from only three classes (Isoquinolines, Phenoxyarylacetamides and Pyrazolopyrimidines) are currently being used or developed for imaging purposes.

The first, and still most widely used substance is the [R]-enantiomer of the isoquinolinecarboxamide PK11195 [15]. This ligand has been labelled with [^{3}H] for autoradiography and [^{11}C] for PET imaging and most microglia imaging studies to date have been performed with PK11195. A disadvantage of PK11195 is its relatively low sensitivity due to high unspecific ligand binding as well as its limited availability as in vivo imaging agent due the short radioactive half-life of [^{11}C] (20 min). The latter requiring a radiochemistry facility with cyclotron at the imaging site. These limitations have prompted the development of alternative TSPO ligands. Among those, [^{18}F]-labelled compounds are of special interest because of the longer half-life of the fluorine isotope (110 min). Especially, the ^{18}F-labelled Phenoxyarylacetamides PBR06 and FEPPA as well as the [^{11}C]-labelled derivatives PBR28 and DAA1106 have been used for imaging studies in humans [16–19].

Table 13.1 Classes of TSPO-receptor ligands and radioligands with reported use in humans

Chemical class	Radioligand for human in vivo imaging
1. Benzodiazepines	
2. 3-Isoquinolinecarboxamides	[^{11}C]PK11195
3. Indoleacetamides	
4. Vinca alkaloids	[^{11}C]vinpocetine
5. Oxodihydropurines	
6. Phenoxyarylacetamides	[^{11}C]DAA1106, [^{18}F]DAA1106, [^{18}F]PBR06, [^{11}C]PBR28, [^{18}F]FEPPA
7. Pyrazolo-[1,5-a]-pyrimidines	[^{11}C]DPA-713, [^{18}F]DPA-714

Other second-generation TSPO ligands which have so far been characterized in biodistribution studies in humans are the vinca alkaloid [^{11}C]-vinpocetine [20] and the Pyrazolopyrimidines [^{11}C]-DPA713 and [^{18}F]-DPA-714 [21].

The higher in vivo specific binding of these new generation ligands under non-pathological conditions, where AMG activity is low, does however not always seem to translate into higher sensitivity to detect a disease associated increase in TSPO receptors when compared to PK11195 [22, 23]. This discrepancy between the high in vitro affinity of these new ligands and the specific binding properties in vivo may hint towards certain, yet to be fully understood properties of the TSPO receptor rather than the design of the ligands themselves [6].

Genetic Polymorphisms of the TSPO Receptor System

In first human, in vivo imaging studies with second-generation TSPO ligands specific binding was absent in about 10 % of normal control subjects [24]. Further in vitro binding studies on post-mortem human brain tissue identified dual binding sites at the TSPO receptor for these second-generation ligands, one with high (4.0 nM) and one with low (313 nM) affinity. Subjects with two high-affinity binding sites were classified as high-affinity binders (HABS), those with two low-affinity binding sites as low-affinity binders (LABS). About 40 % of subjects expressing one high- and one low-affinity binding site, resulting in an intermediate overall affinity, were thus classified as mixed-affinity binders (MABS) [25]. The existence of two receptor binding sites, which are responsible for three binding affinity classes, seemed to point towards a genetic polymorphism with codominant inheritance.

Subsequent genetic association studies with known TSPO polymorphisms confirmed this hypothesis [26]. The rs6971 polymorphism in the TSPO gene on chromosome 22 leads to substitution of the major allele Alanine at position 147 by the minor allele Threonine. Subjects homozygous for Threonine are LABS, HABS are homozygous for Alanine and heterozygous subjects (Ala147Thr) are MABS. Prevalence of this polymorphism appears to depend on ethnicity with 30 % Caucasians carrying the minor allele but only 4–5 % of the Asian population. These variations in binding affinity were observed for all second-generation TSPO ligands to varying degrees and necessitate identification of binding class prior to imaging using either genotyping or thrombocyte binding assays.

Imaging the TSPO-Receptor System Using PET

Measuring receptor concentrations in different tissues quantitatively and non-invasively has always been the domain of nuclear medicine methods like positron emissions tomography (PET) and—to a lesser extent—that of single photon emission tomography (SPECT). In PET, very small concentrations (traces) of a receptor ligand labelled with a positron emitting isotope (radiotracer) are injected into the blood. The temporal and spatial distribution of the radioactivity in the target tissue (e.g. brain) is recorded using tomographic principles and three-dimensional images of the radioactivity distribution in the tissue are reconstructed allowing the measurement of regional radioactivity concentrations which represent the tissue concentration of the radiotracer (usually in units of kBq/ccm).

The imaging process itself exploits the fact that positrons emitted during beta-decay of the isotope label annihilate when encountering their antiparticles the electrons. During this annihilation event, a pair of gamma photons is emitted with both photons travelling in opposite directions under a 180° angle. A coincidence detector pair records this event only if both detectors register the photons at the same time. In modern PET scanners, thousands of such detectors are aligned in detector rings around the scanner aperture and each detector is connected to several detectors in the opposite side of the ring and the neighbouring rings in a fan-like geometry. All parallel detector lines thus record a profile of the radioactivity distribution within a transaxial section of the scanned object. Using filtered back-projection algorithms or iterative methods, a three-dimensional image of the radioactivity distribution in the scanned object can be reconstructed.

From the activity concentration in a certain brain region, quantitative information about the receptor occupancy can be derived by relating tissue activity concentrations of the ligand to its plasma concentration in steady state [27]. Such quantitative models to assess the specific binding of e.g. [^{11}C]-PK11195 to TSPO require repeated arterial blood sampling for the time of the scan to generate a time–activity curve of the radioligand concentration in the plasma. This plasma time–activity curve describes the delivery of the radiotracer to the brain tissue and is thus also called "input function," while the tissue radioactivity concentration measured with the PET camera comprises the concentration of unbound (free) radiotracer, radiotracer bound to the TSPO (specific binding) and radiotracer that binds to other non-receptor carrying cells or cell structures (unspecific binding) [28].

Applying a model to these data allows recovering the specifically bound radiotracer concentration from the input function and the measured tissue activity concentration. Since the requirement for arterial blood sampling limits a broader applicability of the method, analytic models have been developed which estimate the combined free and unspecifically bound radiotracer concentrations from brain regions where no specific binding is supposed to occur (so-called reference region) [29]. This method, however, poses problems to TSPO ligands with high unspecific binding (like [^{11}C]-PK11195) or in case of non-focal spread out inflammatory pathology (such as experimental encephalitis), where a region devoid of specific binding might be difficult to define. In case of unilateral focal pathologies (like stroke) where only receptor binding in a defined target region is of interest, reference tissue methods may still be applicable (e.g. identical mirror region in the contralateral hemisphere) depending on the research question [30]. In such cases, the sensitivity limit to detected increased TSPO binding is given by the physiological TSPO binding in unaffected normal tissue.

Imaging Neuroinflammation in Stroke

Next to tissue necrosis which constitutes the hallmark of acute focal ischaemic brain injury, neuroinflammation is the typical tissue reaction in the late acute and throughout the sub-acute phase and has been one of the first and most widely studied neuroinflammatory processes using TSPO-based imaging methods [31]. Autoradiography [^{3}H]-PK11195 was first used to study the temporal dynamic of the neuroinflammatory response to ischaemia. Ligand binding first occurred in the borderzone of the infarct where microglia and macrophages invaded the necrotic regions from neighbouring blood vessels, with peak activity about 4–8 days post-ischaemia [32]. The intensity of the inflammatory reaction was found to be related to the duration of the ischaemia. Comparing studies of transient and permanent middle cerebral artery occlusion (MCAO), the evolution of microglia activation was found to follow the development of the ischaemic lesions. In transient MCAO, microglia activation starts in the ischaemic core. If ischaemia becomes permanent, the inflammatory processes spread into the non-ischaemic borderzone but leaving the necrotic core devoid of inflammatory cells [33, 34].

The development of Micro-PET cameras facilitated this kind of experiments. In contrast to autoradiography, PET allowed to perform true longitudinal studies in the same animal as well as the non-invasive assessment of related parameters like CBF or CMRGlu in close temporal proximity. Transient MCAO induced ischaemia for 2 h causes a significant increase in TSPO binding for 21 days with a peak around 11 days post-ischaemia [35]. Multitracer studies, using [^{18}F]-fluor-desoxyglucose in conjunction with [^{11}C]-PK11195 also demonstrate an increased metabolic demand in the inflammatory borderzone [4]. This increased demand in an already undersupplied infarct borderzone, may actually proof detrimental to surviving neurons and cause secondary infarct growth beyond the originally ischaemic region thus giving a rational to anti-inflammatory treatment strategies.

While this centrifugal spread of highly increased TSPO binding within the first days after ischaemia could clearly be attributed to microglia activity, a centripetal migration of TSPO expressing astrocytes was observed later in the course of the disease consistent with the formation of a peri-infarct scar. TSPO binding during this period however was considerably lower than in the early phase, dominated by microglia activity [11].

The temporal dynamics of a spreading neuroinflammatory response observed in animal models of focal ischaemia also seems to apply to some extent to human ischaemic stroke [36]. Activated microglia can be found in the infarct core as early as 24–48 h post-ischaemia and spreads into the borderzone over the following 20 days [37]. Microglia cells in the area of the infarct have in common that they mostly express the major histocompatibility complex antigen type 1 (MHC1) and seem to act as phagocytes in removing debris from the site of the infarct. Additional microglia activity has been observed in human ischaemia also in regions remote from the

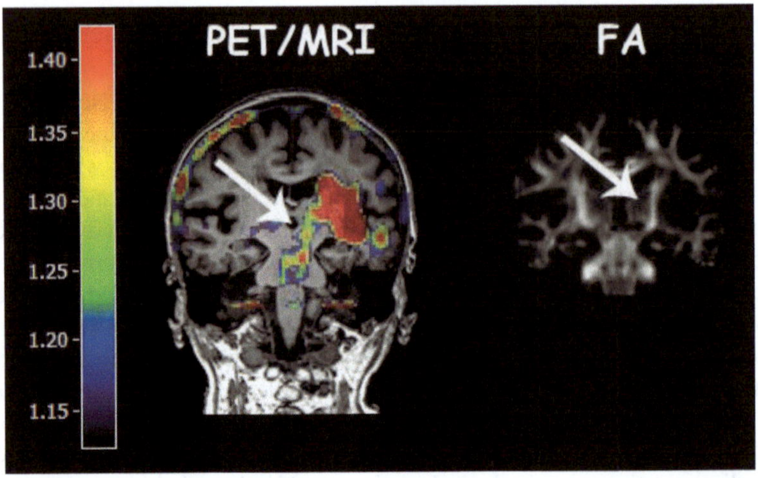

Fig. 13.3 Activated microglia in a patient with left subcortical stroke imaged with [^{11}C]-PK11195 PET. Tracer binding is high in the infarct but extends along the pyramidal tract into the midbrain (*arrow*). The tract is visualized in the fractional anisotropy image from diffusion tensor MRI (*right*). From Thiel and Heiss, Stroke 2011 Feb 1;42(2):507–12

infarct which share common fibre tract connections with the ischaemic area. Activated microglia has been reported in the ipsilateral thalamus following ischaemia of the somatosensory cortex, or along the pyramidal tract down to the spinal cord in primary motor cortex ischaemia. These remote microglia cells differ from their counterparts surrounding the infarct in that they mainly express MHC2 antigens [38, 39].

The advent of diffusion tensor imaging (DTI), an MRI method to measure water diffusion, has played an important role in clarifying the differential roles of local and remote microglia. This MRI technique, based on the restricted diffusion of water molecules along fibre bundles, allows the exact delineation of fibre tracts in an individual brain and thus the measurement of microglia activity along defined tract portions. In addition, altered diffusion properties of fibre tracts are surrogate markers of direct or indirect tract damage. Combined DTI and [^{11}C]-PK11195 PET imaging studies thus suggest that ischaemic tract damage (e.g. subcortical ischaemic infarct at the level of the internal capsule) has an immediate impact on the microstructure of the entire tract [40]. Microglia activity however is only detected in the infarct and along the portion of the tract distal to the lesion (Fig. 13.3). While in most patients inflammation around the infarct usually subsides over the following weeks and months, remote microglia activity can persist (Fig. 13.4). Only in patients where the inflammatory activity in the infarct itself persists over several weeks and months, a further degeneration of tract portions distal to the infarct was observed and those patients tended to have a poorer recovery of motor function [41]. This indicates that persistent inflammatory activity around the infarct seems to drive

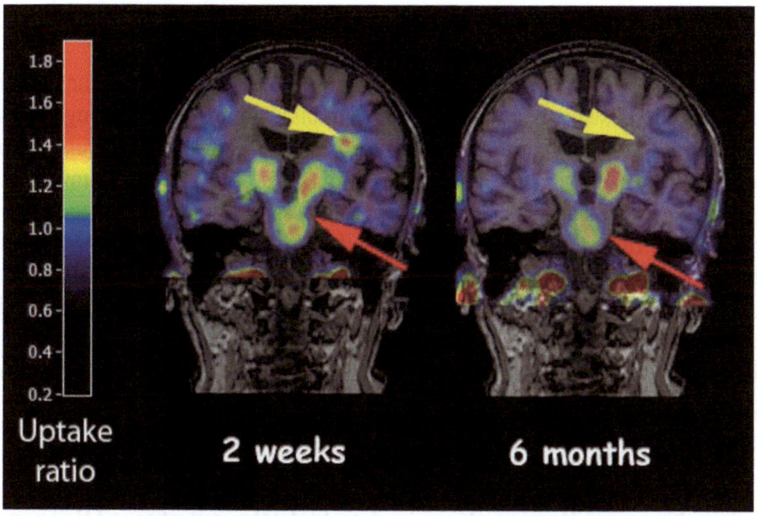

Fig. 13.4 Time course of microglia activation after stroke. In the subacute phase (*left*) tracer uptake is seen in the infarct (*yellow arrow*) as well as along the pyramidal tract (*red arrow*). After 6 months (*right*) inflammation in the infarct subsides but persists along the pyramidal tract. From Thiel and Heiss, Stroke 2011 Feb 1;42(2):507–12

anterograde (i.e. Wallerian) degeneration of fibre tracts while remote microglia activity may be associated with a protective or repair function.

Interestingly, motor rehabilitation after experimental ischaemia in rats does not only improve motor function but also appears to reduce peri-infarct microglia activity compared to control animals not receiving rehabilitation after stroke [42]. Translation of these results to human post-stroke recovery however remains subject to further investigations.

Imaging Neuroinflammation in TBI

Despite the wealth of neuroinflammation research and especially microglia research in TBI, in vivo imaging technologies have not yet been as widely used as in ischaemic stroke or neurodegenerative diseases. In contrast to ischaemic stroke, where necrotic neuronal death precedes the peak of the inflammatory reaction, autoradiography with [^{3}H]-PK11195 indicated an increase of tracer binding in the ipsilateral thalamus between 3 and 14 days after injury. Towards the end of the observation period, around 14 days, thalamic neurons died. Again tracer binding was mainly caused by microglia and macrophages and to a tenfold lesser degree by astrocytes. Similar results were obtained with [^{3}H]-DAA1106 in a controlled cortical impact model of TBI where increased tracer binding was observed in the vicinity of the injury [43–45]. This neuroinflammatory response came along with excitatory neurotransmitter release leading to neuronal apoptosis.

The limitations of in vivo microglia imaging discussed above so far prevented a more widespread use of the technology in humans. Human TBI, although caused by a more or less focal injury, affects the entire brain and can cause a diffuse disruption of the blood–brain barrier, especially in the first days following a trauma. This prohibits the use of reference tissue models in the acute and sub-acute phase after injury because they tend to overestimate specific binding [46]. However, arterial blood sampling, as required for plasma-input function based models, is invasive and can often not be completed in acute TBI patients. In the chronic phase, these limitations do not apply but microglia response, even 6 months after TBI appears to be relatively widespread [47] and reference regions devoid of specific binding may be difficult to define.

Imaging Neuroinflammation in Acute Inflammatory Diseases

CNS infections are the prototype of acute neuroinflammatory CNS diseases and experimental encephalitis models have been used for developing new imaging agents [48]. In patients, MRI showing typical alterations of signal behaviour in T2-weighted and FLAIR imaging sequences is routinely used for the clinical diagnosis of HSV or VZV encephalitis. Direct quantification of the cellular response with in vivo imaging methods in those cases is more challenging. In patients with HSV encephalitis, an increase of microglia activation in limbic brain regions has been demonstrated with [^{11}C]-PK1195 PET. Microglia activity persisted for several months following the treatment and spread into other brain regions [49]. Those brain regions with initially high microglia activity underwent atrophy later in the course of the disease pointing towards a possible neurodegenerative process sparked by the inflammatory reaction. The relatively widespread tracer uptake, indicating a more generalized inflammatory response, may however limit the use of reference tissue methods for quantification for similar reasons as discussed above for traumatic brain injury.

Future Developments: MR Imaging of Neuroinflammation and New Imaging Targets

Since imaging of the TSPO-receptor system for quantification of neuroinflammation is subject to certain limitations and requires a thorough understanding of the imaging methods, recent work in the field has focused on alternative, TSPO-independent surrogate markers of microglia activity. A possible target could be Cyclooxygenase 1 (COX1) which is also expressed in activated microglia and macrophages. A recent study using [^{11}C]-Ketoprofen as selective COX1 ligand was able to demonstrate a tracer accumulation 6 h post-intracerebral injection of quinolinic acid in a rat model which was associated with accumulation of COX1 expressing microglia at the injection site [50].

Another imaging approach aims at neurohumoral components of the inflammatory cascade. β-glucuronidase is a lysosomal enzyme which is released from microglia into extracellular space and hydrolyses glycosaminoglycans on the cell surface. It thus participates in the degradation of the extracellular matrix and is thought to constitute a biomarker for inflammation-associated neurodegeneration. [^{18}F]-FEAnGA is a prodrug carrying the radioactively labelled [^{18}F]-fluoroethylamine ([^{18}F]-FEA) group. [^{18}F]-FEA is cleaved from the molecule by β-glucuronidase and because it is less hydrophilic than [^{18}F]-FEAnGA, its clearance from the tissue is slower than for its parent compound. Uptake in experimental HSV1 encephalitis has been demonstrated which was correlated with the uptake of [^{11}C]-PK11195; however, uptake of [^{18}F]-FEAnGA over the intact BBB was low because of its relatively high hydrophilicity [51].

MRI has also been considered for in vivo imaging of specific neuroinflammatory processes. Most MR strategies rely on labelling monocytes ex vivo with iron oxide particles and then inject those labelled monocytes into the blood. These monocytes will eventually circulate and leave the blood stream at the inflammation site as macrophages. The problem with this ex vivo labelling approach however is that iron introduced into monocytes might not necessarily stay there in vivo. If the labelled cells die, iron is released and may be taken up by other cells or the labelled cells themselves may be subject to phagocytosis. To overcome those problems, ultrasmall superparamagnetic iron oxide nanoparticles (USPIO) can be injected directly and are taken up by the reticuloendothelial system and circulating cells. USPIO however do not cross the BBB and are subject to non-specific vascular egress (in case of disrupted BBB) or endothelial uptake and may thus give more information about inflammation-related changes of the vasculature rather than the inflammatory process in the tissue. Newer MRI probes with microparticles of iron oxide (MPIO) which aim at targeting the vascular (endothelial) component of inflammation are under investigation [52–54], but a reliable MRI marker for cellular components of neuroinflammation remains to be developed.

Conclusion

At present TSPO-based imaging of microglia activity with PK11195 or a second-generation TSPO ligand is the method of choice for in vivo quantification of neuroinflammatory processes. Selection of the appropriate research question and target pathology (focal rather than generalized), knowledge about the specifics of the radioligand (necessity of genotyping for TSPO polymorphisms) as well as the appropriate model for quantification of specific tracer binding (reference tissue model versus plasma input function) are key to a valid application of these methods and proper interpretation of study results.

References

1. Wang X, Feuerstein GZ (2004) The Janus face of inflammation in ischemic brain injury. Acta Neurochir Suppl 89:49–54
2. Jacobs AH, Tavitian B (2012) Noninvasive molecular imaging of neuroinflammation. J Cereb Blood Flow Metab 32(7):1393–1415
3. Schroeter M, Franke C, Stoll G, Hoehn M (2001) Dynamic changes of magnetic resonance imaging abnormalities in relation to inflammation and glial responses after photothrombotic cerebral infarction in the rat brain. Acta Neuropathol 101(2):114–122
4. Schroeter M, Dennin MA, Walberer M, Backes H, Neumaier B, Fink GR et al (2009) Neuroinflammation extends brain tissue at risk to vital peri-infarct tissue: a double tracer [11C] PK1. J Cereb Blood Flow Metab 29(6):1216–1225
5. Banati RB (2002) Visualising microglial activation in vivo. Glia 40(2):206–217
6. Venneti S, Lopresti BJ, Wiley CA (2012) Molecular imaging of microglia/macrophages in the brain. Glia 61(1):10–23
7. Schilling M, Besselmann M, Muller M, Strecker JK, Ringelstein EB, Kiefer R (2005) Predominant phagocytic activity of resident microglia over hematogenous macrophages following transient focal cerebral ischemia: an investigation using green fluorescent protein transgenic bone marrow chimeric mice. Exp Neurol 196(2):290–297
8. Papadopoulos V, Baraldi M, Guilarte TR, Knudsen TB, Lacapore JJ, Lindemann P et al (2006) Translocator protein (18kDa): new nomenclature for the peripheral-type benzodiazepine receptor based on its structure and molecular function. Trends Pharmacol Sci 27(8):402–409
9. McEnery MW, Snowman AM, Trifiletti RR, Snyder SH (1992) Isolation of the mitochondrial benzodiazepine receptor: association with the voltage-dependent anion channel and the adenine nucleotide carrier. Proc Natl Acad Sci USA 89(8):3170–3174
10. Cosenza-Nashat M, Zhao ML, Suh HS, Morgan J, Natividad R, Morgello S et al (2009) Expression of the translocator protein of 18 kDa by microglia, macrophages and astrocytes based on immunohistochemical localization in abnormal human brain. Neuropathol Appl Neurobiol 35(3):306–328
11. Martin A, Boisgard R, Theze B, Van CN, Kuhnast B, Damont A et al (2010) Evaluation of the PBR/TSPO radioligand [(18)F]DPA-714 in a rat model of focal cerebral ischemia. J Cereb Blood Flow Metab 30(1):230–241
12. Ji B, Maeda J, Sawada M, Ono M, Okauchi T, Inaji M et al (2008) Imaging of peripheral benzodiazepine receptor expression as biomarkers of detrimental versus beneficial glial responses in mouse models of Alzheimer's and other CNS pathologies. J Neurosci 28(47): 12255–12267
13. Johansen HL, Wielgosz AT, Nguyen K, Fry RN (2010) Incidence, comorbidity, case fatality and readmission of hospitalized stroke patients in Canada. Can J Cardiol 22(1):65–71
14. Chauveau F, Boutin H, Van CN, Dolle F, Tavitian B (2008) Nuclear imaging of neuroinflammation: a comprehensive review of [11C]PK11195 challengers. Eur J Nucl Med Mol Imaging 35(12):2304–2319
15. Le FG, Guilloux F, Rufat P, Benavides J, Uzan A, Renault C et al (1983) Peripheral benzodiazepine binding sites: effect of PK 11195, 1-(2-chlorophenyl)-N-methyl-(1-methylpropyl)-3 isoquinolinecarboxamide. II. In vivo studies. Life Sci 32(16):1849–1856
16. Fujimura Y, Zoghbi SS, Simeon FG, Taku A, Pike VW, Innis RB et al (2009) Quantification of translocator protein (18 kDa) in the human brain with PET and a novel radioligand, (18) F-PBR06. J Nucl Med 50(7):1047–1053
17. Rusjan PM, Wilson AA, Bloomfield PM, Vitcu I, Meyer JH, Houle S et al (2011) Quantitation of translocator protein binding in human brain with the novel radioligand [lsqb]18F[rsqb]-FEPPA and positron emission tomography. J Cereb Blood Flow Metab 31(8):1807–1816
18. Maeda J, Suhara T, Zhang MR, Okauchi T, Yasuno F, Ikoma Y et al (2004) Novel peripheral benzodiazepine receptor ligand [11C]DAA1106 for PET: an imaging tool for glial cells in the brain. Synapse 52(4):283–291

19. Brown AK, Fujita M, Fujimura Y, Liow JS, Stabin M, Ryu YH et al (2007) Radiation dosimetry and biodistribution in monkey and man of 11C-PBR28: a PET radioligand to image inflammation. J Nucl Med 48(12):2072–2079
20. Gulyazs B, Toth M, Schain M, Airaksinen A, Vas A, Kostulas K et al (2012) Evolution of microglial activation in ischaemic core and peri-infarct regions after stroke: a PET study with the TSPO molecular imaging biomarker [11C]vinpocetine. J Neurol Sci 320(1):110–117
21. Doorduin J, Klein H, Dierckx R, James M, Kassiou M, Vries E (2009) [11C]-DPA-713 and [18F]-DPA-714 as new PET tracers for TSPO: a comparison with [11C]-(R)-PK11195 in a rat model of herpes encephalitis. Mol Imaging Biol 11(6):386–398
22. Dickstein L, Zoghbi S, Fujimura Y, Imaizumi M, Zhang Y, Pike V et al (2011) Comparison of 18F- and 11C-labeled aryloxyanilide analogs to measure translocator protein in human brain using positron emission tomography. Eur J Nucl Med Mol Imaging 38(2):352–357
23. Gulyazs B, Toth M, Vas A, Shchukin E, Kostulas K, Hillert J et al (2012) Visualising neuroinflammation in post-stroke patients: a comparative PET study with the TSPO molecular imaging biomarkers [11C]PK11195 and [11C]vinpocetine. Curr Radiopharm 5(1):19–28
24. Owen DR, Howell OW, Tang SP, Wells LA, Bennacef I, Bergstrom M et al (2010) Two binding sites for [(3)H]PBR28 in human brain: implications for TSPO PET imaging of neuroinflammation. J Cereb Blood Flow Metab 30(9):1608–1618
25. Owen DRJ, Gunn RN, Rabiner EA, Bennacef I, Fujita M, Kreisl WC et al (2011) Mixed-affinity binding in humans with 18-kDa translocator protein ligands. J Nucl Med 52(1):24–32
26. Owen DR, Yeo AJ, Gunn RN, Song K, Wadsworth G, Lewis A et al (2012) An 18-kDa translocator protein (TSPO) polymorphism explains differences in binding affinity of the PET radioligand PBR28. J Cereb Blood Flow Metab 32(1):1–5
27. Lammertsma AA (2002) Radioligand studies: imaging and quantitative analysis. Eur Neuropsychopharmacol 12(6):513–516
28. Kropholler MA, Boellaard R, Schuitemaker A, van Berckel BN, Luurtsema G, Windhorst AD et al (2005) Development of a tracer kinetic plasma input model for (R)-[11C]PK11195 brain studies. J Cereb Blood Flow Metab 25(7):842–851
29. Lammertsma AA, Hume SP (1996) Simplified reference tissue model for PET receptor studies. Neuroimage 4(3):153–158
30. Kropholler MA, Boellaard R, Schuitemaker A, Folkersma H, van Berckel BN, Lammertsma AA (2006) Evaluation of reference tissue models for the analysis of [(11)C](R)-PK11195 studies. J Cereb Blood Flow Metab 25(11):1431–1441
31. Thiel A, Heiss WD (2011) Imaging of microglia activation in stroke. Stroke 42(2):507–512
32. Dubois A, Benavides J, Peny B, Duverger D, Fage D, Gotti B et al (1988) Imaging of primary and remote ischaemic and excitotoxic brain lesions. An autoradiographic study of peripheral type benzodiazepine binding sites in the rat and cat. Brain Res 445(1):77–90
33. Stephenson DT, Schober DA, Smalstig EB, Mincy RE, Gehlert DR, Clemens JA (1995) Peripheral benzodiazepine receptors are colocalized with activated microglia following transient global forebrain ischemia in the rat. J Neurosci 15(7 Pt 2):5263–5274
34. Demerle-Pallardy C, Duverger D, Spinnewyn B, Pirotzky E, Braquet P (1991) Peripheral type benzodiazepine binding sites following transient forebrain ischemia in the rat: effect of neuroprotective drugs. Brain Res 565(2):312–320
35. Rojas S, Martin A, Arranz MJ, Pareto D, Purroy J, Verdaguer E et al (2007) Imaging brain inflammation with [(11)C]PK11195 by PET and induction of the peripheral-type benzodiazepine receptor after transient focal ischemia in rats. J Cereb Blood Flow Metab 27(12): 1975–1986
36. Gerhard A, Schwarz J, Myers R, Wise R, Banati RB (2005) Evolution of microglial activation in patients after ischemic stroke: a [11C](R)-PK11195 PET study. Neuroimage 24(2):591–595
37. Weinstein J, Koerner I, Moller T (2010) Microglia in ischemic brain injury. Future Neurol 5:227–246
38. Schroeter M, Jander S, Witte OW, Stoll G (1999) Heterogeneity of the microglial response in photochemically induced focal ischemia of the rat cerebral cortex. Neuroscience 89(4): 1367–1377

39. Schmitt AB, Brook GA, Buss A, Nacimiento W, Noth J, Kreutzberg GW (1998) Dynamics of microglial activation in the spinal cord after cerebral infarction are revealed by expression of MHC class II antigen. Neuropathol Appl Neurobiol 24(3):167–176
40. Radlinska B, Ghinani S, Lyon P, Jolly D, Soucy JP, Minuk J et al (2009) Multi-modal microglia imaging of fiber tracts in acute sub-cortical stroke. Ann Neurol 66:825–832
41. Thiel A, Radlinska B, Paquette C, Sidel M, Soucy JP, Schirrmacher R et al (2010) The temporal dynamics of post-stroke neuroinflammation: a longitudinal DTI-guided PET-study with [^{11}C]-PK11195 in acute sub-cortical stroke. J Nucl Med 51:1404–1412
42. Keiner S, Wurm F, Kunze A, Witte OW, Redecker C (2008) Rehabilitative therapies differentially alter proliferation and survival of glial cell populations in the perilesional zone of cortical infarcts. Glia 56(5):516–527
43. Rao VL, Dogan A, Bowen KK, Dempsey RJ (2000) Traumatic brain injury leads to increased expression of peripheral-type benzodiazepine receptors, neuronal death, and activation of astrocytes and microglia in rat thalamus. Exp Neurol 161(1):102–114
44. Venneti S, Wagner AK, Wang G, Slagel SL, Chen X, Lopresti BJ et al (2007) The high affinity peripheral benzodiazepine receptor ligand DAA1106 binds specifically to microglia in a rat model of traumatic brain injury: implications for PET imaging. Exp Neurol 207(1):118–127
45. Ramlackhansingh AF, Brooks DJ, Greenwood RJ, Bose SK, Turkheimer FE, Kinnunen KM et al (2011) Inflammation after trauma: microglial activation and traumatic brain injury. Ann Neurol 70(3):374–383
46. Folkersma H, Boellaard R, Vandertop WP, Kloet RW, Lubberink M, Lammertsma AA et al (2009) Reference tissue models and blood–brain barrier disruption: lessons from (R)-[11C] PK11195 in traumatic brain injury. J Nucl Med 50(12):1975–1979
47. Folkersma H, Boellaard R, Yaqub M, Kloet RW, Windhorst AD, Lammertsma AA et al (2011) Widespread and prolonged increase in (R)-11C-PK11195 binding after traumatic brain injury. J Nucl Med 52(8):1235–1239
48. Venneti S, Lopresti BJ, Wang G, Slagel SL, Mason NS, Mathis CA et al (2007) A comparison of the high-affinity peripheral benzodiazepine receptor ligands DAA1106 and (R)-PK11195 in rat models of neuroinflammation: implications for PET imaging of microglial activation. J Neurochem 102(6):2118–2131
49. Cagnin A, Myers R, Gunn RN, Lawrence AD, Stevens T, Kreutzberg GW et al (2001) In vivo visualization of activated glia by [11C] (R)-PK11195-PET following herpes encephalitis reveals projected neuronal damage beyond the primary focal lesion. Brain 124(Pt 10): 2014–2027
50. Shukuri M, Takashima-Hirano M, Tokuda K, Takashima T, Matsumura K, Inoue O et al (2011) In vivo expression of cyclooxygenase-1 in activated microglia and macrophages during neuroinflammation visualized by PET with 11C-ketoprofen methyl ester. J Nucl Med 52(7): 1094–1101
51. Antunes IF, Doorduin J, Haisma HJ, Elsinga PH, van Waarde A, Willemsen ATM et al (2012) 18F-FEAnGA for PET of ß-glucuronidase activity in neuroinflammation. J Nucl Med 53(3): 451–458
52. McAteer M, von Zur Muhlen C, Anthony D, Sibson N, Choudhury R (2011) Magnetic resonance imaging of brain inflammation using microparticles of iron oxide. In: Shah K (ed) Molecular imaging, 680 edn. Humana Press, p 103–15
53. Rosenberg JT, Sachi-Kocher A, Davidson MW, Grant SC (2012) Intracellular SPIO labeling of microglia: high field considerations and limitations for MR microscopy. Contrast Media Mol Imaging 7(2):121–129
54. Nighoghossian N, Wiart M, Cakmak S, Berthezene Y, Derex L, Cho TH et al (2007) Inflammatory response after ischemic stroke. Stroke 38(2):303–307

Chapter 14
Inflammation as a Therapeutic Target after Subarachnoid Hemorrhage: Advances and Challenges

Mutsumi Fujii, Sheng Chen, Damon Klebe, Yoshiteru Soejima, Alexander Vakhmyanin, and John H. Zhang

Abstract Subarachnoid hemorrhage (SAH) results from the rupture of an intracranial aneurysm, and the first consequent events are increased intracranial pressure (ICP), reduced cerebral perfusion pressure (CPP), and decreased cerebral blood flow (CBF). The resultant hypoxic state alters autoregulation, ionic homeostasis, and excitotoxicity as well as initiates secondary injuries such as cytotoxic edema, blood-brain barrier (BBB) disruption, inflammation, and apoptotic cell death. Inflammation persists through hemorrhage degradation in the subarachnoid space. Several different aspects of the inflammatory response have been demonstrated in stroke pathogenesis, including cellular response (e.g., leukocyte adherence and microglia activation), expression of adhesion molecules (e.g., selectins, integrins, and immunoglobulin superfamily), production of inflammatory mediators (e.g., cytokines, nitric oxide/nitric oxide synthase (NO/NOS), and free radicals), and accumulation of platelet aggregates. Since all of these inflammatory aspects lead to brain edema and cell death, inflammation could be a particularly important target for designing therapeutic strategies against secondary injuries after SAH. Given these inflammatory contributions could be seen in large vessels, a plethora of research has been intended to reduce cerebral vasospasm (CVS) after SAH. The main research field, however, is moving toward studying early brain injury (EBI) because some human research demonstrated the morphological alleviation of CVS alone might not improve the functional recovery in patients after SAH. This chapter provides the

M. Fujii • S. Chen • D. Klebe • Y. Soejima • A. Vakhmyanin
Department of Physiology, Loma Linda University, Loma Linda, CA, USA

J.H. Zhang (✉)
Department of Physiology, Pharmacology, and Neurosurgery, Loma Linda University, Loma Linda, CA, USA
e-mail: jhzhang@llu.edu

J. Chen et al. (eds.), *Immunological Mechanisms and Therapies in Brain Injuries and Stroke*, Springer Series in Translational Stroke Research, DOI 10.1007/978-1-4614-8915-3_14,

current knowledge of the inflammatory response, translational research, and human clinical trials in SAH as well as discusses emerging opportunities for novel therapeutic strategies for clinical management of SAH.

Introduction

Numerous studies have recently accumulated supporting the role of the innate and adaptive immune mechanisms in acute brain injuries, such as neurotrauma [1, 2], ischemic stroke [3–6], and hemorrhagic stroke [5, 7–9]. Different inflammatory mechanisms are involved in post-stroke pathogenesis, including cellular responses (e.g., leukocyte adherence, microglia/macrophage activation, and astrocyte activation), adhesion molecule expression (e.g., selectins, integrins, and immunoglobulin superfamily), pro-inflammatory mediator production (e.g., cytokines, chemokines, nitric oxide/nitric oxide synthase, and free radicals), and platelet aggregate accumulation. Since these responses lead to brain edema and cell death, inflammation could be a particularly important target for designing therapeutic strategies against secondary injuries.

Stroke has enormous clinical, social, and economic implications, and it demands a significant effort from both the basic and clinical sciences in searching for successful therapies. Subarachnoid hemorrhage (SAH) is a common and frequently devastating condition, accounting for 5 % of stroke subtypes [10]. Each year, approximately 1 in 10,000 North Americans suffer from aneurysmal SAH with a greater than 50 % combined morbidity and mortality rate [11]. Despite advances in diagnosis and surgical treatment of SAH, effective therapeutic interventions are still limited and clinical outcomes remain disappointingly unimproved. Increasing evidence suggests inflammatory mechanisms are some of the pivotal pathological events during both delayed cerebral vasospasm and acute brain injury after SAH although the inflammatory response is an important pathophysiologic processes after stroke [12]. This chapter provides a thorough review of the current knowledge in the inflammatory response, translational research, and human clinical trials in SAH as well as discusses the emerging opportunities for novel therapeutic strategies in clinical management of SAH.

The Pathological Mechanism in SAH

Cerebral Vasospasm and Early Brain Injury after SAH

Cerebral vasospasm (CVS) after SAH usually occurs on day 3, peaks between days 6–8, and lasts for 2–3 weeks in SAH patients [13]. Delayed cerebral ischemia is thought to be induced by CVS because radiologically confirmed vasospasm is strongly associated with delayed ischemic neurological deficits (DIND) [14–16]. Hence, CVS is widely assumed to be a major cause of the high mortality and poor outcomes after an otherwise successful treatment of a ruptured intracranial aneurysm [17]. On the other hand, the recently coined term early brain injury (EBI) is defined as the period that spans from the moment of initial bleeding to the onset of CVS. EBI describes the immediate injury to the brain after aneurysmal SAH as a whole [18]. Although clazosentan, a selective endothelin receptor type A antagonist, decreased angiographic vasospasm in SAH patients, mortality and clinical outcomes were not improved in the CONSCIOUS-1 trial (clazosentan to overcome neurological ischemia and infarction occurring after SAH) [19]. The CONSCIOUS-1 trial failure indicated the multifactorial pathophysiology underlying CVS and other pathophysiological factors, independent of angiographic vasospasm, contributes to clinical outcomes [20]. Additionally, pathological mechanisms, which are activated within minutes after SAH and lead to EBI, play an important role in CVS development. For instance, vascular injury from acute ischemia, inflammation, and blood products may result in damage of NO-releasing neurons [21]. Therefore, recent intensive research efforts aimed to reveal mechanisms of EBI.

Acute Physiological Events after SAH

EBI was reported as a primary cause of mortality in SAH patients [22], and many important pathological mechanisms are initiated within minutes after aneurysmal SAH [23, 24]. The most immediate event following an intracranial aneurysm rupture is an arrest in intracranial circulation caused by a peak of intracranial pressure (ICP), which rises as high as mean arterial blood pressure within 1 min of ictus. The ICP then falls over several minutes to a much lower baseline but remains higher than normal [25]. The temporary intracranial circulatory arrest promotes hemostasis and contributes to severe global ischemic injury, resulting in loss of autoregulation, reduction in cerebral perfusion pressure (CPP), secondary raised ICP, and decreased cerebral blood flow (CBF) [23, 24, 26]. This hypoxic state culminates in energy failure in neurons and glia and initiates a cascade of events leading to cytotoxic edema [23, 24]. Ischemia also results in apoptosis of cells constituting the blood-brain barrier (BBB) [27]. Death of endothelial cells and perivascular astrocytes causes increased diffusion of serum from the vascular lumen into cerebral tissue (vasogenic edema). These secondary injuries also cause inflammatory responses mentioned below.

Therefore, factors stemming from the initial bleeding in SAH include: raised ICP, decreased CBF and CPP, BBB disruption, brain swelling, brain edema, inflammation, and dysfunction of autoregulation. All factors result in cell death and dysfunction following SAH [23, 24, 28] (Fig. 14.1).

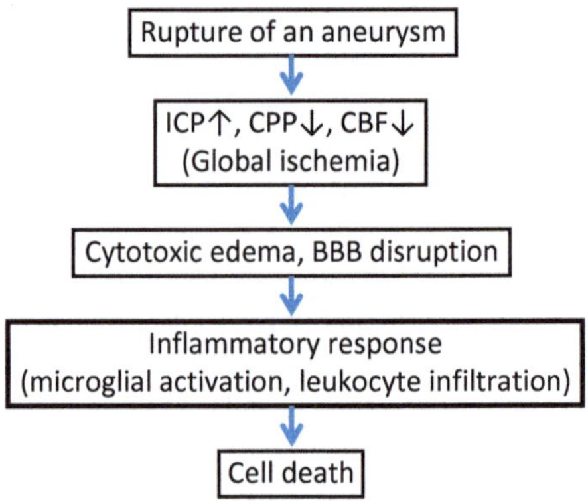

Fig. 14.1 The inflammation plays a crucial role in the pathophysiology of cerebral injury following subarachnoid hemorrhage. *ICP* intracranial pressure, *CPP* cerebral perfusion pressure, *CBF* cerebral blood flow, *BBB* blood–brain barrier

The Inflammatory Response in SAH

The Main Cellular Participants

Blood-derived leukocytes, macrophages, and resident microglia are activated and accumulate in the brain after hemorrhagic stroke [9]. All leukocytes respond to activation signals by altering the composition, expression, and/or functional activity of their trafficking molecules [29]. Neutrophils are the most abundant blood-borne leukocytes, and they infiltrate the brain as early as 10 min after SAH [30]. Neutrophils express abundant adhesion molecules for rapid binding to inflammation-induced counter-receptors on activated endothelial cells, and their chemoattractant receptors sense the release of tissue "distress signals" and pathogen-derived or pathogen-induced molecules. Neutrophils, in particular, mediate secondary tissue damage by releasing cytotoxic mediators. Neutrophil infiltration persists for 3 days after SAH, implicating neutrophil infiltration as a factor leading to delayed CVS development [31, 32]. Monocytes, which express a broad range of adhesion molecules and chemoattractant receptors, are long lived and differentiate into tissue-resident macrophages or dendritic cells [29]. Microglias are resident macrophages of the brain and constitute approximately 12 % of the cells in the central nerve system (CNS). Microglial activation via CD14 induces release of a variety of substances, many of which are cytotoxic and/or cytoprotective [33, 34]. Microglias also induce heme oxygenase-1 (HO-1) expression, which metabolizes heme. Heme contributes to vasospasm and increased oxidative stress after SAH [35]. Dendritic cells are abundant in lymphoid and certain nonlymphonoid tissues, and they are the quintessential antigen-presenting cells [36]. Mast cells have immunomodulatory properties. They are activated by neuropeptides and cytokines, and they release pro-inflammatory mediators with or without degranulation [37], which enhance venular permeability and leukocyte recruitment [29]. Although the role of lymphocytes in SAH is largely unknown, autoreactive and antigen-dependent regulatory T-cell activation may suppress the release of pro-inflammatory cytokines and microglial activation, preventing secondary infarct growth in acute cerebral ischemia as a result [38].

Cell Adhesion Molecules at the Inflamed Sites

Cell adhesion molecules are grouped into selectin, integrin, or immunoglobulin superfamily members. Selectins are type I transmembrane glycoproteins that bind to sialylated carbohydrate structures in a calcium-dependent manner [39]. The selectins, E-selectin (CD62E), P-selectin (CD62P), and L-selectin (CD62L), are involved in the tethering and rolling of leukocytes at inflamed sites on the vessel lumen in microcirculation, and this process is a prerequisite for firm adhesion and subsequent transendothelial migration of leukocytes into tissues [40]. With few exceptions, P- and E-selectin expressions are inducible in endothelial beds and are important determinants for leukocyte recruitment [41]. L-selectin, expressed on circulating leukocytes, can also bind to leukocyte ligands, particularly P-selectin glycoprotein ligand 1 (PSGL-1), and this interaction enhances capturing of leukocytes by intravascular adherent leukocytes [42].

Integrins are transmembrane adhesion receptors that mediate cell-cell and cell-extracellular matrix adhesion as well as induce bidirectional signaling across the cell membrane to regulate cell proliferation, activation, migration, and homeostasis [43]. Each integrin contains one α subunit and one β subunit. The integrins include lymphocyte function-associated antigen-1 (LFA-1; $\alpha_L\beta_2$; or CD11a/CD18), expressed on lymphocytes, granulocytes, and monocytes [44], and Macrophage antigen-1 (Mac-1; $\alpha_M\beta_2$; CR3; CD11b/CD18), expressed on granulocytes, monocytes, and, with the highest expression levels, neutrophils [45]. A humanized monoclonal antibody (mAb), Hu23F2G, targeting CD11/CD18 prevents vasospasm by inhibiting leukocyte adhesion to endothelial cells in the rabbit model of SAH [46]. Since integrins mediate both leukocyte crossing of the basement membrane underlying blood vessels and interstitial migration into the extracellular matrix (ECM), they play an important role in adhesion strengthening and diapedesis across the vessel wall. The immunoglobulin superfamily, such as intercellular cell adhesion molecules (ICAMs) and vascular cell adhesion molecule 1 (VCAM-1), is expressed on endothelial cells. LFA-1 and Mac-1 expressed on circulating leukocytes bind to ICAM-1 and cause transmigration of leukocytes and macrophages across the endothelium into the periadventitial space [47]. A known ligand for LFA-1, ICAM-1 appears to be involved in acute inflammation and is expressed on normal cerebral vessel endothelium in both humans and rodents. ICAM-1 has been implicated in vasospasm; experimental studies determined ICAM-1 is upregulated in the cerebral vasculature after SAH [48] and in blood-exposed vessels that subsequently develop chronic vasospasm [49–51]. Human clinical studies indicated elevated levels of soluble ICAM-1 in patients after aneurysmal SAH [52, 53].

Chemoattractants provide directional cues for the movement of leukocytes in development, homeostasis, and inflammation. Their molecular diversity, selective action on distinct leukocyte subsets, and restricted temporal and spatial expression patterns provide a key mechanism for "fine-tuning" cellular immune responses [29]. Cytokines are a diverse group of soluble short acting proteins, glycoproteins, and peptides produced by various immune cells and vascular cells. Cytokines act in picomolar to nanomolar concentrations to activate specific receptors and modulate the functions of many cells and tissues [54]. Generally, cytokines can be classified into the following categories: (1) tumor necrosis factors (TNFs), (2) interleukins (ILs; cytokines made by one leukocyte and acting on other leukocytes), (3) lymphokines, (4) monokines, (5) interferons (IFNs; include IFN-α, -β, -γ), (6) colony-stimulating factors (CSFs), (7) transforming growth factors (TGFs), (8) chemokines [thought to be involved in chemotaxis and divided into four subfamilies (XC, CC, CXC, and CX3C)], and (9) other proteins. Chemokines comprise macrophage inflammatory protein (MIP)-1α (CCL3), MIP-1β (CCL4), RANTES (regulated on activation, normal T-expressed and secreted, CCL5), and monocyte chemoattractant proteins (MCP)-1 (CCL2) [54]. In an experimental SAH model, increased expression levels of IL-8 and MCP-1 in the basilar artery (BA) were associated with CVS development [33, 34, 55]. SAH induces transcription of inflammatory genes, such as IL-6, TNFα, IL-1β, CXCL1, CXCL2, and CCL20 [56], and increases mRNA expression of IL-1α, IL-6, IL-8, and ICAM-1 in the BA [57].

Multistep Leukocyte Recruitment

The temporary intracranial circulatory arrest resulting from increased ICP promotes hemostasis from the ruptured aneurysm, which contributes to severe global ischemic injury and leads to EBI. Inflammation is characterized by the accumulation of inflammatory cells, such as blood-derived leukocytes and microglia that secrete cytokines. Leukocyte extravasation and transmigration from the blood into the tissues is a regulated multistep process involving a series of coordinated interactions between leukocytes and endothelial cells [29, 58].

After an aneurysm rupture, global ischemia and subarachnoid blood induce the inflammatory response. Inflammation causes the release of cytokines and inflammatory chemoattractants by resident tissue cells, resident and recruited leukocytes, and cytokine-activated endothelial cells. These signals upregulate expression of endothelial selectins and immunoglobulin superfamily members (e.g., ICAM-1 and/or VCAM-1). Chemokine signaling activates leukocyte integrins, such as LFA-1 and Mac-1. After initial tethering, leukocytes roll along the vascular wall with greatly reduced velocity, leading to adhesion and eventual arrest of movement. Subsequently, leukocytes polarize and move by diapedesis across the venular wall. The recruited leukocytes are activated by local pro-inflammatory cytokines and may become desensitized to further chemokine signaling because of high local chemokine concentrations. These inflammatory signals also induce maturation of tissue-resident dendritic cells. In lymph nodes, antigen-loaded mature dendritic cells activate naïve T-cells and expand pools of effector lymphocytes, which enter the blood and migrate back to the inflamed site. Enhanced phagocytosis, cell death, and subsequent degranulation cause these inflammatory cells to release endothelins [59–61] and inflammatory mediators [31, 57, 61], which cause delayed CVS. Leukocytes also die within a few days, releasing endothelins and oxygen free radicals that inactivate nitric oxide, resulting in CVS in 4–14 after SAH [49].

Nitric Oxide/Nitric Oxide Synthase Pathway

The nitric oxide (NO)/nitric oxide synthase (NOS) pathway plays a major role in regulating cerebral hemodynamics. After SAH, the subarachnoid clot is lysed into erythrocyte breakdown products, such as hemoglobin and bilirubin. A time-dependent alteration in the NO/NOS pathway occurs after a hemorrhagic event. Hemoglobin, which has a high affinity for NO [62], scavenges free NO [63] and disrupts many components of NO-mediated vasodilation. Pluta explained the putative involvement of NO in delayed vasospasm development [21]. First, oxyhemoglobin (oxyHb) destroys NO-releasing neurons, leading to diminished availability of NO in the vessel wall and consequent constriction of the vessels. Next, increased shear stress, evoked by the narrowing of the arterial lumen, stimulates endothelial nitric oxide synthase (eNOS). Further metabolism of hemoglobin to bilirubin-oxidized fragments (BOXes) increases asymmetric dimethylarginine, an endogenous inhibitor of eNOS, in the arterial vicinity, further decreasing NO availability and sustaining vasospasm. Finally, when elimination of BOXes commences, increased NO production by eNOS leads to the recovery of endothelium dilatory activity [21]. NO, in the form of peroxynitrite (powerful oxidant), also attacks cell membranes, which damage mitochondria, vascular endothelium, and smooth muscle cells [64], resulting in cell death [65].

Endogenous NO is generated from the precursor amino acid L-arginine, NADPH, and oxygen by a family of three distinct isoforms of NOS [66]. The Ca^{2+}-dependent eNOS and neuronal NOS (nNOS) produce small quantities of NO, while inducible NOS (iNOS), which is expressed by leukocytes and vascular smooth muscle cells in response to inflammatory stimuli and cytokines, generates large amounts of NO in a Ca-independent manner [54, 67]. Previous studies in ischemic brain injury models showed n- and iNOS isozymes are detrimental due to NO-induced neurotoxicity, while eNOS activity is protective because of vasodilatory effects, at least in the early stages of ischemia [68–70]. Although iNOS inhibition did not ameliorate brain edema and neuronal cell death 24 h after SAH in rats [71], eNOS upregulation was beneficial to the brain after SAH [72, 73]. Recently, monomer formation (uncoupling) of eNOS expression is reportedly increased after SAH, which results in decreased NO levels and superoxide anion radical production [74, 75].

Free Radical and Oxidative Stress

The brain becomes exposed to hemoglobin and erythrocyte lysis after SAH because of the extravasated blood in the subarachnoid space. Clot-derived hemoglobin liberates oxyHb, which generates superoxide anions ($O_2^{-\bullet}$) upon its conversion to methemoglobin (metHb) [76]. Superoxide anions react with NO to form a very powerful oxidant, peroxynitrite ($ONOO^-$) [77]. Peroxynitrite is a major mediator of nitric oxide ($NO^\bullet$) toxicity and breaks down into hydroxyl radicals

(OH•) [78]. As oxyHb undergoes auto-oxidation to metHb, O_2^- is also produced, followed by dismutation into hydrogen peroxide (H_2O_2) [79]. OxyHb reduces Fe^{3+} to Fe^{2+}, catalyzing the formation of OH• from H_2O_2, which is called the superoxide-driven Fenton reaction [80, 81]. Enhanced production of $O_2^{-\bullet}$ started at day 2 after SAH in a double hemorrhage dog model [82]. After brain injury, these reactive oxygen species (ROS) are also released by neutrophils, vascular endothelium, and activated microglia/macrophages [83, 84]. Accordingly, ROS damage endothelial cells, vascular smooth muscle cells, astrocytes, and neurons through lipid peroxidation promotion, protein breakdown, DNA damage, and hemoglobin auto-oxidation. Additionally, ROS induce oxidative stress that contributes to ischemic injuries [23, 24, 85]. These events lead to neuronal apoptosis, BBB breakdown, and CVS after SAH.

There are several protective enzymatic systems against free radical production. Superoxide dismutases (SODs), glutathione peroxidases, and catalases are significant enzymatic scavengers in brain tissue [86]. These enzymatic activities, however, are downregulated 6–48 h after SAH in rats, which is associated with increased enzymatic lipid peroxidation via the lipoxygenase pathway [87]. NADPH oxidase expression, an enzymatic source of $O_2^{-\bullet}$ production, increased 24 h after SAH in rats [23, 24]. Additionally, mutant mice deficient in Mn-SOD dismutase had more abundant cytosolic cytochrome *c* and increased cell death within 24 h after subarachnoid hemolysate injection when compared to wild-type mice [88].

Intraluminal Platelet Aggregates

Platelet granules contain many inflammatory and adhesion molecules that are either released or expressed upon activation [89]. After SAH, a variety of pathophysiological stimuli trigger endothelial reorganization, expression of different prothrombotic factors, and activation of platelets and leukocytes that lead to blood cell adhesion to the endothelial monolayer, aggregation as thrombi, and formation of numerous spasmogenic substances [90]. Platelet aggregates are broadly abundant in the microvasculature within 10 min after SAH in animals, leading to mechanical obstruction [91]. Intraluminal platelet aggregates also cause biochemical constriction via release of platelet-derived serotonin, adenosine diphosphate (ADP), and platelet-derived growth factor (PDGF), leading to decreased cerebral blood flow and ischemic injury promotion [92–94]. Hence, platelet aggregates contribute to endothelial cell antigen loss, vascular matrix metalloproteinase (MMP)-9 activation, and collagen IV (the major vascular basal lamina protein) degradation after SAH in rats [91, 95, 96], leading to increased vascular permeability and access to the brain parenchyma [92, 93]. Thus, platelets activate additional inflammatory mechanisms in the parenchyma that further aggravate brain injury after SAH [28].

MMP Increase and BBB Disruption

MMPs are zinc-dependent endopeptidases that are secreted in response to both exogenous insults and inflammatory cytokines, such as TNF-α and IL-1β, and degrade many components of the extracellular matrix. MMPs are normally found in the cytosol in a pro- or inactivated state, and they are activated when cleaved by proteases, such as plasmin or other MMPs [97]. Active MMP expression disrupts the BBB by degrading junctional complex proteins [98]. Thus, MMPs increase capillary permeability. Immunostaining determined collagen IV decreases and MMP-9 increases at 3 h after SAH in rats [95].

Endothelin (ET) is a potent, long-lasting, endogenous vasoconstrictor and is implicated in vasospasm pathogenesis [99, 100]. ET-1 is also a pro-inflammatory factor [101]. ET-1 directly activates neutrophils and endothelial cells, stimulates MCP-1 production, and increases soluble ICAM-1 synthesis [102]. ET-1 also opens the BBB by inducing cyclooxygenase (COX-2) expression [103]. High levels of free radicals also damage the brain endothelium and affect BBB permeability [104, 105]. BBB disruption, increased cytokine expression, and subsequent endothelial and neutrophil adhesion molecule upregulation lead to leukocyte transmigration across the endothelium and BBB [106, 107]. Activated leukocytes enhance BBB permeability by expressing and secreting inflammatory cytokines, soluble factors, ROS, and MMPs. Hence, BBB dysfunction amplifies inflammation, leading to further parenchymal damage and increased brain edema that contribute to ischemic secondary injuries after SAH [23, 24].

Translational Trials in Experimental SAH and Clinical Trials

Inflammation causes tissue injury by diverse mechanisms (noted above) and has detrimental effects on cerebral vessels and brain parenchymal cells after SAH. Since delayed CVS alone is considered the most important cause of DIND and poor outcomes, the aim of basic and clinical research on SAH focused on finding strategies to prevent cerebral vasospasm. Major interventions generated involve leukocyte trafficking inhibition, vessel contraction prevention, oxidative stress reduction by antioxidants (free radical scavengers), NO/NOS pathway intervention, and antiplatelet agent administration in experimental SAH. Clinical studies in human SAH patients produced some promising drugs that showed vasospasm attenuation after SAH. Here, we focus on clinical human trials with the use of anti-inflammatory drugs, the putative effects of which are based on translational animal trials in experimental SAH.

Inhibition of Leukocyte Trafficking

Since the magnitude and character of the inflammatory process are determined in part by leukocyte trafficking into injured and infected sites, leukocyte trafficking prevention may be a promising anti-inflammatory treatment. Patients with poorer outcomes have higher soluble ICAM-1 levels than patients with better outcomes over the first 2 weeks after SAH [52]. Intrathecal anti-ICAM-1 antibody delivery decreased CVS in rabbit BA [108]. Blockage of endothelial ICAM-1 receptors with anti-ICAM-1 monoclonal antibody (mAb) inhibited delayed CVS, which correlated with reduced periadventitial infiltrated macrophages and granulocytes in rats [47, 50]. Systemic administration of anti-LFA-1 mAb prevented vasospasm and reduced periadventitial leukocytes in a rat femoral artery model [47]. Therefore, inhibition of leukocyte infiltration into the periadventitial space ameliorates CVS after SAH. Administration of an E-selectin mAb or blockage of neutrophil/macrophage adhesion molecule CD11/CD18 decreased vasospasm severity after SAH by inhibiting neutrophil and macrophage adhesion and migration into the periadventitial space [46, 109, 110]. Ly6G/C is a surface marker found on cells of myeloid lineage. Recently, Provencio et al. reported administration of myeloid cell-depleting mAb against Ly6G/C ameliorated angiographic vasospasm and normalized neurological behavior in a murine SAH model, suggesting myeloid cells are involved in delayed CVS development [111]. Wang et al demonstrated blockage of CD34, a key adhesion molecule responsible for monocyte/macrophage recruitment and leukocyte attachment to endothelial cells, by a CD34 monoclonal antibody attenuated vasospasm in BA after SAH in rats [112]. Recently, Wu et al. found inhibition of Toll-like receptor 4 by peroxisome proliferator-activated receptor (PPAR) gamma agonist, rosiglitazone, ameliorated CVS by suppressing SAH-induced ICAM-1 and myeloperoxidase (MPO) activation in BA after SAH [113].

Statins are inhibitors of 3-hydroxy-3-methylglutaryl CoA (HMG-CoA) reductase, which possesses pleiotropic activity, including anti-inflammatory properties beyond lipid-lowering. Lovastatin reportedly inhibits CD11b expression and CD11b-dependent monocyte adhesion to the endothelium [114]. Lovastatin also blocks LFA-1 and ICAM-1 interaction by binding to the highly conserved LFA-1 I-domain [115]. McGirt et al demonstrated subcutaneous simvastatin administration ameliorated CVS in BA by attenuating perivascular granulocyte migration after experimental SAH in rabbits [116]. Ibuprofen is also an anti-inflammatory agent that inhibits expression of certain cell adhesion molecules and, therefore, disrupts leukocyte–endothelial cell interactions. In experimental animal models, ibuprofen application within 6 h of adventitia inhibited vasospasm in femoral arteries after blood exposure by, possibly, decreasing the number of macrophages and granulocytes in the periadventitial space [117].

Prevention of Vessel Contraction

One treatment modality for CVS is aimed to improve cerebral blood flow by dilating affected vessels. One potent vasoconstrictor related to inflammation is ET-1 as previously described above. ET-A receptors are found on smooth muscle cells and mediate vasoconstriction. Many experiments showed endothelin receptor antagonists ameliorated CVS [118–132]. Roux et al. reported clazosentan, a selective endothelin receptor type A antagonist, effectively decreased CVS after experimental SAH [125]. Endothelin converting enzyme inhibitors also ameliorated CVS [133–137].

To date, clinical studies were conducted to ascertain if ET-1 inhibition ameliorates CVS and improves outcomes after SAH. Macdonald et al. demonstrated SAH patients treated with clazosentan had a 65 % reduced relative risk in angiographic vasospasm. However, mortality and clinical outcomes were not improved in this randomized, double-blind, placebo-controlled phase II trial named CONSCIOUS-1 (clazosentan to overcome neurological ischemia and infarction occurring after SAH) [19]. CONSCIOUS-2 trial, a randomized, double-blind, placebo-controlled, phase III study, demonstrated clazosentan administration at 5 mg/h had no significant effects on mortality, vasospasm-related morbidity, and functional outcomes [138]. One possible reason for the negative results is successful CVS amelioration is not enough to attenuate poor outcomes in SAH patient. These observations indicate the pathophysiology underlying delayed cerebral ischemia is multifactorial and other pathophysiological factors contribute to poor outcomes that are independent of angiographic vasospasm [20]. Additionally, pathological mechanisms activated within minutes after SAH and lead to EBI may play an important role in delayed ischemic stroke pathogenesis and poor outcome [26].

Reduction of Oxidative Stress by antioxidants

Macdonald et al. showed intrathecal SOD (a potent endogenous antioxidant) and catalase administration failed to attenuate MCA vasospasm in the oxyHb-induced vasospasm model that simulates erythrocyte lysis after SAH without any difference in malondialdehyde (MDA), which indicates lipid peroxidation [139]. However, Shishido et al. demonstrated intracisternal Cu–Zn SOD administration did not reduce BA diameter between 2 and 11 days after SAH in the rabbit intracisternal blood injection model [140]. They observed endothelial injuries in the SAH group were minimized with SOD treatment in the BA. Mori et al. demonstrated Cu–Zn SOD reduced talc-induced contraction of the BA in a beagle dog model. Pathological changes seen in talc injection group, such as myonecrosis, cytoplasmic vacuolations, and detached intercellular junctions, were almost abolished in the Cu–Zn SOD treatment group [141]. Kamii et al. showed Cu–Zn SOD overexpressing transgenic mice had significant amelioration of SAH-induced decrease in MCA diameter at day 3 in the endovascular perforation of the left anterior cerebral artery (ACA) model [142]. N-acetylcysteine, an antioxidant, administration inhibited vasospasm development after SAH in rabbits at day 5 and increased SOD enzymatic activity levels [143]. Moreover, curcumin-treated SAH rats had less mortality and BA contraction as well as higher neurological scores when compared to saline-treated SAH with increased SOD activity in a double hemorrhage rat model [144].

Systemic antioxidant administration in experimental SAH reduced oxidative stress, protected the BBB, and improved neurological scores [145–147]. Kanamaru et al. demonstrated 21-aminosteroid U-74006 F reduced vasospasm and MDA content of the clot in the subarachnoid clot-placement monkey model at day 7 [148]. Lipid peroxidation inhibitor, U74006F (21-aminosteroid tirilazad mesylate) or U74389G, administration also ameliorated vasospasm after SAH in other animal models [122, 123, 149–151]. Suzuki et al. demonstrated Tirilazad mesylate attenuated SAH-induced vasospasm by eliminating phosphatidylcholine hydroperoxide in the cerebral arteries in an SAH monkey model [152]. Hall reported a major neuroprotective mechanism of methylprednisolone is free radical-induced lipid peroxidation inhibition [153]. In vitro production of eicosanoids by brain slices, which is normally increased after SAH, was reduced in rats with high-dose methylprednisolone treatment every 8 h after SAH [154]. Ecdysterone, an insect steroid hormone, attenuated CVS and neurological deficits in SAH rabbits by suppressing vascular adventitial fibroblast proliferation [155].

Moreover, Ebselen, an antioxidant that inhibits arachidonic acid lipoxygenase activity through glutathione peroxidase-like action, ameliorated delayed vasospasm in experimental SAH models [156, 157]. Nakagomi et al. demonstrated Edaravone, a free radical scavenger, attenuated BA narrowing in a canine double hemorrhage model [158]. Nicaraven, a hydroxyl radical scavenger, ameliorated CVS after SAH in rats via synergistic HO-1 induction, which is a protective oxidative stress-inducible enzyme [159]. Aladag et al. demonstrated caffeic acid phenethyl ester, a nontoxic oxygen free radical scavenger, prevented BA vasospasm at day 5 while both reducing lipid peroxidation and increasing NO bioavailability in the double hemorrhage rat SAH model [160].

To date, clinical studies have been conducted to ascertain if antioxidants or free radical scavengers ameliorated SAH outcomes. Saito et al. evaluated the effects of oral administration of Ebselen for 2 weeks in 286 SAH patients in a multicenter, double-blind, clinical trial. Ebselen treatment did not reduce clinically diagnosed DIND, but it improved Glasgow neuroscores when compared to placebo treatment [161]. Edaravone treatment for vasospasm was also clinically examined in 91 SAH patients. Cerebral infarction incidence and poorer outcomes caused by vasospasm were significantly reduced in the Edaravone group when compared to the control group [162]. In larger randomized, double-blind, vehicle-controlled trials in SAH patients, Kassell et al. demonstrated Tirilazad mesylate (6 mg/kg/day) reduced mortality and improved outcomes without reducing symptomatic vasospasm [163], but Haley et al. showed Tirilazad mesylate did not improve the mortality rate, vasospasm incidence and severity, and outcomes at 3 months [164].

On the contrary, Nicaraven, a hydroxyl radical scavenger, reduced DIND incidence, improved outcomes at 1 month, and reduced the mortality rate at 3 months after SAH in a prospective, placebo-controlled, double-blind, multicenter trial [165].

No evidence supports the routine use of steroids in SAH patients. Hashi et al. demonstrated glucocorticoid administration after DIND emergence improved mental, speech, and motor functions rapidly, but it did not ameliorate mortality and outcomes [166]. Although Chyatte et al. showed high-dose methylprednisolone reduced DIND in a retrospective case–control study of 21 patients [167], a small, double-blind, placebo-controlled, randomized trial revealed methylprednisolone (16 mg/kg/day, iv, started within 24–48 h after SAH for 3 days) did not reduce symptomatic vasospasm incidence, but it improved functional outcomes after 1 year [168]. This study supports the idea that starting treatment before DIND emergence may improve outcomes in SAH patients.

Nitric Oxide/Nitric Oxide Synthase Pathway

Since irregular change in the NO/NOS pathway may be detrimental to the brain after SAH, another treatment strategy for CVS or EBI is NO donor administration, iNOS inhibition, or eNOS upregulation. Sehba et al. showed an NO donor, N-nitrosoglutathione, injected 5 min after SAH ameliorated SAH-induced vessel diameter decrease and increased vessel wall thickness 60 min after SAH in an endovascular perforation rat model [169]. Pluta et al. demonstrated L-arginine, the substrate of the NO-producing enzyme NOS, administration increased CBF but did not affect the incidence or degree of CVS in a primate SAH model [170]. Sehba et al. found intravenous NO donor S-nitrosoglutathione infusion attenuated brain parenchymal microvessel permeability by preserving collagen IV and endothelial barrier antigens as well as decreasing collagenase activity when compared to saline-treated rats [96]. Marbacher et al. showed continuous NO donor glyceroltrinitrate infusion into the cisterna magna from days 0–5 did not significantly reduce vasospasm in an autologous blood injection rabbit model [171].

In an endovascular perforation rat model, Yatsushige et al. demonstrated iNOS inhibition did not improve neurological scores and mortality as well as reduce BBB breakdown, brain edema, and neuronal cell death 24 h after SAH [71]. However, Zheng et al. showed aminoguanidine, an iNOS inhibitor, administration reversed CVS after SAH via eNOS upregulation, indicating a regulatory cross talk between eNOS and iNOS in SAH pathogenesis [172]. One action of statins is eNOS upregulation and NO production. Simvastatin treatment before or after SAH attenuated cerebral vasospasm and neurological deficits in experimental animal models, and the mechanisms involved may in part be attributed to eNOS upregulation [72, 73] as well as phosphatidylinositol 3-kinase/Akt (PI3K/Akt) pathway activation [73]. However, attempted NO generation by eNOS gene transfection into subarachnoid space cells using an adenovirus-expressing eNOS vector did not prevent vasospasm [173]. Recently, monomer formation (uncoupling) of eNOS expression is reportedly increased after SAH, resulting in decreased NO levels and reduced superoxide anion radical production [74, 75]. To date, a few randomized clinical trials reported statins attenuated SAH-induced vasospasm, but results from a larger trial are unknown [174–176]. One randomized clinical trial, however, reported statins did not ameliorate transcranial Doppler vasospasm and DIND incidences as well as improve outcomes [177].

Antiplatelet Agents

Since platelet aggregation plays a role in secondary ischemic pathogenesis, experimental studies were performed to evaluate the effectiveness of antiplatelet agents. Aspirin inhibits platelet function and thromboxane production. Linder and Alksne demonstrated aspirin ameliorated persistent BA contraction in dogs [178]. One clinical retrospective study showed aspirin reduced delayed cerebral ischemia risk after SAH [179]. Bilginer et al. demonstrated Cilostazol significantly attenuated cerebral BA vasospasm after SAH in rabbits [180]. Nishino et al. showed Cilostazol decreased perivascular macrophage/dendritic cell infiltration 7 days after SAH in rats [181]. Since platelet-activating factor (PAF) is an inflammatory mediator and has been demonstrated to promote CVS genesis after SAH, Hirashima et al. examined the effects of platelet-activating factor antagonist, E5880, administration after SAH in rabbits [182]. E5880 administration reduced BA constriction by reducing plasma thromboxane B2 concentrations [182]. Meta-analysis in randomized controlled trials comparing any antiplatelet agent to controls in aneurysmal SAH patients concluded antiplatelet treatment prevented secondary ischemia and improved outcomes. Antiplatelet treatment, however, cannot be recommended on the basis of current evidence because trial results did not reach statistical significance, although the results showed a strong trend towards better outcomes [183].

Anti-inflammation Therapy in Early Brain Injury After SAH

The term EBI refers to the period that spans from the moment of the initial bleed to the onset of CVS [184]. Since EBI is influenced by the immediate injury to the brain from aneurysmal rupture in SAH, the main pathophysiological stage is not in the large arteries, but in the brain parenchyma. Because EBI experimental modeling began simulating the intracranial arterial rupture, the common experimental model changed to the rodent "endovascular puncture" model. This model was independently described by Bederson et al. and Veelken et al. [185, 186], and the surgical procedure aims to perforate the internal carotid bifurcation without craniotomy by means of a sharp-ended suture inserted through the external carotid artery. Maximal CVS is observed in rats at 48 h after SAH, and the 24-h time point after SAH seems to be best for EBI analysis [71].

Many reports focused on acute brain edema (shown by brain water content or Evans blue extravasation assay, for example) after EBI was first reported by Kusaka et al. in 2004 because brain edema is a major symptom of EBI [18]. However, the mechanisms involved between brain edema attenuation and anti-inflammatory treatment have not been sufficiently demonstrated. Sozen et al. administered an IL-1β converting enzyme inhibitor, N-Ac-Tyr-Val-Ala-Asp-chloromethyl ketone (Ac-YVAD-CMK), in SAH mice, which ameliorated neurological deficits, mortality, and brain edema by inhibiting c-Jun N-terminal kinase (JNK) phosphorylation, MMP-9 induction, and tight junction protein degradation [187]. SAH-induced neurological deficits and upregulated IL-1β were also ameliorated by argatroban (a direct thrombin inhibitor) or osteopontin (extracellular matrix protein) administration in SAH rats [188, 189]. Suzuki et al. also showed recombinant osteopontin prevented post-SAH BBB disruption by suppressing nuclear factor-kappaB activation SAH rats [189]. Endo et al. demonstrated Cu–Zn SOD transgenic rats had decreased mortality and apoptotic cell death at 24 h. Moreover, survival signals, such as phosphorylation of Akt and glycogen synthase kinase-3beta (GSK3beta), were more enhanced in the cerebral cortex of Cu–Zn SOD transgenic rats after SAH [190]. Gao et al. demonstrated Edaravone ameliorated neurological deficits, decreased MDA levels and apoptotic cell death, and increased SOD activity in the brains of SAH rats [191]. Erşahin et al. demonstrated SAH-induced MDA upregulation and MPO activity were inhibited at 48 h after SAH by Ghrelin treatment into forebrain tissue in a single blood injection SAH rat model [192]. Although many preclinical animal experiments demonstrated anti-inflammatory treatments for SAH had positive outcomes, these treatments have not been translated to randomized, blinded, human clinical trials for EBI management after SAH.

Future Directions of SAH Research

Since it is widely accepted that CVS is a major cause of high mortality and poor outcomes after an otherwise successful treatment of a ruptured intracranial aneurysm [17], the majority of research performed worldwide has focused on strategies limiting arterial narrowing and delayed cerebral ischemia following SAH [193]. Hence, most experiments examined restoration of narrowed large arteries using pharmacological agents. The most common SAH-induced vasospasm model for observing pathophysiology or morphological changes in large arteries is the canine "two-hemorrhage" model, where two blood injections into the dog's basal cistern are performed 48 h apart [194].

Based on CVS amelioration by Endothelin receptor antagonists, CONSCIOUS-1 and CONSCIOUS-2 have been conducted. These studies demonstrated clazosentan did not attenuate mortality or functional outcomes, although it reduced angiographic vasospasm risk. One possible interpretation is successful intervention preventing CVS is not enough for attenuating outcomes in SAH patients. Delayed cerebral ischemia may be multifactorial and other pathophysiological factors may contribute to outcomes that are independent of angiographic vasospasm [20]. Additionally, animal modeling of CVS by injecting blood to make large vessels constrict in order to observe DIND may not be adequate. Furthermore, measures for neurological outcomes in experimental studies may be insufficient. The discrepancy amongst SAH survivors is 21 % develop delayed ischemic injury without vasospasm and only 20–30 % who do develop vasospasm suffer from delayed ischemic injury [195], which suggests researchers should seek new concepts toward the treatment of SAH patients instead of targeting CVS alone [196]. Currently, targeting EBI has great potential for implementing new treatment modalities in SAH patients by attenuating some of the observed long-term, devastating secondary injuries [26, 184]. Mortality and neurological functions in preclinical, animal SAH models should be examined to estimate treatment effectiveness because this information is very important for translating preclinical studies into a clinical application. The endovascular perforation model is more suitable for acute SAH research than the double blood injection model because the endovascular perforation model produces more severe pathophysiological changes and a comparable insult to a ruptured aneurysm [197].

So far, anti-inflammatory research after SAH is limited, and further studies are needed to clarify the exact mechanisms involved. First, natural courses after SAH, such as spatial and temporal histories, are unclear. Utilizing an effective therapeutic window may be one of the biggest challenges. Second, additional basic work is required to investigate SAH-induced changes in the peripheral immune system. Because cytokines, neutrophils, T cells, and macrophages in the peripheral immune system may, theoretically, enter the CNS via an impaired BBB after brain ischemia, changes in the peripheral immune system may significantly affect ischemic brain injury [3]. Third, appropriate target organs must be selected for inhibiting excessive pro-inflammatory reactions. Pro-inflammatory reactions are critically important for removing pathogenic substances and promoting growth, repair, and functional recovery. However, extreme inflammatory responses may be deleterious to tissues injured by ROS and cytokines. Fourth, pharmacokinetic and pathogenic differences between genders must also be considered. Finally, further investigation into the inflammatory responses in stroke patients must be made because of the differences between the immune systems of humans and experimental animals [3]. Considering the potential anti-inflammatory therapeutic development for SAH, further experimental studies are needed to discern key mechanisms underlying the role of inflammation in brain damage.

Conclusion

Since inflammation plays an important role in SAH, many experimental studies targeting anti-inflammation successfully reversed vessel constrictions. However, the reversal of CVS did not improve outcomes in human clinical trials. EBI treatment could be argued to successfully attenuate some of the devastating secondary injuries following SAH. Further studies targeting anti-inflammation may lead to development of novel therapies that improve outcomes for SAH patients.

Acknowledgments This study was supports by the National Institutes of Health NS053407 to J.H. Zhang.

Conflict of Interest Statement We declare that we have no conflicts of interest.

References

1. David S, Kroner A (2011) Repertoire of microglial and macrophage responses after spinal cord injury. Nat Rev Neurosci 12:388–99
2. Loane DJ, Byrnes KR (2010) Role of microglia in neurotrauma. Neurotherapeutics 7: 366–77
3. Barone FC, Feuerstein GZ (1999) Inflammatory mediators and stroke: new opportunities for novel therapeutics. J Cereb Blood Flow Metab 19:819–34
4. Chamorro A, Hallenbeck J (2006) The harms and benefits of inflammatory and immune responses in vascular disease. Stroke 37:291–3
5. Kleinig TJ, Vink R (2009) Suppression of inflammation in ischemic and hemorrhagic stroke: therapeutic options. Curr Opin Neurol 22:294–301
6. Xia W, Han J, Huang G, Ying W (2010) Inflammation in ischaemic brain injury: current advances and future perspectives. Clin Exp Pharmacol Physiol 37:253–8
7. Aronowski J, Hall CE (2005) New horizons for primary intracerebral hemorrhage treatment: experience from preclinical studies. Neurol Res 27:268–79
8. Ducruet AF, Zacharia BE, Hickman ZL, Grobelny BT, Yeh ML, Sosunov SA et al (2009) The complement cascade as a therapeutic target in intracerebral hemorrhage. Exp Neurol 219: 398–403
9. Wang J, Doré S (2007) Inflammation after intracerebral hemorrhage. J Cereb Blood Flow Metab 27:894–908
10. van Gijn J, Kerr RS, Rinkel GJ (2007) Subarachnoid haemorrhage. Lancet 369:306–18
11. King JT Jr (1997) Epidemiology of aneurysmal subarachnoid hemorrhage. Neuroimaging Clin N Am 7:659–68
12. Sehba FA, Hou J, Pluta RM, Zhang JH (2012) The importance of early brain injury after subarachnoid hemorrhage. Prog Neurobiol 97:14–37
13. Wilkins RH (1990) Cerebral vasospasm. Crit Rev Neurobiol 6:51–77
14. Fergusen S, Macdonald RL (2007) Predictors of cerebral infarction in patients with aneurysmal subarachnoid hemorrhage. Neurosurgery 60:658–67
15. Fisher CM, Roberson GH, Ojemann RG (1977) Cerebral vasospasm with ruptured saccular aneurysm–the clinical manifestations. Neurosurgery 1:245–8
16. Rabinstein AA, Friedman JA, Weigand SD, McClelland RL, Fulgham JR, Manno EM et al (2004) Predictors of cerebral infarction in aneurysmal subarachnoid hemorrhage. Stroke 35: 1862–6
17. Dorsch NW (1995) Cerebral arterial spasm–a clinical review. Br J Neurosurg 9:403–12
18. Kusaka G, Ishikawa M, Nanda A, Granger DN, Zhang JH (2004) Signaling pathways for early brain injury after subarachnoid hemorrhage. J Cereb Blood Flow Metab 24:916–25
19. Macdonald RL, Kassell NF, Mayer S, Ruefenacht D, Schmiedek P, Weidauer S et al (2008) Clazosentan to overcome neurological ischemia and infarction occurring after subarachnoid hemorrhage (CONSCIOUS-1): randomized, double-blind, placebo-controlled phase 2 dose-finding trial. Stroke 39:3015–21
20. Macdonald RL, Pluta RM, Zhang JH (2007) Cerebral vasospasm after subarachnoid hemorrhage: the emerging revolution. Nat Clin Pract Neurol 3:256–63
21. Pluta RM (2005) Delayed cerebral vasospasm and nitric oxide: review, new hypothesis, and proposed treatment. Pharmacol Ther 105:23–56

22. Broderick JP, Brott TG, Duldner JE, Tomsick T, Leach A (1994) Initial and recurrent bleeding are the major causes of death following subarachnoid hemorrhage. Stroke 25:1342–7
23. Ostrowski RP, Colohan AR, Zhang JH (2006) Molecular mechanisms of early brain injury after subarachnoid hemorrhage. Neurol Res 28:399–414
24. Ostrowski RP, Tang J, Zhang JH (2006) Hyperbaric oxygen suppresses NADPH oxidase in a rat subarachnoid hemorrhage model. Stroke 37:1314–8
25. Grote E, Hassler W (1988) The critical first minutes after subarachnoid hemorrhage. Neurosurgery 22:654–61
26. Cahill J, Calvert JW, Zhang JH (2006) Mechanisms of early brain injury after subarachnoid hemorrhage. J Cereb Blood Flow Metab 26:1341–53
27. Keep RF, Andjelkovic AV, Stamatovic SM, Shakui P, Ennis SR (2005) Ischemia-induced endothelial cell dysfunction. Acta Neurochir Suppl 95:399–402
28. Sehba FA, Pluta RM, Zhang JH (2011) Metamorphosis of subarachnoid hemorrhage research: from delayed vasospasm to early brain injury. Mol Neurobiol 43:27–40
29. Luster AD, Alon R, von Andrian UH (2005) Immune cell migration in inflammation: present and future therapeutic targets. Nat Immunol 6:1182–90
30. Friedrich V, Flores R, Muller A, Bi W, Peerschke EI, Sehba FA (2011) Reduction of neutrophil activity decreases early microvascular injury after subarachnoid haemorrhage. J Neuroinflammation 8:103
31. Dumont AS, Dumont RJ, Chow MM, Lin CL, Calisaneller T, Ley KF et al (2003) Cerebral vasospasm after subarachnoid hemorrhage: putative role of inflammation. Neurosurgery 53:123–35
32. Provencio JJ, Fu X, Siu A, Rasmussen PA, Hazen SL, Ransohoff RM (2010) CSF neutrophils are implicated in the development of vasospasm in subarachnoid hemorrhage. Neurocrit Care 12:244–51
33. Wang Q, Tang XN, Yenari MA (2007) The inflammatory response in stroke. J Neuroimmunol 184:53–68
34. Wang Y, Zhong M, Tan XX, Yang YJ, Chen WJ, Liu W et al (2007) Expression change of interleukin-8 gene in rabbit basilar artery after subarachnoid hemorrhage. Neurosci Bull 23:151–5
35. Matz P, Turner C, Weinstein PR, Massa SM, Panter SS, Sharp FR (1996) Heme-oxygenase-1 induction in glia throughout rat brain following experimental subarachnoid hemorrhage. Brain Res 713:211–22
36. Banchereau J, Briere F, Caux C, Davoust J, Lebecque S, Liu YJ et al (2000) Immunobiology of dendritic cells. Annu Rev Immunol 1(8):767–811
37. Theoharides TC, Alysandratos KD, Angelidou A, Delivanis DA, Sismanopoulos N, Zhang B et al (2012) Mast cells and inflammation. Biochim Biophys Acta 2012:21–33
38. Liesz A, Suri-Payer E, Veltkamp C, Doerr H, Sommer C, Rivest S et al (2009) Regulatory T cells are key cerebroprotective immunomodulators in acute experimental stroke. Nat Med 15: 192–9
39. Rossi B, Constantin G (2008) Anti-selectin therapy for the treatment of inflammatory diseases. Inflamm Allergy Drug Targets 7:85–93
40. Moore KL (1998) Structure and function of P-selectin glycoprotein ligand-1. Leuk Lymphoma 29:1–15
41. McEver RP (2002) Selectins: lectins that initiate cell adhesion under flow. Curr Opin Cell Biol 14:581–6
42. Sperandio M, Smith ML, Forlow SB, Olson TS, Xia L, McEver RP et al (2003) P-selectin glycoprotein ligand-1 mediates L-selectin-dependent leukocyte rolling in venules. J Exp Med 197:1355–63
43. Hynes RO (2002) Integrins: bidirectional, allosteric signaling machines. Cell 110:673–87
44. Springer TA, Dustin ML, Kishimoto TK, Marlin SD (1987) The lymphocyte function-associated LFA-1, CD2, and LFA-3 molecules: cell adhesion receptors of the immune system. Annu Rev Immunol 5:223–52
45. Miller LJ, Schwarting R, Springer TA (1986) Regulated expression of the Mac-1, LFA-1, p150, 95 glycoprotein family during leukocyte differentiation. J Immunol 137:2891–900

46. Pradilla G, Wang PP, Legnani FG, Ogata L, Dietsch GN, Tamargo RJ (2004) Prevention of vasospasm by anti-CD11/CD18 monoclonal antibody therapy following subarachnoid hemorrhage in rabbits. J Neurosurg 101:88–92
47. Clatterbuck RE, Oshiro EM, Hoffman PA, Dietsch GN, Pardoll DM, Tamargo RJ (2002) Inhibition of vasospasm with lymphocyte function-associated antigen-1 monoclonal antibody in a femoral artery model in rats. J Neurosurg 97:676–82
48. Handa Y, Kubota T, Kaneko M, Tsuchida A, Kobayashi H, Kawano H et al (1995) Expression of intercellular adhesion molecule 1 (ICAM-1) on the cerebral artery following subarachnoid haemorrhage in rats. Acta Neurochir (Wien) 132:92–7
49. Gallia GL, Tamargo RJ (2006) Leukocyte-endothelial cell interactions in chronic vasospasm after subarachnoid hemorrhage. Neurol Res 28:750–8
50. Oshiro EM, Hoffman PA, Dietsch GN, Watts MC, Pardoll DM, Tamargo RJ (1997) Inhibition of experimental vasospasm with anti-intercellular adhesion molecule-1 monoclonal antibody in rats. Stroke 28:2031–8
51. Sills AK Jr, Clatterbuck RE, Thompson RC, Cohen PL, Tamargo RJ (1997) Endothelial cell expression of intercellular adhesion molecule 1 in experimental posthemorrhagic vasospasm. Neurosurgery 41:453–61
52. Mack WJ, Mocco J, Hoh DJ, Huang J, Choudhri TF, Kreiter KT et al (2002) Outcome prediction with serum intercellular adhesion molecule-1 levels after aneurysmal subarachnoid hemorrhage. J Neurosurg 96:71–5
53. Polin RS, Bavbek M, Shaffrey ME, Billups K, Bogaev CA, Kassell NF et al (1998) Detection of soluble E-selectin, ICAM-1, VCAM-1, and L-selectin in the cerebrospinal fluid of patients after subarachnoid hemorrhage. J Neurosurg 89:559–67
54. Sprague AH, Khalil RA (2009) Inflammatory cytokines in vascular dysfunction and vascular disease. Biochem Pharmacol 78:539–52
55. Lu H, Shi JX, Chen HL, Hang CH, Wang HD, Yin HX (2009) Expression of monocyte chemoattractant protein-1 in the cerebral artery after experimental subarachnoid hemorrhage. Brain Res 1262:73–80
56. Vikman P, Ansar S, Edvinsson L (2007) Transcriptional regulation of inflammatory and extracellular matrix-regulating genes in cerebral arteries following experimental subarachnoid hemorrhage in rats. Laboratory investigation. J Neurosurg 107:1015–22
57. Aihara Y, Kasuya H, Onda H, Hori T, Takeda J (2001) Quantitative analysis of gene expressions related to inflammation in canine spastic artery after subarachnoid hemorrhage. Stroke 32:212–7
58. Luster AD (1998) Chemokines–chemotactic cytokines that mediate inflammation. N Engl J Med 338:436–45
59. Ehrenreich H, Anderson RW, Fox CH, Rieckmann P, Hoffman GS, Travis WD et al (1990) Endothelins, peptides with potent vasoactive properties, are produced by human macrophages. J Exp Med 172:1741–8
60. Sessa WC, Kaw S, Hecker M, Vane JR (1991) The biosynthesis of endothelin-1 by human polymorphonuclear leukocytes. Biochem Biophys Res Commun 174:613–8
61. Tierney TS, Clatterbuck RE, Lawson C, Thai QA, Rhines LD, Tamargo RJ (2001) Prevention and reversal of experimental posthemorrhagic vasospasm by the periadventitial administration of nitric oxide from a controlled-release polymer. Neurosurgery 49:945–53
62. Macdonald RL, Weir BK (1991) A review of hemoglobin and the pathogenesis of cerebral vasospasm. Stroke 22:971–82
63. Sehba FA, Schwartz AY, Chereshnev I, Bederson JB (2000) Acute decrease in cerebral nitric oxide levels after subarachnoid hemorrhage. J Cereb Blood Flow Metab 20:604–11
64. Moro MA, Almeida A, Bolanos JP, Lizasoain I (2005) Mitochondrial respiratory chain and free radical generation in stroke. Free Radic Biol Med 39:1291–304
65. Szabó C, Dawson VL (1998) Role of poly(ADP-ribose) synthetase in inflammation and ischaemia-reperfusion. Trends Pharmacol Sci 19:287–98
66. Calvert JW, Zhang JH (2005) Pathophysiology of an hypoxic-ischemic insult during the perinatal period. Neurol Res 27:246–60

67. Weller R (2003) Nitric oxide: a key mediator in cutaneous physiology. Clin Exp Dermatol 28:511–4
68. Huang Z, Huang PL, Panahian N, Dalkara T, Fishman MC, Moskowitz MA (1994) Effects of cerebral ischemia in mice deficient in neuronal nitric oxide synthase. Science 265:1883–5
69. Huang Z, Huang PL, Ma J, Meng W, Ayata C, Fishman MC et al (1996) Enlarged infarcts in endothelial nitric oxide synthase knockout mice are attenuated by nitro-L-arginine. J Cereb Blood Flow Metab 16:981–7
70. Iadecola C, Zhang F, Casey R, Clark HB, Ross ME (1996) Inducible nitric oxide synthase gene expression in vascular cells after transient focal cerebral ischemia. Stroke 27:1373–80
71. Yatsushige H, Calvert JW, Cahill J, Zhang JH (2006) Limited role of inducible nitric oxide synthase in blood–brain barrier function after experimental subarachnoid hemorrhage. J Neurotrauma 23:1874–82
72. McGirt MJ, Lynch JR, Parra A, Sheng H, Pearlstein RD, Laskowitz DT et al (2002) Simvastatin increases endothelial nitric oxide synthase and ameliorates cerebral vasospasm resulting from subarachnoid hemorrhage. Stroke 33:2950–6
73. Sugawara T, Ayer R, Jadhav V, Chen W, Tsubokawa T, Zhang JH (2008) Simvastatin attenuation of cerebral vasospasm after subarachnoid hemorrhage in rats via increased phosphorylation of Akt and endothelial nitric oxide synthase. J Neurosci Res 86:3635–43
74. Sabri M, Ai J, Knight B, Tariq A, Jeon H, Shang X et al (2011) Uncoupling of endothelial nitric oxide synthase after experimental subarachnoid hemorrhage. J Cereb Blood Flow Metab 31:190–9
75. Sabri M, Ai J, Marsden PA, Macdonald RL (2011) Simvastatin re-couples dysfunctional endothelial nitric oxide synthase in experimental subarachnoid hemorrhage. PLoS One 6: e17062
76. Misra HP, Fridovich I (1972) The generation of superoxide radical during the autoxidation of hemoglobin. J Biol Chem 247:6960–2
77. Rubbo H, Trostchansky A, O'Donnell VB (2009) Peroxynitrite-mediated lipid oxidation and nitration: mechanisms and consequences. Arch Biochem Biophys 484:167–72
78. Sehba FA, Bederson JB (2006) Mechanisms of acute brain injury after subarachnoid hemorrhage. Neurol Res 28:381–98
79. Giulivi C, Hochstein P, Davies KJA (1994) Hydrogen peroxide production by red blood cells. Free Radic Biol Med 16:123–9
80. Bruce Van Dykea R, Saltama P (1996) Hemoglobin: a mechanism for the generation of hydroxyl radicals. Free Radic Biol Med 20:985–9
81. Gutteridge JM (1986) Iron promoters of the Fenton reaction and lipid peroxidation can be released from haemoglobin by peroxides. FEBS Lett 201:291–5
82. Mori T, Nagata K, Town T, Tan J, Matsui T, Asano T (2001) Intracisternal increase of superoxide anion production in a canine subarachnoid hemorrhage model. Stroke 32:636–42
83. Facchinetti F, Dawson VL, Dawson TM (1998) Free radicals as mediators of neuronal injury. Cell Mol Neurobiol 18:667–82
84. Weiss SJ (1989) Tissue destruction by neutrophils. N Engl J Med 320:365–76
85. Lin CL, Hsu YT, Lin TK, Morrow JD, Hsu JC, Hsu YH et al (2006) Increased levels of F2-isoprostanes following aneurysmal subarachnoid hemorrhage in humans. Free Radic Biol Med 40:1466–73
86. Lewen A, Matz P, Chan PH (2000) Free radical pathways in CNS injury. J Neurotrauma 17: 871–90
87. Gaetani P, Lombardi D (1992) Brain damage following subarachnoid hemorrhage: the imbalance between anti-oxidant systems and lipid peroxidative processes. J Neurosurg Sci 36:1–10
88. Matz PG, Fujimura M, Lewen A, Morita-Fujimura Y, Chan PH (2001) Increased cytochrome c-mediated DNA fragmentation and cell death in manganese-superoxide dismutase-deficient mice after exposure to subarachnoid hemolysate. Stroke 32:506–15
89. Weyrich AS, Schwertz H, Kraiss LW, Zimmerman GA (2009) Protein synthesis by platelets: historical and new perspectives. J Thromb Haemost 7:241–6

90. Akopov S, Sercombe R, Seylaz J (1996) Endothelium–plateletleukocyte interactions in the cerebral circulation. Cerebrovasc Brain Metab Rev 8:11–94
91. Sehba FA, Mostafa G, Friedrich V Jr, Bederson JB (2005) Acute microvascular platelet aggregation after subarachnoid hemorrhage. J Neurosurg 102:1094–100
92. Friedrich V, Flores R, Muller A, Sehba FA (2010) Escape of intraluminal platelets into brain parenchyma after subarachnoid hemorrhage. Neuroscience 165:968–75
93. Friedrich V, Flores R, Muller A, Sehba FA (2010) Luminal platelet aggregates in functional deficits inparenchymal vessels after subarachnoid hemorrhage. Brain Res 1354:179–87
94. Okada Y, Copeland BR, Mori E, Tung MM, Thomas WS, del Zoppo GJ (1994) P-selectin and intercellular adhesion molecule-1 expression after focal brain ischemia and reperfusion. Stroke 25:202–11
95. Sehba FA, Mostafa G, Knopman J, Friedrich V Jr, Bederson JB (2004) Acute alterations in microvascular basal lamina after subarachnoid hemorrhage. J Neurosurg 101:633–40
96. Sehba FA, Friedrich V Jr, Makonnen G, Bederson JB (2007) Acute cerebral vascular injury after subarachnoid hemorrhage and its prevention by administration of a nitric oxide donor. J Neurosurg 106:321–9
97. Rosenberg GA, Cunningham LA, Wallace J, Alexander S, Estrada EY, Grossetete M et al (2001) Immunohistochemistry of matrix metalloproteinases in reperfusion injury to rat brain: activation of MMP-9 linked to stromelysin-1 and microglia in cell cultures. Brain Res 893: 104–12
98. Liu W, Hendren J, Qin XJ, Shen J, Liu KJ (2009) Normobaric hyperoxia attenuates early blood-brain barrier disruption by inhibiting MMP-9-mediated occludin degradation in focal cerebral ischemia. J Neurochem 108:811–20
99. Asano T, Ikegaki I, Suzuki Y, Satoh S, Shibuya M (1989) Endothelin and the production of cerebral vasospasm in dogs. Biochem Biophys Res Commun 159:1345–51
100. Zimmermann M, Seifert V (2004) Endothelin receptor antagonists and cerebral vasospasm. Clin Auton Res 14:143–5
101. Sercombe R, Dinh YR, Gomis P (2002) Cerebrovascular inflammation following subarachnoid hemorrhage. Jpn J Pharmacol 88:227–49
102. Saleh MA, Pollock DM (2011) Endothelin in renal inflammation and hypertension. Contrib Nephrol 172:160–70
103. Narushima I, Kita T, Kubo K, Yonetani Y, Momochi C, Yoshikawa I et al (1999) Contribution of endothelin-1 to disruption of blood-brain barrier permeability in dogs. Naunyn Schmiedebergs Arch Pharmacol 360:639–45
104. Haorah J, Knipe B, Leibhart J, Ghorpade A, Persidsky Y (2005) Alcohol-induced oxidative stress in brain endothelial cells causes blood-brain barrier dysfunction. J Leukoc Biol 78: 1223–32
105. Imaizumi S, Kondo T, Deli MA, Gobbel G, Joó F, Epstein CJ et al (1996) The influence of oxygen free radicals on the permeability of the monolayer of cultured brain endothelial cells. Neurochem Int 29:205–11
106. Ballabh P, Braun A, Nedergaard M (2004) The blood–brain barrier: an overview: structure, regulation, and clinical implications. Neurobiol Dis 16:1–13
107. van der Goes A, Wouters D, van Der Pol SM, Huizinga R, Ronken E, Adamson P et al (2001) Reactive oxygen species enhance the migration of monocytes across the blood–brain barrier in vitro. FASEB J 15:1852–4
108. Bavbek M, Polin R, Kwan AL, Arthur AS, Kassell NF, Lee KS (1998) Monoclonal antibodies against ICAM-1 and CD18 attenuate cerebral vasospasm after experimental subarachnoid hemorrhage in rabbits. Stroke 29:1930–6
109. Clatterbuck RE, Gailloud P, Ogata L, Gebremariam A, Dietsch GN, Murphy KJ et al (2003) Prevention of cerebral vasospasm by a humanized anti-CD11/CD18 monoclonal antibody administered after experimental subarachnoid hemorrhage in nonhuman primates. J Neurosurg 99:376–82

110. Lin CL, Dumont AS, Calisaneller T, Kwan AL, Hwong SL, Lee KS (2005) Monoclonal antibody against E selectin attenuates subarachnoid hemorrhage-induced cerebral vasospasm. Surg Neurol 64:201–6
111. Provencio JJ, Altay T, Smithason S, Moore SK, Ransohoff RM (2011) Depletion of Ly6G/C(+) cells ameliorates delayed cerebral vasospasm in subarachnoid hemorrhage. J Neuroimmunol 232:94–100
112. Wang Z, Wang KY, Wu Y, Zhou P, Sun XO, Chen G (2010) Potential role of CD34 in cerebral vasospasm after experimental subarachnoid hemorrhage in rats. Cytokine 52:245–51
113. Wu Y, Tang K, Huang RQ, Zhuang Z, Cheng HL, Yin HX et al (2011) Therapeutic potential of peroxisome proliferator-activated receptor γ agonist rosiglitazone in cerebral vasospasm after a rat experimental subarachnoid hemorrhage model. J Neurol Sci 305:85–91
114. Weber C, Erl W, Weber KS, Weber PC (1997) HMG-CoA reductase inhibitors decrease CD11b expression and CD11b-dependent adhesion of monocytes to endothelium and reduce increased adhesiveness of monocytes isolated from patients with hypercholesterolemia. J Am Coll Cardiol 30:1212–7
115. Kallen J, Welzenbach K, Ramage P, Geyl D, Kriwacki R, Legge G et al (1999) Structural basis for LFA-1 inhibition upon lovastatin binding to the CD11a I-domain. J Mol Biol 292:1–9
116. McGirt MJ, Pradilla G, Legnani FG, Thai QA, Recinos PF, Tamargo RJ et al (2006) Systemic administration of simvastatin after the onset of experimental subarachnoid hemorrhage attenuates cerebral vasospasm. Neurosurgery 58:945–51
117. Thai QA, Oshiro EM, Tamargo RJ (1999) Inhibition of experimental vasospasm in rats with the periadventitial administration of ibuprofen using controlled-release polymers. Stroke 30:140–7
118. Hino A, Weir BK, Macdonald RL, Thisted RA, Kim CJ, Johns LM (1995) Prospective, randomized, double-blind trial of BQ-123 and bosentan for prevention of vasospasm following subarachnoid hemorrhage in monkeys. J Neurosurg 83:503–9
119. Itoh S, Sasaki T, Ide K, Ishikawa K, Nishikibe M, Yano M (1993) A novel endothelin ETA receptor antagonist, BQ-485, and its preventive effect on experimental cerebral vasospasm in dogs. Biochem Biophys Res Commun 195:969–75
120. Itoh S, Sasaki T, Asai A, Kuchino Y (1994) Prevention of delayed vasospasm by an endothelin ETA receptor antagonist, BQ-123: change of ETA receptor mRNA expression in a canine subarachnoid hemorrhage model. J Neurosurg 81:759–64
121. Kita T, Kubo K, Hiramatsu K, Sakaki T, Yonetani Y, Sato S et al (1998) Profiles of an intravenously available endothelin A-receptor antagonist, S-0139, for preventing cerebral vasospasm in a canine two-hemorrhage model. Life Sci 63:305–15
122. Macdonald RL, Bassiouny M, Johns L, Sajdak M, Marton LS, Weir BK et al (1998) U74389G prevents vasospasm after subarachnoid hemorrhage in dogs. Neurosurgery 42:1339–46
123. Macdonald RL, Johns L, Lin G, Marton LS, Hallak H, Marcoux F et al (1998) Prevention of vasospasm after subarachnoid hemorrhage in dogs by continuous intravenous infusion of PD156707. Neurol Med Chir (Tokyo) 38(Suppl):138–45
124. Nirei H, Hamada K, Shoubo M, Sogabe K, Notsu Y, Ono T (1993) An endothelin ETA receptor antagonist, FR139317, ameliorates cerebral vasospasm in dogs. Life Sci 52:1869–74
125. Roux S, Breu V, Giller T, Neidhart W, Ramuz H, Coassolo P et al (1997) Ro 61-1790, a new hydrosoluble endothelin antagonist: general pharmacology and effects on experimental cerebral vasospasm. J Pharmacol Exp Ther 283:1110–8
126. Shigeno T, Clozel M, Sakai S, Saito A, Goto K (1995) The effect of bosentan, a new potent endothelin receptor antagonist, on the pathogenesis of cerebral vasospasm. Neurosurgery 37:87–91
127. Wanebo JE, Louis HG, Arthur AS, Zhou J, Kassell NF, Lee KS, et al (1997) Attenuation of cerebral vasospasm by systemic administration of an endothelin-A receptor antagonist, TBC 11251, in a rabbit model of subarachnoid hemorrhage. Neurosurg Focus 3:article
128. Wanebo JE, Arthur AS, Louis HG, West K, Kassell NF, Lee KS et al (1998) Systemic administration of the endothelin-A receptor antagonist TBC 11251 attenuates cerebral vasospasm after experimental subarachnoid hemorrhage: dose study and review of endothelin-based therapies in the literature on cerebral vasospasm. Neurosurgery 43:1409–18

129. Willette RN, Zhang H, Mitchell MP, Sauermelch CF, Ohlstein EH, Sulpizio AC (1994) Nonpeptide endothelin antagonist. Cerebrovascular characterization and effects on delayed cerebral vasospasm. Stroke 25:2450–6
130. Zimmermann M, Seifert V, Löffler BM, Stolke D, Stenzel W (1996) Prevention of cerebral vasospasm after experimental subarachnoid hemorrhage by RO 47-0203, a newly developed orally active endothelin receptor antagonist. Neurosurgery 38:115–20
131. Zuccarello M, Soattin GB, Lewis AI, Breu V, Hallak H, Rapoport RM (1996) Prevention of subarachnoid hemorrhage-induced cerebral vasospasm by oral administration of endothelin receptor antagonists. J Neurosurg 84:503–7
132. Zuccarello M, Boccaletti R, Romano A, Rapoport RM (1998) Endothelin B receptor antagonists attenuate subarachnoid hemorrhage-induced cerebral vasospasm. Stroke 29:1924–9
133. Caner HH, Kwan AL, Arthur A, Jeng AY, Lappe RW, Kassell NF et al (1996) Systemic administration of an inhibitor of endothelin-converting enzyme for attenuation of cerebral vasospasm following experimental subarachnoid hemorrhage. J Neurosurg 85:917–22
134. Kwan AL, Bavbek M, Jeng AY, Maniara W, Toyoda T, Lappe RW et al (1997) Prevention and reversal of cerebral vasospasm by an endothelin-converting enzyme inhibitor, CGS 26303, in an experimental model of subarachnoid hemorrhage. J Neurosurg 87:281–6
135. Kwan AL, Lin CL, Chang CZ, Wu HJ, Hwong SL, Jeng AY et al (2001) Continuous intravenous infusion of CGS 26303, an endothelin-converting enzyme inhibitor, prevents and reverses cerebral vasospasm after experimental subarachnoid hemorrhage. Neurosurgery 49:422–9
136. Kwan AL, Lin CL, Yen CP, Winardi W, Su YF, Winardi D, Dai ZK, Jeng AY, Kassell NF, Howng SL, Wang CJ (2006) Prevention and reversal of vasospasm and ultrastructural changes in basilar artery by continuous infusion of CGS 35066 following subarachnoid hemorrhage. Exp Biol Med (Maywood) 231:1069–74
137. Lin CL, Kwan AL, Dumont AS, Su YF, Kassell NF, Wang CJ et al (2007) Attenuation of experimental subarachnoid hemorrhage-induced increases in circulating intercellular adhesion molecule-1 and cerebral vasospasm by the endothelin-converting enzyme inhibitor CGS 26303. J Neurosurg 106:442–8
138. Macdonald RL, Higashida RT, Keller E, Mayer SA, Molyneux A, Raabe A et al (2011) Clazosentan, an endothelin receptor antagonist, in patients with aneurysmal subarachnoid haemorrhage undergoing surgical clipping: a randomised, double-blind, placebo-controlled phase 3 trial (CONSCIOUS-2). Lancet Neurol 10:618–25
139. Macdonald RL, Weir BK, Runzer TD, Grace MG, Poznansky MJ (1992) Effect of intrathecal superoxide dismutase and catalase on oxyhemoglobin-induced vasospasm in monkeys. Neurosurgery 30:529–39
140. Shishido T, Suzuki R, Qian L, Hirakawa K (1994) The role of superoxide anions in the pathogenesis of cerebral vasospasm. Stroke 25:864–8
141. Mori T, Nagata K, Matsui T, Ishida T, Ohami H, Asano T (1995) Superoxide anions in the pathogenesis of talc-induced cerebral vasocontraction. Neuropathol Appl Neurobiol 21:378–85
142. Kamii H, Kato I, Kinouchi H, Chan PH, Epstein CJ, Akabane A et al (1999) Amelioration of vasospasm after subarachnoid hemorrhage in transgenic mice overexpressing CuZn-superoxide dismutase. Stroke 30:867–72
143. Güney O, Erdi F, Esen H, Kiyici A, Kocaogullar Y (2010) N-acetylcysteine prevents vasospasm after subarachnoid hemorrhage. World Neurosurg 73(42–9):e3
144. Kuo CP, Lu CH, Wen LL, Cherng CH, Wong CS, Borel CO et al (2011) Neuroprotective effect of curcumin in an experimental rat model of subarachnoid hemorrhage. Anesthesiology 115:1229–38
145. Germano A, Imperatore C, d'Avella D, Costa G, Tomasello F (1998) Antivasospastic and brain-protective effects of a hydroxyl radical scavenger (AVS) after experimental subarachnoid hemorrhage. J Neurosurg 88:1075–81
146. Imperatore C, Germano A, d'Avella D, Tomasello F, Costa G (2000) Effects of the radical scavenger AVS on behavioral and BBB changes after experimental subarachnoid hemorrhage. Life Sci 66:779–90

147. Turner CP, Panter SS, Sharp FR (1999) Anti-oxidants prevent focal rat brain injury as assessed by induction of heat shock proteins (HSP70, HO-1/HSP32, HSP47) following subarachnoid injections of lysed blood. Brain Res Mol Brain Res 65:87–102
148. Kanamaru K, Weir BK, Simpson I, Witbeck T, Grace M (1991) Effect of 21-aminosteroid U-74006F on lipid peroxidation in subarachnoid clot. J Neurosurg 74:454–9
149. Matsui T, Asano T (1994) Effects of new 21-aminosteroid tirilazad mesylate (U74006F) on chronic cerebral vasospasm in a "two-hemorrhage" model of beagle dogs. Neurosurgery 34: 1035–9
150. Vollmer DG, Kassell NF, Hongo K, Ogawa H, Tsukahara T (1989) Effect of the nonglucocorticoid 21-aminosteroid U74006F experimental cerebral vasospasm. Surg Neurol 31:190–4
151. Zuccarello M, Marsch JT, Schmitt G, Woodward J, Anderson DK (1989) Effect of the 21-aminosteroid U-74006F on cerebral vasospasm following subarachnoid hemorrhage. J Neurosurg 71:98–104
152. Suzuki H, Kanamaru K, Kuroki M, Sun H, Waga S, Miyazawa T (1999) Effects of tirilazad mesylate on vasospasm and phospholipid hydroperoxides in a primate model of subarachnoid hemorrhage. Stroke 30:450–6
153. Hall ED (1992) The neuroprotective pharmacology of methylprednisolone. J Neurosurg 76: 13–22
154. Gaetani P, Marzatico F, Renault B, Fulle I, Lombardi D, Ferlenga P et al (1990) High-dose methylprednisolone and 'ex vivo' release of eicosanoids after experimental subarachnoid haemorrhage. Neurol Res 12:111–6
155. Tang WH, Chen Z, Liu Z, Zhang JH, Xi G, Feng H (2008) The effect of ecdysterone on cerebral vasospasm following experimental subarachnoid hemorrhage in vitro and in vivo. Neurol Res 30:571–80
156. Handa Y, Kaneko M, Takeuchi H, Tsuchida A, Kobayashi H, Kubota T (2000) Effect of an antioxidant, ebselen, on development of chronic cerebral vasospasm after subarachnoid hemorrhage in primates. Surg Neurol 53:323–9
157. Watanabe T, Nishiyama M, Hori T, Asano T, Shimizu T, Masayasu H (1997) Ebselen (DR3305) ameliorates delayed cerebral vasospasm in a canine two-hemorrhage model. Neurol Res 19:563–5
158. Nakagomi T, Yamakawa K, Sasaki T, Saito I, Takakura K (2003) Effect of edaravone on cerebral vasospasm following experimental subarachnoid hemorrhage. J Stroke Cerebrovasc Dis 12:17–21
159. Shimada Y, Tsunoda H, Zang L, Hirano M, Oka T, Tanaka T (2009) Synergistic induction of heme oxygenase-1 by nicaraven after subarachnoid hemorrhage to prevent delayed cerebral vasospasm. Eur J Pharmacol 620:16–20
160. Aladag MA, Turkoz Y, Ozcan C, Sahna E, Parlakpinar H, Akpolat N et al (2006) Caffeic acid phenethyl ester (CAPE) attenuates cerebral vasospasm after experimental subarachnoidal haemorrhage by increasing brain nitric oxide levels. Int J Dev Neurosci 24:9–14
161. Saito I, Asano T, Sano K, Takakura K, Abe H, Yoshimoto T et al (1998) Neuroprotective effect of an antioxidant, ebselen, in patients with delayed neurological deficits after aneurysmal subarachnoid hemorrhage. Neurosurgery 42:269–78
162. Munakata A, Ohkuma H, Nakano T, Shimamura N, Asano K, Naraoka M (2009) Effect of a free radical scavenger, edaravone, in the treatment of patients with aneurysmal subarachnoid hemorrhage. Neurosurgery 64:423–9
163. Kassell NF, Haley EC Jr, Apperson-Hansen C, Alves WM (1996) Randomized, double-blind, vehicle-controlled trial of tirilazad mesylate in patients with aneurysmal subarachnoid hemorrhage: a cooperative study in Europe, Australia, and New Zealand. J Neurosurg 84:221–8
164. Haley EC Jr, Kassell NF, Apperson-Hansen C, Maile MH, Alves WM (1997) A randomized, double-blind, vehicle-controlled trial of tirilazad mesylate in patients with aneurysmal subarachnoid hemorrhage: a cooperative study in North America. J Neurosurg 86:467–74
165. Asano T, Takakura K, Sano K, Kikuchi H, Nagai H, Saito I et al (1996) Effects of a hydroxyl radical scavenger on delayed ischemic neurological deficits following aneurysmal subarachnoid hemorrhage: results of a multicenter, placebo-controlled double-blind trial. J Neurosurg 84: 792–803

166. Hashi K, Takakura K, Sano K, Ohta T, Saito I, Okada K (1988) Intravenous hydrocortisone in large doses in the treatment of delayed ischemic neurological deficits following subarachnoid hemorrhage—results of a multi-center controlled double-blind clinical study. No to Shinkei—Brain & Nerve 40:373–82 [Japanese]
167. Chyatte D, Fode NC, Nichols DA, Sundt TM Jr (1987) Preliminary report: effects of high dose methylprednisolone on delayed cerebral ischemia in patients at high risk for vasospasm after aneurysmal subarachnoid hemorrhage. Neurosurgery 21:157–160
168. Gomis P, Graftieaux JP, Sercombe R, Hettler D, Scherpereel B, Rousseaux P (2010) Randomized, double-blind, placebo-controlled, pilot trial of high-dose methylprednisolone in aneurysmal subarachnoid hemorrhage. J Neurosurg 112:681–8
169. Sehba FA, Ding WH, Chereshnev I, Bederson JB (1999) Effects of S-nitrosoglutathione on acute vasoconstriction and glutamate release after subarachnoid hemorrhage. Stroke 30: 1955–61
170. Pluta RM, Afshar JK, Thompson BG, Boock RJ, Harvey-White J, Oldfield EH (2000) Increased cerebral blood flow but no reversal or prevention of vasospasm in response to L-arginine infusion after subarachnoid hemorrhage. J Neurosurg 92:121–6
171. Marbacher S, Neuschmelting V, Graupner T, Jakob SM, Fandino J (2008) Prevention of delayed cerebral vasospasm by continuous intrathecal infusion of glyceroltrinitrate and nimodipine in the rabbit model in vivo. Intensive Care Med 34:932–8
172. Zheng B, Zheng T, Wang L, Chen X, Shi C, Zhao S (2010) Aminoguanidine inhibition of iNOS activity ameliorates cerebral vasospasm after subarachnoid hemorrhage in rabbits via restoration of dysfunctional endothelial cells. J Neurol Sci 295:97–103
173. Stoodley M, Weihl CC, Zhang ZD, Lin G, Johns LM, Kowalczuk A et al (2000) Effect of adenovirus-mediated nitric oxide synthase gene transfer on vasospasm after experimental subarachnoid hemorrhage. Neurosurgery 46:1193–1203
174. Chou SH, Smith EE, Badjatia N, Nogueira RG, Sims JR 2nd, Ogilvy CS et al (2008) A randomized, double-blind, placebo-controlled pilot study of simvastatin in aneurysmal subarachnoid hemorrhage. Stroke 39:2891–3
175. Lynch JR, Wang H, McGirt MJ, Floyd J, Friedman AH, Coon AL et al (2005) Simvastatin reduces vasospasm after aneurysmal subarachnoid hemorrhage: results of a pilot randomized clinical trial. Stroke 36:2024–6
176. Tseng MY, Czosnyka M, Richards H, Pickard JD, Kirkpatrick PJ (2005) Effects of acute treatment with pravastatin on cerebral vasospasm, autoregulation, and delayed ischemic deficits after aneurysmal subarachnoid hemorrhage: a phase II randomized placebo-controlled trial. Stroke 36:1627–32
177. Vergouwen MD, Meijers JC, Geskus RB, Coert BA, Horn J, Stroes ES et al (2009) Biologic effects of simvastatin in patients with aneurysmal subarachnoid hemorrhage: a double-blind, placebo-controlled randomized trial. J Cereb Blood Flow Metab 29:1444–53
178. Linder M, Alksne JF (1978) Prevention of persistent cerebral smooth muscle contraction in response to whole blood. Stroke 9:472–7
179. Juvela S (1995) Aspirin and delayed cerebral ischemia after aneurysmal subarachnoid hemorrhage. J Neurosurg 82:945–52
180. Bilginer B, Onal MB, Narin F, Soylemezoglu F, Ziyal IM, Ozgen T (2009) The effects of intravenous cilostazol and nimodipine on cerebral vasospasm after subarachnoid hemorrhage in an experimental rabbit model. Turk Neurosurg 19:374–9
181. Nishino A, Umegaki M, Fujinaka T, Yoshimine T (2010) Cilostazol attenuates cerebral vasospasm after experimental subarachnoid hemorrhage. Neurol Res 32:873–8
182. Hirashima Y, Endo S, Kato R, Takaku A (1996) Prevention of cerebrovasospasm following subarachnoid hemorrhage in rabbits by the platelet-activating factor antagonist, E5880. J Neurosurg 84:826–30
183. Dorhout Mees SM, van den Bergh WM, Algra A, Rinkel GJ (2007) Antiplatelet therapy for aneurysmal subarachnoid haemorrhage. Cochrane Database Syst Rev 17, CD006184
184. Fujii M, Yan J, Rolland WB, Soejima Y, Caner B, Zhang JH (2013) Early brain injury, an evolving frontier in subarachnoid hemorrhage research. Transl Stroke Res 4:432–46

185. Bederson JB, Germano IM, Guarino L (1995) Cortical blood flow and cerebral perfusion pressure in a new noncraniotomy model of subarachnoid hemorrhage in the rat. Stroke 26: 1086–92
186. Veelken JA, Laing RJ, Jakubowski J (1995) The Sheffield model of subarachnoid hemorrhage in rats. Stroke 26:1279–84
187. Sozen T, Tsuchiyama R, Hasegawa Y, Suzuki H, Jadhav V, Nishizawa S et al (2009) Role of interleukin-1beta in early brain injury after subarachnoid hemorrhage in mice. Stroke 40:2519–25
188. Sugawara T, Jadhav V, Ayer R, Chen W, Suzuki H, Zhang JH (2009) Thrombin inhibition by argatroban ameliorates early brain injury and improves neurological outcomes after experimental subarachnoid hemorrhage in rats. Stroke 40:1530–2
189. Suzuki H, Ayer R, Sugawara T, Chen W, Sozen T, Hasegawa Y et al (2010) Protective effects of recombinant osteopontin on early brain injury after subarachnoid hemorrhage in rats. Crit Care Med 38:612–8
190. Endo H, Nito C, Kamada H, Yu F, Chan PH (2007) Reduction in oxidative stress by superoxide dismutase overexpression attenuates acute brain injury after subarachnoid hemorrhage via activation of Akt/glycogensynthase kinase-3beta survival signaling. J Cereb Blood Flow Metab 27:975–982
191. Gao Y, Ding XS, Xu S, Wang W, Zuo QL, Kuai F (2009) Neuroprotective effects of edaravone on early brain injury in rats after subarachnoid hemorrhage. Chin Med J (Engl) 122:1935–40
192. Erşahin M, Toklu HZ, Erzik C, Cetinel S, Akakin D, Velioğlu-Oğünç A et al (2010) The anti-inflammatory and neuroprotective effects of ghrelin in subarachnoid hemorrhage-induced oxidative brain damage in rats. J Neurotrauma 27:1143–55
193. Kassell NF, Sasaki T, Colohan AR, Nazar G (1985) Cerebral vasospasm following aneurysmal subarachnoid hemorrhage. Stroke 16:562–72
194. Megyesi JF, Vollrath B, Cook DA, Findlay JM (2000) In vivo animal models of cerebral vasospasm: a review. Neurosurgery 46:448–61
195. Alaraj A, Charbel FT, Amin-Hanjani S (2009) Peri-operative measures for treatment and prevention of cerebral vasospasm following subarachnoid hemorrhage. Neurol Res 31: 651–9
196. Caner B, Hou J, Altay O, Fujii M, Zhang JH (2012) Transition of research focus from vasospasm to early brain injury after subarachnoid hemorrhage. J Neurochem 123(Suppl 2): 12–21
197. Lee JY, Sagher O, Keep R, Hua Y, Xi G (2009) Comparison of experimental rat models of early brain injury after subarachnoid hemorrhage. Neurosurgery 65:331–43

Chapter 15
Minocycline, A Tetracycline Derivative, as a Potential Protective Agent for Acute Stroke

Jari Koistinaho and Milla Koistinaho

Abstract Minocycline and doxycycline, second-generation tetracyclines that have superior tissue penetration into the brain and cerebrospinal fluid, were reported to provide neuroprotection in global brain ischemia in 1998. Since then these compounds, especially minocycline has been widely studied in numerous in vivo and in vitro models of chronic and acute brain diseases. While the exact mechanism of minocycline's neuroprotective effect is not clear, minocycline has been shown to have anti-inflammatory, anti-apoptotic, and anti-oxidative effects. Currently, minocycline is in clinical trials for several indications, including ischemic stroke. Here, we review the mechanisms found to be behind minocycline's beneficial effect so far in models relevant for stroke. We also discuss the importance of using a wide range of stroke models and addressing the comorbidity and gender issues when evaluating minocycline's potential for treating patients with acute stroke. The chapter also covers the current status of clinical trials of minocycline for treating ischemic stroke.

Minocycline as a Neuroprotective Tetracycline Derivative

Minocycline is a second-generation, broad-spectrum tetracycline antibiotic, which has widely been used for the treatment of various types of bacterial infections for decades [1]. Minocycline is distinguished by its lipophilicity, leading to superior blood–brain barrier permeability [1, 2]. Because minocycline does not cause severe toxicity even

J. Koistinaho (✉)
A.I. Virtanen Institute for Molecular Sciences, University of Eastern Finland, Kuopio, Finland
e-mail: Jari.Koistinaho@uef.fi

M. Koistinaho (✉)
Medeia Therapeutics, Kuopio, Finland
e-mail: milla.koistinaho@uef.fi

J. Chen et al. (eds.), *Immunological Mechanisms and Therapies in Brain Injuries and Stroke*, Springer Series in Translational Stroke Research, DOI 10.1007/978-1-4614-8915-3_15,

in a long-term use, it is commonly used chronically for nonlife-threatening conditions such as acne vulgaris [3] and rheumatoid arthritis [4] in many countries.

Minocycline is known to have many biological effects that potentially possess neuroprotective features in acute brain injuries such as stroke [5–7]. While minocycline's antimicrobial activity is helpful in reducing infections, such as urinary tract infections and pneumonia, resulting from stroke or other acute brain injury-induced immunosuppression, the major beneficial effects of minocycline in stroke are independent of its antimicrobial properties. Minocycline's neuroprotective action is rather explained by various other effects on biological processes, including apoptosis, oxidative stress, iron toxicity, and neuroinflammation [5–9]. The cellular and molecular mechanisms behind these beneficial effects have been extensively explored, yet the exact mechanisms of action of minocycline are not clear.

Past studies have revealed that its *anti-inflammatory action* is due to the modulation or prevention of microglial activity, immune cell activation, and subsequent release of cytokines, chemokines, lipid mediators of inflammation, matrix metalloproteinases, and nitric oxide (NO) [10–19]. Microglial, astrocytic, neutrophilic, and macrophagic production of proinflammatory cytokines, such as TNF-α, IL-β, and IL-6, which are major contributory factors to inflammation and subsequent immune response, are reported to be depressed by minocycline at least partially by inhibition of p38 MAPK and transcription factor nuclear factor kappa-light-chain-enhancer of activated B cells (NFκB) [16, 20–23]. Minocycline reduces production of NO, prostaglandins, and eicosanoid inflammatory mediators by inhibiting expression or activity of the corresponding enzymes responsible for these products, namely inducible nitric oxide synthase (iNOS), cyclooxygenase-2 (COX-2), and 5-lipoxygenase [11, 18, 19]. Finally, minocycline reduces infiltration of blood-derived inflammatory cells, such as monocytes, neutrophils, and T cells, into the CNS upon brain injury [24–28]. Infiltration of leukocytes to the injured CNS is associated with increased expression of matrix metalloproteases 2 and 9 (MMP2 and MMP9), which are efficiently inhibited by minocycline [10, 14, 21, 24, 29].

Both cell culture and in vivo studies have revealed a clear anti-apoptotic effect for minocycline in context relevant for stroke. Minocycline reduces the number of apoptotic neurons in experimental injury model [5, 7, 30, 31]. Again, inhibition of MAP kinases, especially p38 MAPK, and alterations in the activity of extracellular signal-related kinase (Erk), phosphatidylinositol 3-kinase (PI 3-kinase)/Akt, as well as c-jun N-terminal kinase (JNK) signaling pathways appear to directly contribute to neuronal survival [16, 22, 32, 33]. Reported anti-apoptotic in vitro effects of minocycline include caspase-1 inhibition, upregulation of anti-apoptotic Bcl-2 protein, blockade of pro-apoptotic cytochrome c (CytC) and SMAC/Diablo release from the mitochondria [34–36], and inhibition of poly (ADP-ribose) polymerase (PARP) PARP-1 at least in certain conditions [37]. Some in vitro studies suggest that minocycline may have a dual effect on the apoptotic role of CytC: first by decreasing the peroxidase activity of CytC in the early stages of apoptosis and second, by competing with CytC-Apaf-1 binding interaction in the cytosol [38]. Moreover, minocycline has been reported to be able to upregulate nuclear factor erythroid 2-related factor 2 (Nrf2), a transcription factor regulating the expression of a number of antioxidant genes, and induce release of paracrine factors, including brain-derived neurotrophic factor (BDNF), nerve growth factor (NGF), glial cell-derived neurotrophic factor (GDNF), and vascular endothelial growth factor (VEGF), all of them being relevant for neuronal survival in acute brain injury [39]. Besides these specific activities, minocycline may act more generally as an antioxidant protecting against oxidative stress and possessing direct radical-scavenging activity [40, 41].

As acute brain injury such as cerebral stroke involves considerable inflammation and robust cell death, which to at least some extent includes apoptosis or apoptotic-like neuronal death [34, 35, 42–44], it is evident that minocycline can be considered as an excellent drug candidate for the treatment of acute stroke [6, 34, 35, 43–46]. This mechanism-of-action-based conclusion is further supported by minocycline's ability to chelate iron that may be released from hemoglobin to brain tissue upon brain injury [8, 9], as well as minocycline's ability to pass the blood–brain barrier and reach sufficient brain concentrations [1, 2, 7].

Effect of Minocycline in Commonly Used Preclinical Models of Stroke

Neuroprotective effect of minocycline was first described in a gerbil model of global cerebral ischemia in 1998 [19]. Since then more than 30 publications have supported its neuroprotective properties in various rodent models of stroke, including suture or mechanical models of transient and permanent occlusion of the middle cerebral artery (MCAo) and thromboembolic models [45]. Overall, according to a systemic review in 2011 [45], 31 of 33 studies have yielded positive effect for minocycline in rodent models of stroke. Importantly, the efficacy of minocycline has been proven when administered intraperitoneally or intravenously several hours after the onset of ischemia. Dose–response and therapeutic time window experiments on rat models of transient MCAo have demonstrated that an intravenous 3 mg/kg dose provides over 30 % reduction in the infarct size and improves neurological function even when administered 4–5 h after the onset of brain ischemia [47]. It is of great interest that minocycline may be beneficial even with far wider therapeutic time window. Using a mouse model of MCAo, Hayakawa et al. [48] found that long-term minocycline treatment started at 24 h after the ischemic insult resulted in reduced mortality, ischemic damage and improved neurological function and survival when followed up to 14 days. Two studies on rat model of transient MCAo have demonstrated that minocycline administration started 4 days after ischemic insult improved neurological recovery when assessed at 1 month time point [31, 49]. The improved recovery was associated with enhanced ischemia-induced neurogenesis and reduced microgliosis. Thus, the preclinical data on common animal models of stroke is robust.

Minocycline, Gender Issues, and Comorbidities in Stroke

The vast majority of stroke studies have been carried out in healthy, young, male rodents. As stroke obviously is a common disorder among both sexes in the elderly and typically coexists with atherosclerosis, diabetes, or infection, it is evident that preclinical studies of promising drug candidates for stroke should be extended to include studies on both genders as well as clinically relevant risk factors [50–53]. The ischemic infarct size is consistently found to be smaller in young female than young male mice when using transient MCAo model and this difference correlates with overall outcome [54]. Because ovariectomy and post-reproductive age results in the loss of gender difference in brain infarct size and outcome, the use of aged mice in stroke models appears to be more translational and clinically relevant than the use of young mice [55, 56]. Moreover, some studies indicate that the gender difference is not seen in stroke outcome when using permanent MCAo model, suggesting that the impact of gender is related to the pathomechanisms involved with the reperfusion [57]. These observations become crucial when evaluating the efficacy of drug candidates for treating stroke. Minocycline was reported to have beneficial effect in male, but not in female mice after transient MCAo [58, 59]. This gender difference in the efficacy of minocycline appears to be due to the dominant role of NO and PARP in female but not male young mice in transient MCAo model when abrupt reperfusion is taking place [60]. These results emphasize the importance of not only including both male and female rodent but also various ways of inducing the ischemic infarct in experimental stroke studies as recommended also by the Stroke Therapy Academic Industry Roundtable (STAIR) [60, 61].

While stroke studies on aged rodents predisposed to comorbid conditions are gradually increasing, minocycline has not yet been thoroughly investigated in such models. An interesting study was carried out by Hoda et al. [60], who tested the effect of minocycline in both sexes and aged, 18-month-old mice using a thromboembolic stroke model. As pointed out by Hoda et al. [60], thromboembolic models can be seen to mimic quite well the clinical situation, since reperfusion is gradually and partially restored spontaneously but not until 6–12 h when the ischemic cascade is already well advanced. Such gradual and spontaneous reperfusion may represent the vast majority of human MCA territory strokes where treatment with recombinant tissue plasminogen activator (rt-PA) or mechanical removal is not performed. Importantly, minocycline was found to be effective in reducing infarct size and improving short-term neurological outcome in young male and female mice, ovariectomized female mice and aged male and female mice [60]. Moreover, stroke-induced mortality in aged mice was significantly reduced by minocycline treatment. Even though the translationality of this study was limited because of administration of minocycline immediately after stroke onset and a single endpoint at 24 post-stroke hours, this study increases clinically relevant evidence for minocycline's validity as a neuroprotectant for acute stroke.

So far, there are two studies on young spontaneously hypertensive [62, 63]. One of them was carried out using a thromboembolic model and showed neuroprotection by minocycline when administered intravenously 4 h after the onset of stroke

(Murata et al. 2008). There are also several studies showing beneficial effect of minocycline on stroke outcome in either drug-induced hyperglycemic or genetic type 2 diabetic rats [64–66]. Studies on aged mice predisposed to infection or atherosclerosis are still lacking.

Minocycline and Tissue Plasminogen Activator

Tissue plasminogen activator (tPA) is an effective therapy for acute ischemic stroke. Time from the stroke onset is the major determinant for selecting patients eligible for thrombolytic therapy and the efficacy of intravenous tPA is established only up to 4.5 h after symptom onset [67, 68]. The benefit of intravenous tPA gradually declines with longer durations between symptom onset and tPA therapy in part due to increased risks of brain edema, excitotoxicity, and hemorrhagic conversion [69, 70]. Therefore, reduction of tPA-associated BBB injury with other neuroprotective compounds may extend the time window for safe and effective thrombolysis.

Using a thromboembolic clot model in male spontaneously hypertensive rats, Murata et al. (2008) were able to show that minocycline 3 mg/kg intravenously combined with delayed tPA treatment starting at 6 h reduced infarct size, ameliorated tPA-related intracerebral hemorrhage, and reduced plasma MMP-9 levels. In another study, Machado et al. [71] used a suture model of transient 3-h MCAo to test the effect of combined minocycline and tPA treatment. They reported that minocycline, again at 3 mg/kg intravenously, did not affect tPA fibrinolysis, but decreased the incidence of hemorrhages, improved neurological outcome, and appeared to decrease mortality. In addition, minocycline decreased protein expression of MMP-2 and MMP-9 and these expression changes were associated with decreased degradation in collagen IV and laminin-α1. As both of these studies of minocycline—tPA combination therapy were limited to 24-h follow-up time, it remains to be investigated whether the benefit of minocycline administration to patients treated with tPA results in long-term or permanent improvement of stroke outcome. Nevertheless, these studies provide evidence that minocycline protects the BBB during thrombolysis with tPA and could extend the therapeutic time window for safe reperfusion therapy of acute stroke.

Minocycline Towards Clinical Use in Stroke

While the efficacy and neuroprotective potential of minocycline in acute ischemic stroke still needs to be established, this antibiotic has been used in clinical practice for many decades without serious safety concerns. The first study on the effect of minocycline treatment in human acute ischemic stroke was published by Lampl et al. [42]. In their open-label, evaluator-blinded study minocycline at a dose of 200 mg per day was administered orally for 5 days to 74 patients with the therapeutic time window of 6–24 h after onset of stroke. Seventy-seven patients received placebo. Patients were screened for any prior infections before recruitment for the study and no post-stroke complications were observed. In this study, NIH Stroke Scale (NIHSS) and modified Rankin Scale (mRS) scores that measure symptomatic impairment and the degree of disability or dependence in the daily activities, respectively, were significantly lower in minocycline group and Barthel Index (BI) scores that indicate a person's daily activities of daily living and mobility were significantly higher in minocycline-treated patients. These improvements were observed as early as 7 and 30 days during the 90-day follow-up period. On the other hand, deaths, myocardial infarctions, recurrent strokes, and hemorrhagic transformations during the follow-up did not differ between the treatment groups.

Another, smaller randomized single-blinded open-label study including 23 patients receiving oral minocycline 200 mg/day for 5 days and 27 control patients receiving oral vitamin B capsules was reported by Padma Srivastava et al. [72]. The clinical outcome was assessed using NIHSS, modified BI and mRS scores, as well as Magnetic Resonance Imaging (MRI) of the brain at days 1, 7, 30, and 90. Also, in this study, NIHSS score showed statistically significant improvement in patients receiving minocycline at days 30 and 90 as compared with the controls. Similarly, mRS scores and BI showed significant improvement in patients receiving minocycline at 3 months as compared to the control group. No mortality, myocardial infarctions, recurrent strokes, and hemorrhagic transformations were noted in either group. In spite of the small sample size, lack of stroke subtype classification, oral delivery instead of intravenous delivery, and unblinded nature of these studies, these pilot trials indicate tolerance of minocycline at the prescribed dosage and suggest favorable effects of minocycline treatment in acute stroke.

Based on the promising preclinical and pilot clinical studies of minocycline in acute stroke, Fagan et al. [73] evaluated its safety, tolerability, and pharmacokinetics in a dose escalation trial in patients with acute ischemic stroke. Minocycline was administered intravenously within 6 h of stroke symptom onset to 60 patients daily over 72 h. The results indicated that minocycline infusion is safe and well tolerated up to doses of 10 mg/kg alone and in combination with tPA. Intravenous minocycline at doses between 3 and 10 mg/kg daily achieved concentrations in the serum that have been shown to be neuroprotective in experimental stroke models. Because the half-life of minocycline was confirmed to be approximately 24 h in humans, dosing once a day would be sufficient and convenient treatment protocol even with the use of tPA. The encouraging results from these trials have led to the ongoing double-blind, multicenter Neuroprotection with Minocycline Therapy for Acute Stroke Recovery Trial (NeuMAST) sponsored by Singhealth Foundation.

Conclusion

Considering numerous cell culture studies and more than 40 studies on rodent models of stroke, minocycline is apparently one of the most promising drug candidates for treating patients with acute stroke. In addition, both preclinical and small open-label clinical studies indicate that minocycline treatment is safe when combined with tPA in the acute phase of the disease. One of the challenges in clinical development of neuroprotective agents for acute stroke is the lack of predictive, validated response biomarkers for evaluating the proof-of-relevance and therapeutic effects. While clinicians are capable of assessing stroke and reliably score recovery after stroke, identification of a biomarker that measures minocycline's effect on biological function responsible for its beneficial effect in acute stroke is difficult as long as the exact mechanism behind minocycline-mediated neuroprotection remains unknown. However, it is of great interest that plasma levels of MMP-9 and interleukin-6 were found to be decreased in stroke patients receiving minocycline [46]. MMP-9 is an important mediator of BBB disruption, edema, and hemorrhage in acute ischemic stroke. The expression of MMP-9 is increased after cerebral ischemia and its levels are further amplified by tPA treatment [74–77]. In addition, elevations in plasma MMP-9 correlate with stroke severity and are predictive of tPA-related intracerebral hemorrhage [78]. Considering that MMP-9 is a well-characterized biological target of minocycline, plasma levels of MMP-9 might well have value as a predictive response biomarker. Similar to MMP-9, plasma levels of IL-6 are elevated within hours following an acute stroke and remain upregulated up to 7 days. Importantly, increased IL-6 correlates with larger infarct volume, greater stroke severity, and worse clinical outcome. Preclinical studies have also demonstrated minocycline's ability to inhibit IL-6 production among its other anti-inflammatory effects. Whether the ongoing or future phase III trials of minocycline in acute stroke proves the drug to be beneficial or not, it is likely that minocycline remains as a prototype for further development of novel candidate molecules for treating acute stroke and other brain injuries.

References

1. Klein NC, Cunha BA (1995) Third-generation cephalosporins. Med Clin North Am 79(4):705–19
2. Aronson AL (1980) Pharmacotherapeutics of the newer tetracyclines. J Am Vet Med Assoc 176:1061–8
3. Zouboulis CC, Piquero-Martin J (2003) Update and future of systemic acne treatment. Dermatology 206(1):37–53
4. Furst DE (1998) Update on clinical trials in the rheumatic diseases. Curr Opin Rheumatol 10(2):123–8
5. Domercq M, Matute C (2004) Neuroprotection by tetracyclines. Trends Pharmacol Sci 25(12):609–12
6. Fagan SC, Cronic LE, Hess DC (2011) Minocycline development for acute ischemic stroke. Transl Stroke Res 2(2):202–8
7. Koistinaho J, Yrjänheikki J, Kauppinen T, Koistinaho M (2004) Tetracycline derivatives as anti-inflammatory agents and potential agents in stroke treatment. Ernst Schering Res Found Workshop 47:101–15
8. Chen-Roetling J, Chen L, Regan RF (2009) Minocycline attenuates iron neurotoxicity in cortical cell cultures. Biochem Biophys Res Commun 386(2):322–6

9. Zhao F, Hua Y, He Y, Keep RF, Xi G (2011) Minocycline-induced attenuation of iron overload and brain injury after experimental intracerebral hemorrhage. Stroke 42(12):3587–93
10. Chen X, Chen S, Jiang Y, Zhu C, Wu A, Ma X, et al. (2012) Minocycline reduces oxygen-glucose deprivation-induced PC12 cell cytotoxicity via matrix metalloproteinase-9, integrin β1 and phosphorylated Akt modulation. Neurol Sci. (Epub ahead of print)
11. Chu LS, Fang SH, Zhou Y, Yin YJ, Chen WY, Li JH, Sun J, Wang ML, Zhang WP, Wei EQ (2010) Minocycline inhibits 5-lipoxygenase expression and accelerates functional recovery in chronic phase of focal cerebral ischemia in rats. Life Sci 86(5–6):170–7
12. Huang TY, Chu HC, Lin YL, Lin CK, Hsieh TY, Chang WK et al (2009) Minocycline attenuates experimental colitis in mice by blocking expression of inducible nitric oxide synthase and matrix metalloproteinases. Toxicol Appl Pharmacol 237(1):69–82
13. Liu S, Zhen G, Meloni BP, Campbell K, Winn HR (2009) Rodent stroke model guidelines for preclinical stroke trials (1st edition). J Exp Stroke Transl Med 2(2):2–27
14. Machado LS, Kozak A, Ergul A, Hess DC, Borlongan CV, Fagan SC (2006) Delayed minocycline inhibits ischemia-activated matrix metalloproteinases 2 and 9 after experimental stroke. BMC Neurosci 7:56
15. Tikka TM, Koistinaho JE (2001) Minocycline provides neuroprotection against N-methyl-D-aspartate neurotoxicity by inhibiting microglia. J Immunol 166(12):7527–33
16. Tikka T, Fiebich BL, Goldsteins G, Keinanen R, Koistinaho J (2001) Minocycline, a tetracycline derivative, is neuroprotective against excitotoxicity by inhibiting activation and proliferation of microglia. J Neurosci 21(8):2580–8
17. Wilkins A, Nikodemova M, Compston A, Duncan I (2004) Minocycline attenuates nitric oxide-mediated neuronal and axonal destruction in vitro. Neuron Glia Biol 1(3):297–305
18. Yrjänheikki J, Tikka T, Keinänen R, Goldsteins G, Chan PH, Koistinaho J (1999) A tetracycline derivative, minocycline, reduces inflammation and protects against focal cerebral ischemia with a wide therapeutic window. Proc Natl Acad Sci USA 96(23):13496–500
19. Yrjänheikki J, Keinänen R, Pellikka M, Hökfelt T, Koistinaho J (1998) Tetracyclines inhibit microglial activation and are neuroprotective in global brain ischemia. Proc Natl Acad Sci USA 95(26):15769–74
20. Corsaro A, Thellung S, Chiovitti K, Villa V, Simi A, Raggi F et al (2009) Dual modulation of ERK1/2 and p38 MAP kinase activities induced by minocycline reverses the neurotoxic effects of the prion protein fragment 90-231. Neurotox Res 15(2):138–54
21. Nikodemova M, Duncan ID, Watters JJ (2006) Minocycline exerts inhibitory effects on multiple mitogen-activated protein kinases and IkappaBalpha degradation in a stimulus-specific manner in microglia. J Neurochem 96(2):314–23
22. Pi R, Li W, Lee NT, Chan HH, Pu Y, Chan LN et al (2004) Minocycline prevents glutamate-induced apoptosis of cerebellar granule neurons by differential regulation of p38 and Akt pathways. J Neurochem 91(5):1219–30
23. Won SJ, Kim JH, Yoo BH, Sohn M, Kauppinen TM, Park MS et al (2012) Prevention of hypoglycemia-induced neuronal death by minocycline. J Neuroinflammation 9:225
24. Brundula V, Rewcastle NB, Metz LM, Bernard CC, Yong VW (2002) Targeting leukocyte MMPs and transmigration: minocycline as a potential therapy for multiple sclerosis. Brain 125(Pt 6):1297–308
25. Enose-Akahata Y, Matsuura E, Tanaka Y, Oh U, Jacobson S (2012) Minocycline modulates antigen-specific CTL activity through inactivation of mononuclear phagocytes in patients with HTLV-I associated neurologic disease. Retrovirology 9:16
26. Kloppenburg M, Verweij CL, Miltenburg AM, Verhoeven AJ, Daha MR, Dijkmans BA, Breedveld FC (1995) The influence of tetracyclines on T cell activation. Clin Exp Immunol 102(3):635–41
27. Kloppenburg M, Brinkman BM, de Rooij-Dijk HH, Miltenburg AM, Daha MR, Breedveld FC, Dijkmans BA, Verweij C (1996) The tetracycline derivative minocycline differentially affects cytokine production by monocytes and T lymphocytes. Antimicrob Agents Chemother 40(4):934–40

28. Malm TM, Magga J, Kuh GF, Vatanen T, Koistinaho M, Koistinaho J (2008) Minocycline reduces engraftment and activation of bone marrow-derived cells but sustains their phagocytic activity in a mouse model of Alzheimer's disease. Glia 56(16):1767–79
29. Koistinaho M, Malm TM, Kettunen MI, Goldsteins G, Starckx S, Kauppinen RA, Opdenakker G, Koistinaho J (2005) Minocycline protects against permanent cerebral ischemia in wild type but not in matrix metalloprotease-9-deficient mice. J Cereb Blood Flow Metab 25(4):460–7
30. Arvin KL, Han BH, Du Y, Lin SZ, Paul SM, Holtzman DM (2002) Minocycline markedly protects the neonatal brain against hypoxic-ischemic injury. Ann Neurol 52(1):54–61
31. Liu Z, Fan Y, Won SJ, Neumann M, Hu D, Zhou L et al (2007) Chronic treatment with minocycline preserves adult new neurons and reduces functional impairment after focal cerebral ischemia. Stroke 38(1):146–52
32. Garcia-Martinez EM, Sanz-Blasco S, Karachitos A, Bandez MJ, Fernandez-Gomez FJ, Perez-Alvarez S et al (2010) Mitochondria and calcium flux as targets of neuroprotection caused by minocycline in cerebellar granule cells. Biochem Pharmacol 79(2):239–50
33. Lin S, Zhang Y, Dodel R, Farlow MR, Paul SM, Du Y (2001) Minocycline blocks nitric oxide-induced neurotoxicity by inhibition p38 MAP kinase in rat cerebellar granule neurons. Neurosci Lett 315(1–2):61–4
34. Wang CX, Yang T, Shuaib A (2003) Effects of minocycline alone and in combination with mild hypothermia in embolic stroke. Brain Res 963(1–2):327–9
35. Wang X, Zhu S, Drozda M, Zhang W, Stavrovskaya IG, Cattaneo E et al (2003) Minocycline inhibits caspase-independent and -dependent mitochondrial cell death pathways in models of Huntington's disease. Proc Natl Acad Sci USA 100(18):10483–7
36. Zhu S, Stavrovskaya IG, Drozda M, Kim BY, Ona V, Li M et al (2002) Minocycline inhibits cytochrome c release and delays progression of amyotrophic lateral sclerosis in mice. Nature 417(6884):74–8
37. Alano CC, Kauppinen TM, Valls AV, Swanson RA (2006) Minocycline inhibits poly(ADP-ribose) polymerase-1 at nanomolar concentrations. Proc Natl Acad Sci USA 103(25): 9685–90
38. Patriarca A, Polticelli F, Piro MC, Sinibaldi F, Mei G, Bari M et al (2012) Conversion of cytochrome c into a peroxidase: inhibitory mechanisms and implication for neurodegenerative diseases. Arch Biochem Biophys 522(1):62–9
39. Sakata H, Niizuma K, Yoshioka H, Kim GS, Jung JE, Katsu M et al (2012) Minocycline-preconditioned neural stem cells enhance neuroprotection after ischemic stroke in rats. J Neurosci 32(10):3462–73
40. Kraus RL, Pasieczny R, Lariosa-Willingham K, Turner MS, Jiang A, Trauger JW (2005) Antioxidant properties of minocycline: neuroprotection in an oxidative stress assay and direct radical-scavenging activity. J Neurochem 94:819–27
41. Morimoto N, Shimazawa M, Yamashima T, Nagai H, Hara H (2005) Minocycline inhibits oxidative stress and decreases in vitro and in vivo ischemic neuronal damage. Brain Res 1044(1):8–15
42. Lampl Y, Boaz M, Gilad R, Lorberboym M, Dabby R, Rapoport A et al (2007) Minocycline treatment in acute stroke: an open-label, evaluator-blinded study. Neurology 69(14):1404–10
43. Sutherland BA, Minnerup J, Balami JS, Arba F, Buchan AM, Kleinschnitz C (2012) Neuroprotection for ischaemic stroke: translation from the bench to the bedside. Int J Stroke 7(5):407–18
44. Zhang L, Zhang ZG, Chopp M (2012) The neurovascular unit and combination treatment strategies for stroke. Trends Pharmacol Sci 33(8):415–22
45. O'Collins VE, Macleod MR, Cox SF, Van Raay L, Aleksoska E, Donnan GA et al (2011) Preclinical drug evaluation for combination therapy in acute stroke using systematic review, meta-analysis, and subsequent experimental testing. J Cereb Blood Flow Metab 31(3): 962–75
46. Switzer JA, Hess DC, Ergul A, Waller JL, Machado LS, Portik-Dobos V et al (2011) Matrix metalloproteinase-9 in an exploratory trial of intravenous minocycline for acute ischemic stroke. Stroke 42(9):2633–5

47. Xu L, Fagan SC, Waller JL, Edwards D, Borlongan CV, Zheng J et al (2004) Low dose intravenous minocycline is neuroprotective after middle cerebral artery occlusion-reperfusion in rats. BMC Neurol 4:7
48. Hayakawa K, Mishima K, Nozako M, Hazekawa M, Mishima S, Fujioka M et al (2008) Delayed treatment with minocycline ameliorates neurologic impairment through activated microglia expressing a high-mobility group box1-inhibiting mechanism. Stroke 39(3):951–8
49. Hewlett KA, Corbett D (2006) Delayed minocycline treatment reduces long-term functional deficits and histological injury in a rodent model of focal ischemia. Neuroscience 141(1): 27–33
50. Dandona P, Aljada A, Bandyopadhyay A (2004) Inflammation: the link between insulin resistance, obesity and diabetes. Trends Immunol 25(1):4–7
51. Emsley HC, Hopkins SJ (2008) Acute ischaemic stroke and infection: recent and emerging concepts. Lancet Neurol 7(4):341–53
52. Ross R (1999) Atherosclerosis is an inflammatory disease. Am Heart J 138(5 Pt 2):S419–20
53. Savoia C, Schiffrin EL (2006) Inflammation in hypertension. Curr Opin Nephrol Hypertens 15(2):152–8
54. Alkayed NJ, Harukuni I, Kimes AS, London ED, Traystman RJ, Hurn PD (1998) Gender-linked brain injury in experimental stroke. Stroke 29(1):159–65
55. Alkayed NJ, Murphy SJ, Traystman RJ, Hurn PD, Miller VM (2000) Neuroprotective effects of female gonadal steroids in reproductively senescent female rats. Stroke 31(1):161–8
56. Liu F, McCullough LD (2011) Middle cerebral artery occlusion model in rodents: methods and potential pitfalls. J Biomed Biotechnol 2011:464701
57. Brait VH, Jackman KA, Walduck AK, Selemidis S, Diep H, Mast AE et al (2010) Mechanisms contributing to cerebral infarct size after stroke: gender, reperfusion, T lymphocytes, and Nox2-derived superoxide. J Cereb Blood Flow Metab 30(7):1306–17
58. Li J, McCullough LD (2009) Sex differences in minocycline-induced neuroprotection after experimental stroke. J Cereb Blood Flow Metab 29(4):670–4
59. Zeynalov E, Nemoto M, Hurn PD, Koehler RC, Bhardwaj A (2006) Neuroprotective effect of selective kappa opioid receptor agonist is gender specific and linked to reduced neuronal nitric oxide. J Cereb Blood Flow Metab 26(3):414–20
60. Hoda MN, Li W, Ahmad A, Ogbi S, Zemskova MA, Johnson MH et al (2011) Sex-independent neuroprotection with minocycline after experimental thromboembolic stroke. Exp Transl Stroke Med 3(1):16
61. Fisher M, Feuerstein G, Howells DW, Hurn PD, Kent TA, Savitz SI et al (2009) Update of the stroke therapy academic industry roundtable preclinical recommendations. Stroke 40(6): 2244–50
62. Guan W, Kozak A, Fagan SC (2011) Drug repurposing for vascular protection after acute ischemic stroke. Acta Neurochir Suppl 111:295–8
63. Bhatt LK, Addepalli V (2012) Potentiation of aspirin-induced cerebroprotection by minocycline: a therapeutic approach to attenuate exacerbation of transient focal cerebral ischaemia. Diab Vasc Dis Res 9(1):25–34
64. Elgebaly MM, Ogbi S, Li W, Mezzetti EM, Prakash R, Johnson MH et al (2011) Neurovascular injury in acute hyperglycemia and diabetes: a comparative analysis in experimental stroke. Transl Stroke Res 2(3):391–8
65. Kelly-Cobbs AI, Prakash R, Li W, Pillai B, Hafez S, Coucha M et al (2013) Targets of vascular protection in acute ischemic stroke differ in type 2 diabetes. Am J Physiol Heart Circ Physiol 304:H806–15
66. Fisher M, Albers GW (2013) Advanced imaging to extend the therapeutic time window of acute ischemic stroke. Ann Neurol 73(1):4–9
67. Lees KR, Bluhmki E, von Kummer R, Brott TG, Toni D, Grotta JJ et al (2010) Time to treatment with intravenous alteplase and outcome in stroke: an updated pooled analysis of ECASS, ATLANTIS, NINDS, and EPITHET trials. Lancet 375(9727):1695–703
68. Abu Fanne R, Nassar T, Yarovoi S, Rayan A, Lamensdorf I, Karakoveski M, Vadim P, Jammal M, Cines DB, Higazi AA (2010) Blood–brain barrier permeability and tPA-mediated neurotoxicity. Neuropharmacology 58(7):972–80

69. Lopez-Atalaya JP, Roussel BD, Levrat D, Parcq J, Nicole O, Hommet Y, Benchenane K, Castel H, Leprince J, Van To D, Bureau R, Rault S, Vaudry H, Petersen KU, Santos JS, Ali C, Vivien D (2008) Toward safer thrombolytic agents in stroke: molecular requirements for NMDA receptor-mediated neurotoxicity. J Cereb Blood Flow Metab 28(6):1212–21
70. Machado LS, Sazonova IY, Kozak A, Wiley DC, El-Remessy AB, Ergul A et al (2009) Minocycline and tissue-type plasminogen activator for stroke: assessment of interaction potential. Stroke 40(9):3028–33
71. Padma Srivastava MV, Bhasin A, Bhatia R, Garg A, Gaikwad S, Prasad K et al (2012) Efficacy of minocycline in acute ischemic stroke: a single-blinded, placebo-controlled trial. Neurol India 60(1):23–8
72. Fagan SC, Waller JL, Nichols FT, Edwards DJ, Pettigrew LC, Clark WM et al (2010) Minocycline to improve neurologic outcome in stroke (MINOS): a dose-finding study. Stroke 41(10):2283–7
73. Fukuda S, Fini CA, Mabuchi T, Koziol JA, Eggleston LL Jr, del Zoppo GJ (2004) Focal cerebral ischemia induces active proteases that degrade microvascular matrix. Stroke 35(4): 998–1004
74. Gasche Y, Fujimura M, Morita-Fujimura Y, Copin J-C, Kawase M, Massengale J et al (1999) Early appearance of activated matrix metalloproteinase-9 after focal cerebral ischemia in mice: a possible role in blood–brain barrier dysfunction. J Cereb Blood Flow Metab 19(9):1020–8
75. Heo JH, Lucero J, Abumiya T, Koziol JA, Copeland BR, del Zoppo GJ (1999) Matrix metalloproteinases increase very early during experimental focal cerebral ischemia. J Cereb Blood Flow Metab 19(6):624–33
76. Rosenberg GA, Navratil M, Barone F, Feuerstein G (1996) Proteolytic cascade enzymes increase in focal cerebral ischemia in rat. J Cereb Blood Flow Metab 16(3):360–6
77. Montaner J, Molina CA, Monasterio J, Abilleira S, Arenillas JF, Ribo M et al (2003) Matrix metalloproteinase-9 pretreatment level predicts intracranial hemorrhagic complications after thrombolysis in human stroke. Circulation 107(4):598–603
78. Murata Y, Rosell A, Scannevin RH, Rhodes KJ, Wang X, Lo EH (2008) Extension of the thrombolytic time window with minocycline in experimental stroke. Stroke 39(12):3372–7

Chapter 16
Tolerization to Brain and Vascular Antigens: Targeting Autoimmunity After Acute Brain Injuries and Preventing Stroke

Kyra J. Becker and John Hallenbeck

Abstract The immune response can contribute to the risk of stroke as well as to brain injury following stroke. Directed modulation of the immune response to attenuate risk or injury is thus a potential therapeutic strategy for treating stroke. Inducing a population of antigen-specific regulatory T cells is one such strategy and can be used to locally modulate the immune response to the organs in which the antigen is present, thereby limiting potential side effects. This chapter addresses the use of mucosal tolerance to both prevent and treat ischemic stroke.

Introduction

Following stroke there is compromise of the blood–brain barrier (BBB) that allows for cells of the peripheral immune system to enter brain [1–4]. In addition, this compromise of the BBB allows for dying brain cells to release their contents into the systemic circulation. For instance, proteins associated with astrocytes (S100, glial fibrillary acidic protein [GFAP]), oligodendrocytes (myelin basic protein [MBP]), and neurons (neuron-specific enolase [NSE]) are all found in the in the bloodstream after stroke [5, 6]. It is thus possible for central nervous system (CNS) antigens to be presented to lymphocytes in either the brain or in peripheral lymphoid organs following stroke. Indeed, there is evidence of CNS antigen presentation in the cervical lymph nodes within hours after experimental stroke and in the palatine tonsils of stroke patients within days after stroke onset [7, 8]. The possibility of developing autoimmune response to brain antigens thus exists, and the type of

K.J. Becker (✉)
Neurology and Neurological Surgery, University of Washington School of Medicine, Seattle, WA, USA
e-mail: kjb@u.washington.edu

J. Hallenbeck
Stroke Branch, NINDS, NIH, Bethesda, MD, USA

J. Chen et al. (eds.), *Immunological Mechanisms and Therapies in Brain Injuries and Stroke*, Springer Series in Translational Stroke Research, DOI 10.1007/978-1-4614-8915-3_16,

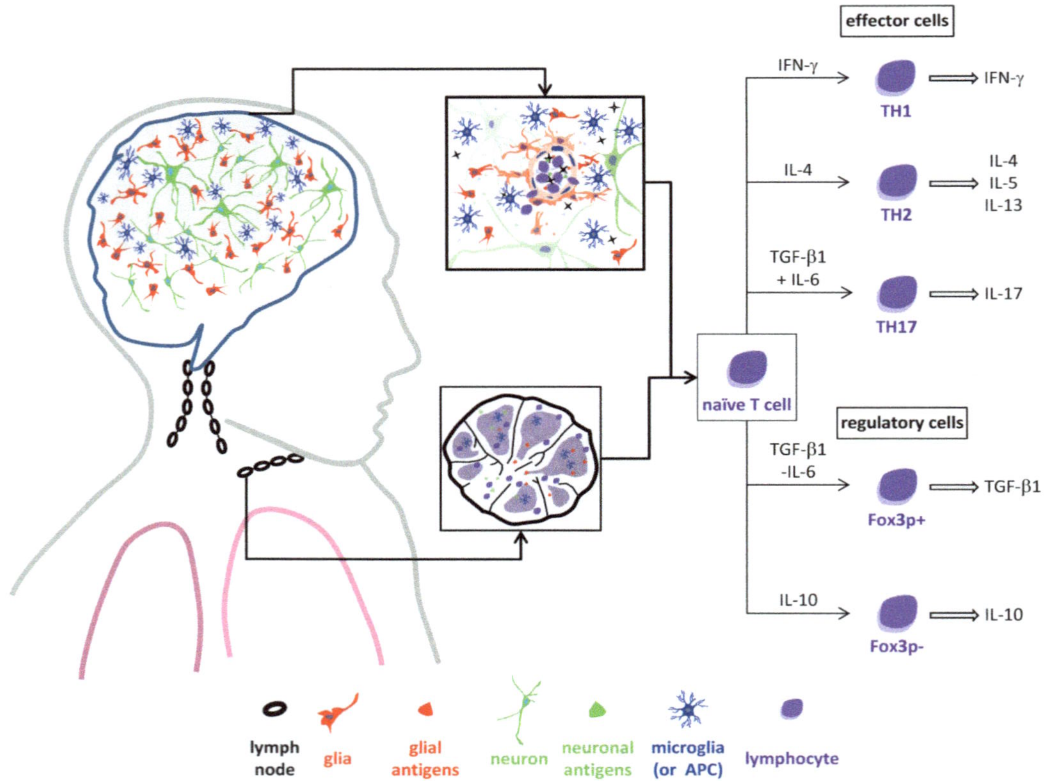

Fig. 16.1 Following stroke, the immune system can encounter novel CNS antigens both in the brain and in the periphery (including the lymph nodes). The type of T-cell response (effector or regulatory) that develops following successful antigen presentation depends on the microenvironment at the site of antigen encounter

immune response that develops to these antigens (TH1, TH2, TH17, or TREG) depends upon the microenvironment at the site of antigen presentation (Fig. 16.1).

That autoimmune responses to CNS antigens occur following stroke was first documented decades ago when patients with stroke were recruited as "other neurologic disease controls" for studies of classic autoimmune diseases like multiple sclerosis (MS) and acute inflammatory demyelinating polyneuropathy (AIDP) [9–11]. These initial studies focused on the cellular responses to MBP and "nervous tissue antigens" and found increased lymphocyte reactivity to these antigens among patients with a history of stroke. The possibility that these responses could contribute to ischemic brain injury or affect outcome from stroke was not considered at the time, and the fact that lymphocytes could have such profoundly different phenotypes was not yet appreciated.

In recent decades, however, there has been increased interest in the post-ischemic immune response with efforts to better understand the effects of the immune response on ischemic brain injury and how it can be manipulated to improve outcome. Importantly, it is now known that not only does the immune response contribute to ischemic brain injury but that stroke itself alters the systemic immune response. Through activation of the sympathetic nervous system, stroke induces dysfunction in the systemic immune response, and this sympathetically mediated immunodepression is associated with increased risk of infection [12, 13]. Given this immunodepression, it is not surprising that TH1 type immune responses to CNS antigens appear to be distinctly uncommon in both experimental models of stroke as well as in patients who experience stroke [14, 15]. Despite what appears to be a protective mechanism to prevent CNS autoimmunity after stroke, however, TH1 type immune responses to CNS antigens (MBP, PLP, and NSE) are seen in some animals after experimental stroke [16]. The predisposition to developing TH1 type immune responses to these antigens (and MBP in particular) can be increased by systemic administration of lipopolysaccharide (LPS). LPS, a component of the Gram-negative bacterial cell wall, contains a highly conserved "pathogen-associated molecular pattern" (PAMP) that potently initiates the innate immune response by stimulating dendritic/antigen-presenting cells (APCs) through toll-like receptor (TLR) 4. (Other PAMPS are able to initiate the innate immune response by stimulating other TLRs.) Similar to the increase in TH1 type immune responses seen in animals following treatment with LPS, the predisposition to developing TH1 responses to MBP in patients with stroke is increased in patients who develop systemic infections (especially severe infections like pneumonia) [17]. Notably, TH1 type immune responses to MBP are associated with worse outcome from experimental stroke [15, 16, 18, 19] as well as clinical stroke [17].

It takes time for a clinically meaningful immune response to develop to an antigen, as is illustrated by the delay in protection afforded by immunization. Generally, the antigen must be appropriately presented to lymphocytes and there must be persistence of the antigen to promote expansion of these lymphocytes to detectable levels. As might be expected, then, we were unable to detect a TH1 response to MBP until 1 month after experimental stroke [15]. And in patients, clinically important antigen-specific TH1 responses were not seen until 3 months after stroke onset [17]. In speaking about the contribution of *antigen-specific* immune responses to stroke outcome, it is thus the effect on long-term outcome that is of interest. For patients with a prior history of stroke, however, it is possible that TH1 responses induced by the previous stroke(s) could contribute to early ischemic brain injury following a recurrent stroke through the phenomenon of immunological memory.

The data thus suggest that stroke-induced immunodepression serves to limit the development of TH1 type immune responses to MBP (and other antigens) after stroke, but that induction of systemic inflammation by infection (or LPS) circumvents this immunodepressive effect and supports the development of autoimmunity. Similarly, onset of autoimmune diseases following infectious illnesses is commonly reported and often attributed to "bystander activation" of the immune response to self-antigens [20]. Given that infections occur in roughly 30% of patients who experience stroke, "bystander activation" of the immune response in patients with a compromised BBB and circulating brain antigens could contribute to the development of CNS autoimmunity [21].

Post-stroke infection is associated with an increased risk of death and disability at long-term follow-up [22–24]. Since stroke severity is the most important risk factor for post-stroke infection, it could be assumed that the relationship between infection and stroke outcome is confounded [25–27]. Patients with the most severe strokes are thus at the highest risk of infection and, subsequently, at the highest risk of developing autoimmune responses to CNS antigens exposed by the stroke. Multivariate models, however, show that the detrimental effect of infection (especially pneumonia) persists even after controlling for stroke severity and other prognostic variables [23, 27–29]. These autoimmune responses may explain (at least in part) how infection mediates a long-lasting detrimental effect on stroke outcome.

Despite the potent effect of infection (or infection mimics) on the development of TH1 responses to brain antigens after stroke, TH1 type immune responses to brain antigens are also seen in patients without infection (and in animals that do not receive LPS) [15, 17]. Apart from infection (and infection mimics), the major risk factor for developing such responses seems to be stroke severity/infarct volume [17]. Necrotic cells release molecules known as alarmins, and similar to LPS, alarmins are able to activate the innate immune response through stimulation of TLRs [30]. Common alarmins include heat shock proteins (HSPs), uric acid, and high-mobility group box protein (HMGB)-1. PAMPs and alarmins are collectively referred to as danger-associated molecular patterns (DAMPs). By initiation of the innate immune response, DAMPs increase the likelihood of successful antigen presentation; when these antigens are self-antigens, autoimmunity may develop [31, 32].

Prevention of Autoimmune Post-ischemic Autoimmune Responses

One strategy to prevent the development of post-ischemic autoimmune responses to brain would be to prevent infection. Animal studies suggest that prophylactic antibiotic therapy prevents infection and improves outcome after stroke [33].

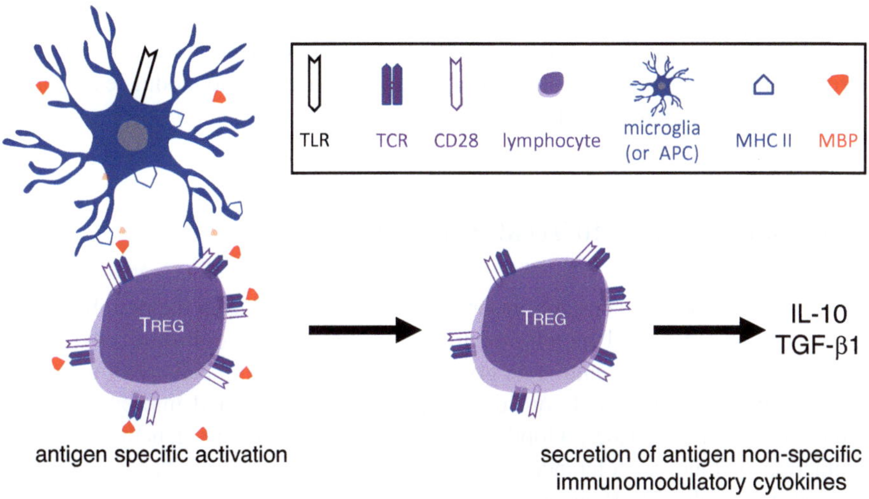

Fig. 16.2 TREG cells are activated in an antigen-specific fashion, but the cytokines they secrete (IL-10 and TGF-β1) act in an antigen-nonspecific fashion. This fact means that induction of TREG responses to a given antigen can also affect the immune response to antigens in the nearby vicinity, a phenomenon referred to as "bystander suppression"

There have been three prospective randomized controlled trials of prophylactic antibiotic therapy aimed at preventing post-stroke infection; these studies have had very mixed results [34–36]. Whether preventing infection with prophylactic antibiotics would prevent the development of CNS-specific immune responses after stroke is unknown. Another potential strategy to prevent the development of CNS autoimmunity following stroke is the use of immunomodulatory therapies. Many traditional approaches to immunomodulation increase the risk of infection under the best of circumstances; their use in a population prone to infection could be particularly dangerous. A more selective immunomodulatory strategy would thus be needed—one that could limit the development of CNS autoimmune responses without affecting the ability of the immune system to respond to infectious pathogens. One such selective approach would be to expand the compartment of *CNS antigen-specific* TREG cells. There are several different methods by which this expansion could be accomplished, including treatment with immunomodulatory peptides like α-melanocyte-stimulating hormone (MSH) [37]. Another method to expand the compartment of antigen-specific TREG cells is through the process of mucosal tolerance. By inducing TREG responses to MBP prior to experimental stroke (and treatment with LPS), we found that animals could be prevented from developing TH1 type immune response to MBP [18]. The concept behind induction of mucosal tolerance is that the presence of the antigen to which tolerance is induced will lead to activation of antigen-specific TREG cells that secrete immunomodulatory cytokines which lack antigen specificity (Fig. 16.2). Thus, wherever the antigen is present, there will be a local response that suppresses the activation of lymphocytes to that antigen as well as to neighboring/unrelated antigens ("bystander suppression") [38]. Induction of tolerance to MBP prior to stroke, for instance, could theoretically prevent the development of autoimmune responses to MBP and modulate responses to other CNS antigens as well.

Mucosal Tolerance to Treat Acute Stroke

The fact that CNS antigens are present in the brain as well as in the periphery after stroke can be capitalized upon for treatment of acute stroke. For instance, expansion of either MBP or oligodendrocyte glycoprotein (MOG) TREG cells prior to experimental stroke is associated with decreased infarct volume and improved clinical outcome [39–41]. Further, adoptive transfer of the TREG cells generated through induction of mucosal tolerance leads to similar neuroprotective benefits in naïve recipient animals [40]. Regulatory T cells secrete cytokines like TGF-β1 and IL-10, both of which are neuroprotective [42–47]. Thus, independent of the effect of TREGS in preventing the development of detrimental TH1 type responses after stroke, induction of TREG responses can be used to treat acute stroke. TREGS primed to CNS antigens will secrete TGF-β1 and IL-10 when encountering their cognate antigen, an effect which is most likely to happen in brain where the antigen concentration is the highest.

Mucosal Tolerance to Prevent Stroke

Exposure of the blood–endothelial interface of the CNS vascular tree to DAMPS can also participate in stroke initiation. The localized Shwartzman reaction is an intriguing model of focal blood vessel activation initially observed in rabbits [48]. In the *preparatory step* of the Shwartzman reaction, endotoxin, a PAMP, is injected intradermally where it acts through TLRs to release proinflammatory cytokines that activate skin blood vessels [49]. This process leads to local erythema that subsides unless there is further stimulation. If, however, there is a *provocative step* induced by a small nontoxic dose of intravenous endotoxin 18–24 h after the *preparatory step*, hemorrhagic necrosis develops in the *prepared* skin [50]. This dramatic effect is restricted to the *prepared* skin; all other tissues in the body appear to be spared. In rat models, established risk factors for stroke can act locally to prepare blood vessels in the brain for a "modified Shwartzman reaction" in response to a provocative inflammatory stimulus, such as endotoxin, injected intracisternally or intravenously [48]. Stroke risk factors such as hypertension, diabetes, advanced age, and genetic predisposition to stroke are also sufficient to prepare rat brainstem vessels so that a single injection of endotoxin can provoke the "modified Shwartzman" reaction [48]. Affected rats manifest neurologic deficits accompanied by pathologic lesions while brain infarcts develop in only a small proportion of risk factor-free rats. The fraction of animals that develop infarcts in response to the provocative step varies directly with the relative importance of the stroke risk factors. These data suggest that one role of stroke risk factors is to prime vessels for activation and promote endothelial dysfunction in conformity with the local Shwartzman reaction, but that a second time critical systemic activation of inflammation and/or hemostatic potential is necessary to precipitate local thrombosis or hemorrhage in the vessel.

Related studies show that Spontaneously Hypertensive, genetically Stroke Prone (SHR-SP) and Spontaneously Hypertensive (SHR) rats have elevated numbers of activated blood monocytes (as detected by nitroblue tetrazolium staining indicating the production of superoxide) when compared to normotensive Wistar-Kyoto (WKY) and Sprague-Dawley (SD) rats [51]. In addition, carotid arteries from the hypertensive strains show patchy endothelial expression of ICAM-1 with endothelium-adherent monocytes/macrophages, while those from normotensive strains do not. Further, incubation of carotid rings with endotoxin induces a concentration-dependent expression of interleukin (IL)-1β mRNA and release of tumor necrosis factor (TNF)-α to a significantly greater degree in SHR than WKY rats. Hypertension thus appears to be associated with activation of both monocytes and endothelium, increased monocyte/macrophage adhesion to endothelium, and an increased capacity for blood vessels to produce proinflammatory cytokines. In aggregate, these studies implicate immune and inflammatory mechanisms as integral to the stroke initiation process. Indeed, Rosenberg et al. characterized endothelium in discrete vascular segments within an organ and found it to be continuously integrating signals from the blood, blood vessel wall, and surrounding parenchymal tissue to modify the hemostatic potential of each vessel segment in a cyclic and asynchronous fashion [52].

Based on the foregoing findings and analyses, it was reasoned that because inflammatory and immune mediators drive cyclic changes of local hemostatic potential that stroke in stroke-prone individuals might also occur when a homeostatic threshold is exceeded and local thrombosis or hemorrhage occurs. If this hypothesis is true, immunomodulation of blood vessel activation might be an effective approach to stroke prevention. The challenge of this approach to stroke prevention is directing the immunomodulatory therapy to the activating blood vessels. As described above, induction of mucosal tolerance to a priming antigen could potentially be used to target these activating segments. The expression of E-selectin, for instance, is confined to the luminal surface of endothelium; it is not constitutively expressed but is upregulated in vessel segments that are becoming activated [53]. Using the paradigm of mucosal tolerance to induce E-selectin-specific Treg cells, a robust suppression of spontaneous strokes was found in stroke-prone rats (SHR-SP/Izm) [54]. In comparison to control groups, the SHR-SP rats repeatedly tolerized to recombinant human E-selectin (rhES) over the course of 56 weeks had a marked reduction in the incidence of ischemic strokes and the absence of parenchymal hemorrhages [54].

In addition to stroke prevention, induction of E-selectin Treg cells is also neuroprotective leading to smaller infarcts and better outcome following MCAO [55]. Further, E-selectin Treg cells reduce vascular cognitive impairment, prevent vasospasm following subarachnoid hemorrhage, suppress experimental autoimmune encephalomyelitis (EAE), and inhibit the development of atherosclerosis in ApoE null mice fed a hyperlipidemic diet (reviewed in [56]). Interestingly, post-ischemic adult neurogenesis is enhanced by repetitive intranasal instillation of E-selectin in SHR rats subjected to permanent middle cerebral artery occlusion [57]. On the basis of this laboratory work, a strategy for induction of TREG cells to E-selectin is being developed for prevention of recurrent strokes in patients at risk.

Potential Dangers of Mucosal Tolerance

The data suggest that induction of TREGs prior to stroke onset is associated with decreased infarct volume and improved neurological outcome immediately after stroke onset [39, 41, 55]. By preventing the induction of TH1 responses, improved outcomes are also seen 1 month after experimental stroke [18]. Additional follow-up, however, showed at least some of the animals undergoing induction of mucosal tolerance to MBP had developed TH1 responses to MBP at 3 months after experimental stroke [19]. As it turns out, there is not much that separates the requirements for induction of TREG cells and TH17 cells, the latter of which secrete IL-17 and are associated with autoimmune disease. TGF-β1 is needed for the development of both cell lines; in the presence of IL-6, lymphocytes will become TH17 cells, and in the absence of IL-6, they will become TREG cells (Fig. 16.1) [58]. Further, it is becoming increasingly apparent that the TREG phenotype is not stable, and cells originally committed to a TREG phenotype can convert to TH1 or TH17 phenotypes [59].

Efforts to move immunomodulatory therapies into clinical trials must thus address potential adverse effects (like those that might be related to unstable TREG cells) thorough careful preclinical studies. For instance, there are reports of mucosal administration of antigen-exacerbating inflammation in EAE and oral administration of insulin-inducing autoimmune diabetes [60, 61]. Mucosal administration of antigen can also promote a humoral response to the administered antigen, as seen in a nonhuman primate model of EAE where animals treated with soluble MOG developed detrimental antibodies to MOG [62]. Minimal requirements for preclinical studies include prolonged monitoring for evidence of rebound immune activation after termination of procedures to induce TREG responses (or after cessation of other immunomodulatory therapies). Further, the risks of long term immunosuppression, even if localized to the target organ, are largely unknown and a wide range of possible effects must be considered.

Human studies of mucosal tolerance have been conducted for treatment of a variety of autoimmune diseases, including multiple sclerosis, rheumatoid arthritis, autoimmune diabetes, uveitis, and inflammatory bowel disease [63–72]. The efficacy of mucosal tolerance for the diseases treated thus far is unclear, but the safety of the approach does not appear to be an issue in any of the studies. Based on these safety data and the extensive preclinical studies of E-selectin in a variety of stroke/ animal models, an NINDS workshop on immunomodulation supported the early translation of mucosal tolerance to E-selectin from animal to clinical studies (predicated on the successful completion of preclinical toxicology and immunotoxicology studies) [73]. The primary goals of the early human E-selectin studies are to determine the safety and appropriate dosing of E-selectin in humans and to explore potential biomarkers that would indicate success of the immunologic manipulation (i.e., induction of E-selectin Treg cells) as well as the success of the overall strategy for preventing/treating vascular disease and stroke. These initial studies will be small with very closely monitored. And importantly, as mentioned, these clinical studies will only occur after adequate preclinical toxicology and immunotoxicology have been completed. If unanticipated problems arise, the hope is to minimize the harm done and to not repeat the well-publicized disaster of the TGN1412 study. In the Phase I study of TGN1412, a novel antibody that acts as an agonist of CD28, all 6 treated patients rapidly developed a systemic inflammatory syndrome with multi-organ failure [74]. Because preclinical animal studies did not predict this systemic inflammatory response, it is important to consider species differences in the immune system and to be certain that initial human dosing is done one individual at a time to prevent multiple volunteers from being unnecessarily exposed to harm.

In addition to the "escape of tolerance" that can occur following induction of Treg cells, there are other potential detrimental long-term effects of protracted immunomodulation for which patients will need to be monitored closely. These long-term effects include susceptibility to infection and malignancies. While mucosal tolerance is not expected to have profound effects on systemic immunity, the experience with natalizumab, a monoclonal antibody directed towards the α subunit of very late antigen (VLA)-4 for the treatment of multiple sclerosis highlights the potential risks. This antibody blocks lymphocyte adhesion to and transit from the vasculature and is associated with an increased likelihood of developing progressive multifocal leukoencephalopathy (PML), a central nervous infection caused by the JC virus [75–77].

Conclusion

Induction of antigen-specific regulatory T-cell responses is an immunomodulatory strategy that can be used to both prevent and treat stroke. While these Treg cells are activated in an antigen-specific fashion, the cytokines secreted by these cells act in an antigen nonspecific fashion, modulation of the immune response is thus limited to wherever the relevant antigen is present, irrespective of whether that antigen provoked an immune response. Antigen-specific Treg cells can therefore be used to target immunomodulatory responses in a site and time-dependent manner, hopefully limiting potential side effects associated with systemic immunomodulatory therapies. The utility of expanding the Treg compartment with mucosal administration of antigen for prevention and treatment of stroke in humans is unknown, but preclinical data suggest reasons for optimism.

References

1. Jander S, Kraemer M, Schroeter M, Witte OW, Stoll G (1995) Lymphocytic infiltration and expression of intercellular adhesion molecule-1 in photochemically induced ischemia of the rat cortex. J Cereb Blood Flow Metab 15(1):42–51
2. Schroeter M, Jander S, Witte OW, Stoll G (1994) Local immune responses in the rat cerebral cortex after middle cerebral artery occlusion. J Neuroimmunol 55(2):195–203
3. Braun JS, Jander S, Schroeter M, Witte OW, Stoll G (1996) Spatiotemporal relationship of apoptotic cell death to lymphomonocytic infiltration in photochemically induced focal ischemia of the rat cerebral cortex. Acta Neuropathol (Berl) 92(3):255–63
4. Campanella M, Sciorati C, Tarozzo G, Beltramo M (2002) Flow cytometric analysis of inflammatory cells in ischemic rat brain. Stroke 33(2):586–92
5. Herrmann M, Vos P, Wunderlich MT, de Bruijn CH, Lamers KJ (2000) Release of glial tissue-specific proteins after acute stroke: a comparative analysis of serum concentrations of protein S-100B and glial fibrillary acidic protein. Stroke 31(11):2670–7
6. Jauch EC, Lindsell C, Broderick J, Fagan SC, Tilley BC, Levine SR (2006) Association of serial biochemical markers with acute ischemic stroke: the National Institute of Neurological Disorders and Stroke recombinant tissue plasminogen activator Stroke Study. Stroke 37(10): 2508–13
7. van Zwam M, Huizinga R, Melief MJ, Wierenga-Wolf AF, van Meurs M, Voerman JS et al (2009) Brain antigens in functionally distinct antigen-presenting cell populations in cervical lymph nodes in MS and EAE. J Mol Med 87(3):273–86
8. Planas AM, Gomez-Choco M, Urra X, Gorina R, Caballero M, Chamorro A (2012) Brain-derived antigens in lymphoid tissue of patients with acute stroke. J Immunol 188(5):2156–63
9. Youngchaiyud U, Coates AS, Whittingham S, Mackay IR (1974) Cellular-immune response to myelin protein: absence in multiple sclerosis and presence in cerebrovascular accidents. Aust N Z J Med 4(6):535–8
10. Kallen B, Nilsson O, Thelin C (1977) Effect of encephalitogenic protein on migration in agarose of leukoytes from patients with multiple sclerosis. A longitudinal study of patients with relapsing multiple sclerosis or with cerebral infarction. Acta Neurol Scand 55(1):47–56
11. Rocklin RE, Sheremata WA, Feldman RG, Kies MW, David JR (1971) The Guillain-Barre syndrome and multiple sclerosis. In vitro cellular responses to nervous-tissue antigens. N Engl J Med 284(15):803–8
12. Prass K, Meisel C, Hoflich C, Braun J, Halle E, Wolf T et al (2003) Stroke-induced immunodeficiency promotes spontaneous bacterial infections and is mediated by sympathetic activation reversal by poststroke T helper cell type 1-like immunostimulation. J Exp Med 198(5): 725–36
13. Wong CH, Jenne CN, Lee WY, Leger C, Kubes P (2011) Functional innervation of hepatic iNKT cells is immunosuppressive following stroke. Science 334(6052):101–5
14. Becker K (2012) Autoimmune responses to brain following stroke. Transl Stroke Res
15. Becker KJ, Kindrick DL, Lester MP, Shea C, Ye ZC (2005) Sensitization to brain antigens after stroke is augmented by lipopolysaccharide. J Cereb Blood Flow Metab 25(12):1634–44
16. Zierath D, Thullbery M, Hadwin J, Gee JM, Savos A, Kalil A et al (2010) CNS immune responses following experimental stroke. Neurocrit Care 12(2):274–84
17. Becker KJ, Kalil AJ, Tanzi P, Zierath DK, Savos AV, Gee JM et al (2011) Autoimmune responses to the brain after stroke are associated with worse outcome. Stroke 42:2763–9
18. Gee JM, Kalil A, Thullbery M, Becker KJ (2008) Induction of immunologic tolerance to myelin basic protein prevents central nervous system autoimmunity and improves outcome after stroke. Stroke 39(5):1575–82
19. Gee JM, Zierath D, Hadwin J, Savos A, Kalil A, Thullbery M et al (2009) Long term immunologic consequences of experimental stroke and mucosal tolerance. Exp Transl Stroke Med 1:3
20. Sfriso P, Ghirardello A, Botsios C, Tonon M, Zen M, Bassi N et al (2010) Infections and autoimmunity: the multifaceted relationship. J Leukoc Biol 87(3):385–95

21. Westendorp WF, Nederkoorn PJ, Vermeij JD, Dijkgraaf MG, van de Beek D (2011) Post-stroke infection: a systematic review and meta-analysis. BMC Neurol 11:110
22. Johnston KC, Li JY, Lyden PD, Hanson SK, Feasby TE, Adams RJ et al (1998) Medical and neurological complications of ischemic stroke: experience from the RANTTAS trial. RANTTAS Investigators. Stroke 29(2):447–53
23. Aslanyan S, Weir CJ, Diener HC, Kaste M, Lees KR (2004) Pneumonia and urinary tract infection after acute ischaemic stroke: a tertiary analysis of the GAIN International trial. Eur J Neurol 11(1):49–53
24. Katzan IL, Cebul RD, Husak SH, Dawson NV, Baker DW (2003) The effect of pneumonia on mortality among patients hospitalized for acute stroke. Neurology 60(4):620–5
25. Minnerup J, Wersching H, Brokinkel B, Dziewas R, Heuschmann PU, Nabavi DG et al (2010) The impact of lesion location and lesion size on poststroke infection frequency. J Neurol Neurosurg Psychiatr 81(2):198–202
26. Hug A, Dalpke A, Wieczorek N, Giese T, Lorenz A, Auffarth G et al (2009) Infarct volume is a major determiner of post-stroke immune cell function and susceptibility to infection. Stroke 40(10):3226–32
27. Tanzi P, Cain K, Kalil A, Zierath D, Savos A, Gee JM et al (2011) Post-stroke infection: a role for IL-1ra? Neurocrit Care 14(2):244–52
28. Hong KS, Kang DW, Koo JS, Yu KH, Han MK, Cho YJ et al (2008) Impact of neurological and medical complications on 3-month outcomes in acute ischaemic stroke. Eur J Neurol 15(12):1324–31
29. Vermeij FH, Scholte op Reimer WJ, de Man P, van Oostenbrugge RJ, Franke CL, de Jong G (2009) Stroke-associated infection is an independent risk factor for poor outcome after acute ischemic stroke: data from the Netherlands Stroke Survey. Cerebrovasc Dis 27(5):465–71
30. Oppenheim JJ, Tewary P, de la Rosa G, Yang D (2007) Alarmins initiate host defense. Adv Exp Med Biol 601:185–94
31. Yang D, de la Rosa G, Tewary P, Oppenheim JJ (2009) Alarmins link neutrophils and dendritic cells. Trends Immunol 30(11):531–7
32. Bianchi ME, Manfredi AA (2007) High-mobility group box 1 (HMGB1) protein at the crossroads between innate and adaptive immunity. Immunol Rev 220:35–46
33. Meisel C, Prass K, Braun J, Victorov I, Wolf T, Megow D et al (2004) Preventive antibacterial treatment improves the general medical and neurological outcome in a mouse model of stroke. Stroke 35(1):2–6
34. Chamorro A, Horcajada JP, Obach V, Vargas M, Revilla M, Torres F et al (2005) The early systemic prophylaxis of infection after stroke study: a randomized clinical trial. Stroke 36(7): 1495–500
35. Harms H, Prass K, Meisel C, Klehmet J, Rogge W, Drenckhahn C et al (2008) Preventive antibacterial therapy in acute ischemic stroke: a randomized controlled trial. PLoS One 3(5): e2158
36. Schwarz S, Al-Shajlawi F, Sick C, Meairs S, Hennerici MG (2008) Effects of prophylactic antibiotic therapy with mezlocillin plus sulbactam on the incidence and height of fever after severe acute ischemic stroke: the Mannheim infection in stroke study (MISS). Stroke 39(4):1220–7
37. Savos AV, Gee JM, Zierath D, Becker KJ (2011) alpha-MSH: A potential neuroprotective and immunomodulatory agent for the treatment of stroke. J Cereb Blood Flow Metab 31(2): 606–13
38. Miller A, Lider O, Weiner HL (1991) Antigen-driven bystander suppression after oral administration of antigens. J Exp Med 174(4):791–8
39. Becker KJ, McCarron RM, Ruetzler C, Laban O, Sternberg E, Flanders KC et al (1997) Immunologic tolerance to myelin basic protein decreases stroke size after transient focal cerebral ischemia. Proc Natl Acad Sci U S A 94(20):10873–8
40. Becker K, Kindrick D, McCarron R, Hallenbeck J, Winn R (2003) Adoptive transfer of myelin basic protein-tolerized splenocytes to naive animals reduces infarct size: a role for lymphocytes in ischemic brain injury? Stroke 34(7):1809–15

41. Frenkel D, Huang Z, Maron R, Koldzic DN, Moskowitz MA, Weiner HL (2005) Neuroprotection by IL-10-producing MOG CD4+ T cells following ischemic stroke. J Neurol Sci 233(1–2): 125–32
42. Spera PA, Ellison JA, Feuerstein GZ, Barone FC (1998) IL-10 reduces rat brain injury following focal stroke. Neurosci Lett 251(3):189–92
43. Grilli M, Barbieri I, Basudev H, Brusa R, Casati C, Lozza G et al (2000) Interleukin-10 modulates neuronal threshold of vulnerability to ischaemic damage. Eur J Neurosci 12(7):2265–72
44. Zhou Z, Peng X, Insolera R, Fink DJ, Mata M (2009) Interleukin-10 provides direct trophic support to neurons. J Neurochem 110(5):1617–27
45. Justicia C, Perez-Asensio FJ, Burguete MC, Salom JB, Planas AM (2001) Administration of transforming growth factor-alpha reduces infarct volume after transient focal cerebral ischemia in the rat. J Cereb Blood Flow Metab 21(9):1097–104
46. Henrich-Noack P, Prehn JH, Krieglstein J (1996) TGF-beta 1 protects hippocampal neurons against degeneration caused by transient global ischemia. Dose-response relationship and potential neuroprotective mechanisms. Stroke 27(9):1609–14, discussion 15
47. Ma M, Ma Y, Yi X, Guo R, Zhu W, Fan X et al (2008) Intranasal delivery of transforming growth factor-beta1 in mice after stroke reduces infarct volume and increases neurogenesis in the subventricular zone. BMC Neurosci 9:117
48. Hallenbeck JM, Dutka AJ, Kochanek PM, Siren A, Pezeshkpour GH, Feuerstein G (1988) Stroke risk factors prepare rat brainstem tissues for modified local Shwartzman reaction. Stroke 19(7):863–9
49. Akira S, Takeda K, Kaisho T (2001) Toll-like receptors: critical proteins linking innate and acquired immunity. Nat Immunol 2(8):675–80
50. Thomas L (1983) The youngest science: notes of a medicine-watcher. Viking, New York
51. Liu Y, Liu T, McCarron RM, Spatz M, Feuerstein G, Hallenbeck JM et al (1996) Evidence for activation of endothelium and monocytes in hypertensive rats. Am J Physiol 270(6 Pt 2): H2125–31
52. Rosenberg RD, Aird WC (1999) Vascular-bed-specific hemostasis and hypercoagulable states. N Engl J Med 340(20):1555–64
53. Bevilacqua MP, Stengelin S, Gimbrone MA Jr, Seed B (1989) Endothelial leukocyte adhesion molecule 1: an inducible receptor for neutrophils related to complement regulatory proteins and lectins. Science 243(4895):1160–5
54. Takeda H, Spatz M, Ruetzler C, McCarron R, Becker K, Hallenbeck J (2002) Induction of mucosal tolerance to E-selectin prevents ischemic and hemorrhagic stroke in spontaneously hypertensive genetically stroke-prone rats. Stroke 33(9):2156–63
55. Chen Y, Ruetzler C, Pandipati S, Spatz M, McCarron RM, Becker K et al (2003) Mucosal tolerance to E-selectin provides cell-mediated protection against ischemic brain injury. Proc Natl Acad Sci U S A 100(25):15107–12
56. Hallenbeck J (2010) How inflammation modulates central nervous system vessel activation and provides targets for intervention–a personal perspective. Ann N Y Acad Sci 1207:1–7
57. Ishibashi S, Maric D, Mou Y, Ohtani R, Ruetzler C, Hallenbeck JM (2009) Mucosal tolerance to E-selectin promotes the survival of newly generated neuroblasts via regulatory T-cell induction after stroke in spontaneously hypertensive rats. J Cereb Blood Flow Metab 29(3):606–20
58. Vila J, Isaacs JD, Anderson AE (2009) Regulatory T cells and autoimmunity. Curr Opin Hematol 16(4):274–9
59. Lee YK, Mukasa R, Hatton RD, Weaver CT (2009) Developmental plasticity of Th17 and Treg cells. Curr Opin Immunol 21(3):274–80
60. Xiao BG, Link H (1997) Mucosal tolerance: a two-edged sword to prevent and treat autoimmune diseases. Clin Immunol Immunopathol 85(2):119–28
61. Blanas E, Carbone FR, Allison J, Miller JF, Heath WR (1996) Induction of autoimmune diabetes by oral administration of autoantigen. Science 274(5293):1707–9
62. Genain CP, Abel K, Belmar N, Villinger F, Rosenberg DP, Linington C et al (1996) Late complications of immune deviation therapy in a nonhuman primate. Science 274(5295): 2054–7

63. Hafler DA, Kent SC, Pietrusewicz MJ, Khoury SJ, Weiner HL, Fukaura H (1997) Oral administration of myelin induces antigen-specific TGF-beta 1 secreting T cells in patients with multiple sclerosis. Ann N Y Acad Sci 835:120–31
64. Fukaura H, Kent SC, Pietrusewicz MJ, Khoury SJ, Weiner HL, Hafler DA (1996) Induction of circulating myelin basic protein and proteolipid protein-specific transforming growth factor-beta1-secreting Th3 T cells by oral administration of myelin in multiple sclerosis patients. J Clin Invest 98(1):70–7
65. Wei W, Zhang LL, Xu JH, Xiao F, Bao CD, Ni LQ et al (2009) A multicenter, double-blind, randomized, controlled phase III clinical trial of chicken type II collagen in rheumatoid arthritis. Arthritis Res Ther 11(6):R180
66. Choy EH, Scott DL, Kingsley GH, Thomas S, Murphy AG, Staines N et al (2001) Control of rheumatoid arthritis by oral tolerance. Arthritis Rheum 44(9):1993–7
67. Barnett ML, Kremer JM, St Clair EW, Clegg DO, Furst D, Weisman M et al (1998) Treatment of rheumatoid arthritis with oral type II collagen. Results of a multicenter, double-blind, placebo-controlled trial. Arthritis Rheum 41(2):290–7
68. Fourlanos S, Perry C, Gellert SA, Martinuzzi E, Mallone R, Butler J et al (2011) Evidence that nasal insulin induces immune tolerance to insulin in adults with autoimmune diabetes. Diabetes 60(4):1237–45
69. Nussenblatt R (2004) Orally and nasally induced tolerance studies in ocular inflammatory disease: guidance for future interventions. Ann N Y Acad Sci 1029:278–85
70. Kraus TA, Mayer L (2005) Oral tolerance and inflammatory bowel disease. Curr Opin Gastroenterol 21(6):692–6
71. Margalit M, Israeli E, Shibolet O, Zigmond E, Klein A, Hemed N et al (2006) A double-blind clinical trial for treatment of Crohn's disease by oral administration of Alequel, a mixture of autologous colon-extracted proteins: a patient-tailored approach. Am J Gastroenterol 101(3): 561–8
72. Ilan Y (2004) Oral immune regulation toward disease-associated antigens: results of phase I clinical trials in Crohn's disease and chronic hepatitis. Ann N Y Acad Sci 1029:286–98
73. Hallenbeck J, Del Zoppo G, Jacobs T, Hakim A, Goldman S, Utz U et al (2006) Immunomodulation strategies for preventing vascular disease of the brain and heart: workshop summary. Stroke 37(12):3035–42
74. Suntharalingam G, Perry MR, Ward S, Brett SJ, Castello-Cortes A, Brunner MD et al (2006) Cytokine storm in a phase 1 trial of the anti-CD28 monoclonal antibody TGN1412. N Engl J Med 355(10):1018–28
75. Kleinschmidt-DeMasters BK, Tyler KL (2005) Progressive multifocal leukoencephalopathy complicating treatment with natalizumab and interferon beta-1a for multiple sclerosis. N Engl J Med 353(4):369–74
76. Langer-Gould A, Atlas SW, Green AJ, Bollen AW, Pelletier D (2005) Progressive multifocal leukoencephalopathy in a patient treated with natalizumab. N Engl J Med 353(4):375–81
77. Clifford DB, De Luca A, Simpson DM, Arendt G, Giovannoni G, Nath A (2010) Natalizumab-associated progressive multifocal leukoencephalopathy in patients with multiple sclerosis: lessons from 28 cases. Lancet Neurol 9(4):438–46

Chapter 17
The Role of PPARγ in Stroke

Xiurong Zhao and Jaroslaw Aronowski

Abstract Over the last decade, the transcription factor PPARγ, previously known for its essential role in regulation of metabolic processes in adipose tissue, emerged as highly promising new target for the treatment of many neurological conditions, including ischemic and hemorrhagic stroke. Based on many cell culture and animal studies, activation of PPARγ was demonstrated to be associated with a broad range of biological effects (via genomic and non-genomic mode of action in virtually all brain cell types) which could effectively ameliorate pathogenic processes triggered by stroke, including inflammation, oxidative damage, edema, BBB preservation, and excitotoxicity, as well as help in the post-stroke recovery process by modulating the macrophage-mediated brain cleanup process. Some key aspects of PPARγ as target for stroke treatment are reviewed in this chapter.

Introduction

The peroxisome proliferator-activated receptors (PPARs), including α, γ, and δ/β, are encoded by separate genes and are members of the nuclear hormone receptor superfamily of ligand-activated nuclear transcription factors. *PPARγ*, also known as NR1C3 (nuclear receptor subfamily 1, group C, member 3), is a pleiotropic type II nuclear receptor, which was termed for its ability to induce proliferation of hepatic peroxisomes in response to xenobiotic stimuli in mice [1]. Three different PPARγ transcripts (PPARγ 1, 2, and 3), each a derivative of the PPARγ gene through differential promoter usage and alternative splicing, have been identified [2, 3].

X. Zhao
Department of Neurology, The University of Texas Medical School at Houston,
Houston, TX, USA

J. Aronowski (✉)
The University of Texas Medical School at Houston, Houston, TX, USA
e-mail: J.Aronowski@uth.tmc.edu

J. Chen et al. (eds.), *Immunological Mechanisms and Therapies in Brain Injuries and Stroke*, Springer Series in Translational Stroke Research, DOI 10.1007/978-1-4614-8915-3_17,

While PPARγ 2 is the form primarily expressed in adipose tissue, PPARγ 1 has broader tissue distribution including presence in the brain [2]. As a transcription factor that regulates target gene expression through binding to the conserved DNA sequence termed *peroxisome-proliferator response element* (*PPRE*) [2, 4, 5], PPARγ was initially described in adipose tissue as a key regulator of metabolic processes [6–9]. Soon after, PPARγ was shown to be a unique therapeutic target for the treatment of metabolic disorders, e.g., diabetes (insulin resistance), obesity, hyperlipidemia, and hyperglycemia [10–12]. Among many compounds, ligands for PPARγ activation include fatty acids (especially the oxidized form) [13–15], cyclopentanone prostaglandins (e.g., 15-deoxy-Δ12,14-prostaglandin J_2; -15d-PGJ_2) [16], lipoxygenase products [17, 18], the nonsteroidal anti-inflammatory drugs (NSAIDs) [19, 20], and a class of clinically relevant compounds, the thiazolidinediones (*TZDs*) [10, 21]; of which pioglitazone and rosiglitazone are used to treat the type 2 diabetes mellitus [22–25]. In addition, PPARγ transactivation is regulated by its phosphorylation [26, 27]. Specifically, phosphorylation of PPARγ by the extracellular signal-regulated kinase (*ERK1/2*) and C-Jun N-terminal kinase (*JNK*) reduces PPARγ activity [26, 27]. Since JNK is activated by H_2O_2, oxygen–glucose deprivation (*OGD*), NMDA or ischemic stroke and acts as pro-death signal [28–32], the deleterious JNK functions may be secondary to the phosphorylation-mediated PPARγ inhibition.

Later studies on the mechanism of PPARγ action in other than fat tissue demonstrated its important role in regulation of anti-oxidative and anti-inflammatory processes [33–35]. It is primarily the anti-inflammatory properties of PPARγ ligands that ultimately brought the closer attention to PPARγ and PPARγ-activating agents to vascular diseases process [36–38]. PPARγ (and primarily PPARγ1) expression is ubiquitous regarding the type of tissues and cells it is expressed. In terms of neurological conditions, PPARγ in preclinical studies was shown to act as potential target for the treatment of ischemic stroke [39–51], intracerebral hemorrhage [52], neurotrauma [53–58], Alzheimer's and neurodegenerative diseases [59–69], autoimmune encephalomyelitis (*EAE*), a model for multiple sclerosis [70–72]. In this chapter, our focus is mainly on the role of PPARγ in ischemic stroke, attempting to discuss the interactions of PPARγ with the NF-E2-related factor 2 (*Nrf2*) and the nuclear factor kappa B (*NF-κB*) signaling pathways in regulating pro- and anti-inflammatory responses in the brain.

Pleiotropic Effect of PPARγ Agonists in Ischemic Stroke

Based on the known function of gene targets, PPARγ acts as a key regulator in a broad range of processes virtually in all brain cells including neurons [45, 73], astroglia [74–76], oligodendroglia [77–79], microglia [54, 80, 81], and endothelial cells [82, 83]. Primarily through the use of various PPARγ agonists but also through the use of cell-specific PPARγ knockouts, PPARγ was demonstrated to protect brain from

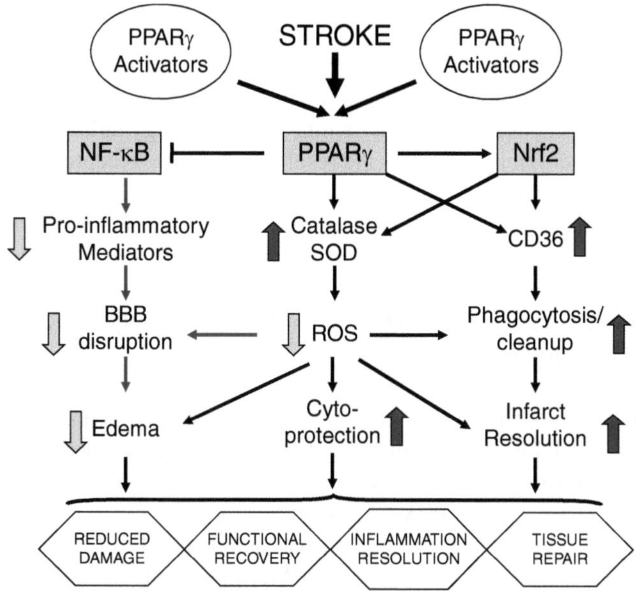

Fig. 17.1 PPARγ regulated pathways after stroke—role of PPARγ activators. PPARγ transcriptionally controls expression of numerous genes including the anti-oxidative enzymes, such as catalase and superoxide dismutase (SOD), as well as the transcription factor Nrf2. Nrf2 plays a key role in amplifying the expression of many anti-oxidative genes including catalase and SOD, similar to PPARγ. This anti-oxidative feature of PPARγ is critical in combating oxidative damage imposed by cerebral ischemia. Importantly, since PPARγ and Nrf2 are ubiquitously expressed, this anti-oxidative mechanism may apply to all brain cell types affected by stroke. In addition, both PPARγ and Nrf2 regulate expression of CD36, a scavenger receptor that is abundant on microglia/macrophages. CD36 plays important role in endocytosis of oxidized lipids and phagocytosis of dead (including apoptotic) cells and other cellular debris, thereof aiding in cleanup—process allowing for a faster inflammation resolution and more efficient tissue repair. Another important task of PPARγ is to inhibit NF-κB, a proinflammatory transcription factor implicated in BBB disruption and brain edema formation. Ultimately, augmented PPARγ activation improves inflammation resolution, tissue repair, and functional recovery after stroke

damages caused by ischemic [41, 42, 45, 84–87] and hemorrhagic stroke [53–55]. The beneficial effects of PPARγ activation was linked to (1) repression of pro-inflammatory mediators production (at least in part through inhibition of NF-κB either directly or by upregulation of endogenous NF-κB inhibitor, IκB [33, 34, 53, 88–95]), (2) upregulation of antioxidant enzymes including CuZn-superoxide dismutase (*SOD*) and catalase [41, 54], (3) inhibition of excitotoxicity [96, 97], and (4) activation of phagocytotic activities by microglia and macrophages via mechanism involving the PPARγ-target gene—scavenger receptor CD36, the molecule that assists in cleanup of damaged brain tissue, a process necessary for efficient recovery and the termination of deleterious pro-inflammatory cascade (Fig. 17.1) [54, 98–101].

PPARγ and Neuroprotection

In response to the prolonged ischemia, neurons that are localized in the ischemic core die rapidly as consequence of ischemia-induced energy failure, anoxic depolarization, and excitotoxicity, which is the result of glutamate receptors overactivation, calcium overload, and a breakdown of ion homeostasis [102–109]. Using oxygen–glucose deprivation (OGD) or glutamate/NMDA toxicity (in vitro models of ischemia) to study the neuroprotective capacity of PPARγ agonists [including pioglitazone, rosiglitazone, or cyclopentanone prostaglandins (*CyPG*)], we and other groups demonstrated that activation of PPARγ potently reduces the neuronal death in the primary neurons [96, 97, 110], implying that PPARγ may act as pro-survival factor for neurons under the ischemic/excitotoxic stress. The anti-excitotoxic effect of PPARγ agonists was observed not only in cultured neurons but also in the animal injury model that assess the extent of brain damage caused by intracortical injection of NMDA [97]. Finally, we have established that neurons derived from animals engineered to lack PPARγ, selectively in neurons, demonstrated significantly increased susceptibility to excitotoxic damage and to OGD [84]. In agreement with the in vitro data, mice lacking PPARγ in neurons were significantly more susceptible to the ischemic damage caused by focal cerebral ischemia [84].

Reactive oxygen species (*ROS*) are well known to represent one of the most important components of brain injure in response to ischemia/reperfusion insult. ROS are generated by the ischemia-affected brain cells, the activated microglia, and infiltrating neutrophils that collectively impose oxidative stress to cells located in proximity to the ischemia [111–114]. To combat the oxidative stress, cells have developed a number of self-defense mechanisms including upregulation of enzymes with anti-oxidative functions. Superoxide dismutase along with catalase and glutathione peroxidase plays key roles in eliminating ROS through catalytic decomposition of superoxide or H_2O_2 [84, 115, 116]. Catalase is a large homotetrameric protein that is usually localized in peroxisomes (the membrane-bound organelles that house β-oxidation of very long chains of fatty acids, in which toxic peroxides are generated as side products) [117], where it acts to protect the cells from the toxic effects of H_2O_2 by catalyzing its decomposition. As a ubiquitous enzyme to most cells in our body including neuroglia and neurons [118], catalase expression is regulated by PPARγ and Nrf2 [115, 119]. The distribution pattern of catalase-immunopositive neurons throughout the brain inversely corresponds to increased susceptibility to damage induced by global cerebral ischemia [118], suggesting that catalase plays important role in cell survival. Overexpression of catalase in rat striatum through virus-mediated gene transfer decreases the vulnerability to ischemic stroke [120]. In response to PPARγ activation, expression of catalase rapidly increased in the ischemia-affected brain [118, 121] and in the OGD-injured neurons [122], which likely reflect an adaptive response aiming at improving the antioxidant buffering capacity under the pathological scenarios. In agreement with this notion, treatment with catalase of neurons in culture subjected to H_2O_2-induced injury provided a robust cytoprotection [123, 124]. Thus, catalase upregulation by PPARγ may reflect a self-protective mechanism to combat oxidative stress in stroke. It is important to point out that in addition to catalase, PPARγ regulates expression of superoxide dismutase (including in neurons), an enzyme well recognized for decades as a key player in mitigating oxidative injury and brain damage after cerebral ischemia [41, 84, 125, 126].

PPARγ-Induced CD36 Expression on Phagocytes and the Endogenous Cleanup Mechanism

After cerebral ischemia, the infarcted/dead tissue not only acts as a reservoir of various cytotoxic and pro-inflammatory molecules that harm the adjacent healthy brain tissue, but it also forms a biological and physical barrier hampering neural reorganization, repair, and ultimately, neurological recovery. Thus, in order to minimize such detrimental effects, infarcted tissue needs to be removed to facilitate recovery. Microglia and hematogenous macrophages (*MMΦ*) are the cells primarily responsible for such cleanup and repair processes. Successful removal of the disintegrated and apoptotic brain cells or debris (including the neutrophils that accumulate in brain in response to injury and consequently die through apoptosis) by MMΦ is also essential in achieving resolution of inflammation. While apoptotic cells appear to be considerably benign to the surrounding brain tissue, an apoptotic cells non-phagocytosed in a timely manner may undergo secondary necrosis causing spill of the intracellular toxic content, leading to the damage to the neighboring cells and causing inflammation. Several macrophage scavenger receptors that mediate cleanup process have been identified. These include not only CD36 but also CD91, SR-A, and several others [54, 127–133]. Regarding apoptotic cell efferocytosis by macrophages, the phosphatidyl serine on the sickle red blood cells, symmetric red cell ghosts [134–136], or apoptotic neutrophils was suggested to act as the recognition molecule for CD36, a class II scavenger receptor on macrophages [137–139]. Expression of CD36 on macrophages (*MΦ*) is transcriptionally regulated by both PPARγ [98, 140, 141] and Nrf2 [142–145]. Although CD36 has various functions, one of its primary roles is to mediate endocytosis of (oxidized) fatty acids and phagocytosis of dead/apoptotic cells [129, 137, 146–148]. Deficiency of CD36 in macrophages due to genetic deletion of PPARγ leads to delayed uptake of oxidized LDL by macrophages and aggravation of atherosclerotic lesions [149]. In CD36-KO mice, aberrant phagocytotic capacity of macrophages was proposed to explain the deficiency in remyelination in response to sciatic nerve crush injury [150]. In addition, transfection of non-phagocytic cells with CD36 renders these cells capable of ingesting apoptotic neutrophils, lymphocytes, and fibroblasts [138], further confirming the important role of CD36 in phagocytosis. As pointed above, since CD36 transcription is under control of Nrf2 and PPARγ, the upregulation of CD36 by MMΦ in response to Nrf2 and/or PPARγ activators may ensure a more efficient interaction between the MMΦ and their targets for phagocytosis. This may allow for more efficient phagocytosis-mediated clearance of dead cells/tissues from the ischemic brain. However, despite its beneficial role in the cleanup process, CD36 may have detrimental effect which is normally characterized by increased oxidative stress and pro-inflammatory responses, as adult animals deficient in CD36 suffer from the less profound damage in response to cerebral ischemia [151, 152]. The nature of these responses is not known; however, the likelihood is that upon engulfment of cellular debris including oxidized lipids, the MMΦ generate damaging levels of oxidative stress during degradation of debris in the phagolysosomes. Interestingly, CD36 knockout neonates subjected to cerebral ischemia experienced more damage (suggesting beneficial function of CD36), which was suggested to be in part due to the impaired cleanup mechanism [153]. Independent of the natural responses that were tested in experiments using CD36 knockout mice, we suggest that under conditions using pharmacologic agents to activate PPARγ, MMΦ not only express higher levels of CD36 for a more efficient phagocytosis but also produce more anti-oxidative enzymes (e.g., catalase) that are regulated by PPARγ. Recently, we provided the evidence that MMΦ in culture challenged with PPARγ or Nrf2 activators, despite expressing CD36 at much higher level and demonstrating the augmented phagocytosis, experienced less oxidative damage and showed reduced pro-inflammatory gene expression [54].

Thus, in response to PPARγ in activated microglia, the upregulation of the antioxidant enzymes (in addition to CD36) may play a protective role allowing for effective and safe phagocytosis. Consequently, cleaning the apoptotic/dislocated/damaged cells or debris will help to reestablish the nurturing environment necessary for restoring tissue structure and neurological function recovery [154, 155].

PPARγ Activation and the Interaction of PPARγ and RXR

PPARγ regulates target gene expression by binding to PPRE as heterodimers with the retinoic acid receptor (*RXR*). Interestingly, existing studies indicate that activation of PPARγ–RXR complex can be achieved with either PPARγ and/or by RXR ligand (e.g., 9-*cis* retinoic acid), indicating some level of the promiscuity in activation of PPARγ [156, 157]. Although each ligand can initiate transactivation independently, the effect of co-activation appears to be stronger [9], suggesting that the occupancy of both PPARγ and RXR ligand (e.g., 15d-PGJ_2 plus 9-*cis* retinoic acid) is needed for the maximal receptor activity [9, 158–160]. In agreement with this notion, we found that co-treatment of cultured neurons with 15d-PGJ_2 and 9-*cis* retinoic acid was more effective in reducing the OGD-induced damage, as compared to each ligand alone [53]. This beneficial interaction between PPARγ and RXR ligands in our neuroprotection assay is consistent with an earlier report showing that combination use of 15d-PGJ_2 and 9-*cis* retinoic acid was superior to each drug alone in reducing behavioral dysfunction in a mouse model of experimental autoimmune encephalomyelitis [161].

Interaction of PPARγ and Nrf2 and NF-κB

The pro-survival role of PPARγ includes the non-genomic inhibition of deleterious pro-inflammatory transcription factor, nuclear factor kappa B, NF-κB. In the ischemia-injured brain, the delayed cell death is in part triggered by the overproduction of pro-inflammatory molecules including pro-inflammatory cytokines (such as tumor necrosis factor alpha, *TNF-α* or interleukin-1 beta, *IL-1β*), adhesion molecules (such as intercellular adhesion molecule 1, *ICAM-1* or vascular cell adhesion molecule, *VCAM*), matrix metalloproteinases (including *MMP9*) or the pro-oxidative inducible form of nitric oxide synthase (*iNOS*) capable of generating large quantities of nitric oxide, that in presence of superoxide generated by NADPH oxidase is converted to a highly cytotoxic peroxynitrites [109, 162–165]. Once perpetuated by ischemia, these potentially deleterious factors act in concert to damage blood–brain barrier (BBB) and cause edema and/or hemorrhage [166–168]. Interestingly, the expression of all these factors is tightly regulated by NF-κB. The activation of PPARγ can antagonize these harmful effects through inhibition of NF-κB [33, 34], which may be achieved by at least three independent mechanisms (Fig. 17.1) [33, 34, 53, 88–95]. First, PPARγ may directly bind to the NF-κB subunits, p50 and p65, resulting in NF-κB inactivation [169]; second, PPARγ may indirectly inhibit NF-κB by sequestering the common transcription co-activators such as SRC-1 [170] and p300/CBP (CREB-binding protein) [88–90]; and third, PPARγ may upregulate the production of inhibitor kappa B (*IκB*) [91, 93–95], the protein that directly inhibit NF-κB activation. Inhibition of NF-κB by PPARγ agonists may reduce generation of pro-inflammatory mediators involved in the secondary brain damage.

Nrf2 is a ubiquitous pleiotropic transcription factor and a key genomic homeostatic regulator of intracellular stress [171]. By combining with Mif family proteins, Nrf2 forms heterodimeric complexes capable of transactivating the antioxidant response elements *(ARE)* within the regulatory region of many cytoprotective target genes including catalase, superoxide dismutase, glutathione-*S*-transferase, thioredoxin, NQO1, and many other proteins with important role in neutralization of oxidative stress and detoxification [172]. In most cells, Nrf2 is present at low concentrations due to continuous Nrf2 degradation through the proteasome pathway [173, 174]. Nrf2 contributes to cytoprotection and amelioration of tissue damage through reducing the oxidative stress in many pathogenic conditions including cerebral ischemia [175–181], neurodegenerative diseases [182], and mitochondrial metabolic stress [183]. The growing body of evidence suggests that PPARγ may play important role in regulation of Nrf2 and thus Nrf2 target genes (Fig. 17.1). The interaction between PPARγ and Nrf2 may involve several layers of interaction. Most importantly, PPARγ was demonstrated to regulate Nrf2 gene expression and Nrf2-regulated genes containing putative PPREs [184]. Interestingly, it appears that Nrf2 also regulates PPARγ and PPARγ-regulated genes containing the ARE [185]. Next, PPRE and ARE coexist in the same genes, such as CD36 and catalase, suggesting an interactive function of Nrf2 and PPARγ in expression of these genes.

Finally, an interaction between PPARγ and Nrf2 may be through NF-κB inhibition. Since NF-κB activation requires the presence of oxidative stress [186], the effect of Nrf2 in ameliorating oxidative stress was proposed to inhibit NF-κB [187]. As different mechanisms are used by Nrf2 and PPARγ in inhibiting NF-κB, it is likely that the mutual effect may lead to a synergistic role [188–190].

Adverse Effects of PPARγ Agonists

There is a small number of observations reporting the dose-dependent neurotoxic effects of the endogenous PPARγ ligand 15d-PGJ_2 in cerebellar granule cells [191], primary cortical neurons [192], and spinal cord motor neurons [193]. The mechanism that underlies this neurotoxicity is unclear and some reports indicate that these harmful actions are probably not directly linked to PPARγ [191]. In our studies using mouse and rat neurons in culture, we have not observed neurotoxicity using PPARγ activating ligands to date. In fact, all the tested PPARγ agonists including 15d-PGJ_2, 15d-PGD_2, ciglitazone, rosiglitazone, and pioglitazone demonstrated potent cytoprotective effects in models of OGD and excitotoxicity [45, 50, 97]. The only instance showing toxicity was when the doses of the agonists were higher than these needed for the cytoprotection. Unlike synthetic TZDs that display rather significant levels of PPARγ specificity, prostaglandin D_2 derivatives, including 15d-PGJ2, have a limited selectivity toward PPARγ and many of their biological activities are independent of PPARγ [92, 194–198]. However, the clinical use of PPARγ ligands, and primarily rosiglitazone, was associated with hemodilution, peripheral edema, increase in body weight, as well as cardiomyopathies and heart failure [46, 199–201]. Again, these are the known side effects of long-term use of these medications and as such should not necessarily influence the safety of patients subjected to short-term treatment. The study evaluating the safety of pioglitazone in patients with hemorrhagic stroke is currently ongoing [52].

PPARγ Agonists and Clinical Trials

Two of the thiazolidinediones (TZDs), pioglitazone and rosiglitazone, are currently approved by the FDA for treatment of type 2 diabetes mellitus. These insulin-sensitizing PPARγ agonists are unique among all the glucose-lowering agents as they act independent of secretion of insulin from pancreas (TZDs do not change blood insulin levels, rather make cells more sensitive to its effect) [22, 202]. The glucose-lowering effect of TZDs is of clinical importance since hyperglycemia during ischemia/reperfusion may worsens the brain damage and neurological outcome, including by increasing incidence of hemorrhage in patients subjected to thrombolysis with rt-PA [203–206]. A first case-matched controlled study reporting improved functional recovery in stroke patients with type 2 diabetes receiving pioglitazone or rosiglitazone (vs. control type 2 diabetes patients not receiving TZDs) yields a promising outlook [207]. Subsequently, PROACTIVE (*PRO*spective pioglit*A*zone *C*linical *T*rial *I*n macro*V*ascular *E*vents; NCT00174993), a randomized, double-blinded, placebo-controlled study looked at the impact of pioglitazone on total mortality and macrovascular morbidity in 5,238 patients with diabetes and macrovascular disease. This secondary prevention study showed safety and a macrovascular benefit with pioglitazone in terms of major adverse cardiovascular events including all-cause mortality, nonfatal myocardial infarction, acute coronary syndrome, cardiac intervention (including coronary artery bypass graft or percutaneous coronary intervention), and stroke [208–210]. The higher beneficial rates were observed in patients with prior stroke compared with those without prior stroke [211, 212]. A meta-analysis of 19 randomized clinical trials with pioglitazone revealed a statistical difference regarding the favorable outcome including mortality, nonfatal MI, and stroke when using pioglitazone [201]. However, a recent study suggests that use of rosiglitazone may impose 1.4-fold increase in risk of acute MI and death from cardiovascular diseases compared with non-TZDs therapies [213]. As compared to pioglitazone, rosiglitazone significantly increased the risk of stroke, heart failure, and death in elderly patients [214]. In contrast, from the stroke prevention point, pioglitazone has shown significant protection from both micro- and macrovascular cardiovascular events and plaque progression [215–217].

References

1. Dreyer C, Keller H, Mahfoudi A, Laudet V, Krey G, Wahli W (1993) Positive regulation of the peroxisomal beta-oxidation pathway by fatty acids through activation of peroxisome proliferator-activated receptors (PPAR). Biol Cell 77:67–76
2. Elbrecht A, Chen Y, Cullinan CA, Hayes N, Leibowitz M, Moller DE et al (1996) Molecular cloning, expression and characterization of human peroxisome proliferator activated receptors gamma 1 and gamma 2. Biochem Biophys Res Commun 224:431–437
3. Fajas L, Auboeuf D, Raspe E, Schoonjans K, Lefebvre AM, Saladin R et al (1997) The organization, promoter analysis, and expression of the human PPARgamma gene. J Biol Chem 272:18779–18789
4. Greene ME, Blumberg B, McBride OW, Yi HF, Kronquist K, Kwan K et al (1995) Isolation of the human peroxisome proliferator activated receptor gamma cDNA: expression in hematopoietic cells and chromosomal mapping. Gene Expr 4:281–299
5. Adams M, Montague CT, Prins JB, Holder JC, Smith SA, Sanders L et al (1997) Activators of peroxisome proliferator-activated receptor gamma have depot-specific effects on human preadipocyte differentiation. J Clin Invest 100:3149–3153
6. Mangelsdorf DJ, Thummel C, Beato M, Herrlich P, Schutz G, Umesono K et al (1995) The nuclear receptor superfamily: the second decade. Cell 83:835–839
7. Lemberger T, Desvergne B, Wahli W (1996) Peroxisome proliferator-activated receptors: a nuclear receptor signaling pathway in lipid physiology. Annu Rev Cell Dev Biol 12:335–363
8. Berger J, Wagner JA (2002) Physiological and therapeutic roles of peroxisome proliferator-activated receptors. Diabetes Technol Ther 4:163–174
9. Berger J, Moller DE (2002) The mechanisms of action of PPARs. Annu Rev Med 53: 409–435

10. Lehmann JM, Moore LB, Smith-Oliver TA, Wilkison WO, Willson TM, Kliewer SA (1995) An antidiabetic thiazolidinedione is a high affinity ligand for peroxisome proliferator-activated receptor gamma (PPAR gamma). J Biol Chem 270:12953–12956
11. Moller DE, Berger JP (2003) Role of PPARs in the regulation of obesity-related insulin sensitivity and inflammation. Int J Obes Relat Metab Disord 27(Suppl 3):S17–S21
12. Li Y, Zhang J, Schopfer FJ, Martynowski D, Garcia-Barrio MT, Kovach A et al (2008) Molecular recognition of nitrated fatty acids by PPAR gamma. Nat Struct Mol Biol 15: 865–867
13. Yamamoto K, Itoh T, Abe D, Shimizu M, Kanda T, Koyama T et al (2005) Identification of putative metabolites of docosahexaenoic acid as potent PPARgamma agonists and antidiabetic agents. Bioorg Med Chem Lett 15:517–522
14. Itoh T, Murota I, Yoshikai K, Yamada S, Yamamoto K (2006) Synthesis of docosahexaenoic acid derivatives designed as novel PPARgamma agonists and antidiabetic agents. Bioorg Med Chem 14:98–108
15. Li H, Ruan XZ, Powis SH, Fernando R, Mon WY, Wheeler DC et al (2005) EPA and DHA reduce LPS-induced inflammation responses in HK-2 cells: evidence for a PPAR-gamma-dependent mechanism. Kidney Int 67:867–874
16. Kliewer SA, Lenhard JM, Willson TM, Patel I, Morris DC, Lehmann JM (1995) A prostaglandin J2 metabolite binds peroxisome proliferator-activated receptor gamma and promotes adipocyte differentiation. Cell 83:813–819
17. Yuan H, Li MY, Ma LT, Hsin MK, Mok TS, Underwood MJ et al (2010) 15-Lipoxygenases and its metabolites 15(S)-HETE and 13(S)-HODE in the development of non-small cell lung cancer. Thorax 65:321–326
18. Cimen I, Astarci E, Banerjee S (2011) 15-Lipoxygenase-1 exerts its tumor suppressive role by inhibiting nuclear factor-kappa B via activation of PPAR gamma. J Cell Biochem 112:2490–2501
19. Jaradat MS, Wongsud B, Phornchirasilp S, Rangwala SM, Shams G, Sutton M et al (2001) Activation of peroxisome proliferator-activated receptor isoforms and inhibition of prostaglandin H(2) synthases by ibuprofen, naproxen, and indomethacin. Biochem Pharmacol 62:1587–1595
20. Sastre M, Dewachter I, Landreth GE, Willson TM, Klockgether T, van Leuven F et al (2003) Nonsteroidal anti-inflammatory drugs and peroxisome proliferator-activated receptor-gamma agonists modulate immunostimulated processing of amyloid precursor protein through regulation of beta-secretase. J Neurosci 23:9796–9804
21. Lambe KG, Tugwood JD (1996) A human peroxisome-proliferator-activated receptor-gamma is activated by inducers of adipogenesis, including thiazolidinedione drugs. Eur J Biochem 239:1–7
22. Grossman SL, Lessem J (1997) Mechanisms and clinical effects of thiazolidinediones. Expert Opin Investig Drugs 6:1025–1040
23. Nattrass M, Bailey CJ (1999) New agents for type 2 diabetes. Baillieres Best Pract Res Clin Endocrinol Metab 13:309–329
24. Gillies PS, Dunn CJ (2000) Pioglitazone. Drugs 60:333–343, discussion 344–335
25. Goldstein BJ (2000) Rosiglitazone. Int J Clin Pract 54:333–337
26. Adams M, Reginato MJ, Shao D, Lazar MA, Chatterjee VK (1997) Transcriptional activation by peroxisome proliferator-activated receptor gamma is inhibited by phosphorylation at a consensus mitogen-activated protein kinase site. J Biol Chem 272:5128–5132
27. Camp HS, Tafuri SR, Leff T (1999) C-jun n-terminal kinase phosphorylates peroxisome proliferator-activated receptor-gamma1 and negatively regulates its transcriptional activity. Endocrinology 140:392–397
28. Mielke K, Damm A, Yang DD, Herdegen T (2000) Selective expression of JNK isoforms and stress-specific JNK activity in different neural cell lines. Brain Res Mol Brain Res 75:128–137
29. Barbin G, Roisin MP, Zalc B (2001) Tumor necrosis factor alpha activates the phosphorylation of ERK, SAPK/JNK, and p38 kinase in primary cultures of neurons. Neurochem Res 26: 107–112

30. Borsello T, Clarke PG, Hirt L, Vercelli A, Repici M, Schorderet DF et al (2003) A peptide inhibitor of c-Jun N-terminal kinase protects against excitotoxicity and cerebral ischemia. Nat Med 9:1180–1186
31. Borsello T, Croquelois K, Hornung JP, Clarke PG (2003) N-methyl-d-aspartate-triggered neuronal death in organotypic hippocampal cultures is endocytic, autophagic and mediated by the c-Jun N-terminal kinase pathway. Eur J Neurosci 18:473–485
32. Wang Q, Wang X, Studzinski GP (2003) Jun N-terminal kinase pathway enhances signaling of monocytic differentiation of human leukemia cells induced by 1,25-dihydroxyvitamin D3. J Cell Biochem 89:1087–1101
33. Jiang C, Ting AT, Seed B (1998) PPAR-gamma agonists inhibit production of monocyte inflammatory cytokines. Nature 391:82–86
34. Ricote M, Li AC, Willson TM, Kelly CJ, Glass CK (1998) The peroxisome proliferator-activated receptor-gamma is a negative regulator of macrophage activation. Nature 391:79–82
35. Ohara M, Sawa T (1999) Current topics in the regulation of prostanoids-4. The feedback regulation by PPAR-gamma. Masui 48:146–151
36. Barbier O, Torra IP, Duguay Y, Blanquart C, Fruchart JC, Glineur C et al (2002) Pleiotropic actions of peroxisome proliferator-activated receptors in lipid metabolism and atherosclerosis. Arterioscler Thromb Vasc Biol 22:717–726
37. Blanquart C, Barbier O, Fruchart JC, Staels B, Glineur C (2003) Peroxisome proliferator-activated receptors: regulation of transcriptional activities and roles in inflammation. J Steroid Biochem Mol Biol 85:267–273
38. Kiaei M, Kipiani K, Chen J, Calingasan NY, Beal MF (2005) Peroxisome proliferator-activated receptor-gamma agonist extends survival in transgenic mouse model of amyotrophic lateral sclerosis. Exp Neurol 191:331–336
39. Sundararajan S, Landreth GE (2004) Antiinflammatory properties of PPARgamma agonists following ischemia. Drug News Perspect 17:229–236
40. Pereira MP, Hurtado O, Cardenas A, Alonso-Escolano D, Bosca L, Vivancos J et al (2005) The nonthiazolidinedione PPARgamma agonist l-796,449 is neuroprotective in experimental stroke. J Neuropathol Exp Neurol 64:797–805
41. Shimazu T, Inoue I, Araki N, Asano Y, Sawada M, Furuya D et al (2005) A peroxisome proliferator-activated receptor-gamma agonist reduces infarct size in transient but not in permanent ischemia. Stroke 36:353–359
42. Sundararajan S, Gamboa JL, Victor NA, Wanderi EW, Lust WD, Landreth GE (2005) Peroxisome proliferator-activated receptor-gamma ligands reduce inflammation and infarction size in transient focal ischemia. Neuroscience 130:685–696
43. Zhao Y, Patzer A, Gohlke P, Herdegen T, Culman J (2005) The intracerebral application of the PPARgamma-ligand pioglitazone confers neuroprotection against focal ischaemia in the rat brain. Eur J Neurosci 22:278–282
44. Bordet R, Ouk T, Petrault O, Gele P, Gautier S, Laprais M et al (2006) PPAR: a new pharmacological target for neuroprotection in stroke and neurodegenerative diseases. Biochem Soc Trans 34:1341–1346
45. Ou Z, Zhao X, Labiche LA, Strong R, Grotta JC, Herrmann O et al (2006) Neuronal expression of peroxisome proliferator-activated receptor-gamma (PPARgamma) and 15d-prostaglandin j2-mediated protection of brain after experimental cerebral ischemia in rat. Brain Res 1096:196–203
46. Chu K, Lee ST, Koo JS, Jung KH, Kim EH, Sinn DI et al (2006) Peroxisome proliferator-activated receptor-gamma-agonist, rosiglitazone, promotes angiogenesis after focal cerebral ischemia. Brain Res 1093:208–218
47. Luo Y, Yin W, Signore AP, Zhang F, Hong Z, Wang S et al (2006) Neuroprotection against focal ischemic brain injury by the peroxisome proliferator-activated receptor-gamma agonist rosiglitazone. J Neurochem 97:435–448
48. Collino M, Aragno M, Mastrocola R, Gallicchio M, Rosa AC, Dianzani C et al (2006) Modulation of the oxidative stress and inflammatory response by PPAR-gamma agonists in

the hippocampus of rats exposed to cerebral ischemia/reperfusion. Eur J Pharmacol 530: 70–80
49. Lin TN, Cheung WM, Wu JS, Chen JJ, Lin H, Liou JY et al (2006) 15d-prostaglandin J2 protects brain from ischemia-reperfusion injury. Arterioscler Thromb Vasc Biol 26:481–487
50. Victor NA, Wanderi EW, Gamboa J, Zhao X, Aronowski J, Deininger K et al (2006) Altered PPARgamma expression and activation after transient focal ischemia in rats. Eur J Neurosci 24:1653–1663
51. Culman J, Nguyen-Ngoc M, Glatz T, Gohlke P, Herdegen T, Zhao Y (2012) Treatment of rats with pioglitazone in the reperfusion phase of focal cerebral ischemia: a preclinical stroke trial. Exp Neurol 238:243–253
52. Gonzales NR, Shah J, Sangha N, Sosa L, Martinez R, Shen L et al (2012) Design of a prospective, dose-escalation study evaluating the safety of pioglitazone for hematoma resolution in intracerebral hemorrhage (SHRINC). Int J Stroke 8(5):388–396
53. Zhao X, Zhang Y, Strong R, Grotta JC, Aronowski J (2006) 15d-prostaglandin J2 activates peroxisome proliferator-activated receptor-gamma, promotes expression of catalase, and reduces inflammation, behavioral dysfunction, and neuronal loss after intracerebral hemorrhage in rats. J Cereb Blood Flow Metab 26:811–820
54. Zhao X, Sun G, Zhang J, Strong R, Song W, Gonzales N et al (2007) Hematoma resolution as a target for intracerebral hemorrhage treatment: role for peroxisome proliferator-activated receptor gamma in microglia/macrophages. Ann Neurol 61:352–362
55. Hyong A, Jadhav V, Lee S, Tong W, Rowe J, Zhang JH et al (2008) Rosiglitazone, a PPAR gamma agonist, attenuates inflammation after surgical brain injury in rodents. Brain Res 1215:218–224
56. Yi JH, Park SW, Brooks N, Lang BT, Vemuganti R (2008) PPARgamma agonist rosiglitazone is neuroprotective after traumatic brain injury via anti-inflammatory and anti-oxidative mechanisms. Brain Res 1244:164–172
57. Zhang Q, Hu W, Meng B, Tang T (2010) PPARgamma agonist rosiglitazone is neuroprotective after traumatic spinal cord injury via anti-inflammatory in adult rats. Neurol Res 32: 852–859
58. Sauerbeck A, Gao J, Readnower R, Liu M, Pauly JR, Bing G et al (2011) Pioglitazone attenuates mitochondrial dysfunction, cognitive impairment, cortical tissue loss, and inflammation following traumatic brain injury. Exp Neurol 227:128–135
59. Combs CK, Johnson DE, Karlo JC, Cannady SB, Landreth GE (2000) Inflammatory mechanisms in Alzheimer's disease: inhibition of beta-amyloid-stimulated proinflammatory responses and neurotoxicity by PPARgamma agonists. J Neurosci 20:558–567
60. Landreth GE, Heneka MT (2001) Anti-inflammatory actions of peroxisome proliferator-activated receptor gamma agonists in Alzheimer's disease. Neurobiol Aging 22:937–944
61. Heneka MT, Landreth GE, Feinstein DL (2001) Role for peroxisome proliferator-activated receptor-gamma in Alzheimer's disease. Ann Neurol 49:276
62. Breidert T, Callebert J, Heneka MT, Landreth G, Launay JM, Hirsch EC (2002) Protective action of the peroxisome proliferator-activated receptor-gamma agonist pioglitazone in a mouse model of Parkinson's disease. J Neurochem 82:615–624
63. Kielian T, Drew PD (2003) Effects of peroxisome proliferator-activated receptor-gamma agonists on central nervous system inflammation. J Neurosci Res 71:315–325
64. Duvanel CB, Honegger P, Pershadsingh H, Feinstein D, Matthieu JM (2003) Inhibition of glial cell proinflammatory activities by peroxisome proliferator-activated receptor gamma agonist confers partial protection during antimyelin oligodendrocyte glycoprotein demyelination in vitro. J Neurosci Res 71:246–255
65. Dehmer T, Heneka MT, Sastre M, Dichgans J, Schulz JB (2004) Protection by pioglitazone in the MPTP model of Parkinson's disease correlates with I kappa B alpha induction and block of NF kappa B and INOS activation. J Neurochem 88:494–501
66. Gasparini L, Ongini E, Wenk G (2004) Non-steroidal anti-inflammatory drugs (NSAIDs) in Alzheimer's disease: old and new mechanisms of action. J Neurochem 91:521–536

67. d'Abramo C, Massone S, Zingg JM, Pizzuti A, Marambaud P, Dalla Piccola B et al (2005) Role of peroxisome proliferator-activated receptor gamma in amyloid precursor protein processing and amyloid beta-mediated cell death. Biochem J 391:693–698
68. Heneka MT, Landreth GE, Hull M (2007) Drug insight: effects mediated by peroxisome proliferator-activated receptor-gamma in CNS disorders. Nat Clin Pract Neurol 3:496–504
69. Heneka MT, Reyes-Irisarri E, Hull M, Kummer MP (2011) Impact and therapeutic potential of PPARs in Alzheimer's disease. Curr Neuropharmacol 9:643–650
70. Feinstein DL, Galea E, Gavrilyuk V, Brosnan CF, Whitacre CC, Dumitrescu-Ozimek L et al (2002) Peroxisome proliferator-activated receptor-gamma agonists prevent experimental autoimmune encephalomyelitis. Ann Neurol 51:694–702
71. Storer PD, Xu J, Chavis J, Drew PD (2005) Peroxisome proliferator-activated receptor-gamma agonists inhibit the activation of microglia and astrocytes: implications for multiple sclerosis. J Neuroimmunol 161:113–122
72. Loria F, Petrosino S, Hernangomez M, Mestre L, Spagnolo A, Correa F et al (2010) An endocannabinoid tone limits excitotoxicity in vitro and in a model of multiple sclerosis. Neurobiol Dis 37:166–176
73. Luna-Medina R, Cortes-Canteli M, Alonso M, Santos A, Martinez A, Perez-Castillo A (2005) Regulation of inflammatory response in neural cells in vitro by thiadiazolidinones derivatives through peroxisome proliferator-activated receptor gamma activation. J Biol Chem 280:21453–21462
74. Dello Russo C, Gavrilyuk V, Weinberg G, Almeida A, Bolanos JP, Palmer J et al (2003) Peroxisome proliferator-activated receptor gamma thiazolidinedione agonists increase glucose metabolism in astrocytes. J Biol Chem 278:5828–5836
75. Cristiano L, Bernardo A, Ceru MP (2001) Peroxisome proliferator-activated receptors (PPARs) and peroxisomes in rat cortical and cerebellar astrocytes. J Neurocytol 30:671–683
76. Janabi N (2002) Selective inhibition of cyclooxygenase-2 expression by 15-deoxy-delta(12,14)(12,14)-prostaglandin J(2) in activated human astrocytes, but not in human brain macrophages. J Immunol 168:4747–4755
77. Roth AD, Leisewitz AV, Jung JE, Cassina P, Barbeito L, Inestrosa NC et al (2003) PPAR gamma activators induce growth arrest and process extension in B12 oligodendrocyte-like cells and terminal differentiation of cultured oligodendrocytes. J Neurosci Res 72:425–435
78. Bernardo A, Bianchi D, Magnaghi V, Minghetti L (2009) Peroxisome proliferator-activated receptor-gamma agonists promote differentiation and antioxidant defenses of oligodendrocyte progenitor cells. J Neuropathol Exp Neurol 68:797–808
79. De Nuccio C, Bernardo A, De Simone R, Mancuso E, Magnaghi V, Visentin S et al (2011) Peroxisome proliferator-activated receptor gamma agonists accelerate oligodendrocyte maturation and influence mitochondrial functions and oscillatory Ca(2+) waves. J Neuropathol Exp Neurol 70:900–912
80. Petrova TV, Akama KT, Van Eldik LJ (1999) Cyclopentenone prostaglandins suppress activation of microglia: down-regulation of inducible nitric-oxide synthase by 15-deoxy-delta12,14-prostaglandin J2. Proc Natl Acad Sci USA 96:4668–4673
81. Bernardo A, Minghetti L (2006) PPAR-gamma agonists as regulators of microglial activation and brain inflammation. Curr Pharm Des 12:93–109
82. Hamblin M, Chang L, Fan Y, Zhang J, Chen YE (2009) PPARs and the cardiovascular system. Antioxid Redox Signal 11:1415–1452
83. Wu QQ, Wang Y, Senitko M, Meyer C, Wigley WC, Ferguson DA et al (2011) Bardoxolone methyl (BARD) ameliorates ischemic AKI and increases expression of protective genes Nrf2, PPARgamma, and HO-1. Am J Physiol Renal Physiol 300:F1180–F1192
84. Zhao X, Strong R, Zhang J, Sun G, Tsien JZ, Cui Z et al (2009) Neuronal PPARgamma deficiency increases susceptibility to brain damage after cerebral ischemia. J Neurosci 29:6186–6195
85. Vemuganti R (2008) Therapeutic potential of PPARgamma activation in stroke. PPAR Res 2008:461981

86. Culman J, Zhao Y, Gohlke P, Herdegen T (2007) PPAR-gamma: therapeutic target for ischemic stroke. Trends Pharmacol Sci 28:244–249
87. Fatehi-Hassanabad Z, Tasker RA (2011) Peroxisome proliferator-activated receptor-gamma (PPAR-gamma) activation confers functional neuroprotection in global ischemia. Neurotox Res 19:462–471
88. Dowell P, Ishmael JE, Avram D, Peterson VJ, Nevrivy DJ, Leid M (1997) P300 functions as a coactivator for the peroxisome proliferator-activated receptor alpha. J Biol Chem 272:33435–33443
89. Gelman L, Zhou G, Fajas L, Raspe E, Fruchart JC, Auwerx J (1999) P300 interacts with the N- and C-terminal part of PPARgamma2 in a ligand-independent and -dependent manner, respectively. J Biol Chem 274:7681–7688
90. Li M, Pascual G, Glass CK (2000) Peroxisome proliferator-activated receptor gamma-dependent repression of the inducible nitric oxide synthase gene. Mol Cell Biol 20:4699–4707
91. Straus DS, Pascual G, Li M, Welch JS, Ricote M, Hsiang CH et al (2000) 15-deoxy-delta 12,14-prostaglandin J2 inhibits multiple steps in the NF- kappa B signaling pathway. Proc Natl Acad Sci USA 97:4844–4849
92. Rossi A, Kapahi P, Natoli G, Takahashi T, Chen Y, Karin M et al (2000) Anti-inflammatory cyclopentenone prostaglandins are direct inhibitors of ikappaB kinase. Nature 403:103–108
93. Delerive P, Gervois P, Fruchart JC, Staels B (2000) Induction of ikappaBalpha expression as a mechanism contributing to the anti-inflammatory activities of peroxisome proliferator-activated receptor-alpha activators. J Biol Chem 275:36703–36707
94. Cernuda-Morollon E, Rodriguez-Pascual F, Klatt P, Lamas S, Perez-Sala D (2002) PPAR agonists amplify inos expression while inhibiting NF-kappaB: implications for mesangial cell activation by cytokines. J Am Soc Nephrol 13:2223–2231
95. Heneka MT, Gavrilyuk V, Landreth GE, O'Banion MK, Weinberg G, Feinstein DL (2003) Noradrenergic depletion increases inflammatory responses in brain: effects on ikappaB and HSP70 expression. J Neurochem 85:387–398
96. Uryu S, Harada J, Hisamoto M, Oda T (2002) Troglitazone inhibits both post-glutamate neurotoxicity and low-potassium-induced apoptosis in cerebellar granule neurons. Brain Res 924:229–236
97. Zhao X, Ou Z, Grotta JC, Waxham N, Aronowski J (2006) Peroxisome-proliferator-activated receptor-gamma (PPARgamma) activation protects neurons from NMDA excitotoxicity. Brain Res 1073–1074:460–469
98. Moore KJ, Rosen ED, Fitzgerald ML, Randow F, Andersson LP, Altshuler D et al (2001) The role of PPAR-gamma in macrophage differentiation and cholesterol uptake. Nat Med 7:41–47
99. Asada K, Sasaki S, Suda T, Chida K, Nakamura H (2004) Antiinflammatory roles of peroxisome proliferator-activated receptor gamma in human alveolar macrophages. Am J Respir Crit Care Med 169:195–200
100. Patel SN, Serghides L, Smith TG, Febbraio M, Silverstein RL, Kurtz TW et al (2004) Cd36 mediates the phagocytosis of Plasmodium falciparum-infected erythrocytes by rodent macrophages. J Infect Dis 189:204–213
101. Majai G, Sarang Z, Csomos K, Zahuczky G, Fesus L (2007) PPARgamma-dependent regulation of human macrophages in phagocytosis of apoptotic cells. Eur J Immunol 37:1343–1354
102. Zheng Z, Lee JE, Yenari MA (2003) Stroke: molecular mechanisms and potential targets for treatment. Curr Mol Med 3:361–372
103. Danton GH, Dietrich WD (2003) Inflammatory mechanisms after ischemia and stroke. J Neuropathol Exp Neurol 62:127–136
104. Wen YD, Zhang HL, Qin ZH (2006) Inflammatory mechanism in ischemic neuronal injury. Neurosci Bull 22:171–182
105. Mehta SL, Manhas N, Raghubir R (2007) Molecular targets in cerebral ischemia for developing novel therapeutics. Brain Res Rev 54:34–66
106. Nakka VP, Gusain A, Mehta SL, Raghubir R (2008) Molecular mechanisms of apoptosis in cerebral ischemia: multiple neuroprotective opportunities. Mol Neurobiol 37:7–38

107. Brea D, Sobrino T, Ramos-Cabrer P, Castillo J (2009) Inflammatory and neuroimmunomodulatory changes in acute cerebral ischemia. Cerebrovasc Dis 27(Suppl 1):48–64
108. Candelario-Jalil E (2009) Injury and repair mechanisms in ischemic stroke: considerations for the development of novel neurotherapeutics. Curr Opin Investig Drugs 10:644–654
109. Guo MF, Yu JZ, Ma CG (2011) Mechanisms related to neuron injury and death in cerebral hypoxic ischaemia. Folia Neuropathol 49:78–87
110. Zhang YQ, Zhang YN, Wu J, Zhu XY, Xu CQ (2005) Effect of peroxisome proliferation activated receptor-gamma on neuronal cell death induced by hypoxia and ischemia in rats in vitro and in vivo. Zhonghua Yi Xue Za Zhi 85:684–688
111. Clemens JA (2000) Cerebral ischemia: gene activation, neuronal injury, and the protective role of antioxidants. Free Radic Biol Med 28:1526–1531
112. Starkov AA, Chinopoulos C, Fiskum G (2004) Mitochondrial calcium and oxidative stress as mediators of ischemic brain injury. Cell Calcium 36:257–264
113. Tuttolomondo A, Di Sciacca R, Di Raimondo D, Arnao V, Renda C, Pinto A et al (2009) Neuron protection as a therapeutic target in acute ischemic stroke. Curr Top Med Chem 9:1317–1334
114. Chen SD, Yang DI, Lin TK, Shaw FZ, Liou CW, Chuang YC (2011) Roles of oxidative stress, apoptosis, PGC-1alpha and mitochondrial biogenesis in cerebral ischemia. Int J Mol Sci 12:7199–7215
115. Girnun GD, Domann FE, Moore SA, Robbins ME (2002) Identification of a functional peroxisome proliferator-activated receptor response element in the rat catalase promoter. Mol Endocrinol 16:2793–2801
116. Jung TW, Lee JY, Shim WS, Kang ES, Kim SK, Ahn CW et al (2007) Rosiglitazone protects human neuroblastoma SH-SY5Y cells against MPP+ induced cytotoxicity via inhibition of mitochondrial dysfunction and ROS production. J Neurol Sci 253:53–60
117. Chance B, Sies H, Boveris A (1979) Hydroperoxide metabolism in mammalian organs. Physiol Rev 59:527–605
118. Moreno S, Mugnaini E, Ceru MP (1995) Immunocytochemical localization of catalase in the central nervous system of the rat. J Histochem Cytochem 43:1253–1267
119. Kwak MK, Itoh K, Yamamoto M, Sutter TR, Kensler TW (2001) Role of transcription factor Nrf2 in the induction of hepatic phase 2 and antioxidative enzymes in vivo by the cancer chemoprotective agent, 3H-1, 2-dimethiole-3-thione. Mol Med 7:135–145
120. Gu W, Zhao H, Yenari MA, Sapolsky RM, Steinberg GK (2004) Catalase over-expression protects striatal neurons from transient focal cerebral ischemia. Neuroreport 15:413–416
121. Mahadik SP, Makar TK, Murthy JN, Ortiz A, Wakade CG, Karpiak SE (1993) Temporal changes in superoxide dismutase, glutathione peroxidase, and catalase levels in primary and peri-ischemic tissue. Monosialoganglioside (GM1) treatment effects. Mol Chem Neuropathol 18:1–14
122. Ye R, Li N, Han J, Kong X, Cao R, Rao Z et al (2009) Neuroprotective effects of ginsenoside RD against oxygen-glucose deprivation in cultured hippocampal neurons. Neurosci Res 64:306–310
123. Ricart KC, Fiszman ML (2001) Hydrogen peroxide-induced neurotoxicity in cultured cortical cells grown in serum-free and serum-containing media. Neurochem Res 26:801–808
124. Amantea D, Marrone MC, Nistico R, Federici M, Bagetta G, Bernardi G et al (2009) Oxidative stress in stroke pathophysiology validation of hydrogen peroxide metabolism as a pharmacological target to afford neuroprotection. Int Rev Neurobiol 85:363–374
125. Kondo T, Reaume AG, Huang TT, Carlson E, Murakami K, Chen SF et al (1997) Reduction of cuzn-superoxide dismutase activity exacerbates neuronal cell injury and edema formation after transient focal cerebral ischemia. J Neurosci 17:4180–4189
126. Chan PH (1994) Oxygen radicals in focal cerebral ischemia. Brain Pathol 4:59–65
127. Sambrano GR, Steinberg D (1995) Recognition of oxidatively damaged and apoptotic cells by an oxidized low density lipoprotein receptor on mouse peritoneal macrophages: role of membrane phosphatidylserine. Proc Natl Acad Sci USA 92:1396–1400

128. Fadok VA, Warner ML, Bratton DL, Henson PM (1998) Cd36 is required for phagocytosis of apoptotic cells by human macrophages that use either a phosphatidylserine receptor or the vitronectin receptor (alpha v beta 3). J Immunol 161:6250–6257
129. Yamada Y, Doi T, Hamakubo T, Kodama T (1998) Scavenger receptor family proteins: roles for atherosclerosis, host defence and disorders of the central nervous system. Cell Mol Life Sci 54:628–640
130. Aderem A, Underhill DM (1999) Mechanisms of phagocytosis in macrophages. Annu Rev Immunol 17:593–623
131. Stolzing A, Grune T (2004) Neuronal apoptotic bodies: phagocytosis and degradation by primary microglial cells. FASEB J 18:743–745
132. Nicholson AC (2004) Expression of cd36 in macrophages and atherosclerosis: the role of lipid regulation of PPARgamma signaling. Trends Cardiovasc Med 14:8–12
133. Vinals M, Bermudez I, Llaverias G, Alegret M, Sanchez RM, Vazquez-Carrera M et al (2005) Aspirin increases CD36, SR-BI, and ABCA1 expression in human THP-1 macrophages. Cardiovasc Res 66:141–149
134. Hebbel RP, Miller WJ (1984) Phagocytosis of sickle erythrocytes: immunologic and oxidative determinants of hemolytic anemia. Blood 64:733–741
135. Schwartz RS, Tanaka Y, Fidler IJ, Chiu DT, Lubin B, Schroit AJ (1985) Increased adherence of sickled and phosphatidylserine-enriched human erythrocytes to cultured human peripheral blood monocytes. J Clin Invest 75:1965–1972
136. Connor J, Pak CC, Schroit AJ (1994) Exposure of phosphatidylserine in the outer leaflet of human red blood cells. Relationship to cell density, cell age, and clearance by mononuclear cells. J Biol Chem 269:2399–2404
137. Haslett C, Savill JS, Whyte MK, Stern M, Dransfield I, Meagher LC (1994) Granulocyte apoptosis and the control of inflammation. Philos Trans R Soc Lond B Biol Sci 345:327–333
138. Ren Y, Silverstein RL, Allen J, Savill J (1995) Cd36 gene transfer confers capacity for phagocytosis of cells undergoing apoptosis. J Exp Med 181:1857–1862
139. Navazo MD, Daviet L, Savill J, Ren Y, Leung LL, McGregor JL (1996) Identification of a domain (155–183) on CD36 implicated in the phagocytosis of apoptotic neutrophils. J Biol Chem 271:15381–15385
140. Tontonoz P, Nagy L, Alvarez JG, Thomazy VA, Evans RM (1998) PPARgamma promotes monocyte/macrophage differentiation and uptake of oxidized LDL. Cell 93:241–252
141. Babaev VR, Yancey PG, Ryzhov SV, Kon V, Breyer MD, Magnuson MA et al (2005) Conditional knockout of macrophage PPARgamma increases atherosclerosis in C57BL/6 and low-density lipoprotein receptor-deficient mice. Arterioscler Thromb Vasc Biol 25: 1647–1653
142. Ishii T, Itoh K, Ruiz E, Leake DS, Unoki H, Yamamoto M et al (2004) Role of Nrf2 in the regulation of CD36 and stress protein expression in murine macrophages: activation by oxidatively modified LDL and 4-hydroxynonenal. Circ Res 94:609–616
143. Sussan TE, Jun J, Thimmulappa R, Bedja D, Antero M, Gabrielson KL et al (2008) Disruption of Nrf2, a key inducer of antioxidant defenses, attenuates ApoE-mediated atherosclerosis in mice. PLoS One 3:e3791
144. Maruyama A, Tsukamoto S, Nishikawa K, Yoshida A, Harada N, Motojima K et al (2008) Nrf2 regulates the alternative first exons of CD36 in macrophages through specific antioxidant response elements. Arch Biochem Biophys 477:139–145
145. Olagnier D, Lavergne RA, Meunier E, Lefevre L, Dardenne C, Aubouy A et al (2011) Nrf2, a PPARgamma alternative pathway to promote CD36 expression on inflammatory macrophages: implication for malaria. PLoS Pathog 7:e1002254
146. Savill J (1997) Recognition and phagocytosis of cells undergoing apoptosis. Br Med Bull 53:491–508
147. Boullier A, Bird DA, Chang MK, Dennis EA, Friedman P, Gillotre-Taylor K et al (2001) Scavenger receptors, oxidized LDL, and atherosclerosis. Ann N Y Acad Sci 947:214–222, discussion 222–213

148. Husemann J, Loike JD, Anankov R, Febbraio M, Silverstein SC (2002) Scavenger receptors in neurobiology and neuropathology: their role on microglia and other cells of the nervous system. Glia 40:195–205
149. Nagy L, Tontonoz P, Alvarez JG, Chen H, Evans RM (1998) Oxidized ldl regulates macrophage gene expression through ligand activation of PPARgamma. Cell 93:229–240
150. Eto M, Yoshikawa H, Fujimura H, Naba I, Sumi-Akamaru H, Takayasu S et al (2003) The role of cd36 in peripheral nerve remyelination after crush injury. Eur J Neurosci 17:2659–2666
151. Janabi M, Yamashita S, Hirano K, Sakai N, Hiraoka H, Matsumoto K et al (2000) Oxidized LDL-induced NF-kappa b activation and subsequent expression of proinflammatory genes are defective in monocyte-derived macrophages from cd36-deficient patients. Arterioscler Thromb Vasc Biol 20:1953–1960
152. Cho S, Park EM, Febbraio M, Anrather J, Park L, Racchumi G et al (2005) The class B scavenger receptor CD36 mediates free radical production and tissue injury in cerebral ischemia. J Neurosci 25:2504–2512
153. Woo MS, Wang X, Faustino JV, Derugin N, Wendland MF, Zhou P et al (2012) Genetic deletion of CD36 enhances injury after acute neonatal stroke. Ann Neurol 72:961–970
154. Majno G (1975) The healing hand: man and wound in the ancient world. Harvard University Press, Cambridge, MA
155. Giulian D, Chen J, Ingeman JE, George JK, Noponen M (1989) The role of mononuclear phagocytes in wound healing after traumatic injury to adult mammalian brain. J Neurosci 9:4416–4429
156. Marcus SL, Miyata KS, Zhang B, Subramani S, Rachubinski RA, Capone JP (1993) Diverse peroxisome proliferator-activated receptors bind to the peroxisome proliferator-responsive elements of the rat hydratase/dehydrogenase and fatty acyl-coa oxidase genes but differentially induce expression. Proc Natl Acad Sci USA 90:5723–5727
157. Aperlo C, Pognonec P, Saladin R, Auwerx J, Boulukos KE (1995) Cdna cloning and characterization of the transcriptional activities of the hamster peroxisome proliferator-activated receptor happar gamma. Gene 162:297–302
158. Mukherjee R, Davies PJ, Crombie DL, Bischoff ED, Cesario RM, Jow L et al (1997) Sensitization of diabetic and obese mice to insulin by retinoid x receptor agonists. Nature 386:407–410
159. Mukherjee R, Jow L, Croston GE, Paterniti JR Jr (1997) Identification, characterization, and tissue distribution of human peroxisome proliferator-activated receptor (PPAR) isoforms PPARgamma2 versus PPARgamma1 and activation with retinoid x receptor agonists and antagonists. J Biol Chem 272:8071–8076
160. Tontonoz P, Singer S, Forman BM, Sarraf P, Fletcher JA, Fletcher CD et al (1997) Terminal differentiation of human liposarcoma cells induced by ligands for peroxisome proliferator-activated receptor gamma and the retinoid x receptor. Proc Natl Acad Sci U S A 94:237–241
161. Diab A, Hussain RZ, Lovett-Racke AE, Chavis JA, Drew PD, Racke MK (2004) Ligands for the peroxisome proliferator-activated receptor-gamma and the retinoid x receptor exert additive anti-inflammatory effects on experimental autoimmune encephalomyelitis. J Neuroimmunol 148:116–126
162. Okuma Y, Uehara T, Miyazaki H, Miyasaka T, Nomura Y (1998) The involvement of cytokines, chemokines and inducible nitric oxide synthase (INOS) induced by a transient ischemia in neuronal survival/death in rat brain. Nihon Yakurigaku Zasshi 111:37–44
163. Forster C, Clark HB, Ross ME, Iadecola C (1999) Inducible nitric oxide synthase expression in human cerebral infarcts. Acta Neuropathol 97:215–220
164. Wang Q, Tang XN, Yenari MA (2007) The inflammatory response in stroke. J Neuroimmunol 184:53–68
165. Iadecola C, Alexander M (2001) Cerebral ischemia and inflammation. Curr Opin Neurol 14:89–94
166. Stowe AM, Adair-Kirk TL, Gonzales ER, Perez RS, Shah AR, Park TS et al (2009) Neutrophil elastase and neurovascular injury following focal stroke and reperfusion. Neurobiol Dis 35:82–90

167. Ishrat T, Sayeed I, Atif F, Hua F, Stein DG (2010) Progesterone and allopregnanolone attenuate blood–brain barrier dysfunction following permanent focal ischemia by regulating the expression of matrix metalloproteinases. Exp Neurol 226:183–190
168. Jin R, Yang G, Li G (2010) Molecular insights and therapeutic targets for blood–brain barrier disruption in ischemic stroke: critical role of matrix metalloproteinases and tissue-type plasminogen activator. Neurobiol Dis 38:376–385
169. Chung SW, Kang BY, Kim SH, Pak YK, Cho D, Trinchieri G et al (2000) Oxidized low density lipoprotein inhibits interleukin-12 production in lipopolysaccharide-activated mouse macrophages via direct interactions between peroxisome proliferator-activated receptor-gamma and nuclear factor-kappa B. J Biol Chem 275:32681–32687
170. Nolte RT, Wisely GB, Westin S, Cobb JE, Lambert MH, Kurokawa R et al (1998) Ligand binding and co-activator assembly of the peroxisome proliferator-activated receptor-gamma. Nature 395:137–143
171. Moi P, Chan K, Asunis I, Cao A, Kan YW (1994) Isolation of NF-E2-related factor 2 (Nrf2), a NF-E2-like basic leucine zipper transcriptional activator that binds to the tandem NF-E2/AP1 repeat of the beta-globin locus control region. Proc Natl Acad Sci USA 91:9926–9930
172. Ishii T, Itoh K, Yamamoto M (2002) Roles of Nrf2 in activation of antioxidant enzyme genes via antioxidant responsive elements. Methods Enzymol 348:182–190
173. Itoh K, Wakabayashi N, Katoh Y, Ishii T, O'Connor T, Yamamoto M (2003) Keap1 regulates both cytoplasmic-nuclear shuttling and degradation of Nrf2 in response to electrophiles. Genes Cells 8:379–391
174. Giudice A, Montella M (2006) Activation of the Nrf2-are signaling pathway: a promising strategy in cancer prevention. Bioessays 28:169–181
175. Lee JM, Shih AY, Murphy TH, Johnson JA (2003) NF-E2-related factor-2 mediates neuroprotection against mitochondrial complex i inhibitors and increased concentrations of intracellular calcium in primary cortical neurons. J Biol Chem 278:37948–37956
176. Kraft AD, Johnson DA, Johnson JA (2004) Nuclear factor E2-related factor 2-dependent antioxidant response element activation by tert-butylhydroquinone and sulforaphane occurring preferentially in astrocytes conditions neurons against oxidative insult. J Neurosci 24:1101–1112
177. Shih AY, Li P, Murphy TH (2005) A small-molecule-inducible Nrf2-mediated antioxidant response provides effective prophylaxis against cerebral ischemia in vivo. J Neurosci 25: 10321–10335
178. Leonard MO, Kieran NE, Howell K, Burne MJ, Varadarajan R, Dhakshinamoorthy S et al (2006) Reoxygenation-specific activation of the antioxidant transcription factor Nrf2 mediates cytoprotective gene expression in ischemia-reperfusion injury. FASEB J 20:2624–2626
179. Kraft AD, Lee JM, Johnson DA, Kan YW, Johnson JA (2006) Neuronal sensitivity to kainic acid is dependent on the Nrf2-mediated actions of the antioxidant response element. J Neurochem 98:1852–1865
180. Satoh T, Okamoto SI, Cui J, Watanabe Y, Furuta K, Suzuki M et al (2006) Activation of the keap1/Nrf2 pathway for neuroprotection by electrophilic [correction of electrophillic] phase II inducers. Proc Natl Acad Sci USA 103:768–773
181. Zhao J, Kobori N, Aronowski J, Dash PK (2006) Sulforaphane reduces infarct volume following focal cerebral ischemia in rodents. Neurosci Lett 393:108–112
182. van Muiswinkel FL, Kuiperij HB (2005) The nrf2-are signalling pathway: promising drug target to combat oxidative stress in neurodegenerative disorders. Curr Drug Targets CNS Neurol Disord 4:267–281
183. Shih AY, Imbeault S, Barakauskas V, Erb H, Jiang L, Li P et al (2005) Induction of the Nrf2-driven antioxidant response confers neuroprotection during mitochondrial stress in vivo. J Biol Chem 280:22925–22936
184. Park EY, Cho IJ, Kim SG (2004) Transactivation of the PPAR-responsive enhancer module in chemopreventive glutathione s-transferase gene by the peroxisome proliferator-activated receptor-gamma and retinoid x receptor heterodimer. Cancer Res 64:3701–3713

185. Cho HY, Gladwell W, Wang X, Chorley B, Bell D, Reddy SP et al (2010) Nrf2-regulated PPAR{gamma} expression is critical to protection against acute lung injury in mice. Am J Respir Crit Care Med 182:170–182
186. Bowie A, O'Neill LA (2000) Oxidative stress and nuclear factor-kappab activation: a reassessment of the evidence in the light of recent discoveries. Biochem Pharmacol 59:13–23
187. Thimmulappa RK, Lee H, Rangasamy T, Reddy SP, Yamamoto M, Kensler TW et al (2006) Nrf2 is a critical regulator of the innate immune response and survival during experimental sepsis. J Clin Invest 116:984–995
188. Reuter S, Gupta SC, Chaturvedi MM, Aggarwal BB (2010) Oxidative stress, inflammation, and cancer: how are they linked? Free Radic Biol Med 49:1603–1616
189. Higgins LG, Hayes JD (2011) Mechanisms of induction of cytosolic and microsomal glutathione transferase (GST) genes by xenobiotics and pro-inflammatory agents. Drug Metab Rev 43:92–137
190. Polvani S, Tarocchi M, Galli A (2012) PPARgamma and oxidative stress: con(beta) catenating Nrf2 and foxo. PPAR Res 2012:641087
191. Smith SA, Monteith GR, Holman NA, Robinson JA, May FJ, Roberts-Thomson SJ (2003) Effects of peroxisome proliferator-activated receptor gamma ligands ciglitazone and 15-deoxy-delta 12,14-prostaglandin J2 on rat cultured cerebellar granule neuronal viability. J Neurosci Res 72:747–755
192. Rohn TT, Wong SM, Cotman CW, Cribbs DH (2001) 15-deoxy-delta12,14-prostaglandin J2, a specific ligand for peroxisome proliferator-activated receptor-gamma, induces neuronal apoptosis. Neuroreport 12:839–843
193. Kondo M, Shibata T, Kumagai T, Osawa T, Shibata N, Kobayashi M et al (2002) 15-deoxy-delta(12,14)-prostaglandin J(2): the endogenous electrophile that induces neuronal apoptosis. Proc Natl Acad Sci USA 99:7367–7372
194. Castrillo A, Diaz-Guerra MJ, Hortelano S, Martin-Sanz P, Bosca L (2000) Inhibition of ikappaB kinase and ikappaB phosphorylation by 15-deoxy- delta(12,14)-prostaglandin J(2) in activated murine macrophages. Mol Cell Biol 20:1692–1698
195. Cernuda-Morollon E, Pineda-Molina E, Canada FJ, Perez-Sala D (2001) 15-deoxy-delta 12,14-prostaglandin J2 inhibition of nf-kappaB-DNA binding through covalent modification of the p50 subunit. J Biol Chem 276:35530–35536
196. Campo PA, Das S, Hsiang CH, Bui T, Samuel CE, Straus DS (2002) Translational regulation of cyclin d1 by 15-deoxy-delta(12,14)-prostaglandin J(2). Cell Growth Differ 13:409–420
197. Oliva JL, Perez-Sala D, Castrillo A, Martinez N, Canada FJ, Bosca L et al (2003) The cyclopentenone 15-deoxy-delta 12,14-prostaglandin J2 binds to and activates H-Ras. Proc Natl Acad Sci USA 100:4772–4777
198. Park EJ, Park SY, Joe EH, Jou I (2003) 15d-PGJ2 and rosiglitazone suppress Janus kinase-stat inflammatory signaling through induction of suppressor of cytokine signaling 1 (SOCS1) and SOCS3 in glia. J Biol Chem 278:14747–14752
199. Diamant M, Heine RJ (2003) Thiazolidinediones in type 2 diabetes mellitus: current clinical evidence. Drugs 63:1373–1405
200. Belcher G, Lambert C, Edwards G, Urquhart R, Matthews DR (2005) Safety and tolerability of pioglitazone, metformin, and gliclazide in the treatment of type 2 diabetes. Diabetes Res Clin Pract 70:53–62
201. Lincoff AM, Wolski K, Nicholls SJ, Nissen SE (2007) Pioglitazone and risk of cardiovascular events in patients with type 2 diabetes mellitus: a meta-analysis of randomized trials. JAMA 298:1180–1188
202. Dana SL, Hoener PA, Bilakovics JM, Crombie DL, Ogilvie KM, Kauffman RF et al (2001) Peroxisome proliferator-activated receptor subtype-specific regulation of hepatic and peripheral gene expression in the zucker diabetic fatty rat. Metabolism 50:963–971
203. Nedergaard M (1987) Transient focal ischemia in hyperglycemic rats is associated with increased cerebral infarction. Brain Res 408:79–85
204. Demchuk AM, Morgenstern LB, Krieger DW, Linda Chi T, Hu W, Wein TH et al (1999) Serum glucose level and diabetes predict tissue plasminogen activator-related intracerebral hemorrhage in acute ischemic stroke. Stroke 30:34–39

205. Thal SC, Engelhard K, Werner C (2005) New cerebral protection strategies. Curr Opin Anaesthesiol 18:490–495
206. Martini SR, Kent TA (2007) Hyperglycemia in acute ischemic stroke: a vascular perspective. J Cereb Blood Flow Metab 27:435–451
207. Lee J, Reding M (2007) Effects of thiazolidinediones on stroke recovery: a case-matched controlled study. Neurochem Res 32:635–638
208. Dormandy JA, Charbonnel B, Eckland DJ, Erdmann E, Massi-Benedetti M, Moules IK et al (2005) Secondary prevention of macrovascular events in patients with type 2 diabetes in the proactive study (prospective pioglitazone clinical trial in macrovascular events): a randomised controlled trial. Lancet 366:1279–1289
209. Betteridge DJ, DeFronzo RA, Chilton RJ (2008) Proactive: time for a critical appraisal. Eur Heart J 29:969–983
210. Scheen AJ (2012) Outcomes and lessons from the proactive study. Diabetes Res Clin Pract 98:175–186
211. Wilcox R, Bousser MG, Betteridge DJ, Schernthaner G, Pirags V, Kupfer S et al (2007) Effects of pioglitazone in patients with type 2 diabetes with or without previous stroke: results from proactive (prospective pioglitazone clinical trial in macrovascular events 04). Stroke 38:865–873
212. Wilcox R, Kupfer S, Erdmann E (2008) Effects of pioglitazone on major adverse cardiovascular events in high-risk patients with type 2 diabetes: results from prospective pioglitazone clinical trial in macro vascular events (proactive 10). Am Heart J 155:712–717
213. Nissen SE, Wolski K (2007) Effect of rosiglitazone on the risk of myocardial infarction and death from cardiovascular causes. N Engl J Med 356:2457–2471
214. Graham DJ, Ouellet-Hellstrom R, MaCurdy TE, Ali F, Sholley C, Worrall C et al (2010) Risk of acute myocardial infarction, stroke, heart failure, and death in elderly medicare patients treated with rosiglitazone or pioglitazone. JAMA 304:411–418
215. Charbonnel B, Dormandy J, Erdmann E, Massi-Benedetti M, Skene A (2004) The prospective pioglitazone clinical trial in macrovascular events (PROactive): can pioglitazone reduce cardiovascular events in diabetes? Study design and baseline characteristics of 5238 patients. Diabetes Care 27:1647–1653
216. Mannucci E, Monami M, Lamanna C, Gensini GF, Marchionni N (2008) Pioglitazone and cardiovascular risk. A comprehensive meta-analysis of randomized clinical trials. Diabetes Obes Metab 10:1221–1238
217. Nissen SE, Nicholls SJ, Wolski K, Nesto R, Kupfer S, Perez A et al (2008) Comparison of pioglitazone vs glimepiride on progression of coronary atherosclerosis in patients with type 2 diabetes: the periscope randomized controlled trial. JAMA 299:1561–1573

Chapter 18
CD36: An Inflammatory Mediator in Acute Brain Injury

Sunghee Cho and Maria Febbraio

Abstract Stroke is a major leading cause of death and disability in the human population. The pathology of stroke-induced brain injury involves multifactorial pro-death processes. Among them, inflammation is an important contributor to stroke pathology as indicated by the close association between excessive inflammation and exacerbation of the disease process. Considerable experimental evidence indicates that disease outcome is modulated by several factors including predisposing clinical conditions. Stroke compromises vascular permeability and leads to breakdown of the blood–brain barrier. While the pathology primarily occurs in the CNS, the presence of peripheral immune cells in the infarcted area suggests their potential role in post-ischemic inflammation. Given recent advances highlighting the heterogeneity of peripheral immune cells and diversity of their function, we review neuroimmune interaction in the setting of acute cerebral ischemia, post-ischemic inflammation, and the trafficking of peripheral immune cells to inflamed tissue, with specific focus on the involvement of the class B scavenger receptor, CD36. We discuss CD36 expression and functions, the contribution of the receptor to stroke pathology, its relevance to peripheral inflammatory conditions, and potential strategies to target the CD36-associated neuroinflammatory pathway.

List of Abbreviations

AGE	Advanced glycation end product
BBB	Blood–brain barrier

S. Cho (✉)
Neuroscience, Weill Cornell Medical College, Burke Medical Research Institute, White Plains, NY, USA
e-mail: suc2002@med.cornell.edu

M. Febbraio
University of Alberta, Edmonton, AB, Canada

J. Chen et al. (eds.), *Immunological Mechanisms and Therapies in Brain Injuries and Stroke*, Springer Series in Translational Stroke Research, DOI 10.1007/978-1-4614-8915-3_18,

CCR2	c–c chemokine receptor type 2
DAMP	Damage-associated molecular pattern
EAE	Experimental autoimmune encephalitis
fAβ	Fibrillar beta-amyloid
ICAM	Intercellular adhesion molecule
MCP-1	Monocyte chemotactic protein-1
mLDL	Modified low-density lipoprotein
oxLDL	Oxidized low-density lipoprotein
$_{ox}PC_{CD36}$	Oxidized choline glycerophospholipid species
PAMP	Pathogen-associated molecular pattern
PPAR-γ	Peroxisome proliferator-activated receptor-γ
PRR	Pattern recognition receptor
TLR	Toll-like receptors
TSPs	Thrombospondins
SAB	Salvianolic acid B

CD36, A Multifunctional Scavenger Receptor

Overview

CD36 is an 88 kDa heavily N-linked glycosylated membrane protein [1, 2]. It has to date defied crystallization, so we can only imagine its structure based on protein prediction and modeling. Short intracellular tails extend from the two transmembrane domains (the N-terminal tail results from an uncleaved signal peptide), anchoring the protein, and exposing a large extracellular loop. There is a hydrophobic region that is predicted to dip back towards or into the plasma membrane, and disulfide bonds of the 6 extracellular cysteines constrain the molecule [3, 4]. Two cysteine residues that are palmitoylated characterize both cytoplasmic domains, and both cytoplasmic tails are necessary for efficient plasma membrane CD36 expression [5, 6]. Posttranslational disulfide bond formation, glycosylation, and palmitoylation are all essential in targeting CD36 to the plasma membrane, and the latter is also required for positioning CD36 in caveolae, detergent-resistant membranes, or lipid rafts [4, 7, 8]. In some cell types, expression of caveolin-1 has been shown to be mandatory for plasma membrane targeting of CD36, and disruption of caveolae may affect some CD36-dependent functions [9–11]. The partitioning of CD36 to specific plasma membrane domains may facilitate interaction with signaling partners and interacting proteins that are a requirement for CD36-dependent responses.

The human CD36 gene (including all variants) extends about 77 kb on chromosome 7q11.2 and encodes a predicted protein of 471 amino acids with a predicted molecular weight of 53 kDa (http://www.ncbi.nlm.nih.gov/gene/948). Human CD36 has ten potential N-linked glycosylation sites, and thus the actual molecular weight varies from ~80–100 kDa [12]. The thick complex carbohydrate coat of

CD36 may protect it from proteolysis in harsh environments. Variant transcripts and a multitude of single nucleotide polymorphisms (SNP) mostly in noncoding regions have been identified [13, 14]. Mutations which result in absence of CD36 expression in platelets (Type II CD36 deficiency) or in monocytes and platelets (Type I CD36 deficiency) have been found at a frequency of 3–10 % in Asian and African populations and may persist as a result of selective pressure by the malaria parasite [15–18]. There is controversy as to whether absence of CD36 leads to or predisposes to human pathology, or is protective against malaria or other disease states. This may relate to whether CD36 is absent from all cells and tissues or some subset that differs depending upon the particular polymorphism, and the presence or absence of other interacting gene products.

Signaling

CD36 binding sites for oxidized low-density lipoprotein (oxLDL), growth hormone-releasing peptide and the family of thrombospondins (TSPs) are well defined, while the site for fatty acid binding is less precise and has not been tested definitively [19–22]. Other ligands have been assigned to the immunodominant domain (amino acids 155–183) by virtue of antibody blockade [23]. Alternatively, antibody binding to this domain may lead to a disruptive conformational change. There are two potential phosphorylation sites, both on the extracellular face of CD36, threonine 92 and serine 237. To date, phosphorylation at serine 237 has not been observed. However, there is data suggesting important biological consequences with regard to the phosphorylation of threonine 92. On platelets, phosphorylation reduces palmitate uptake and inhibits binding of TSP-1 and perhaps platelet activation [24–26]. Although it was long presumed that the “default” status in resting platelets was the phosphorylated state, recent evidence points towards a low basal level of CD36 phosphorylation in both platelets and microvascular endothelial cells [26]. Thus, why there is little platelet TSP binding and activation remains an open question. Threonine 92 is recognized by protein kinase C and to a lesser extent by protein kinase A [25, 27]. Recent work in transfected cell lines suggests that phosphorylation may also occur intracellularly as a posttranslational modification under certain conditions [26].

CD36 facilitates fatty acid uptake, but not by a classical transporter mechanism. Fatty acids probably bind transiently in the hydrophobic domain, and this facilitates flip-flop across the membrane, followed by rapid esterification or sequestration by cytoplasmic fatty acid binding proteins [28, 29]. Alternatively, CD36 may provide a hydrophobic pore facilitating fatty acid entry into membranes, analogous to the hypothesized mechanism by which scavenger receptor B1 facilitates cholesterol exchange between high-density lipoprotein and cells [30]. In contrast, uptake of oxLDL is via a caveolin-independent endocytic pathway and depends upon CD36-mediated lyn activation of the vav family of guanine exchange factors for Rho family GTPases, for subsequent vesicle maturation [31, 32]. There is recent evidence suggesting that CD36 is also expressed on mitochondrial membranes, but its function remains controversial [33–35]. While all groups consistently show that increased fatty acid oxidation is accompanied by increased CD36-mediated fatty acid uptake at the plasma membrane, there has been no definitive evidence that CD36 plays a direct role in fatty acid delivery into mitochondria.

Expression and Function in the CNS and Periphery

Initially characterized as a platelet receptor for thrombospondin-1 (TSP-1, designated glycoprotein IV), CD36 expression has subsequently been found on many types of cells and tissues [36]. CD36 is expressed on blood cells and cells of the vasculature, including platelets, reticulocytes, monocytes and macrophages, dendritic cells, endothelial cells, and smooth muscle cells [37–44]. It is found in specialized epithelium, including mammary epithelium, retinal pigment epithelium, apical enterocytes of the proximal small intestine, and the proximal tubule epithelium of the kidney [45–50]. CD36 is expressed in insulin-sensitive cells and tissues, including adipocytes, hepatocytes, cardiac and skeletal muscle, and pancreatic beta cell granules [28, 51]. CD36 is also found in taste receptors of the circumvallate papillae and steroidogenic cells of the adrenal, testes, and ovary [48, 52, 53].

The functions of CD36 are dictated by the cell type, circumstances, and ligand (Fig. 18.1). For example, CD36 plays a major role in fatty acid uptake required for production of energy or heat, especially in heart, skeletal muscle, and brown adipose tissue, and also in fat storage in white adipose tissue, and pathologically in liver and muscle [54–56]. CD36 also functions in recognition of malaria parasites and plays a role in fatty acid sensing by taste buds in the mouth and enterocytes in the gut [2, 48]. The uptake of oxLDLs by monocytes/macrophages leads to the formation of "foamy" macrophages and is a key step in atherosclerotic lesion development [57]. In binding the matricellular protein TSPs, CD36 may not only function as an adhesion receptor in platelets and between platelets, monocytes, tumor cells, sickled erythrocytes, and endothelium but can also modulate TGF-β activation, inhibit angiogenesis, and facilitate uptake of apoptotic cells [58]. CD36 is classified as a pattern recognition receptor (PRR) because it recognizes pathogen-associated molecular patterns or PAMPs, and danger or damage-associated molecular patterns also known as DAMPs [59]. These are repetitive motifs found on pathogens, modified phospholipids, or oxidatively denatured cytoplasmic or nuclear constituents that present as nonself [59, 60]. Some examples include oxidatively modified phospholipids in rod outer segments, apoptotic cells and low- and high-density lipoproteins, advanced glycation end products (AGEs), neurotoxic prion protein, amyloid-beta (Aβ), and diacylglycerides in the cell walls of Gram-positive bacteria. The diverse responses by interacting with specific CD36 ligands often converge into endothelial dysfunction and inflammation, common pathological features of cardio- and cerebrovascular diseases.

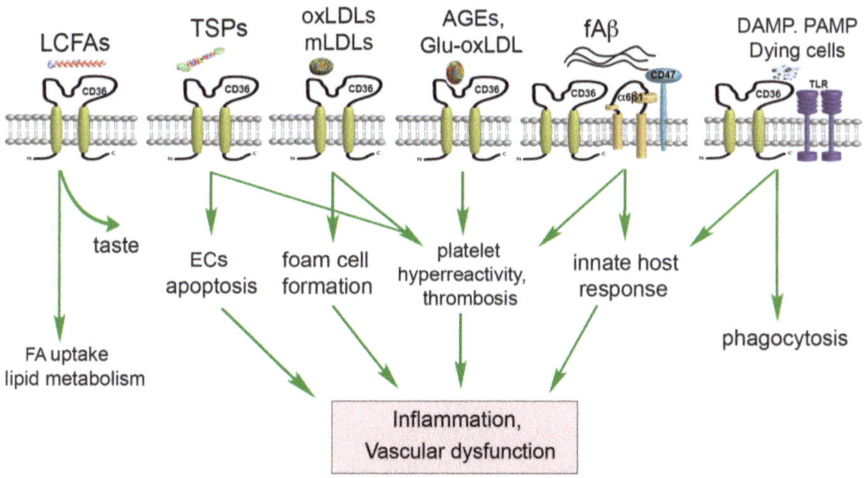

Fig. 18.1 CD36 as a multifunctional receptor. By recognizing a host of ligands, CD36 elicits myriad responses and the interaction between specific ligands and the receptor results in diverse physiological and pathological responses. CD36 forms a complex with α6β1 integrin and CD47 to elicit downstream function and also acts as a coreceptor for TLRs. Many CD36-associated pathways converge in inflammatory responses. *LCFAs* long chain fatty acids, *TSPs* thrombospondins, *EC* endothelial cells, *ox(m)LDLs* oxidized (modified) low-density lipoprotein, *AGEs* advanced glycation end products, *Glu-oxLDL* Glucose-oxLDL, *fAβ* fibrillar β amyloid, *TLR* toll-like receptor

CNS

CD36 is expressed in neurons, microglia, astrocytes, and the endothelium of the blood–brain barrier. CD36 is expressed in neurons found in regions involved in pheromone responses and reproductive behavior. Specifically, it is found in neurons of the pyramidal layer of the ventral hippocampus, CA1 field, amygdalopiriform transition area, the perirhinal cortex, and the ectorhinal cortex [61, 62]. Neurons of the ventromedial hypothalamic nucleus utilize both glucose and fatty acids as signaling molecules and regulate energy homeostasis through central modulation of feeding behavior, hepatic glucose production, and hormone secretion. About 50 % of oleic acid sensing by ventromedial neurons of the hypothalamus was shown to be CD36 dependent, and this sensing was independent of fatty acid metabolism [62]. The mechanism of CD36-dependent fatty acid sensing is presumed to be analogous to that which has been defined in taste buds; CD36 binding by fatty acids is postulated to cause neurotransmitter release by activation of a protein tyrosine kinase and liberation of inositol 1,4,5-triphosphate, leading to calcium-dependent membrane depolarization [62, 63]. Dysfunctional central fatty acid sensing by CD36 may play a role in insulin-resistant states and obesity.

Long chain fatty acid uptake by endothelial cells at the blood–brain barrier is at least partially receptor dependent, and the role of CD36 in this process has been evaluated. The uptake of monounsaturated fatty acids is probably partially CD36 dependent; CD36 knockout mice have a significant decrease in this class of fatty acids, and in cultured blood–brain barrier-derived endothelium, knockdown of CD36 significantly decreased oleic acid uptake [64, 65]. Uptake of polyunsaturated fatty acids is most probably CD36 independent. The role of CD36 in fatty acid uptake within the brain remains to be elucidated.

CD36 expression was found in a subset of astrocytes in the post-ischemic brain [66]. The expression was temporally and spatially limited, only found 3 days following stroke and in the peri-infarct area, where the glial scar forms. Bao and colleagues subsequently showed a close relationship in the expression of CD36 and glial fibrillary acidic protein (GFAP). Inhibition of CD36 expression coincided with decreased GFAP expression and reduced glial scar, suggesting the involvement of CD36 in injury-induced scar formation [67].

CD36 has been found to play a role in microglial activation induced by amyloid-beta (Aβ) in Alzheimer's disease (AD) plaque and by the neurotoxic prion protein, leading to secretion of pro-inflammatory cytokines [68–70]. There are apparently multiple mechanisms of microglial activation and downstream signaling that are CD36 dependent. Studies demonstrated that fibrillar amyloid beta (fAβ) engaged microglia by a complex of receptors that included CD36, scavenger receptor A, α6β1 integrin, CD47, and toll-like receptors (TLRs) 2 and 4. These initiate a signaling cascade that includes p38, src kinase, vav proteins, rac, and reactive oxygen species, leading to cytokine release and phagocytosis [71, 72]. Stewart et al. suggested that TLR 4/6 and CD36 engaged Aβ and activated the inflammasome, leading to release of inflammatory cytokines. This group also found that CD36 was essential to a signaling cascade involving the src kinase fyn phosphorylating p130CAS, a focal adhesion scaffolding protein. This led to recruitment and phosphorylation of pyk2 and paxillin to the leading edge and membrane ruffles, resulting in an increase in microglia migration [73].

Periphery

CD36 is highly expressed in mononuclear phagocytes, including monocytes, macrophages, and dendritic cells [41, 74, 75]. Receptor-mediated uptake of oxidatively modified LDL (mLDL)/oxLDL by monocytes increases transcription of CD36 and several other genes via activation of the nuclear hormone receptor peroxisome proliferator-activated receptor-γ (PPAR-γ) [76]. The activating ligand(s) for PPAR-γ are oxidized phospholipids, such as 9- and 13-HODE, and these and/or their precursor lipids are delivered to the cell within the oxLDL particle. Similarly, in heart and liver, uptake of ligands for other members of the PPAR family is mediated by CD36 and may contribute to cardiac lipotoxicity and hepatic steatosis [77, 78].

On platelets, CD36 expression is variable in the population, and this is attributed to genetic polymorphisms but potentially may also relate to physiological status [79]. For example, both a high-fat diet and insulin resistance upregulate CD36 on monocytes/macrophages and could have similar effect on megakaryocytes.

Humans and mice deficient in platelet CD36 do not have a significant phenotype. However, recent studies using pathophysiological ligands of CD36, specifically, oxidized phospholipids and AGEs, support the hypothesis that platelet CD36 can modulate platelet reactivity by transducing prothrombotic signals [80–82]. The pathway in platelets follows a recurring theme: src kinase activation (in platelets, fyn and lyn are associated with CD36 following oxLDL binding), phosphorylation of vav family proteins, and Map kinase (in this case, MKK4), resulting ultimately in jnk activation. AGEs also trigger platelet CD36-mediated jnk2 activation [83]. In addition to activation of signaling pathways, CD36 may foster exchange of phospholipids and fatty acids between platelets and lipoproteins. The type of fatty acid/phospholipid, and whether it is oxidized, may alter platelet membranes rendering them more susceptible to aggregation [84]. Platelet CD36 has been shown to interact with amyloids, and platelet aggregation was mediated by a p38 MAP kinase and thromboxane A2-dependent pathway [85]. Thus, increased platelet expression and sensitization by pathophysiological CD36 ligands may explain platelet hyperreactivity in inflammation and hyperlipidemic and insulin-resistant states, among others, and lead to appreciable thrombosis in response to subthreshold platelet stimulating agents. In contrast to CD36-expressing mice, CD36 KO mice do not show platelet hyperreactivity in response to high-fat diet feeding or insulin-resistant states and have normal thrombosis profiles in experimental in vivo models that enrich for CD36 ligands [80, 84].

Neuroinflammation

Post-ischemic Inflammation

Post-ischemic inflammation is a contributing negative factor in stroke, exacerbating injury, and influencing outcome [86, 87]. Stroke increases inflammatory mediator: free radical, cytokines/chemokines (IL-1β, TNF-α, IL-6, monocyte chemotactic protein (MCP)-1, macrophage inflammatory protein-1, C-C chemokine receptor type 2 (CCR2), and inflammatory proteins (inducible nitric oxide synthase, cyclooxygenase, and matrix metalloproteinases). These mediators increase endothelial expression of adhesion molecules, such as intercellular adhesion molecule (ICAM)-1 and p-selectin, leading to leukocyte arrest and transendothelial migration at the injured site. These mediators either act in concert or converge into an inflammatory response following activation of temporally and spatially separated cascades [88, 89]. A number of studies have shown that targeting specific neuroinflammatory mediators attenuates stroke-induced brain injury [90–92]. Mice deficient in ICAM-1 or p-selectin displayed smaller infarct size compared to wild type [93–95]. Deficiency in either MCP-1 or its cognate receptor, CCR2, also resulted in protection in murine stroke models [96, 97], while increased expression of MCP-1 exacerbated ischemic outcome with enhanced recruitment of inflammatory cells at the injury sites [98].

Despite apparent benefits of attenuating neuroinflammatory pathways in animal models, clinical trials in human stroke, using antibodies against adhesion molecules and neutrophils, were not effective [99, 100].

Recent studies recognized the complexity in targeting neuroinflammatory pathways. Stroke-induced inflammation is a double-edged sword in that it not only is necessary for containment and repair but can also lead to further damage. It has become increasingly clear that peripheral inflammatory status and comorbidities are important factors in neuroinflammation and outcome [101–103]. Understanding the temporal sequence of activation of inflammatory mediators, spatial localization of the cascade in the infarct (core vs. penumbra), and identification of cell types activated remain to be investigated to selectively reduce adverse while preserving beneficial aspects of the inflammatory response.

CD36: A Modulator of the Innate Immune System

In response to an encounter with microbes, the host elicits a rapid, specific, and self-constraining acute inflammatory response to avoid inflammatory-mediated damage to neighboring tissues [104]. This primordial defense response involves initial recognition of the triggers, so-called PAMPs, through pattern recognition receptors (PRRs), which include scavenger receptors. Subsequently, triggered adaptive immune responses lead to resolution to reinstate tissue homeostasis in a timely manner. Sterile inflammation occurs in post-ischemic tissues in the absence of microbes [105, 106]. The triggers of sterile inflammation are elements of damaged tissue, including oxidized lipids and cytoplasmic proteins, DNA, RNA, and proteolyzed or oxidized extracellular matrix components, which, as previously noted, are collectively termed DAMPs. Regardless of microbial or endogenous in nature, PRRs are believed to be involved in recognizing the triggers and eliciting inflammatory responses (Fig. 18.2).

In addition to playing a role in the endocytic uptake of modified lipoproteins leading to foam cell formation, an important role for monocyte/macrophage CD36 as a PRR in innate immune modulation has emerged [107–109]. This is both in association with and independent of TLRs. CD36 recognizes nonself, and this is one of the oldest and most conserved functions of these receptors, beginning with recognition of apoptotic cells during development as a result of normal homeostasis [110]. The recognition motif, altered fatty acid chains that become hydrophilic and more easily accessed by surface receptors, accounts for the crossover recognition to modified lipoproteins carrying these ligands as a result of oxidative stress [111, 112]. CD36 interaction with apoptotic cells invokes an anti-inflammatory response, consistent with the maintenance of organism status quo. This response involves p38 activation and transcriptional upregulation of the IL-10 promoter by Pbx-1 and Prep-1 [113, 114]. On dendritic cells, CD36 mediates uptake of apoptotic cells and is also involved in cross-presentation of antigens to cytotoxic T cells [41].

In recognition of lipids in cell walls of bacteria, CD36 may play a role not only in endocytosis/phagocytosis but also in TLR signaling responses [73, 115–121]

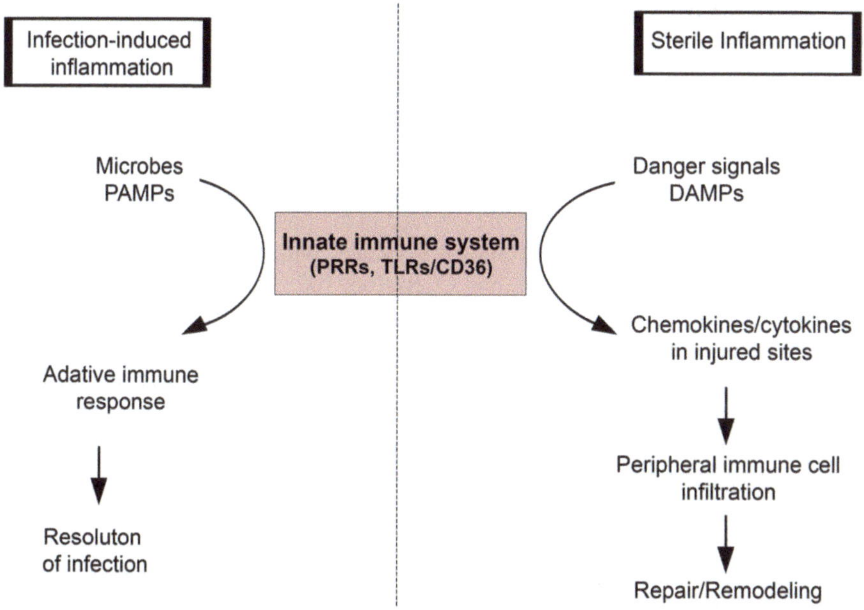

Fig. 18.2 Innate immune receptors resolve pathogen-induced and sterile inflammation. Convergence of innate immune system to resolve inflammation. PAMPs or DAMPs are recognized through pattern recognition receptors such as TLRs and/or CD36. The interactions elicit appropriate downstream responses to resolve infection and repair tissue damage. *TLRs* toll-like receptors, *PAMPs* pathogen-associated molecular patterns, *DAMPs* damage-associated molecular patterns

(Fig. 18.1). This is an emerging topic in CD36 biology, and the exact mechanism is still under investigation. One hypothesis is that CD36 acts an accessory protein for TLRs to deliver ligands, while alternative data suggest that CD36 enhances downstream signaling. This may prove to be important in many of the inflammatory responses mediated by CD36, as there is significant crossover between TLR and CD36 ligands, including fatty acids, amyloid-beta, modified LDL, and other PAMPs/DAMPs. Specifically, CD36 acts as a co-receptor for the recognition of bacteria-derived diacylglyceride through a TLR2/6 complex [115, 122]. OxLDL and Aβ trigger inflammatory signaling through TLR4/6 [73]. Abe and colleagues found that CD36 plays a key role in the inflammatory response and tissue damage mediated by TLR2/1, but not TLR2/6, as a result of cerebral ischemia [121]. This differed from the accessory role of CD36 in monocyte/macrophages in the periphery, where it was instead important for TLR 2/6 signaling. The explanation for this difference was not explored and may relate to differences in expression of other TLR accessory proteins. Nonetheless, these studies suggest that CD36 is involved in the pathogenesis of sterile inflammation through convergence with TLR signaling.

CD36: An Inflammatory Receptor

The pro-inflammatory nature of CD36 has been implicated in atherosclerosis, vascular dysfunction, and neurodegenerative diseases [57, 69, 123, 124]. The concept that CD36 is a prototypic inflammatory receptor and contributes to stroke pathology also has been recognized [66, 102, 125]. In addition to elevated CD36 expression, CD36 ligands such as fAβ, mLDL, oxLDL, and TSPs are elevated in the post-ischemic brain [126–130]. Several studies showed CD36 activation is associated with elevated levels of free radicals, IL-1β, TNF-α, IL-6, MCP-1, and CCR2 [66, 102, 131]. CD36 expression was found predominantly on CD11b+ cells within the infarct territory and the presence of the cells occurs throughout the course of infarct development. However, the identity of the CD11b+ cells as to resident microglia versus infiltrated peripheral mononuclear phagocytes has not been explored. Although CD36 expression was not detected in neurons in the injured tissue, the expression was shown in the subsets of GFAP-expressing astrocytes in the glial scar area [66]. As inflammation is essential in glial scar formation [132], the involvement of CD36 in injury-induced scar formation was addressed. The study by Bao and colleagues reported that CD36 expression covaries with GFAP, an intermediated filament in astrocytes [67]. This study identified CD36 as a novel mediator of GFAP expression and glial scar formation and suggested that targeting CD36 may decrease the barring effect of scar tissue to promote regeneration.

Comorbidities Influencing CD36 Expression/Function

Cardiovascular and cerebrovascular diseases share many prevalent risk factors. These comorbid conditions include hyperlipidemia, insulin resistance associated with metabolic syndrome, obesity and diabetes, impaired vascular function, and hypertension. CD36 expression has been shown to be modulated in comorbid conditions. Since comorbidities increase CD36 expression and a specific set of ligands in a feed-forward manner [76], excessive ligands/receptor interaction associated with risk factors presumably intensify CD36 pathways in disease conditions (Fig. 18.3).

Dyslipidemia

Podrez and colleagues reported increased lipid-based CD36 ligands in ApoE KO mice fed a high-fat diet [80]. They demonstrated a profound upregulation of structurally defined, oxidized choline glycerophospholipid species (${}_{ox}PC_{CD36}$), that serve as high-affinity ligands for CD36 in lipoproteins in the plasma of hyperlipidemic mice and also in humans with low HDL levels [133]. The abundance of ${}_{ox}PC_{CD36}$ in hyperlipidemia led to vascular foam cell formation, a key event in atherosclerotic lesion development [57]. Compared to controls, hyperlipidemic mice subjected to

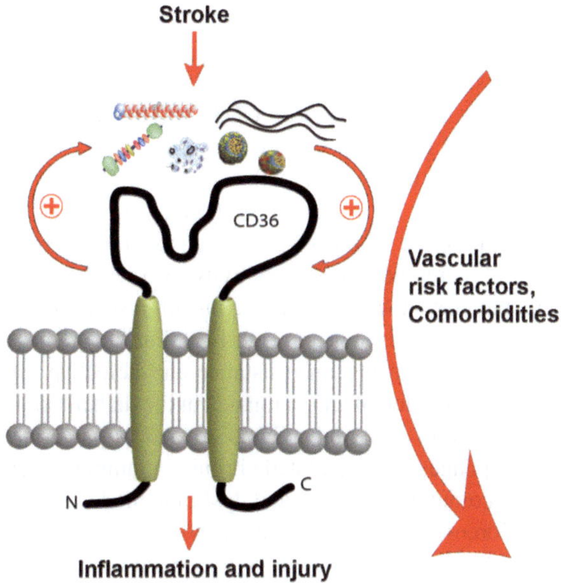

Fig. 18.3 CD36 exacerbates stroke-induced injury as a result of disease modifying risk factors. Following stroke, interaction of CD36 with its ligand occurs in a feed-forward manner. Vascular risk factors and cormodities such as hyperlipidemia, insulin resistance, vascular dysfunction, and hypertension enhance the generation of CD36 ligands and intensify CD36 pathways

cerebral ischemia displayed larger infarcts and heightened post-ischemic inflammation [131]. An underlying hypothesis is that "priming" of peripheral mononuclear phagocytes by CD36/ligands prior to stroke might account for the exacerbation. This study also showed higher expression of CD36 in peripheral monocytes/macrophages in hyperlipidemic mice prior to ischemia. Following stroke, the mice displayed elevated CD36 expression, foam cell area, and pro-inflammatory cytokines/chemokines (MCP-1, CCR2, IL-1β, TNFα, and IL-6) in the post-ischemic brain. The absence of CD36 reversed the hyperlipidemia-associated phenotype [131]. Clear indication from the study is CD36's involvement in hyperlipidemia-induced exacerbation of ischemic inflammation and injury and notably, its peripheral influence on CNS injury development. The possibility that CD36-dependent stroke outcomes could be influenced by the presence of other comorbidities at the time of stroke and also involvement of other factors such as developmental stage (neonatal vs. adult) remain to be investigated.

Insulin Resistance

Insulin resistance associated with diabetes is a predisposing risk factor for stroke as indicated by the fact that 70 % of new stroke victims were previously diagnosed with diabetes, occult diabetes, or were prediabetic with impaired insulin sensitivity [134, 135]. Diabetic conditions promote a chronic pro-inflammatory state that increases the burden of CD36 ligands via modifications of LDL and AGEs, and augments CD36 expression and function. Increased expression of CD36 in monocytes/macrophages not only influences the peripheral inflammatory state but also impacts at the localized site of cerebral ischemic injury.

A potential link between CD36 and impaired insulin sensitivity has been found experimentally in mice. However, in humans, studies are less equivocal, and this may depend upon which cells/tissues are affected by the specific CD36 mutation/SNP, whether the mutation/SNP leads to reduced or increased CD36 expression, and expression of other gene products [136–138]. CD36 KO mice not only have overall increased insulin sensitivity, as a result of increased glucose uptake in muscle, but also show liver-specific insulin resistance as a result of reduced capacity to utilize fatty acids in heart [139, 140]. Further studies showed that CD36 is linked to inflammation in insulin resistance and defective insulin signaling [55, 141]. The increased burden of CD36 ligands in the diabetic state was shown to promote a pro-inflammatory state and a CD36-dependent paracrine loop between adipocytes and macrophages that facilitated chronic inflammation and contributed to insulin resistance common in obesity and dyslipidemia [142].

Moreover, there is abundant evidence that glucose/diabetes modulates CD36 expression and thus impacts CD36 downstream effects. For example, glucose administration upregulates CD36 expression on macrophages [143] and in proximal renal tubular epithelia in humans [144]. Other studies have shown that CD36 expression is increased in monocytes from type 2 diabetic patients [145] and in diabetic mouse hearts [146]. In diabetes associated with atherosclerosis, increased plasma MCP-1 levels were associated with increased monocyte CCR2, CD68, and CD36 and increased vessel wall monocyte number [147]. Liang and colleagues reported increased CD36 protein in macrophages as a response to defective insulin signaling [148]. Glucose was also shown to promote LDL oxidation, and the resulting glucose-oxLDLs stimulated macrophage proliferation in a manner that was dependent on CD36 [149]. Human THP-1 macrophages that were exposed to glucoxidized LDLs increased both CD36 gene expression and accumulation of cholesterol ester (an indicator of foam cell formation) to extents greater than those produced by glycated LDLs or oxLDLs [150].

CD36 has been shown to be localized on insulin-containing granules in human pancreatic beta cells and mediates fatty acid effects on insulin secretion [151]. Handberg and colleagues identified CD36 in plasma (later shown to be contained within microparticles) as a novel marker of insulin resistance [152, 153]. Thus, multiple lines of evidence show that CD36 is modulated by insulin and glucose pathways, and that CD36 has effects on these pathways.

Vascular Dysfunction

The deposition of Aβ in the microvasculature, a hallmark of AD, contributes to oxidative stress and compromises blood–brain barrier (BBB) integrity [154, 155]. Due to the nature of the ligand, many studies on CD36 relevant to fAβ were focused on innate host response and inflammation associated with AD [69, 71, 156]. A key role for CD36 was reported, as CD36 deficiency attenuated fAβ-induced secretion of cytokines, chemokines, and reactive oxygen species in microglia. Macrophage or microglia recruitment into the peritoneum or brain, respectively, in response to injection of fAβ was attenuated in CD36 KO mice [69]. A multi-receptor complex comprised of CD36/$\alpha_6\beta_1$-integrin/CD47 stimulates intracellular tyrosine kinase-based signaling cascades and cellular activation, as detailed previously, which leads to the secretion of pro-inflammatory molecules [72]. In animal models of AD, interaction of Aβ with CD36 causes cerebrovascular oxidative stress and neurovascular dysfunction. The dysfunction was abrogated in the absence of CD36, suggesting that a strategy of CD36 inhibition to normalize cerebrovascular dysfunction might be effective [123, 157]. Lee and colleagues showed an increased level of circulating Aβ in patients with acute ischemic stroke and suggested that the ligand is derived from brain as a consequence of ischemic insult [158].

Hypertension

Hypertension is a major risk factor for stroke [159, 160]. Clinical trials employing antihypertensive agents that aim at reducing blood pressure have been effective in management and prevention [161, 162]. CD36 has been implicated in blood pressure control and modulated by hypertension. Pravenec and colleagues showed that CD36 mutation in the kidney can increase blood pressure and identified renal CD36 as a genetical determinant of blood pressure and risk factor for hypertension [163]. In the stroke prone spontaneous hypertensive rat, BBB impairment was associated with increased CD36 expression in the vessel [164]. Similar to what occurs in diabetics, macrophages from hypertensive subjects show significant increase in CD36 expression, and this was associated with enhanced adhesion to endothelial cells and greater production of ROS [165]. Circulating human endothelial cells also show increased CD36 expression in pulmonary hypertensive states [166]. In hypoxia-induced pulmonary hypertension, CD36 expression increases on intrapulmonary arteries [167]. Human gene association studies have been equivocal with respect to CD36, probably for similar reasons described above in the case of insulin resistance/diabetes.

Neuroimmune Interaction

The presence of granulocytes (neutrophils), subsets of T cells, and monocytes/macrophages in the post-stroke brain suggests mobilization of peripheral immune cells to the injured tissue [88, 168]. There has been controversy regarding the order and timing among the types of immune cells for trafficking. An early study reported that neutrophil infiltration occurs prior to macrophages/activated microglia following stroke [169], while others showed that the accumulation of microglia and/or macrophages in the infarct territory precedes neutrophils [170]. Despite disagreement regarding the order of infiltrating cell types, it is believed that the accumulation of peripheral immune cells contributes to injury development during the acute phase of stroke.

MCP-1/CCR2 Axis for Monocyte Trafficking

Experimental autoimmune encephalitis (EAE) in mice is an example of how inflammatory cells impact disease in the CNS and demonstrates the importance of the MCP-1/CCR2 axis in monocyte recruitment. Among the types of infiltrating cells, monocytes was most tightly coupled to neurobehavioral severity in EAE [171]. Specific inhibition of monocyte recruitment reduced EAE lesion progression, while the presence of T cells was independent of disease severity, strongly implicating infiltrating monocytes in EAE pathogenesis, and confirming an earlier finding [172]. Through serial experimental manipulation using parabiosis (suturing a pair of mice to share circulation) and irradiation/bone marrow transplant of stem cells from genetically engineered mice, CCR2, a G-protein-linked membrane receptor, was found to be the essential mediator of monocyte trafficking, as the study showed the absence of monocyte CCR2 profoundly attenuated paralytic progression of the disease [171].

Monocytes exhibit distinct subsets that are reminiscent of macrophage phenotypes [173–175]. The subset that expresses a high level of the hematopoietic cell differentiation antigen Ly-6C (Ly-6C^{hi}) also expresses CCR2. Ly-6C^{hi} (CCR2+) monocytes are specifically recruited to an injury site and become classically activated M1 macrophages. This CCR2+ subset is chemotactic to MCP-1, which is produced in the inflamed tissue. Recruitment of this subset to inflammatory sites is believed to be CCR2 dependent since monocytes from CCR2-deficient mice do not traffic as efficiently into areas of inflammation [176, 177]. The Ly-6C^{low} monocyte subset expresses CX3CR1, a receptor for CX3CL1 (fractalkine), but is devoid of CCR2 expression. This anti-inflammatory Ly-6C^{low} (CCR2-/CX3CR1+) subset is recruited to normal tissues and develops into resident M2 macrophages that function in host defense and repair after injury [174].

Secreted by microvascular endothelial cells, monocytes/macrophages, and astrocytes upon injury [98, 178–180], MCP-1 is a member of the CC chemokine family and functions in the trafficking of CCR2-expressing monocytes into an injury site. Previous work has established the importance of the MCP-1/CCR2 axis in monocyte/macrophage trafficking in cerebral ischemia. Stroke increases MCP-1 expression in the affected hemisphere. The overexpression of MCP-1 increases infarct volume and enhances the recruitment of monocytes to the injury site [98]. The absence of CCR2 or MCP-1 reduces infarct size [96, 97]. In the absence of CD36, CD36 ligands and injury-induced CC and CXC chemokine production were attenuated [102, 131]. In other disease models that involve recruitment of classically activated M1 macrophages, such as obesity and atherosclerosis, the absence of CD36 is associated with decreased monocyte/macrophage migration/ infiltration and reduced overall numbers of macrophages [141, 181–183], suggesting MCP-1/CCR2 as a major chemokine/ receptor axis for immune cell trafficking.

CD36 in Monocyte/Macrophage Trafficking

Studies indicate the involvement of CD36 in cell mortality and mobility, an important function in cell trafficking. CD36 has been shown to signal through the P130Cas complex to the actin cytoskeleton and regulate microglial migration [184]. Harb and colleagues addressed the role of CD36 in regulating mononuclear phagocyte trafficking to pro-inflammatory atherosclerotic lesions. This study showed that inhibition of CD36 attenuated macrophage accumulation in atherosclerotic lesions, and this was associated with reduced expression of MCP-1 [182]. Cell polarization is essential for migration and mobility of leukocytes. Thus, studies by Park et al. showing that oxLDL/CD36 interaction induced loss of cell polarity and reduced macrophage migration through a vav-Rac-myosin II pathway provide a mechanistic framework to consider CD36 actions [185]. This work explains why macrophages become trapped in areas rich in CD36 ligands and promote further inflammation.

CD36 ligands are elevated in hyperlipidemic conditions and in injured tissues where oxidative or damaged products from cells are released [80, 107]. Through the uptake of lipid-based ligands and foam cell formation in hyperlipidemic conditions, monocyte/macrophage CD36 has been shown to play a role in atherosclerosis and stroke pathology [102, 131, 182, 186]. In a recent study, Kim and colleagues showed that infiltrating immune cells from the periphery are the major source of CD36 in the post-ischemic brain and contribute to stroke-induced brain injury in a hyperlipidemic condition. Mice receiving CD36-deficient bone marrow showed attenuated infarct volume and MCP-1 and CCR2 expression in the brain. The reverse transplantation study (transplantation of CD36-expressing bone marrow-derived cells to CD36 KO mice) showed no increase in infarct volume. The study suggested that CD36 in both host and periphery is required for peripheral CD36 to exert its effect on the hyperlipidemia-induced exacerbation in stroke injury. The underlying mechanism of the exacerbation presumably is that CD36 regulates immune cell trafficking via modulation of the expression of MCP-1 and CCR2 [102].

Targeting CD36 to Attenuate Inflammation

In light of the receptor's pro-inflammatory properties, downregulation of CD36 has been suggested as a strategy to reduce inflammation-associated cerebro- and cardiovascular diseases including altherosclerosis and stroke. Several pharmacological agents were identified to attenuate CD36 expression and function. The antioxidant, α-tocopherol, reduces expression of CD36 and the uptake of oxLDL into macrophages [187–190]. Statins downregulate CD36 expression and suppress oxLDL uptake [187, 191, 192] and subsequently prevent oxLDL-induced macrophage foam cell formation [193]. Hexarelin is a member of the hexapeptide growth hormone-releasing peptide family and binds to CD36 and inhibits its expression [20]. Treatment of mice with hexarelin or a structurally related analogue, EP80317, resulted in a marked decrease in atherosclerotic lesions [194]. Using a high-throughput screening approach for CD36 antagonists based on competition in an oxLDL-binding assay, salvianolic acid B (SAB) was identified as a CD36 inhibitor [195]. SAB is a water-soluble polyphenolic antioxidant isolated from *Danshen*, a Chinese herb that has been used for the prevention and treatment of atherosclerosis and stroke in Asian countries. The specificity and efficacy of SAB in the inhibition of CD36-mediated lipid uptake were confirmed by binding studies for the physical interaction of SAB with CD36. SAB reduces oxLDL-induced CD36 gene expression in cultured cell lines and primary macrophages. Moreover, SAB reduces CD36 gene expression and lipid uptake into macrophages in hyperlipidemic ApoE KO mice [196].

Due to the issues regarding developmental compensatory changes with germ line deletions, investigation on the efficacy of CD36 inhibitors has been complemented by genetic approaches. Besides finding from CD36 KO mice that displayed attenuated stroke-induced inflammation and brain injury [66, 125], effects of a new class of antioxidants peptide, SS31, has been tested against cerebral ischemia [197]. Mice treated with SS31 peptides had attenuated ischemia-induced glutathione (GSH) depletion in the cortex and showed smaller infarct size. The absence of stroke-induced glutathione depletion and no effect on infarct volume in CD36 KO mice treated with the peptide suggested that the protection occurred through the downregulation of CD36 pathways. Because CD36 is a multi-ligand and multifunctional receptor and its expression occurs in a positive feed-forward manner that promotes its functions, targeting at the level of the receptor by interrupting the feed-forward loop to downregulate the CD36 pathway (a multimodal approach) has been suggested [125].

Conclusion

CD36 is an inflammatory receptor that is at the junction of cardio and cerebral vascular disease. Defining the role of CD36 in the CNS and periphery through neuroimmune interactions has been an important emerging area in understanding the pathophysiology of brain injury as a result of cerebral ischemia. CD36 expression is altered in peripheral inflammatory conditions, including obesity, insulin resistance, hyperlipidemia, and hypertension, which also increase stroke incidence either singly or through clustering of these risk factors. Accumulating evidence suggests that CD36 regulates injury-induced mobilization of peripheral immune cells and influences the outcome of stroke. Thus, stroke-induced injury is viewed as the summation of intrinsic ischemic insult and peripheral influences through the neuroimmune interaction. As less favorable outcomes are predicted in patients with various risk factors, targeting CD36-associated pathways may modulate the neuroinflammatory responses in comorbid conditions and serve as a potential approach to limit secondary expansion of primary injury in the setting of acute ischemia.

References

1. Gruarin P, De Monte L, Alessio M (2000) CD36 folding revealed by conformational epitope expression is essential for cytoadherence of Plasmodium falciparum-infected red blood cells. Parasite Immunol 22(7):349–60
2. Oquendo P, Hundt E, Lawler J, Seed B (1989) CD36 directly mediates cytoadherence of Plasmodium falciparum parasitized erythrocytes. Cell 58(1):95–101
3. Rasmussen JT, Berglund L, Rasmussen MS, Petersen TE (1998) Assignment of disulfide bridges in bovine CD36. Eur J Biochem 257(2):488–94
4. Gruarin P, Sitia R, Alessio M (1997) Formation of one or more intrachain disulphide bonds is required for the intracellular processing and transport of CD36. Biochem J 328(Pt 2): 635–42
5. Jochen A, Hays J (1993) Purification of the major substrate for palmitoylation in rat adipocytes: N-terminal homology with CD36 and evidence for cell surface acylation. J Lipid Res 34(10):1783–92
6. Tao N, Wagner SJ, Lublin DM (1996) CD36 is palmitoylated on both N- and C-terminal cytoplasmic tails. J Biol Chem 271(37):22315–20
7. Eyre NS, Cleland LG, Tandon NN, Mayrhofer G (2007) Involvement of the C-terminal cytoplasmic domain in the plasma membrane localization of FAT/CD36 and its ability to mediate long-chain fatty acid uptake. J Lipid Res 48:528–42
8. Eyre NS, Cleland LG, Tandon NN, Mayrhofer G (2007) Importance of the carboxyl terminus of FAT/CD36 for plasma membrane localization and function in long-chain fatty acid uptake. J Lipid Res 48(3):528–42
9. Ring A, Le Lay S, Pohl J, Verkade P, Stremmel W (2006) Caveolin-1 is required for fatty acid translocase (FAT/CD36) localization and function at the plasma membrane of mouse embryonic fibroblasts. Biochim Biophys Acta 1761(4):416–23
10. Pohl J, Ring A, Ehehalt R, Schulze-Bergkamen H, Schad A, Verkade P et al (2004) Long-chain fatty acid uptake into adipocytes depends on lipid raft function. Biochemistry 43(14):4179–87
11. Pohl J, Ring A, Korkmaz U, Ehehalt R, Stremmel W (2005) FAT/CD36-mediated long-chain fatty acid uptake in adipocytes requires plasma membrane rafts. Mol Biol Cell 16(1):24–31
12. Hoosdally SJ, Andress EJ, Wooding C, Martin CA, Linton KJ (2009) The Human Scavenger Receptor CD36: glycosylation status and its role in trafficking and function. J Biol Chem 284(24): 16277–88
13. Wyler B, Daviet L, Bortkiewicz H, Bordet JC, McGregor JL (1993) Cloning of the cDNA encoding human platelet CD36: comparison to PCR amplified fragments of monocyte, endothelial and HEL cells. Thromb Haemost 70(3):500–5
14. Taylor KT, Tang Y, Sobieski DA, Lipsky RH (1993) Characterization of two alternatively spliced 5′-untranslated exons of the human CD36 gene in different cell types. Gene 133(2): 205–12
15. Yamamoto N, Akamatsu N, Sakuraba H, Yamazaki H, Tanoue K (1994) Platelet glycoprotein IV (CD36) deficiency is associated with the absence (type I) or the presence (type II) of glycoprotein IV on monocytes. Blood 83(2):392–7

16. Lee K, Godeau B, Fromont P, Plonquet A, Debili N, Bachir D et al (1999) CD36 deficiency is frequent and can cause platelet immunization in Africans. Transfusion 39(8):873–9
17. Aitman TJ, Cooper LD, Norsworthy PJ, Wahid FN, Gray JK, Curtis BR et al (2000) Malaria susceptibility and CD36 mutation. Nature 405(6790):1015–6
18. Serghides L, Smith TG, Patel SN, Kain KC (2003) CD36 and malaria: friends or foes? Trends Parasitol 19(10):461–9
19. Pearce SF, Roy P, Nicholson AC, Hajjar DP, Febbraio M, Silverstein RL (1998) Recombinant glutathione S-transferase/CD36 fusion proteins define an oxidized low density lipoprotein-binding domain. J Biol Chem 273(52):34875–81
20. Demers A, McNicoll N, Febbraio M, Servant M, Marleau S, Silverstein R et al (2004) Identification of the growth hormone-releasing peptide binding site in CD36: a photoaffinity cross-linking study. Biochem J 382(Pt 2):417–24
21. Pearce SF, Wu J, Silverstein RL (1995) Recombinant GST/CD36 fusion proteins define a thrombospondin binding domain. Evidence for a single calcium-dependent binding site on CD36. J Biol Chem 270(7):2981–6
22. Baillie AGS, Coburn CT, Abumrad NA (1996) Reversible binding of long-chain fatty acids to purified FAT, the adipose CD36 homolog. J Membr Biol 153(1):75–81
23. Navazo MD, Daviet L, Savill J, Ren Y, Leung LL, McGregor JL (1996) Identification of a domain (155–183) on CD36 implicated in the phagocytosis of apoptotic neutrophils. J Biol Chem 271(26):15381–5
24. Guthmann F, Maehl P, Preiss J, Kolleck I, Rustow B (2002) Ectoprotein kinase-mediated phosphorylation of FAT/CD36 regulates palmitate uptake by human platelets. Cell Mol Life Sci 59(11):1999–2003
25. Asch AS, Liu I, Briccetti FM, Barnwell JW, Kwakye-Berko F, Dokun A et al (1993) Analysis of CD36 binding domains: ligand specificity controlled by dephosphorylation of an ectodomain. Science (New York, NY 262(5138):1436–40
26. Chu LY, Silverstein RL (2012) CD36 ectodomain phosphorylation blocks thrombospondin-1 binding: structure-function relationships and regulation by protein kinase C. Arterioscler Thromb Vasc Biol 32(3):760–7
27. Ho M, Hoang HL, Lee KM, Liu N, MacRae T, Montes L et al (2005) Ectophosphorylation of CD36 regulates cytoadherence of Plasmodium falciparum to microvascular endothelium under flow conditions. Infect Immun 73(12):8179–87
28. Abumrad NA, el-Maghrabi MR, Amri EZ, Lopez E, Grimaldi PA (1993) Cloning of a rat adipocyte membrane protein implicated in binding or transport of long-chain fatty acids that is induced during preadipocyte differentiation. Homology with human CD36. J Biol Chem 268(24):17665–8
29. Schwenk RW, Holloway GP, Luiken JJ, Bonen A, Glatz JF (2010) Fatty acid transport across the cell membrane: regulation by fatty acid transporters. Prostaglandins Leukot Essent Fatty Acids 82(4–6):149–54
30. Papale GA, Nicholson K, Hanson PJ, Pavlovic M, Drover VA, Sahoo D (2010) Extracellular hydrophobic regions in scavenger receptor BI play a key role in mediating HDL-cholesterol transport. Arch Biochem Biophys 496(2):132–9
31. Sun B, Boyanovsky BB, Connelly MA, Shridas P, van der Westhuyzen DR, Webb NR (2007) Distinct mechanisms for OxLDL uptake and cellular trafficking by class B scavenger receptors CD36 and SR-BI. J Lipid Res 48(12):2560–70
32. Zeng Y, Tao N, Chung KN, Heuser JE, Lublin DM (2003) Endocytosis of oxidized low density lipoprotein through scavenger receptor CD36 utilizes a lipid raft pathway that does not require caveolin-1. J Biol Chem 278(46):45931–6
33. King KL, Stanley WC, Rosca M, Kerner J, Hoppel CL, Febbraio M (2007) Fatty acid oxidation in cardiac and skeletal muscle mitochondria is unaffected by deletion of CD36. Arch Biochem Biophys 467(2):234–8
34. Campbell SE, Tandon NN, Woldegiorgis G, Luiken JJ, Glatz JF, Bonen A (2004) A novel function for fatty acid translocase (FAT)/CD36: involvement in long chain fatty acid transfer into the mitochondria. J Biol Chem 279(35):36235–41

35. Bezaire V, Bruce CR, Heigenhauser GJ, Tandon NN, Glatz JF, Luiken JJ et al (2006) Identification of fatty acid translocase on human skeletal muscle mitochondrial membranes: essential role in fatty acid oxidation. Am J Physiol Endocrinol Metab 290(3):E509–15
36. Tandon NN, Lipsky RH, Burgess WH, Jamieson GA (1989) Isolation and characterization of platelet glycoprotein IV (CD36). J Biol Chem 264(13):7570–5
37. Joneckis CC, Ackley RL, Orringer EP, Wayner EA, Parise LV (1993) Integrin alpha 4 beta 1 and glycoprotein IV (CD36) are expressed on circulating reticulocytes in sickle cell anemia. Blood 82(12):3548–55
38. Endemann G, Stanton LW, Madden KS, Bryant CM, White RT, Protter AA (1993) CD36 is a receptor for oxidized low density lipoprotein. J Biol Chem 268(16):11811–6
39. Huh HY, Lo SK, Yesner LM, Silverstein RL (1995) CD36 induction on human monocytes upon adhesion to tumor necrosis factor-activated endothelial cells. J Biol Chem 270(11): 6267–71
40. Rouabhia M, Jobin N, Doucet R Jr, Bergeron J, Auger FA (1994) The skin immune system: CD36(+)-dendritic epidermal cell–a putative actor in posttransplant immunological events. Transplant Proc 26(6):3482
41. Albert ML, Pearce SF, Francisco LM, Sauter B, Roy P, Silverstein RL et al (1998) Immature dendritic cells phagocytose apoptotic cells via alphavbeta5 and CD36, and cross-present antigens to cytotoxic T lymphocytes. J Exp Med 188(7):1359–68
42. Swerlick RA, Lee KH, Wick TM, Lawley TJ (1992) Human dermal microvascular endothelial but not human umbilical vein endothelial cells express CD36 in vivo and in vitro. J Immunol 148(1):78–83
43. Lim HJ, Lee S, Lee KS, Park JH, Jang Y, Lee EJ et al (2006) PPARgamma activation induces CD36 expression and stimulates foam cell like changes in rVSMCs. Prostaglandins Other Lipid Mediat 80(3–4):165–74
44. Li W, Febbraio M, Reddy SP, Yu DY, Yamamoto M, Silverstein RL (2010) CD36 participates in a signaling pathway that regulates ROS formation in murine VSMCs. J Clin Investig 120(11):3996–4006
45. Clezardin P, Frappart L, Clerget M, Pechoux C, Delmas PD (1993) Expression of thrombospondin (TSP1) and its receptors (CD36 and CD51) in normal, hyperplastic, and neoplastic human breast. Cancer Res 53(6):1421–30
46. Ryeom SW, Sparrow JR, Silverstein RL (1996) CD36 participates in the phagocytosis of rod outer segments by retinal pigment epithelium. J Cell Sci 109(Pt 2):387–95
47. Lobo MV, Huerta L, Ruiz-Velasco N, Teixeiro E, de la Cueva P, Celdran A et al (2001) Localization of the lipid receptors CD36 and CLA-1/SR-BI in the human gastrointestinal tract: towards the identification of receptors mediating the intestinal absorption of dietary lipids. J Histochem Cytochem 49(10):1253–60
48. Laugerette F, Passilly-Degrace P, Patris B, Niot I, Febbraio M, Montmayeur JP et al (2005) CD36 involvement in orosensory detection of dietary lipids, spontaneous fat preference, and digestive secretions. J Clin Investig 115(11):3177–84
49. Zhang X, Fitzsimmons RL, Cleland LG, Ey PL, Zannettino AC, Farmer EA et al (2003) CD36/fatty acid translocase in rats: distribution, isolation from hepatocytes, and comparison with the scavenger receptor SR-B1. Lab Invest 83(3):317–32
50. Baines RJ, Chana RS, Hall M, Febbraio M, Kennedy D, Brunskill NJ (2012) CD36 mediates proximal tubular binding and uptake of albumin and is upregulated in proteinuric nephropathies. Am J Physiol Renal Physiol 303(7):F1006–14
51. Van Nieuwenhoven FA, Verstijnen CP, Abumrad NA, Willemsen PH, Van Eys GJ, Van der Vusse GJ et al (1995) Putative membrane fatty acid translocase and cytoplasmic fatty acid-binding protein are co-expressed in rat heart and skeletal muscles. Biochem Biophys Res Commun [Comparative Study] 207(2):747–52
52. Gillot I, Jehl-Pietri C, Gounon P, Luquet S, Rassoulzadegan M, Grimaldi P et al (2005) Germ cells and fatty acids induce translocation of CD36 scavenger receptor to the plasma membrane of Sertoli cells. J Cell Sci 118(Pt 14):3027–35

53. Zibara K, Malaud E, McGregor JL (2002) CD36 mRNA and protein expression levels are significantly increased in the heart and testis of apoE deficient mice in comparison to wild type (C57BL/6). J Biomed Biotechnol 2(1):14–21
54. Coburn CT, Knapp FF Jr, Febbraio M, Beets AL, Silverstein RL, Abumrad NA (2000) Defective uptake and utilization of long chain fatty acids in muscle and adipose tissues of CD36 knockout mice. J Biol Chem 275(42):32523–9
55. Kennedy DJ, Kuchibhotla S, Westfall KM, Silverstein RL, Morton RE, Febbraio M (2010) A CD36-dependent pathway enhances macrophage and adipose tissue inflammation and impairs insulin signalling. Cardiovasc Res 89(3):604–13
56. Zhou J, Febbraio M, Wada T, Zhai Y, Kuruba R, He J et al (2008) Hepatic fatty acid transporter Cd36 is a common target of LXR, PXR, and PPARgamma in promoting steatosis. Gastroenterology 134(2):556–67
57. Febbraio M, Podrez EA, Smith JD, Hajjar DP, Hazen SL, Hoff HF et al (2000) Targeted disruption of the class B scavenger receptor CD36 protects against atherosclerotic lesion development in mice. J Clin Investig 105(8):1049–56
58. Febbraio M, Hajjar DP, Silverstein RL (2001) CD36: a class B scavenger receptor involved in angiogenesis, atherosclerosis, inflammation, and lipid metabolism. J Clin Investig 108(6): 785–91
59. Gordon S (2002) Pattern recognition receptors: doubling up for the innate immune response. Cell 111(7):927–30
60. Miller YI, Chang MK, Binder CJ, Shaw PX, Witztum JL (2003) Oxidized low density lipoprotein and innate immune receptors. Curr Opin Lipidol 14(5):437–45
61. Glezer I, Bittencourt JC, Rivest S (2009) Neuronal expression of Cd36, Cd44, and Cd83 antigen transcripts maps to distinct and specific murine brain circuits. J Comp Neurol 517(6):906–24
62. Le Foll C, Irani BG, Magnan C, Dunn-Meynell AA, Levin BE (2009) Characteristics and mechanisms of hypothalamic neuronal fatty acid sensing. Am J Physiol Regul Integr Comp Physiol 297(3):R655–64
63. El-Yassimi A, Hichami A, Besnard P, Khan NA (2008) Linoleic acid induces calcium signaling, Src kinase phosphorylation, and neurotransmitter release in mouse CD36-positive gustatory cells. J Biol Chem 283(19):12949–59
64. Mitchell RW, Edmundson CL, Miller DW, Hatch GM (2009) On the mechanism of oleate transport across human brain microvessel endothelial cells. J Neurochem 110(3):1049–57
65. Song BJ, Elbert A, Rahman T, Orr SK, Chen CT, Febbraio M et al (2010) Genetic ablation of CD36 does not alter mouse brain polyunsaturated fatty acid concentrations. Lipids 45(4): 291–9
66. Cho S, Park EM, Febbraio M, Anrather J, Park L, Racchumi G et al (2005) The class B scavenger receptor CD36 mediates free radical production and tissue injury in cerebral ischemia. J Neurosci 25(10):2504–12
67. Bao Y, Qin L, Kim E, Bhosle S, Guo H, Febbraio M et al (2012) CD36 is involved in astrocyte activation and astroglial scar formation. J Cereb Blood Flow Metab 32(8):1567–77
68. El Khoury J, Hickman SE, Thomas CA, Cao L, Silverstein SC, Loike JD (1996) Scavenger receptor-mediated adhesion of microglia to beta-amyloid fibrils. Nature 382(6593):716–9
69. El Khoury JB, Moore KJ, Means TK, Leung J, Terada K, Toft M et al (2003) CD36 mediates the innate host response to beta-amyloid. J Exp Med 197(12):1657–66
70. Kouadir M, Yang L, Tan R, Shi F, Lu Y, Zhang S et al (2012) CD36 participates in PrP (106–126)-induced activation of microglia. PLoS ONE 7(1):e30756
71. Bamberger ME, Harris ME, McDonald DR, Husemann J, Landreth GE (2003) A cell surface receptor complex for fibrillar beta-amyloid mediates microglial activation. J Neurosci 23(7): 2665–74
72. Wilkinson B, Koenigsknecht-Talboo J, Grommes C, Lee CY, Landreth G (2006) Fibrillar beta-amyloid-stimulated intracellular signaling cascades require Vav for induction of respiratory burst and phagocytosis in monocytes and microglia. J Biol Chem 281(30):20842–50

73. Stewart CR, Stuart LM, Wilkinson K, van Gils JM, Deng J, Halle A et al (2010) CD36 ligands promote sterile inflammation through assembly of a Toll-like receptor 4 and 6 heterodimer. Nat Immunol 11(2):155–61
74. Endemann G, Stanton LW, Madden KS, Bryant CM, White RT, Protter AA (1993) CD36 is a receptor for oxidized low density lipoprotein. J Biol Chem 268(16):11811–6
75. Huh HY, Lo SK, Yesner LM, Silverstein RL (1995) CD36 induction on human monocytes upon adhesion to tumor necrosis factor-activated endothelial cells. J Biol Chem 270(11): 6267–71
76. Tontonoz P, Nagy L, Alvarez JG, Thomazy VA, Evans RM (1998) PPARgamma promotes monocyte/macrophage differentiation and uptake of oxidized LDL. Cell 93(2):241–52
77. Yang J, Sambandam N, Han X, Gross RW, Courtois M, Kovacs A et al (2007) CD36 deficiency rescues lipotoxic cardiomyopathy. Circulation Res 100(8):1208–17
78. Ito M, Suzuki J, Tsujioka S, Sasaki M, Gomori A, Shirakura T et al (2007) Longitudinal analysis of murine steatohepatitis model induced by chronic exposure to high-fat diet. Hepatol Res 37(1):50–7
79. Ghosh A, Murugesan G, Chen K, Zhang L, Wang Q, Febbraio M et al (2011) Platelet CD36 surface expression levels affect functional responses to oxidized LDL and are associated with inheritance of specific genetic polymorphisms. Blood 117(23):6355–66
80. Podrez EA, Byzova TV, Febbraio M, Salomon RG, Ma Y, Valiyaveettil M et al (2007) Platelet CD36 links hyperlipidemia, oxidant stress and a prothrombotic phenotype. Nat Med 13(9): 1086–95
81. Chen K, Febbraio M, Li W, Silverstein RL (2008) A specific CD36-dependent signaling pathway is required for platelet activation by oxidized low-density lipoprotein. Circulation Res 102(12):1512–9
82. Chen K, Li W, Major J, Rahaman SO, Febbraio M, Silverstein RL (2011) Vav guanine nucleotide exchange factors link hyperlipidemia and a prothrombotic state. Blood 117(21): 5744–50
83. Zhu W, Li W, Silverstein RL (2012) Advanced glycation end products induce a prothrombotic phenotype in mice via interaction with platelet CD36. Blood 119(25):6136–44
84. Korporaal SJ, Van Eck M, Adelmeijer J, Ijsseldijk M, Out R, Lisman T et al (2007) Platelet activation by oxidized low density lipoprotein is mediated by CD36 and scavenger receptor-A. Arterioscler Thromb Vasc Biol 27(11):2476–83
85. Herczenik E, Bouma B, Korporaal SJ, Strangi R, Zeng Q, Gros P et al (2007) Activation of human platelets by misfolded proteins. Arterioscler Thromb Vasc Biol 27(7):1657–65
86. Iadecola C, Cho S, Feuerstein GZ, Hallenbeck J (2004) Cerebral Ischemia and Inflammation. In Stroke: Pathophysiology, Diagnosis, and Management. 883–94
87. Dirnagl U, Iadecola C, Moskowitz MA (1999) Pathobiology of ischaemic stroke: an integrated view. Trends Neurosci 22(9):391–7
88. Amantea D, Nappi G, Bernardi G, Bagetta G, Corasaniti MT (2009) Post-ischemic brain damage: pathophysiology and role of inflammatory mediators. FEBS J 276(1):13–26
89. Tan KT, Lip GY, Blann AD (2003) Post-stroke inflammatory response: effects of stroke evolution and outcome. Curr Atheroscler Rep 5(4):245–51
90. del Zoppo G, Ginis I, Hallenbeck JM, Iadecola C, Wang X, Feuerstein GZ (2000) Inflammation and stroke: putative role for cytokines, adhesion molecules and iNOS in brain response to ischemia. Brain Pathol (Zurich, Switzerland) 10(1):95–112
91. Barone FC, Feuerstein GZ (1999) Inflammatory mediators and stroke: new opportunities for novel therapeutics. J Cereb Blood Flow Metab 19(8):819–34
92. Becker KJ (1998) Inflammation and acute stroke. Curr Opin Neurol 11(1):45–9
93. Connolly ES Jr, Winfree CJ, Prestigiacomo CJ, Kim SC, Choudhri TF, Hoh BL et al (1997) Exacerbation of cerebral injury in mice that express the P-selectin gene: identification of P-selectin blockade as a new target for the treatment of stroke. Circulation Res 81(3):304–10
94. Connolly ES Jr, Winfree CJ, Springer TA, Naka Y, Liao H, Yan SD et al (1996) Cerebral protection in homozygous null ICAM-1 mice after middle cerebral artery occlusion. Role of neutrophil adhesion in the pathogenesis of stroke. J Clin Invest 97(1):209–16

95. Soriano SG, Lipton SA, Wang YF, Xiao M, Springer TA, Gutierrez-Ramos JC et al (1996) Intercellular adhesion molecule-1-deficient mice are less susceptible to cerebral ischemia-reperfusion injury. Ann Neurol 39(5):618–24
96. Hughes PM, Allegrini PR, Rudin M, Perry VH, Mir AK, Wiessner C (2002) Monocyte chemoattractant protein-1 deficiency is protective in a murine stroke model. J Cereb Blood Flow Metab 22(3):308–17
97. Dimitrijevic OB, Stamatovic SM, Keep RF, Andjelkovic AV (2007) Absence of the chemokine receptor CCR2 protects against cerebral ischemia/reperfusion injury in mice. Stroke 38(4):1345–53
98. Chen Y, Hallenbeck JM, Ruetzler C, Bol D, Thomas K, Berman NE et al (2003) Overexpression of monocyte chemoattractant protein 1 in the brain exacerbates ischemic brain injury and is associated with recruitment of inflammatory cells. J Cereb Blood Flow Metab 23(6):748–55
99. Enlimomab Acute Stroke Trial investigators (2001) Use of anti-ICAM-1 therapy in ischemic stroke: results of the Enlimomab Acute Stroke Trial. Neurology 57(8):1428–34
100. Becker KJ (2002) Anti-leukocyte antibodies: LeukArrest (Hu23F2G) and Enlimomab (R6.5) in acute stroke. Curr Med Res Opin 18(Suppl 2):s18–22
101. Doyle KP, Buckwalter MS (2012) The double-edged sword of inflammation after stroke: what sharpens each edge? Ann Neurol 71(6):729–31
102. Kim E, Febbraio M, Bao Y, Tolhurst AT, Epstein JM, Cho S (2012) CD36 in the periphery and brain synergizes in stroke injury in hyperlipidemia. Ann Neurol 71(6):753–64
103. Gliem M, Mausberg AK, Lee JI, Simiantonakis I, van Rooijen N, Hartung HP et al (2012) Macrophages prevent hemorrhagic infarct transformation in murine stroke models. Ann Neurol 71(6):743–52
104. Akira S, Uematsu S, Takeuchi O (2006) Pathogen recognition and innate immunity. Cell 124(4):783–801
105. Bamboat ZM, Balachandran VP, Ocuin LM, Obaid H, Plitas G, DeMatteo RP (2010) Toll-like receptor 9 inhibition confers protection from liver ischemia-reperfusion injury. Hepatology 51(2):621–32
106. Bamboat ZM, Ocuin LM, Balachandran VP, Obaid H, Plitas G, DeMatteo RP (2010) Conventional DCs reduce liver ischemia/reperfusion injury in mice via IL-10 secretion. J Clin Invest 120(2):559–69
107. Greenberg ME, Sun M, Zhang R, Febbraio M, Silverstein R, Hazen SL (2006) Oxidized phosphatidylserine-CD36 interactions play an essential role in macrophage-dependent phagocytosis of apoptotic cells. J Exp Med 203(12):2613–25
108. Ren Y, Silverstein RL, Allen J, Savill J (1995) CD36 gene transfer confers capacity for phagocytosis of cells undergoing apoptosis. J Exp Med 181(5):1857–62
109. Savill J (1997) Recognition and phagocytosis of cells undergoing apoptosis. Br Med Bull 53(3):491–508
110. Franc NC, Dimarcq JL, Lagueux M, Hoffmann J, Ezekowitz RA (1996) Croquemort, a novel Drosophila hemocyte/macrophage receptor that recognizes apoptotic cells. Immunity 4(5): 431–43
111. Greenberg ME, Li XM, Gugiu BG, Gu X, Qin J, Salomon RG et al (2008) The lipid whisker model of the structure of oxidized cell membranes. J Biol Chem 283(4):2385–96
112. Hazen SL (2008) Oxidized phospholipids as endogenous pattern recognition ligands in innate immunity. J Biol Chem 283:15527–31
113. Voll RE, Herrmann M, Roth EA, Stach C, Kalden JR, Girkontaite I (1997) Immunosuppressive effects of apoptotic cells. Nature 390(6658):350–1
114. Chung EY, Liu J, Homma Y, Zhang Y, Brendolan A, Saggese M et al (2007) Interleukin-10 expression in macrophages during phagocytosis of apoptotic cells is mediated by homeodomain proteins Pbx1 and Prep-1. Immunity 27(6):952–64
115. Hoebe K, Georgel P, Rutschmann S, Du X, Mudd S, Crozat K et al (2005) CD36 is a sensor of diacylglycerides. Nature 433(7025):523–7
116. Stuart LM, Deng J, Silver JM, Takahashi K, Tseng AA, Hennessy EJ et al (2005) Response to Staphylococcus aureus requires CD36-mediated phagocytosis triggered by the COOH-terminal cytoplasmic domain. J Cell Biol 170(3):477–85

117. Drage MG, Pecora ND, Hise AG, Febbraio M, Silverstein RL, Golenbock DT et al (2009) TLR2 and its co-receptors determine responses of macrophages and dendritic cells to lipoproteins of Mycobacterium tuberculosis. Cellular Immunol 258(1):29–37
118. Triantafilou M, Gamper FG, Lepper PM, Mouratis MA, Schumann C, Harokopakis E et al (2007) Lipopolysaccharides from atherosclerosis-associated bacteria antagonize TLR4, induce formation of TLR2/1/CD36 complexes in lipid rafts and trigger TLR2-induced inflammatory responses in human vascular endothelial cells. Cell Microbiol 9(8):2030–9
119. Triantafilou M, Gamper FG, Haston RM, Mouratis MA, Morath S, Hartung T et al (2006) Membrane sorting of toll-like receptor (TLR)-2/6 and TLR2/1 heterodimers at the cell surface determines heterotypic associations with CD36 and intracellular targeting. J Biol Chem 281(41):31002–11
120. Seimon TA, Nadolski MJ, Liao X, Magallon J, Nguyen M, Feric NT et al (2010) Atherogenic lipids and lipoproteins trigger CD36-TLR2-dependent apoptosis in macrophages undergoing endoplasmic reticulum stress. Cell Metab 12(5):467–82
121. Abe T, Shimamura M, Jackman K, Kurinami H, Anrather J, Zhou P et al (2010) Key role of CD36 in Toll-like receptor 2 signaling in cerebral ischemia. Stroke 41(5):898–904
122. Jimenez-Dalmaroni MJ, Xiao N, Corper AL, Verdino P, Ainge GD, Larsen DS et al (2009) Soluble CD36 ectodomain binds negatively charged diacylglycerol ligands and acts as a co-receptor for TLR2. PLoS ONE 4(10):e7411
123. Park L, Wang G, Zhou P, Zhou J, Pitstick R, Previti ML et al (2011) Scavenger receptor CD36 is essential for the cerebrovascular oxidative stress and neurovascular dysfunction induced by amyloid-beta. Proc Natl Acad Sci U S A 108(12):5063–8
124. Su X, Maguire-Zeiss KA, Giuliano R, Prifti L, Venkatesh K, Federoff HJ (2007) Synuclein activates microglia in a model of Parkinson's disease. Neurobiol Aging 29:1690–701
125. Cho S, Kim E (2009) CD36: a multi-modal target for acute stroke therapy. J Neurochem 109(Suppl 1):126–32
126. Kim E, Tolhurst AT, Qin LY, Chen XY, Febbraio M, Cho S (2008) CD36/fatty acid translocase, an inflammatory mediator, is involved in hyperlipidemia-induced exacerbation in ischemic brain injury. J Neurosci 28(18):4661–70
127. Fitch MT, Silver J (1997) Activated macrophages and the blood–brain barrier: inflammation after CNS injury leads to increases in putative inhibitory molecules. Exp Neurol 148(2): 587–603
128. Nihashi T, Inao S, Kajita Y, Kawai T, Sugimoto T, Niwa M et al (2001) Expression and distribution of beta amyloid precursor protein and beta amyloid peptide in reactive astrocytes after transient middle cerebral artery occlusion. Acta Neurochir (Wien) 143(3):287–95
129. Hayashi T, Noshita N, Sugawara T, Chan PH (2003) Temporal profile of angiogenesis and expression of related genes in the brain after ischemia. J Cereb Blood Flow Metab 23(2): 166–80
130. Pilitsis JG, Coplin WM, O'Regan MH, Wellwood JM, Diaz FG, Fairfax MR et al (2003) Measurement of free fatty acids in cerebrospinal fluid from patients with hemorrhagic and ischemic stroke. Brain Res 985(2):198–201
131. Uno M, Kitazato KT, Nishi K, Itabe H, Nagahiro S (2003) Raised plasma oxidised LDL in acute cerebral infarction. J Neurol Neurosurg Psychiatry 74(3):312–6
132. Shie FS, Neely MD, Maezawa I, Wu H, Olson SJ, Jurgens G et al (2004) Oxidized low-density lipoprotein is present in astrocytes surrounding cerebral infarcts and stimulates astrocyte interleukin-6 secretion. Am J Pathol 164(4):1173–81
133. Podrez EA, Poliakov E, Shen Z, Zhang R, Deng Y, Sun M et al (2002) Identification of a novel family of oxidized phospholipids that serve as ligands for the macrophage scavenger receptor CD36. J Biol Chem 277(41):38503–16
134. Kernan WN, Inzucchi SE (2004) Type 2 diabetes mellitus and insulin resistance: stroke prevention and management. Curr Treat Options Neurol 6(6):443–50
135. Kernan WN, Inzucchi SE, Viscoli CM, Brass LM, Bravata DM, Shulman GI et al (2003) Impaired insulin sensitivity among nondiabetic patients with a recent TIA or ischemic stroke. Neurology 60(9):1447–51

136. Gautam S, Banerjee M (2011) The macrophage Ox-LDL receptor, CD36 and its association with type II diabetes mellitus. Mol Genet Metab [Review] 102(4):389–98
137. Yanai H, Chiba H, Morimoto M, Jamieson GA, Matsuno K (2000) Type I CD36 deficiency in humans is not associated with insulin resistance syndrome. Thromb Haemost 83(5):786
138. Wang Y, Zhou XO, Zhang Y, Gao PJ, Zhu DL (2012) Association of the CD36 gene with impaired glucose tolerance, impaired fasting glucose, type-2 diabetes, and lipid metabolism in essential hypertensive patients. Genet Mol Res 11(3):2163–70
139. Goudriaan JR, Dahlmans VE, Teusink B, Ouwens DM, Febbraio M, Maassen JA et al (2003) CD36 deficiency increases insulin sensitivity in muscle, but induces insulin resistance in the liver in mice. J lipid Res 44(12):2270–7
140. Kuang M, Febbraio M, Wagg C, Lopaschuk GD, Dyck JR (2004) Fatty acid translocase/CD36 deficiency does not energetically or functionally compromise hearts before or after ischemia. Circulation 109(12):1550–7
141. Nicholls HT, Kowalski G, Kennedy DJ, Risis S, Zaffino LA, Watson N et al (2011) Hematopoietic cell-restricted deletion of CD36 reduces high-fat diet-induced macrophage infiltration and improves insulin signaling in adipose tissue. Diabetes 60(4):1100–10
142. Kennedy DJ, Kuchibhotla S, Westfall KM, Silverstein RL, Morton RE, Febbraio M (2011) A CD36-dependent pathway enhances macrophage and adipose tissue inflammation and impairs insulin signalling. Cardiovasc Res 89(3):604–13
143. Griffin E, Re A, Hamel N, Fu C, Bush H, McCaffrey T et al (2001) A link between diabetes and atherosclerosis: glucose regulates expression of CD36 at the level of translation. Nat Med 7(7):840–6
144. Susztak K, Ciccone E, McCue P, Sharma K, Bottinger EP (2005) Multiple metabolic hits converge on CD36 as novel mediator of tubular epithelial apoptosis in diabetic nephropathy. PLoS Med 2(2):e45
145. Sampson MJ, Davies IR, Braschi S, Ivory K, Hughes DA (2003) Increased expression of a scavenger receptor (CD36) in monocytes from subjects with Type 2 diabetes. Atherosclerosis 167(1):129–34
146. Greenwalt DE, Scheck SH, Rhinehart-Jones T (1995) Heart CD36 expression is increased in murine models of diabetes and in mice fed a high fat diet. J Clin Investig 96(3):1382–8
147. Mine S, Okada Y, Tanikawa T, Kawahara C, Tabata T, Tanaka Y (2006) Increased expression levels of monocyte CCR2 and monocyte chemoattractant protein-1 in patients with diabetes mellitus. Biochem Biophys Res Commun 344(3):780–5
148. Liang CP, Han S, Okamoto H, Carnemolla R, Tabas I, Accili D et al (2004) Increased CD36 protein as a response to defective insulin signaling in macrophages. J Clin Investig 113(5): 764–73
149. Lamharzi N, Renard CB, Kramer F, Pennathur S, Heinecke JW, Chait A et al (2004) Hyperlipidemia in concert with hyperglycemia stimulates the proliferation of macrophages in atherosclerotic lesions: potential role of glucose-oxidized LDL. Diabetes 53(12):3217–25
150. Lam MC, Tan KC, Lam KS (2004) Glycoxidized low-density lipoprotein regulates the expression of scavenger receptors in THP-1 macrophages. Atherosclerosis 177(2):313–20
151. Noushmehr H, D'Amico E, Farilla L, Hui H, Wawrowsky KA, Mlynarski W et al (2005) Fatty acid translocase (FAT/CD36) is localized on insulin-containing granules in human pancreatic beta-cells and mediates fatty acid effects on insulin secretion. Diabetes 54(2): 472–81
152. Handberg A, Levin K, Hojlund K, Beck-Nielsen H (2006) Identification of the oxidized low-density lipoprotein scavenger receptor CD36 in plasma: a novel marker of insulin resistance. Circulation 114(11):1169–76
153. Alkhatatbeh MJ, Mhaidat NM, Enjeti AK, Lincz LF, Thorne RF (2011) The putative diabetic plasma marker, soluble CD36, is non-cleaved, non-soluble and entirely associated with microparticles. J Thromb Haemost 9(4):844–51
154. Hartz AM, Bauer B, Soldner EL, Wolf A, Boy S, Backhaus R et al (2011) Amyloid-beta contributes to blood–brain barrier leakage in transgenic human amyloid precursor protein mice and in humans with cerebral amyloid angiopathy. Stroke 43:514–23

155. Carrano A, Hoozemans JJ, van der Vies SM, Rozemuller AJ, van Horssen J, de Vries HE (2011) Amyloid Beta induces oxidative stress-mediated blood–brain barrier changes in capillary amyloid angiopathy. Antioxid Redox Signal 15(5):1167–78
156. Moore KJ, El Khoury J, Medeiros LA, Terada K, Geula C, Luster AD et al (2002) A CD36-initiated signaling cascade mediates inflammatory effects of beta-amyloid. J Biol Chem 277(49):47373–9
157. Wilkinson K, Boyd JD, Glicksman M, Moore KJ, El Khoury J (2011) A high content drug screen identifies ursolic acid as an inhibitor of amyloid beta protein interactions with its receptor CD36. J Biol Chem 286(40):34914–22
158. Lee PH, Bang OY, Hwang EM, Lee JS, Joo US, Mook-Jung I et al (2005) Circulating beta amyloid protein is elevated in patients with acute ischemic stroke. J Neural Transm 112(10): 1371–9
159. Amenta F, Di Tullio MA, Tomassoni D (2003) Arterial hypertension and brain damage–evidence from animal models (review). Clin Exp Hypertens 25(6):359–80
160. Droste DW, Ritter MA, Dittrich R, Heidenreich S, Wichter T, Freund M et al (2003) Arterial hypertension and ischaemic stroke. Acta Neurol Scand 107(4):241–51
161. Bornstein N, Silvestrelli G, Caso V, Parnetti L (2006) Arterial hypertension and stroke prevention: an update. Clin Exp Hypertens 28(3–4):317–26
162. Amenta F, Mignini F, Rabbia F, Tomassoni D, Veglio F (2002) Protective effect of anti-hypertensive treatment on cognitive function in essential hypertension: analysis of published clinical data. J Neurol Sci 203–204:147–51
163. Pravenec M, Churchill P, Chuchill MC, Viklicky O, Kazdova L, Aitman TJ (2008) Identification of renal CD36 as a determinant of blood pressure and risk for hypertension. Nature Genetics 40:952–4
164. Ueno M, Nakagawa T, Nagai Y, Nishi N, Kusaka T, Kanenishi K et al (2011) The expression of CD36 in vessels with blood–brain barrier impairment in a stroke-prone hypertensive model. Neuropathol Appl Neurobiol 37(7):727–37
165. Zapolska-Downar D, Siennicka A, Chelstowski K, Widecka K, Goracy I, Halasa M et al (2006) Is there an association between angiotensin-converting enzyme gene polymorphism and functional activation of monocytes and macrophage in young patients with essential hypertension? J Hypertens 24(8):1565–73
166. Bull HA, Brickell PM, Dowd PM (1994) Src-related protein tyrosine kinases are physically associated with the surface antigen CD36 in human dermal microvascular endothelial cells. FEBS Lett 351(1):41–4
167. Kwapiszewska G, Wilhelm J, Wolff S, Laumanns I, Koenig IR, Ziegler A et al (2005) Expression profiling of laser-microdissected intrapulmonary arteries in hypoxia-induced pulmonary hypertension. Respir Res 6:109
168. Schilling M, Besselmann M, Leonhard C, Mueller M, Ringelstein EB, Kiefer R (2003) Microglial activation precedes and predominates over macrophage infiltration in transient focal cerebral ischemia: a study in green fluorescent protein transgenic bone marrow chimeric mice. Exp Neurol 183(1):25–33
169. Stevens SL, Bao J, Hollis J, Lessov NS, Clark WM, Stenzel-Poore MP (2002) The use of flow cytometry to evaluate temporal changes in inflammatory cells following focal cerebral ischemia in mice. Brain Res 932(1–2):110–9
170. Gelderblom M, Leypoldt F, Steinbach K, Behrens D, Choe CU, Siler DA et al (2009) Temporal and spatial dynamics of cerebral immune cell accumulation in stroke. Stroke 40(5): 1849–57
171. Ajami B, Bennett JL, Krieger C, McNagny KM, Rossi FM (2011) Infiltrating monocytes trigger EAE progression, but do not contribute to the resident microglia pool. Nat Neurosci 14(9):1142–9
172. Bauer J, Huitinga I, Zhao W, Lassmann H, Hickey WF, Dijkstra CD (1995) The role of macrophages, perivascular cells, and microglial cells in the pathogenesis of experimental autoimmune encephalomyelitis. Glia 15(4):437–46
173. Gordon S (2007) Macrophage heterogeneity and tissue lipids. J Clin Investig 117(1):89–93

174. Auffray C, Sieweke MH, Geissmann F (2009) Blood monocytes: development, heterogeneity, and relationship with dendritic cells. Ann Rev Immunol 27:669–92
175. Geissmann F, Manz MG, Jung S, Sieweke MH, Merad M, Ley K (2010) Development of monocytes, macrophages, and dendritic cells. Science (New York, NY); 327(5966):656–61
176. Nahrendorf M, Swirski FK, Aikawa E, Stangenberg L, Wurdinger T, Figueiredo JL et al (2007) The healing myocardium sequentially mobilizes two monocyte subsets with divergent and complementary functions. J Exp Med 204(12):3037–47
177. Swirski FK, Nahrendorf M, Etzrodt M, Wildgruber M, Cortez-Retamozo V, Panizzi P et al (2009) Identification of splenic reservoir monocytes and their deployment to inflammatory sites. Science (New York, NY 325(5940):612–6
178. Kim JS, Gautam SC, Chopp M, Zaloga C, Jones ML, Ward PA et al (1995) Expression of monocyte chemoattractant protein-1 and macrophage inflammatory protein-1 after focal cerebral ischemia in the rat. J Neuroimmunol 56(2):127–34
179. Minami M, Satoh M (2003) Chemokines and their receptors in the brain: pathophysiological roles in ischemic brain injury. Life Sci 74(2–3):321–7
180. Che X, Ye W, Panga L, Wu DC, Yang GY (2001) Monocyte chemoattractant protein-1 expressed in neurons and astrocytes during focal ischemia in mice. Brain Res 902(2):171–7
181. Kennedy DJ, Kuchibhotla SD, Guy E, Park YM, Nimako G, Vanegas D et al (2009) Dietary cholesterol plays a role in CD36-mediated atherogenesis in LDLR-knockout mice. Arterioscl Thromb Vasc biol 29(10):1481–7
182. Harb D, Bujold K, Febbraio M, Sirois MG, Ong H, Marleau S (2009) The role of the scavenger receptor CD36 in regulating mononuclear phagocyte trafficking to atherosclerotic lesions and vascular inflammation. Cardiovasc Res 83:42–51
183. Kuchibhotla S, Vanegas D, Kennedy DJ, Guy E, Nimako G, Morton RE et al (2008) Absence of CD36 protects against atherosclerosis in ApoE knock-out mice with no additional protection provided by absence of scavenger receptor A I/II. Cardiovasc Res 78:185–96
184. Stuart LM, Bell SA, Stewart CR, Silver JM, Richard J, Goss JL et al (2007) CD36 signals to the actin cytoskeleton and regulates microglial migration via a p130Cas complex. J Biol Chem 282(37):27392–401
185. Park YM, Drazba JA, Vasanji A, Egelhoff T, Febbraio M, Silverstein RL (2012) Oxidized LDL/CD36 interaction induces loss of cell polarity and inhibits macrophage locomotion. Mol Biol Cell 23(16):3057–68
186. Park YM, Febbraio M, Silverstein RL (2009) CD36 modulates migration of mouse and human macrophages in response to oxidized LDL and may contribute to macrophage trapping in the arterial intima. J Clin Investig 119(1):136–45
187. Fuhrman B, Volkova N, Aviram M (2002) Oxidative stress increases the expression of the CD36 scavenger receptor and the cellular uptake of oxidized low-density lipoprotein in macrophages from atherosclerotic mice: protective role of antioxidants and of paraoxonase. Atherosclerosis 161(2):307–16
188. Venugopal SK, Devaraj S, Jialal I (2004) RRR-alpha-tocopherol decreases the expression of the major scavenger receptor, CD36, in human macrophages via inhibition of tyrosine kinase (Tyk2). Atherosclerosis 175(2):213–20
189. Ricciarelli R, Zingg JM, Azzi A (2000) Vitamin E reduces the uptake of oxidized LDL by inhibiting CD36 scavenger receptor expression in cultured aortic smooth muscle cells. Circulation 102(1):82–7
190. Munteanu A, Zingg JM, Ogru E, Libinaki R, Gianello R, West S et al (2004) Modulation of cell proliferation and gene expression by alpha-tocopheryl phosphates: relevance to atherosclerosis and inflammation. Biochem Biophys Res Commun 318(1):311–6
191. Han J, Zhou X, Yokoyama T, Hajjar DP, Gotto AM Jr, Nicholson AC (2004) Pitavastatin downregulates expression of the macrophage type B scavenger receptor, CD36. Circulation 109(6):790–6
192. Puccetti L, Sawamura T, Pasqui AL, Pastorelli M, Auteri A, Bruni F (2005) Atorvastatin reduces platelet-oxidized-LDL receptor expression in hypercholesterolaemic patients. Eur J Clin Investig 35(1):47–51

193. Akisato Y, Ishii I, Kitahara M, Tamaki T, Saito Y, Kitada M (2008) Effect of pitavastatin on macrophage cholesterol metabolism. Yakugaku Zasshi 128(3):357–63
194. Marleau S, Harb D, Bujold K, Avallone R, Iken K, Wang Y et al (2005) EP 80317, a ligand of the CD36 scavenger receptor, protects apolipoprotein E-deficient mice from developing atherosclerotic lesions. Faseb J 19(13):1869–71
195. Wang L, Bao Y, Yang Y, Wu Y, Chen X, Si S et al (2010) Discovery of antagonists for human scavenger receptor CD36 via an ELISA-like high-throughput screening assay. J Biomol Screen 15(3):239–50
196. Bao Y, Wang L, Xu Y, Yang Y, Si S, Cho S et al (2012) Salvianolic acid B inhibits macrophage uptake of modified low density lipoprotein (mLDL) in a scavenger receptor CD36-dependent manner. Atherosclerosis 223(1):152–9
197. Cho S, Szeto HH, Kim E, Kim H, Tolhurst AT, Pinto JT (2007) A novel cell-permeable antioxidant peptide, SS31, attenuates ischemic brain injury by down-regulating CD36. J Biol Chem 282(7):4634–42

Chapter 19
Cool Down the Inflammation: Hypothermia as a Therapeutic Strategy for Acute Brain Injuries

Yama Akbari, Maximilian Mulder, Ali Razmara, and Romergryko Geocadin

Abstract While hypothermia has historically been appreciated as a possible cause of injury and mortality, its role as a therapeutic agent of choice in special circumstances has been less delineated in its origins. There are reports of ancient Egyptians, Greeks, and Romans possibly implementing it, including Hippocrates specifically recommending that wounded soldiers be surrounded by snow to improve survival (Polderman, Intensive Care Med 30:757–769, 2004). The first clinical usages of hypothermia for brain injury were conducted by the neurosurgeon Temple Fay in the 1930s with subsequent pioneering reports of its therapeutic usage in various brain disorders (Fay, Ann Surg 101:76–132, 1959; Harris et al., Arch Neurol 59:1077–1083, 2002). Over the last few decades, there have been various animal studies and clinical trials that have investigated the various indications and mechanisms of therapeutic hypothermia, many of which implicate a role in curtailing the inflammatory cascade. Only over the past decade, in the setting of adult and pediatric post-anoxic encephalopathy, has hypothermia become an evidence-based therapy. Indications for treatment as well as mitigation of the inflammatory processes in other acute brain injuries such as ischemic and hemorrhagic strokes, traumatic brain injury, and status epilepticus remain controversial. This chapter reviews the role of therapeutic hypothermia, how "cooling the inflammation" may or may not be indicated in a variety of acute brain injuries: ischemic stroke, neonatal hypoxia–ischemia, post-cardiac arrest global ischemia, traumatic brain injury,

Y. Akbari
University of California, Irvine, CA, USA

M. Mulder
Johns Hopkins Hospital, Baltimore, MD, USA

A. Razmara
Massachusetts General Hospital, Boston, MA, USA

R. Geocadin (✉)
The Johns Hopkins University School of Medicine, Baltimore, MD, USA
e-mail: rgeocadi@jhmi.edu

J. Chen et al. (eds.), *Immunological Mechanisms and Therapies in Brain Injuries and Stroke*, Springer Series in Translational Stroke Research, DOI 10.1007/978-1-4614-8915-3_19,

intracerebral hemorrhage, subarachnoid hemorrhage, status epilepticus, meningitis/encephalitis, and acute liver failure.

Introduction

While hypothermia has historically been appreciated as a possible cause of injury and mortality, its role as a therapeutic agent of choice in special circumstances has been less delineated in its origins. There are reports of ancient Egyptians, Greeks, and Romans possibly implementing it, including Hippocrates specifically recommending that wounded soldiers be surrounded by snow to improve survival [1]. The first clinical usages of hypothermia for brain injury were conducted by the neurosurgeon Temple Fay in the 1930s with subsequent pioneering reports of its therapeutic usage in various brain disorders [2, 3]. Over the last few decades, there have been various animal studies and clinical trials that have investigated the various indications and mechanisms of therapeutic hypothermia, many of which implicate a role in curtailing the inflammatory cascade. Only over the past decade, in the setting of adult and pediatric post-anoxic encephalopathy, has hypothermia become an evidence-based therapy. Indications for treatment as well as mitigation of the inflammatory processes in other acute brain injuries such as ischemic and hemorrhagic strokes, traumatic brain injury, and status epilepticus remain controversial. This chapter reviews the role of therapeutic hypothermia, how "cooling the inflammation" may or may not be indicated in a variety of acute brain injuries: ischemic stroke, neonatal hypoxia–ischemia, post-cardiac arrest global ischemia, traumatic brain injury, intracerebral hemorrhage, subarachnoid hemorrhage, status epilepticus, meningitis/encephalitis, and acute liver failure.

Therapeutic Hypothermia: Methods of In Vivo Implementation

There are a few different methods to implement therapeutic hypothermia. External, noninvasive devices focus on surface cooling. Such methods include simple ice packs, cooling blankets, as well as more sophisticated pads that tightly bind a large amount of skin surface area and have feedback controls for continuous circulation of absorbed heat to maintain hypothermia [4]. Other attempted methods of implementing hypothermia include the usage of cool air during ventilation and the simple infusion of cold saline through a peripheral intravenous catheter [4]. Intravascular, core-cooling invasive devices circulate cold liquid through closed circuit central venous catheters [5]. Surface cooling devices and intravascular cooling devices each have their own advantages and disadvantages depending on the scenario. For example, one study found that surface cooling can lead to higher incidence of hyperglycemia, whereas intravascular core cooling can lead to hypomagnesemia [6]. However, this study did not find significant differences in survival outcome or in the time required to achieve target temperature using either method. More recently, there have been studies that demonstrate feasibility of achieving hypothermia using molecules or medication that affect central temperature regulation [7], such as cholecystokinin octapeptide [8] and cannabinoids [9], as well as circulating cool air or evaporative concepts using the intranasal pathway [10]. The optimal depth of hypothermia has traditionally been controversial, where severe hypothermia (15–22 °C), thought to be more neuroprotective at one time, actually led to more complications when compared to mild and moderate hypothermia (30–35 °C) [3]. This underscores the point that adverse effects may be a result of either the depth of hypothermia or the specific cooling method used to reach the targeted temperature. With mild hypothermia, however, complications specifically from cooling devices only involved 29 of 3,133 patients (1 %) and were related to infection, bleeding, pulmonary edema, and deep vein thrombosis [5]. Complications from targeted temperature management (independent of the type of cooling method used), in a recent observational study on post-cardiac arrest patients, included pneumonia (41 %), hyperglycemia (37 %), cardiac arrhythmias (33 %), seizures (24 %), and electrolyte disturbances (19 %) [11]. As these data are specifically from cardiac arrest patients, the profile and frequency of adverse events will vary not only on the depth of hypothermia but likely also on the type of acute brain injury as well.

Ischemic Stroke

Mechanism of Injury: In Vitro and In Vivo Studies

After a series of studies in the 1980s on rodents undergoing brain ischemia demonstrated that small decreases in temperature resulted in the prevention of neuronal death [12, 13], therapeutic hypothermia for ischemic stroke became a topic of interest. Hypothermia impacts neurovascular pathophysiology in the acute, subacute, and chronic stages of ischemia. The mechanisms underlying therapeutic hypothermia's neuroprotection have been investigated over the last few decades, and we have tremendously increased our knowledge of the fundamental effects on a cellular, tissue, and organ level. Hypothermia impacts various molecular cascades in inflammation, mitochondrial dysfunction, oxidative stress, blood flow, energy metabolism, and blood–brain barrier integrity, thus affecting neurogenesis and angiogenesis following an ischemic insult, leading to neurovascular protection [14, 15]. Here, we focus on these underlying pathophysiologic mechanisms in animal models of focal and global ischemia.

It is important to realize that methods of cooling are distinct between animal in vivo and in vitro models of ischemia. Typical cooling experiments are conducted in animals by application of cooling blankets or spraying water or alcohol on anesthetized animals, usually small rodents, for only a few hours, however if longer duration of cooling is pursued (i.e., more than 24 h) in awake animals, then automated systems have been developed [16]. Several studies on animals have demonstrated that relatively small decreases in temperature are neuroprotective without resulting in significant side effects [17]. Although timing and duration of hypothermia are important factors, the degree of hypothermia appears to be less critical. Mild-to-moderate hypothermia to 30–34 °C provided similar protection to severe hypothermia below 25 °C in many settings [17]. Moreover, the effect of hypothermia in different animal stroke models appears to also be an important consideration. Whereas in global ischemia models the time window for therapeutic hypothermia is longer, in focal ischemia the window of opportunity for hypothermia appears to be shorter where it must be started within 2 h of ischemia onset [18]. However, initiation of therapeutic hypothermia after 2 h may still be beneficial if maintained for prolonged periods of time up to 48 h after ischemia onset [18]. Although some studies have shown long-term benefits of therapeutic hypothermia [19], the majority of animal studies have been conducted under shorter duration of ischemia and cooling with shorter follow-up testing for beneficial outcomes.

The Role of Inflammation and Hypothermia in Stroke

While ischemic brain injury leads to many different effects including inflammatory cascades, therapeutic effects of hypothermia remain primarily a result of reductions in metabolism and cerebral blood flow during injurious periods in which the brain is susceptible to energy depletion and reperfusion injury [20]. Cooling results in decreases in brain glucose and oxygen consumption, thus maintaining ATP stores and leading to decreased lactic acidosis and oxidative stress. Cooling also prevents excitotoxicity by modulation of ATP production, thereby preventing the accumulation of excitotoxic amino acids, including glutamate [21] and, consequently, affecting calcium influx through glutamate receptor 2 subunit of AMPA receptor [22]. Moreover, hypothermia appears to affect early gene expression in ischemia by regulating expression of inducible heat shock proteins in various ways [23, 24], though it remains unclear whether modulation of gene expression is involved in the neuroprotective mechanisms of cooling. More recently, microRNA expression changes during acute brain injuries has garnered interest. Cooling appears to modulate a number of microRNAs during traumatic brain injury, and with ischemic stroke models showing a multitude of changes in microRNAs [25], it is reasonable to speculate that hypothermia may affect microRNAs in focal stroke models as well [26, 27]. Nevertheless, these mechanisms during acute ischemic injury do not fully explain the neuroprotective effects and inflammation-mediated effects of hypothermia, and further investigation is necessary.

Inflammation-mediated effects of hypothermia are more apparent in the post-acute phases of focal ischemic strokes. The subacute phase of ischemia, which begins after many hours of the insult and can last up to 1–2 weeks, has a multitude of pro-inflammatory effects. Though some aspects of these inflammatory processes generated by an ischemic stroke may be evolutionarily advantageous, it appears that other aspects may worsen brain injury and outcome, especially during the acute and subacute phases of injury [28]. For example, hypothermia has been shown to significantly reduce microglia and neutrophils in the region of ischemia while also reducing reactive oxygen species [29, 30]. Moreover, various cytokines are reduced by hypothermia, including interleukin-1β (IL-1β), IL-6, and tumor necrosis factor-α (TNFα) [28]. Additionally, hypothermia has been shown in rodents to reduce inducible nitric oxide synthase (iNOS) and reactive nitric oxide, and this reduction in reactive nitrogen species may mitigate ischemic infarct size [31].

Some hypothermia-mediated mechanisms involving the inflammatory cascade appear to show variable effects in different experimental paradigms, suggesting that further studies are needed. For example, while the mitogen-activated protein kinase (MAPK) pathway, also known as the extracellular-signal-regulated kinase (ERK) pathway, is important in modulating a number of inflammatory molecules including intercellular adhesion molecule 1 (ICAM1), hypothermia has opposite effects in vitro versus in vivo: hypothermia reduces MAPK signaling in cultured microglia paradigms while stimulating the MAPK in CNS endothelial cells in in vivo stroke models [32, 33]. Moreover, not all of the effects of hypothermia appear anti-inflammatory. For example, hypothermia appears to decrease IL-10, an anti-inflammatory cytokine [34]. Thus, hypothermia-mediated mechanisms in ischemic stroke models are complex and need further clarification.

Hypothermia also appears to have a multitude of effects on the apoptosis pathway, which can be triggered after ischemic insults. Apoptotic pathways, in turn, may be affected by inflammatory factors, though a review of hypothermia's role with this interplay is beyond the scope of this chapter [14].

Clinical Applications Using Hypothermia in Stroke

Given the laboratory data showing benefit of hypothermia in ischemic stroke models with the various mechanisms being further clarified, significant optimism arose towards implementation of hypothermia during human ischemic stroke, including hope to reduce propagation of inflammatory cascades involved therein. Initially, retrospective analysis of a prior prospective stroke study involving 390 stroke patients demonstrated an important correlation between body temperature and clinical outcome after stroke; for every spontaneous reduction in body temperature of 1 °C, there was a progressively significant improvement in clinical outcome (i.e., relative risk improvement of 2.2), including improvements in infarct size, stroke severity, and mortality [35]. Over the last two decades, multiple prospective clinical trials have been conducted utilizing hypothermia during ischemic stroke, with a wide variety of clinical designs, including randomized and nonrandomized studies, safety and efficacy studies, various time windows after stroke onset, and integration with intravenous (iv) thrombolysis (i.e., tissue plasminogen activator; tPA).

The first clinical trial on the effect of moderate hypothermia to 33 °C in only 25 patients with severe middle cerebral artery (MCA) infarction was reported in 1998 [36]. The authors hoped to mitigate post-stroke cytotoxic edema while monitoring core and brain temperature, intracranial pressure (ICP), cerebral perfusion pressure, and other factors. They found that during the period of hypothermia, better control of ICP was achieved, but that during the rewarming period, there was a rebound increase of the ICP. Overall, the study demonstrated a significant reduction in mortality from roughly 79 %, as expected from other studies at the time, down to 44 % along with a more favorable outcome in the survivors. A follow-up study on a slightly larger sample size of 50 patients demonstrated similar efficacy even though it was mostly concerned with showing safety [37]. These studies demonstrated that side effects of hypothermia can include pneumonia, bradycardia, and thrombocytopenia. While these studies were not controlled, the first case–control safety trial was conducted in 2000 on 17 patients and showed a good safety profile [38]. Since these original studies and other subsequent studies that established safety and planted hope for efficacy, there have been a number of additional clinical trials, including a few randomized controlled trials, which have failed to show statistically significant improvements in stroke-related measures or clinical outcomes. In two of these trials, there did appear to be a trend towards significance, and it is possible that future studies with larger sample sizes and better designs may achieve statistical significance and better establish efficacy [39, 40]. One of these studies which focused on establishing safety, Cooling for Acute Ischemic Brain Damage ("COOL AID"), was an open controlled study on 19 patients, including patients who had received tPA for thrombolysis. COOL AID demonstrated a small (non-statistically significant) trend towards improved clinical function at 3 months after the stroke [39]. However, a follow-up randomized multicenter study, COOL AID II, included 40 total patients who received endovascular cooling along with possible tPA and did not find such a trend favoring hypothermia. Again, however, the main purpose of this study was to further establish feasibility and safety, a goal that was accomplished [41]. The other study that showed a possible non-statistically significant benefit was a prospective randomized study done by Els et al. [40] on 25 patients in which mild hypothermia to 35 °C was started immediately after decompressive hemicraniectomy for malignant ischemic stroke [40]. In this study, there was a trend ($p < 0.08$) towards a very small functional benefit in patients after 6 months when assessed by the National Institutes of Health Stroke Scale, a standardized quantitative clinical scoring scale. Another study on the usage of hypothermia in acute stroke was The Intravascular Cooling for the Treatment of Stroke (ICTuS) trial [42]. This was a multicenter uncontrolled study which involved 18 patients and included some post-thrombolysis patients who had received tPA within 3 h. ICTuS achieved its goal of demonstrating acceptable safety profile [42]. The follow-up ICTuS-L study, a multicenter controlled trial involving 59 patients which allowed for a longer window of time (up to 3–6 h) after stroke, was unable to establish any benefit for hypothermia in acute stroke [43].

Summary of Stroke, Inflammation, and Hypothermia

In summary, even though a wealth of in vitro and in vivo animal studies demonstrate the benefits of hypothermia with regard to inflammatory processes and functional outcome of stroke models, there appears to be no convincing evidence in support of its use in humans with various different functional outcomes and survivability. While the clinical trials utilizing therapeutic hypothermia began only 15 years ago, and trials have established safety, there are many factors that may account for the failed efficacy. Along with variables such as depth, onset, and duration of hypothermia as well as patient selection criteria including size and location of the ischemic stroke, administration of thrombolysis, and presence of decompressive hemicraniectomy, the most critical limitation has been small sample size. Future studies are expected to minimize and control for variability while selecting and choosing optimal factors so as to standardize experimental designs. Then, increasing sample size may allow for any potential benefit to manifest itself in the data. Subsequently, studies should assess the role of the inflammatory cascade in achieving any possible benefit.

Global Cerebral Ischemia: Neonatal Hypoxia–Ischemia

Mechanism of Injury and the Role of Inflammation

While ischemic stroke constitutes a focal area of ischemia, when the insult is more proximal than a single cerebral artery such as bilateral carotid arteries or the heart, it can result in global ischemia. Also, if systemic hypoxia occurs, this can lead to brain hypoxia as well as an eventual reduction in cerebral blood flow, both of which can lead to brain injury recognized as hypoxic–ischemic. The acute brain injury and subsequent clinical syndrome thereafter may be called hypoxic–ischemic encephalopathy. Neonatal hypoxic–ischemic episodes can happen during birth due to a variety of perinatal insults that may originate from the newborn infant, the mother, or environmental factors.

Though the exact pathophysiology of neonatal hypoxic–ischemic encephalopathy is not clear, it appears to involve some of the same cellular and subcellular mechanisms involved in focal ischemic stroke. Inflammation seems to have a significant impact on the pathophysiology of neonatal hypoxia–ischemia and includes many of the same inflammatory cascade components described in focal ischemia with some differences [44, 45].

Clinical Applications in Neonatal Hypoxia–Ischemia

Clinical trials of hypothermia in human neonates suffering from perinatal hypoxia–ischemia have shown a significant benefit. A recent Cochrane Review scrutinized 11 randomized controlled trials of infants suffering from moderate-to-severe hypoxic–ischemic insults, including asphyxia, and demonstrated significant improvement in survival as well as clinical function with lower neurodevelopmental disability up to

18 months of age [46]. In fact, the number needed to treat (NNT) to achieve a clinical improvement in outcome was 7 while the NNT to achieve survivability as the outcome was 11, both of which are impressive when compared to other FDA-approved neurologic treatments such as tPA for acute stroke. With tolerable side effects of bradycardia and thrombocytopenia, therapeutic hypothermia in neonatal hypoxia–ischemia has been embraced as a standard treatment. Much of this was achieved with two landmark studies [47, 48] in recent years, which are both included in the aforementioned Cochrane Review. Evidence for the mechanism of how hypothermia may achieve such favorable outcome in this setting comes mostly from in vitro and in vivo animal studies, and more recently, from human serum markers that demonstrate various reductions in inflammatory molecules [46, 49–51].

Global Cerebral Ischemia: Cardiac Arrest

Mechanism of Injury and Role of Inflammation

Therapeutic hypothermia is the only definitive treatment to date after resuscitation is achieved for cardiac arrest. The literature in support of hypothermia after cardiac arrest is convincing, even though the mechanisms are not clearly understood, and may involve downregulation of the inflammatory cascade.

During cardiac arrest-induced global ischemia, some inflammatory processes in the brain are similar to focal ischemic strokes [14, 52]. However, even more so than neonatal hypoxia–ischemia, cardiac arrest can trigger a more generalized inflammatory process with multiple organ-specific processes that occur alongside a brain-specific inflammatory cascade. The inflammation-related mechanistic effects of hypothermia on cardiac arrest have not been studied extensively, and some of the data that has been reported does not appear fully consistent, with some groups finding significant changes in inflammatory cascades and others finding minimal to none [53–55]. Many of the variation in reports may originate from different animal models and slight differences in the structure of the cardiac arrest and hypothermia design. Overall, as in other brain injury disorders reported in this chapter, hypothermia after cardiac arrest does appear to change the dynamics of inflammation in the brain. More impressively, however, is that hypothermia after cardiac arrest modulates the systemic inflammatory process that is much more robust in cardiac arrest when compared to isolated brain injuries.

In vitro studies in human cell culture systems demonstrate that hypothermia modifies NF-κB activity, a pro-inflammatory transcription factor, including delaying its induction and prolonging its duration while changing downstream cytokine activation [56]. Studies that focus on in vivo human data from serum of patients having undergone cardiac arrest and therapeutic hypothermia seem to show a more systematic inflammatory increase in IL-6 during the latter part of a 24-h hypothermia period, which correlates with a slight rise in body temperature from the coldest nadir, while the subsequent rewarming process led to complement activation [57]. In further support of hypothermia leading to a rise in IL-6, another human study that measured serum inflammatory factors after resuscitation and therapeutic hypothermia also found a significant increase in IL-6 alongside a large increase in bacterial colonization. Outcome data from this study showed that hypothermia-treated patients consistently had lower mortality despite higher bacterial colonization and a possible higher risk of infection [55]. This underscores the impressive improvement in outcome that hypothermia has after cardiac arrest, far outweighing any possible deleterious or adverse effect. More on its specific effects on the inflammatory cascade, a study of hypothermia on rats demonstrated a significant reduction in IL-10 [58]. On the other hand, a recent study in pigs demonstrated that hypothermia after cardiac arrest has minimal change on IL-6 and IL-10, but a significant reduction in TNFα, which is dependent on NF-κB activity. Thus, it is important to realize that hypothermia may have varying effects on the inflammatory cascade on different organs and different species during cardiac arrest. For example, one study in pigs focused on examining cardiac tissue and found that hypothermia after cardiac arrest led to decreased apoptosis, reduced inflammatory markers (i.e., IL-1β and IL-6) and lower matrix metalloproteinase-9 (MM9) within the cardiac tissue itself [59]. Interestingly, although hypothermia appears to have an overall anti-inflammatory response with human trials showing an increased risk of infection, some data exists to suggest that hypothermia may, at times, strengthen the immune response. For example, in a rat model of septic shock using lipopolysaccharide (LPS) injection to achieve endotoxemia, hypothermia improved mortality while increasing IL-10 [60]. Thus, even though conflicting data exists on the mechanistic role of hypothermia after cardiac arrest, the general anti-inflammatory mechanisms of hypothermia may, in some particular circumstances, actually render the immune system stronger rather than weaker. This further demonstrates the complexity of hypothermia's actions on the inflammatory process and the body as a whole.

Post-cardiac arrest periods are further complicated by changes in the blood–brain barrier. The interplay among multiple organs being affected, the release of various factors into the blood, and a possible breakdown in the blood–brain barrier magnifies the complexity of the situation. This breakdown in the blood–brain barrier can significantly contribute to brain edema while also confounding assays for inflammatory factors. Deciphering whether an inflammatory process originated from the brain or other organs becomes challenging. Brain edema, which itself contributes to breakdown of the blood–brain barrier, has been shown to be regulated by aquaporin-4, a membrane protein on glia that regulates water transport, and matrix metalloprotease-9 (MMP9) [61]. Cardiac arrest leads to upregulation of aquaporin-4, contributing to brain edema, while hypothermia can downregulate this process and thereby lessen brain edema [62]. Minimizing brain edema can help to maintain the blood–brain barrier, which in turn protects the brain from systemic inflammatory factors. The rewarming period of therapeutic hypothermia leads to yet another shift in the inflammatory balance. For example, complement activation can occur during this post-hypothermia period, and this activation can contribute further towards breakdown of the blood–brain barrier [57].

Regardless of the variability in studies above, hypothermia appears to be convincingly neuroprotective after cardiac arrest. The mechanistic evidence for this is vast, and a detailed review of this is beyond the scope of this chapter. For further in-depth review of the neuroprotection literature, one can refer to other sources [14, 63]. Briefly, the mechanisms of neuroprotection appear to involve a decrease in overall metabolism of vulnerable neurons, reduction in excitotoxic pathways and inflammatory cascades, less reactive oxygen species, and lower apoptosis.

Clinical Applications of Hypothermia After Cardiac Arrest

Two landmark studies, both prospective randomized controlled trials, have been conducted to date in strong support of therapeutic hypothermia after cardiac arrest [64, 65]. Based on these two trials, the American Heart Association along with the European Resuscitation Council advocated therapeutic hypothermia targeting 32–34 °C for cardiac arrest due to ventricular fibrillation and pulseless ventricular tachycardia. There are also other clinical trials that are in support of these two landmark trials [66]. Though these studies support the use of therapeutic hypothermia for shockable cardiac rhythms, many clinicians go on to implement it for cardiac arrest due to pulseless electrical activity also despite the lack of convincing studies showing benefit in non-shockable rhythms. Given the low side effect profile and good tolerability of therapeutic hypothermia, it is not unreasonable to fathom its potential benefit for cardiac arrest due to pulseless electrical activity. Because we do not fully understand the mechanisms of hypothermia's benefits, this provides a dilemma with future investigations being randomized controlled trials, as it would be unethical not to initiate therapeutic hypothermia for a post-cardiac arrest patient in the setting of a shockable rhythm. Therefore, we may have to rely on animal studies and in vitro studies to allow us better experimental strategies in uncovering the mechanisms involved, including the role of the inflammatory system. Such bench-to-bedside approach will be key towards understanding and advancing neurologic treatments in this area. This also underscores the importance of focusing on the functional outcome and survivability in studies despite at times not clearly understanding the mechanisms.

Traumatic Brain Injury

Mechanism of Injury

Traumatic brain injury (TBI) induces cerebral damage through a variety of methods and can be classified as primary or secondary insults. The primary insults are the direct result of trauma and result in parenchymal and vascular damage. These lead to cell death from direct cell disruption and from cellular energetic dysfunction which may be ischemic and/or hypoxic in nature. Substrate deprivation results in a decrease of ATP production with resulting cellular energy starvation as well as loss of cellular integrity from dysfunction of membrane ATP-dependent Na^+–K^+ pumps. This in turn leads to the uncontrolled release of glutamate, an excitatory neurotransmitter, which leads to injury from excitotoxicity [67–69] mediated principally through *N*-methyl-D-aspartate (NMDA) receptors [70]. As part of this cascade of cellular injury, inhibitory neurotransmitters that normally dampen glutamate excitotoxicity, such as γ-aminobutyric acid (GABA) and glycine [71], are decreased as well [72]. Glutamate excitotoxicity is mediated by activation of NMDA receptors, which create an intracellular calcium influx that in turn activates a number of second messengers that amplify cellular injury by increasing calcium permeability and increasing glutamate release leading to a vicious cycle [73–75]. There is also an activation of neuronal nitric oxide synthase (nNOS), which results in the production of oxygen-free radical species, which are also responsible for cellular injury by direct DNA fragmentation, protein oxidation, lipid peroxidation [76], and disruption of the mitochondrial respiratory chain. Oxidative stress mediated by these mechanisms also leads to further inflammation and injury through complement activation and subsequent degradation [77, 78], cytokine production (IL-1, IL-6, IL-8, and TNF-α), expression of leukocyte adhesion molecules, and microvascular dysfunction [79]. Secondary injury is a delayed phenomenon and is mediated by cerebral edema with resulting elevations in intracranial pressure, non-convulsive seizures, and blood–brain barrier disruption. Cytotoxic edema is a result of previously discussed excitotoxicity and ionic pump failure, as well as cellular water shift across impaired aquaporin function [80]. Matrix metalloproteases (MMPs) are responsible for blood–brain barrier disruption and are decreased by therapeutic hypothermia in human studies [81].

Clinical Applications and Role of Hypothermia in TBI

Human TBI studies have shown that mild therapeutic hypothermia has positive effects on cerebral metabolic imbalance [82]. Energetic dysfunction and cellular metabolic crises have been shown to be important pathophysiologic processes in humans with TBI [83] and are predictors of outcome [84]. Studies in anesthetized patients have established that in humans the cerebral rate of oxygen consumption ($CMRO_2$) is decreased by 6.5 % per °C in temperature reduction [85], alleviating the oxygen deprivation encountered during both primary and secondary injuries.

A 2009 Cochrane systematic review by Sydenham et al. [86] found 23 trials with a total of 1,614 randomized patients in trials that utilized therapeutic hypothermia (<35 °C) for at least 12 h. The authors of this analysis concluded that there is no evidence that hypothermia is beneficial in the treatment of head injury. Hypothermia may be effective in reducing death and unfavorable outcomes for traumatic head-injured patients, but significant benefit was only found in low-quality trials. As evidenced by the number of reviews and studies [15, 87–89], there is great interest in therapeutic hypothermia for the treatment of TBI. Some of the criticisms against the negative trials in hypothermia for TBI have included delays in initiation of hypothermia, as well as inadequate duration of hypothermia which has been shown to decrease ICP in these patients who then had rebound elevations in ICP upon rewarming [90]. Moreover, lack of positive results may also be partly due to heterogeneity in the variables of the treatment groups, in particular the presence and degree of intracranial hypertension. The 2007 Severe Traumatic Brain Injury Guidelines [91] state that there is insufficient data to support level 1 and 2 recommendations for the use of therapeutic hypothermia in TBI. At a level 3, the guidelines mention that while therapeutic hypothermia was not associated with decreased mortality compared to normothermic controls, preliminary findings suggest that a greater decrease in mortality is observed when target temperature is maintained for >48 h. A 2011 multi-society consensus statement on Targeted Temperature Management (TTM) [92] concluded that existing data does not support a recommendation for or against the use of TTM (including TH) to treat TBI. Furthermore, it states that although existing data suggest that TH can decrease ICP in TBI, the relationship between ICP reduction and outcomes is indeterminate. The decision of whether to employ TH and how to do so in TBI must be individualized with particular attention paid to the presence of elevated ICP pending the results of the ongoing Eurotherm 3235 Trial [93], which aims to definitively address the impact of TH on 6-month mortality, outcomes, ICP control, and cost-effectiveness.

Subarachnoid Hemorrhage and Intracerebral Hemorrhage

Mechanism of Injury in SAH and ICH

Subarachnoid hemorrhage (SAH) and intracerebral hemorrhage (ICH) share the main pathophysiologic mechanisms for inflammation and injury with other forms of brain injury. Some unique aspects of the inflammatory process in SAH and ICH include the presence of significant amounts of blood in and around the brain. Exposure of blood to the surface of the brain induces a series of inflammatory changes including cytotoxic edema, breakdown of the blood–brain barrier, and delayed cerebral ischemia (DCI). Animal models have shown that free heme induces IL-1-mediated inflammation [94]. As in other forms of brain injury, CSF and serum levels of interleukins and TNFα are elevated in SAH [95]. Delayed cerebral ischemia in SAH is thought to be mediated via the mitogen-activated protein kinase pathway (MAPK) which ultimately results in increases of IL-1β, IL-6, and TNFα [96]. Vasospasm seems to be partially mediated by dysregulation of endothelin-1 (ET-1) and nitric oxide (NO) levels, as well as by asymmetric D-methyl arginine (ADMA), which mediate endothelial dysfunction as well [97]. Also, matrix metalloprotease-9 likely plays an important role in SAH-mediated blood–brain barrier disruption and inflammatory mechanisms [96].

Preclinical Data on Hypothermia and SAH

Animal studies of therapeutic hypothermia have shown promise and significant effect on reducing inflammation based on quantification of inflammatory markers as well as radiologic and clinical parameters [98]. Early animal studies showed decreases in pro-inflammatory heat-shock proteins (hsp70) after treatment with mild therapeutic hypothermia [99]. Dog models of SAH have shown that hypothermia decreases angiographic vasospasm as well as clinical performance status that correlate with modulation of ET-1 and NO [100]. In a separate rat model of experimental SAH, therapeutic hypothermia decreased metabolic stress and cytotoxic edema as measured by diffusion-weighted MRI and magnetic resonance spectroscopy (MRS) [101]. A laser-Doppler and microdialysis study in a massive SAH rat model again demonstrated that hypothermia decreased early ischemia, improved cerebral vascular autoregulation and cellular metabolic crisis, and also decreased inflammatory mediators [102].

Clinical Applications of Hypothermia in SAH

The Intraoperative Hypothermia for Aneurysm Surgery Trial (IHAST) is the main study of hypothermia in SAH [103]. Results from this large prospective multicenter trial of intraoperative hypothermia in patients with good grade SAH failed to show treatment benefit with intraoperative hypothermia. In another publication stemming from the same trial [104], therapeutic hypothermia failed to show benefit in terms of neuropsychological outcomes between patients treated with and without hypothermia. Furthermore, in contrast to animal studies, data from the IHAST trial showed no difference in outcomes in patients with SAH treated with a combination of hypothermia and barbiturates [105]. Gasser et al. [106] evaluated the feasibility and safety of long-term hypothermia (>72 h) in the treatment of severe brain edema after poor-grade SAH. The functional independence at 3 months, defined as a Glasgow Outcome Scale (GOS) score of 4 or 5, did not differ between the two groups.

In the 2012 AHA Guidelines for the Management of Aneurysmal Subarachnoid Hemorrhage, it states that induced hypothermia during aneurysm surgery may be reasonable in some cases but is not routinely recommended (IIIb) [107]. In the same guidelines, no statement was provided on the use of therapeutic hypothermia outside of surgery. Nguyen et al. have shown, however, that perioperative hypothermia to 33 °C is safe from a cardiovascular standpoint [108]. However in Gasser et al.'s aforementioned study [106], there was a significant increase in infectious complications in patients with hypothermia >72 h. More recent data seems to show that maintenance of normothermia (37 °C) may be safer and result in better outcomes post-SAH [109].

Preclinical Data on Hypothermia and ICH

Animal models of inflammation in ICH as well as human data point to the majority of pathophysiologic mechanisms discussed previously. By the same token, studies in rats of therapeutic hypothermia in ICH models have shown stabilization of the blood–brain barrier as well as decreased inflammatory signaling and oxidative stress [110, 111]. Another study again showed improvement in measures of inflammation with hypothermia in rats, but failed to show clinically significant improvements [112].

Clinical Applications of Hypothermia in ICH

At present, there is little evidence to support the use of therapeutic hypothermia in spontaneous ICH. The 2010 AHA Guidelines for the Management of Spontaneous Intracerebral Hemorrhage do not recommend the use of hypothermia given the lack of evidence in humans at this time [113]. However, since the publication of that guideline, a small study by Staykov et al. [114] reported the use of mild therapeutic hypothermia at 35 °C via endovascular cooling for 8–10 days in patients with ICH volumes >25 cm^3. They demonstrated significant decreases in the development of perihematomal edema between days 3 and 10 in the TH group, as well as a trend towards decreased mortality, though it is unclear if the mechanism of these findings involves inflammation-related changes. These results have led the same group to launch the prospective Cooling in IntraCerebral Hemorrhage (CINCH) trial in Europe [115], which aims to determine whether TH improves survival and decreases lesion volume after large ICH.

Seizures and Status Epilepticus

Mechanism of Injury, the Role of Inflammation, and Preclinical Data

The immune and inflammatory mechanisms of status epilepticus (SE) are in some ways similar to other acute brain injuries. As in other forms of brain injury, inflammatory mediators such as IL-1, IL-6, and TNFα play an essential role in the pathogenesis of SE [116]. These findings are corroborated by amelioration of seizure-induced inflammation in rats treated with IL-1 receptor antagonists [117]. Inhibition of MAPK and cyclooxygenase-2 (COX-2) decreased markers of cellular injury and oxidative stress as well as mortality in a rat model of SE treated with an anti-inflammatory and antioxidant compound [118]. NMDA receptor-1 upregulation has been associated with increased neuronal death in the rat model, and mild hypothermia delayed and decreased the amount of neuronal death [119]. In another rat model, the neuroprotective effect of hypothermia in SE was thought to be mediated by a decrease in cerebral edema. Of note, in this study hypothermia also improved the cognitive function and decreased the recurrence of seizures in treated rats in comparison to controls [120].

Clinical Applications of Hypothermia in Seizures

The body of evidence for TH in status epilepticus is quite limited. In a recent review [121] the outcomes of therapies for refractory and super-refractory status epilepticus of nine patients in four reports were described. All cases achieved initial control and 7/9 fully recovered. In a small case series of patients with refractory status epilepticus, hypothermia of 30–31 °C was combined with barbiturate coma, and successful control of seizures was achieved [122]. The role of hypothermia in management of refractory seizures warrants further investigation [123]. Based on what little evidence exists, reviews [121, 124] recommend using TH as a second-line agent, ideally after a combination of GABAergic and NMDA antagonist drugs is in place [125].

Meningitis and Encephalitis

Mechanism of Injury, Inflammation, and Clinical Considerations

Bacterial meningitis and viral encephalitis share many of the underlying inflammatory processes as the pathologic entities previously discussed in this chapter. However, the pathogen–host interaction adds another layer of complexity from a molecular and immune standpoint. The details of individual bacterial or viral species virulence factors and interactions with the human central nervous system are quite beyond the scope of this chapter. However, detailed information on the molecular mechanisms of inflammation and pathogenicity of bacterial meningitis and viral encephalitis may be sought in other references [126–128]. Bacterial meningitis induces CNS immune activation by a variety of cell wall products that as a first step require recognition by CD-14 membrane receptors, which eventually activate toll-like receptor-2 (TLR-2) among other mechanisms; this is accompanied by migration and proliferation of activated immune cells in the CNS. Bacterial products stimulate IL-1β, IL-6, and TNFα, as well as COX-2, MMP-8 and 9 and dysregulated iNOS, all of which by now constitute a form of common inflammatory pathway for brain injury [129–132]. Viral encephalitis induces CNS inflammation in much the same way as bacterial meningitis. Animal models of meningitis have shown that hypothermia decreases inflammatory markers [129] as well as improving ICP and metabolic measures in rats treated with hypothermia at 32–34 °C [133]. This data led to the IHPOTOTAM (Induced HyPOthermia TO Treat Adult Meningitis; http://www.clinicaltrials.org #NCT00774631) trial of which results are pending at this time.

Acute Liver Failure and Hepatic Encephalopathy

Mechanism of Injury, Role of Inflammation, and Animal Studies

While hypothermia has been investigated as a potential agent to reduce inflammation, cerebral edema, and resulting increases in intracranial pressure (ICP) due to several conditions resulting from direct (i.e. stroke, traumatic brain injury, and meningitis) neurologic injuries, it is also investigated for indirect neurologic injuries, such as in cerebral edema secondary to liver failure. Therapeutic hypothermia (TH) for acute liver failure (ALF) concerns mainly the use of TH as a measure to decrease ICP as a bridge to definitive therapy, usually orthotopic liver transplantation (OLT) [134]. Purported mechanisms by which TH may be useful in ALF are reductions in cerebral edema mediated by decreased ammonia production, brain hyperosmolarity, oxidative stress, and cerebral hyperperfusion [135, 136].

The pathophysiology of central nervous system dysfunction during hepatic encephalopathy, resulting from both acute and chronic liver failure, is not yet fully understood. Cerebral edema is usually absent in cases of chronic liver failure but frequent in acute liver failure. There are a variety of molecular mechanisms that have been implicated in the development of cerebral edema and resulting increased intracranial pressure, as well as in the onset of hepatic encephalopathy. Ammonia is believed to be the main neurotoxin in hepatic encephalopathy, as shown *in vitro* by vacuolization and cellular damage of cortical astrocytes exposed to ammonia [137] and confirmed by human autopsy data [138, 139]. Additionally, serum ammonia levels have been correlated to herniation events and death in humans with acute liver failure [140]. Hepatic encephalopathy and resulting cerebral edema are also mediated by cerebral metabolic imbalances. Animal experimental models have demonstrated decreased brain glucose utilization, resulting in reduced production of adenosine-triphosphate (ATP) [141]. This is compounded by increases in lactate production in hyperammonemic states [141–144] that are at least partially mediated by inhibition of brain α-ketoglutarate dehydrogenase. Elevated levels of brain lactate have been implicated in cytotoxic edema. Oxidative stress is another mechanism by which hepatic encephalopathy induces brain injury. In a rat model, Murthy et al. [145] demonstrated astrocyte derived free radicals' role in the pathophysiology of hyperammonemic brain injury. Other animal studies have shown increased activity of inducible nitric oxide synthase I and II in hepatic encephalopathy [146, 147]. In a rabbit model of acute liver injury [148], blood–brain barrier disruption was reported. This, in turn, can lead to many possible pathways for altered cerebral metabolism and homeostasis leading to secondary brain injury. Matrix metalloproteases have been implicated in blood–brain barrier disruption in several forms of brain injury including hepatic encephalopathy [149]. Other pathophysiologic mechanisms of cerebral dysfunction and injury include increases in pro-inflammatory cytokines (IL-1β, IL-6, and TNFα) and concomitant decreases in anti-inflammatory mediators [136, 150–154].

In a key study, Jiang et al. [155] demonstrated in a rat model that mild hypothermia to 35 °C decreased brain edema as measured by brain water content. They also found reduced serum and cerebrospinal fluid ammonia concentrations as well as blunted serum protein levels and CNS mRNA levels of IL-1β, IL-6, and TNFα. This study shows several important mechanisms by which hypothermia attenuates neuroinflammation in acute liver failure. Barba et al. [156] employed hydrogen nuclear MRI to perform functional metabolic analysis on a rat model of acute liver failure and demonstrated decreased levels of alanine and lactate, demonstrating decreased metabolic dysregulation in the brains of hypothermic rats in comparison to normothermic controls. These results parallel those of Chatauret and colleagues [144] who demonstrated improvement in brain glucose metabolism imbalance with hypothermia.

Clinical Applications of Hypothermia in Acute Liver Failure

A number of human studies regarding therapeutic hypothermia for treating hepatic encephalopathy and resulting cerebral edema comes from Jalan et al. from King's College [157–159]. In these studies, TH was applied as part of a strategy to optimize and bridge patients with acute liver failure to liver transplantation by decreasing intracranial pressure. These small case series have led to enthusiasm for undertaking a prospective study of therapeutic hypothermia in acute liver failure in the form of the "Hypothermia to Prevent High Intracranial Pressure in Patients With Acute Liver Failure" study (http://www.clinicaltrials.org #NCT00670124). Pending publication of the results of this study, there is no compelling evidence to recommend routine application of therapeutic hypothermia to patients with hepatic encephalopathy; therapeutic hypothermia may be considered, however, in high-risk patients with refractory elevations in intracranial pressures [87, 160].

Fever

Mechanism of Injury, Role of Inflammation, and Clinical Relevance

Fever is an inflammatory reaction par excellence, and though its advantageous evolution in animals is arguable, it can lead to deleterious reactions in cases of brain injury. The causes of fever are varied and numerous. Therefore, this section focuses on those where brain injury is implicated: stroke, ICH and SAH, hypoxic–ischemic encephalopathy, seizures, and CNS infections. The preoptic region of the anterior hypothalamus is the principal center of the body's thermoregulatory circuits. Direct lesions or inflammation of this structure as well as its afferent and efferent pathways can result in impaired thermoregulation, including fever. Exogenous pyrogens include bacterial superantigens and lipopolysaccharide (LPS), and endogenous pyrogens include IL-1, IL-6, interferon γ (INFγ), TNFα, and ciliary neurotrophic factor (CNTF) which are normally counterbalanced by endogenous cryogens (interleukin-10, steroid hormones, and others). An imbalance of these factors can lead to fever or hypothermia. Interplay of the humoral (immune) and the neural fever pathways is required for the initiation and maintenance of fever [161–163].

Fever exacerbates brain injury via a number of mechanisms. These include increasing vascular and blood–brain barrier permeability and increasing cellular metabolism which can increase $CMRO_2$. A rise in $CMRO_2$ can precipitate an energy crisis which results in higher lactate/pyruvate ratios and decreases in pH. It can also lead to oxidative stress, increased cerebral edema, and consequently intracranial hypertension. These various steps are mediated by the inflammatory mechanisms discussed over the course of this chapter. The exact mechanism of action of therapeutic hypothermia as a neuroprotectant is unclear but is most likely multifactorial: limiting excitotoxicity [69], reducing cerebral cellular metabolism [18, 164], decreasing tissue nitric oxide and oxidative stress [165], modulating transcription factors and microglial activation [166], decreasing cerebral edema [110, 167], and stabilizing the blood–brain barrier [168]. Fever has been clinically implicated in poor outcomes in neurologic conditions and is recognized as a significant concern for brain-injured patients [15, 169–173]. For these reasons, normothermia or relative hypothermia compared to baseline fever states is often targeted clinically to mitigate inflammatory processes that, in turn, can contribute to direct and indirect worsening of the various brain injuries discussed throughout this chapter.

Conclusion

In vitro and animal studies provide us convincing experimental evidence that inflammation has a central role in many aspects of acute brain injuries. Distinguishing when inflammation may be detrimental versus beneficial in physiologic processes is key. Hypothermia, for the most part, has a generalized anti-inflammatory role. This is supported strongly by *in vitro* studies that show hypothermia curtailing injury and animal studies that demonstrate improved recovery from brain injury. Clinical trials in humans thus far, however, have only shown neurological benefit in neonatal hypoxia–ischemia and post-cardiac arrest brain injury. Given that adverse side effects of therapeutic hypothermia appear tolerable, we await ongoing and future studies on whether hypothermia has a clinical role in other acute brain injuries, including ischemic stroke, subarachnoid hemorrhage, intracerebral hemorrhage, seizure disorders, meningitis/encephalitis, and acute liver failure. Some clinical trials may have failed due to high variance in patient selection, suboptimal design, or small sample size. *In vitro* studies and preclinical studies on animals provide us the bench-to-bedside opportunity to design better human clinical trials, which hold promise towards allowing us potentially to apply therapeutic hypothermia in more types of acute brain injury.

References

1. Polderman KH (2004) Application of therapeutic hypothermia in the intensive care unit. Opportunities and pitfalls of a promising treatment modality–part 2: practical aspects and side effects. Intensive Care Med 30:757–69
2. Fay T (1959) Early experiences refrigeration with local and generalized of the human brain. Ann Surg 101:76–132
3. Harris OA, Colford JM, Good MC, Matz PG (2002) The role of hypothermia in the management of severe brain injury: a meta-analysis. Arch Neurol 59:1077–83
4. Holzer M (2008) Devices for rapid induction of hypothermia. Eur J Anaesthesiol Suppl 42:31–8
5. Holzer M (2010) Targeted temperature management for comatose survivors of cardiac arrest. N Engl J Med 363:1256–64
6. Tømte Ø, Drægni T, Mangschau A, Jacobsen D, Auestad B, Sunde K (2011) A comparison of intravascular and surface cooling techniques in comatose cardiac arrest survivors. Crit Care Med 39:443–9
7. Akbari Y, Geocadin RG (2011) A new generation of therapeutic hypothermia: using a warm syringe to cool. Crit Care Med 39:2558–9
8. Weng Y, Sun S, Song F, Phil Chung S, Park J, Harry Weil M, Tang W (2011) Cholecystokinin octapeptide induces hypothermia and improves outcomes in a rat model of cardiopulmonary resuscitation. Crit Care Med 39:2407–12
9. Sun S, Tang W, Song F, Chung SP, Weng Y, Yu T, Weil MH (2010) Pharmacologically induced hypothermia with cannabinoid receptor agonist WIN55, 212–2 after cardiopulmonary resuscitation. Crit Care Med 38:2282–6
10. Castrén M, Nordberg P, Svensson L et al (2010) Intra-arrest transnasal evaporative cooling: a randomized, prehospital, multicenter study (PRINCE: Pre-ROSC IntraNasal Cooling Effectiveness). Circulation 122:729–36
11. Nielsen N, Hovdenes J, Nilsson F, Rubertsson S, Stammet P, Sunde K, Valsson F, Wanscher M, Friberg H (2009) Outcome, timing and adverse events in therapeutic hypothermia after out-of-hospital cardiac arrest. Acta Anaesthesiol Scand 53:926–34
12. Busto R, Dietrich WD, Globus MY, Valdés I, Scheinberg P, Ginsberg MD (1987) Small differences in intraischemic brain temperature critically determine the extent of ischemic neuronal injury. J Cereb Blood Flow Metab 7:729–38
13. Busto R, Globus MY, Dietrich WD, Martinez E, Valdés I, Ginsberg MD (1989) Effect of mild hypothermia on ischemia-induced release of neurotransmitters and free fatty acids in rat brain. Stroke 20:904–10
14. Yenari MA, Han HS (2012) Neuroprotective mechanisms of hypothermia in brain ischaemia. Nat Rev Neurosci 13:267–78
15. Polderman KH (2008) Induced hypothermia and fever control for prevention and treatment of neurologic injuries. Lancet 371:1955–69
16. Colbourne F, Sutherland GR, Auer RN (1996) An automated system for regulating brain temperature in awake and freely moving rodents. J Neurosci Methods 67:185–90
17. Van der Worp HB, Sena ES, Donnan GA, Howells DW, Macleod MR (2007) Hypothermia in animal models of acute ischaemic stroke: a systematic review and meta-analysis. Brain 130: 3063–74

18. Yenari M, Kitagawa K, Lyden P, Perez-Pinzon M (2008) Metabolic downregulation: a key to successful neuroprotection? Stroke 39:2910–7
19. Maier CM, Sun GH, Kunis D, Yenari MA, Steinberg GK (2001) Delayed induction and long-term effects of mild hypothermia in a focal model of transient cerebral ischemia: neurological outcome and infarct size. J Neurosurg 94:90–6
20. Erecinska M, Thoresen M, Silver IA (2003) Effects of hypothermia on energy metabolism in Mammalian central nervous system. J Cereb Blood Flow Metab 23:513–30
21. Lee JM, Zipfel GJ, Choi DW (1999) The changing landscape of ischaemic brain injury mechanisms. Nature 399:A7–14
22. Colbourne F, Grooms SY, Zukin RS, Buchan AM, Bennett MVL (2003) Hypothermia rescues hippocampal CA1 neurons and attenuates down-regulation of the AMPA receptor GluR2 subunit after forebrain ischemia. Proc Natl Acad Sci USA 100:2906–10
23. Terao Y, Miyamoto S, Hirai K, Kamiguchi H, Ohta H, Shimojo M, Kiyota Y, Asahi S, Sakura Y, Shintani Y (2009) Hypothermia enhances heat-shock protein 70 production in ischemic brains. Neuroreport 20:745–9
24. Kumar K, Wu X, Evans AT, Marcoux F (1995) The effect of hypothermia on induction of heat shock protein (HSP)-72 in ischemic brain. Metab Brain Dis 10:283–91
25. Truettner JS, Alonso OF, Bramlett HM, Dietrich WD (2011) Therapeutic hypothermia alters microRNA responses to traumatic brain injury in rats. J Cereb Blood Flow Metab 31: 1897–907
26. Lee Y, Johnson KR, Hallenbeck JM (2012) Global protein conjugation by ubiquitin-like-modifiers during ischemic stress is regulated by microRNAs and confers robust tolerance to ischemia. PLoS One 7:e47787
27. Weiss JB, Eisenhardt SU, Stark GB, Bode C, Moser M, Grundmann S (2012) MicroRNAs in ischemia-reperfusion injury. Am J Cardiovasc Dis 2:237–47
28. Wang Q, Tang XN, Yenari MA (2007) The inflammatory response in stroke. J Neuroimmunol 184:53–68
29. Maier CM, Sun GH, Cheng D, Yenari MA, Chan PH, Steinberg GK (2002) Effects of mild hypothermia on superoxide anion production, superoxide dismutase expression, and activity following transient focal cerebral ischemia. Neurobiol Dis 11:28–42
30. Wang GJ, Deng HY, Maier CM, Sun GH, Yenari MA (2002) Mild hypothermia reduces ICAM-1 expression, neutrophil infiltration and microglia/monocyte accumulation following experimental stroke. Neuroscience 114:1081–90
31. Han HS, Qiao Y, Karabiyikoglu M, Giffard RG, Yenari MA (2002) Influence of mild hypothermia on inducible nitric oxide synthase expression and reactive nitrogen production in experimental stroke and inflammation. J Neurosci 22:3921–8
32. Choi JS, Park J, Suk K, Moon C, Park Y-K, Han HS (2011) Mild hypothermia attenuates intercellular adhesion molecule-1 induction via activation of extracellular signal-regulated kinase-1/2 in a focal cerebral ischemia model. Stroke Res Treat 2011:846716
33. Schmitt KRL, Diestel A, Lehnardt S, Schwartlander R, Lange PE, Berger F, Ullrich O, Abdul-Khaliq H (2007) Hypothermia suppresses inflammation via ERK signaling pathway in stimulated microglial cells. J Neuroimmunol 189:7–16
34. Matsui T, Kakeda T (2008) IL-10 production is reduced by hypothermia but augmented by hyperthermia in rat microglia. J Neurotrauma 25:709–15
35. Reith J, Jørgensen HS, Pedersen PM, Nakayama H, Raaschou HO, Jeppesen LL, Olsen TS (1996) Body temperature in acute stroke: relation to stroke severity, infarct size, mortality, and outcome. Lancet 347:422–5
36. Schwab S, Schwarz S, Spranger M, Keller E, Bertram M, Hacke W (1998) Moderate hypothermia in the treatment of patients with severe middle cerebral artery infarction. Stroke 29:2461–6
37. Schwab S, Georgiadis D, Berrouschot J, Schellinger PD, Graffagnino C, Mayer SA (2001) Feasibility and safety of moderate hypothermia after massive hemispheric infarction. Stroke 32:2033–5

38. Kammersgaard LP, Rasmussen BH, Jørgensen HS, Reith J, Weber U, Olsen TS (2000) Feasibility and safety of inducing modest hypothermia in awake patients with acute stroke through surface cooling: a case–control study: the Copenhagen Stroke Study. Stroke 31: 2251–6
39. Krieger DW, De Georgia MA, Abou-Chebl A, Andrefsky JC, Sila CA, Katzan IL, Mayberg MR, Furlan AJ (2001) Cooling for acute ischemic brain damage (cool aid): an open pilot study of induced hypothermia in acute ischemic stroke. Stroke 32:1847–54
40. Els T, Oehm E, Voigt S, Klisch J, Hetzel A, Kassubek J (2006) Safety and therapeutical benefit of hemicraniectomy combined with mild hypothermia in comparison with hemicraniectomy alone in patients with malignant ischemic stroke. Cerebrovasc Dis 21:79–85
41. De Georgia MA, Krieger DW, Abou-Chebl A, Devlin TG, Jauss M, Davis SM, Koroshetz WJ, Rordorf G, Warach S (2004) Cooling for Acute Ischemic Brain Damage (COOL AID): a feasibility trial of endovascular cooling. Neurology 63:312–7
42. Lyden PD, Allgren RL, Ng K, Akins P, Meyer B, Al-Sanani F, Lutsep H, Dobak J, Matsubara BS, Zivin J (2005) Intravascular cooling in the treatment of stroke (ICTuS): early clinical experience. J Stroke Cerebrovasc Dis 14:107–14
43. Hemmen TM, Raman R, Guluma KZ, Meyer BC, Gomes JA, Cruz-Flores S, Wijman CA, Rapp KS, Grotta JC, Lyden PD (2010) Intravenous thrombolysis plus hypothermia for acute treatment of ischemic stroke (ICTuS-L): final results. Stroke 41:2265–70
44. Winerdal M, Winerdal ME, Kinn J, Urmaliya V, Winqvist O, Adén U (2012) Long lasting local and systemic inflammation after cerebral hypoxic ischemia in newborn mice. PLoS One 7:e36422
45. Pimentel VC, Pinheiro FV, De Bona KS et al (2011) Hypoxic-ischemic brain injury stimulates inflammatory response and enzymatic activities in the hippocampus of neonatal rats. Brain Res 1388:134–40
46. Jacobs SE, Berg M, Hunt R, Tarnow-Mordi WO, Inder TE, Davis PG (2013) Cooling for newborns with hypoxic ischaemic encephalopathy. Cochrane Database Syst Rev 1, CD003311
47. Shankaran S, Laptook AR, Ehrenkranz RA et al (2005) Whole-body hypothermia for neonates with hypoxic-ischemic encephalopathy. N Engl J Med 353:1574–84
48. Gluckman PD, Wyatt JS, Azzopardi D et al (2005) Selective head cooling with mild systemic hypothermia after neonatal encephalopathy: multicentre randomised trial. Lancet 365: 663–70
49. Jenkins DD, Rollins LG, Perkel JK et al (2012) Serum cytokines in a clinical trial of hypothermia for neonatal hypoxic-ischemic encephalopathy. J Cereb Blood Flow Metab 32:1888–96
50. Higgins RD, Raju T, Edwards AD et al (2011) Hypothermia and other treatment options for neonatal encephalopathy: an executive summary of the Eunice Kennedy Shriver NICHD workshop. J Pediatr 159:851.e1–858.e1
51. Xiong M, Chen L-X, Ma S-M, Yang Y, Zhou W-H (2013) Short-term effects of hypothermia on axonal injury, preoligodendrocyte accumulation and oligodendrocyte myelination after hypoxia-ischemia in the hippocampus of immature rat brain. Dev Neurosci 35:17–27. doi:10.1159/000346324
52. Lakhan SE, Kirchgessner A, Hofer M (2009) Inflammatory mechanisms in ischemic stroke: therapeutic approaches. J Transl Med 7:97
53. Callaway CW, Rittenberger JC, Logue ES, McMichael MJ (2008) Hypothermia after cardiac arrest does not alter serum inflammatory markers. Crit Care Med 36:2607–12
54. Meybohm P, Gruenewald M, Zacharowski KD, Albrecht M, Lucius R, Fösel N, Hensler J, Zitta K, Bein B (2010) Mild hypothermia alone or in combination with anesthetic post-conditioning reduces expression of inflammatory cytokines in the cerebral cortex of pigs after cardiopulmonary resuscitation. Crit Care 14:R21
55. Fries M, Stoppe C, Brücken D, Rossaint R, Kuhlen R (2009) Influence of mild therapeutic hypothermia on the inflammatory response after successful resuscitation from cardiac arrest. J Crit Care 24:453–7

56. Fairchild KD, Singh IS, Patel S, Drysdale BE, Viscardi RM, Hester L, Lazusky HM, Hasday JD (2004) Hypothermia prolongs activation of NF-kappaB and augments generation of inflammatory cytokines. Am J Physiol Cell Physiol 287:C422–31
57. Bisschops LLA, Hoedemaekers CWE, Mollnes TE, Van der Hoeven JG (2012) Rewarming after hypothermia after cardiac arrest shifts the inflammatory balance. Crit Care Med 40:1136–42
58. Sarcia PJ, Scumpia PO, Moldawer LL, DeMarco VG, Skimming JW (2003) Hypothermia induces interleukin-10 and attenuates injury in the lungs of endotoxemic rats. Shock 20:41–5
59. Meybohm P, Gruenewald M, Albrecht M, Zacharowski KD, Lucius R, Zitta K, Koch A, Tran N, Scholz J, Bein B (2009) Hypothermia and postconditioning after cardiopulmonary resuscitation reduce cardiac dysfunction by modulating inflammation, apoptosis and remodeling. PLoS One 4:e7588
60. Huet O, Kinirons B, Dupic L, Lajeunie E, Mazoit JX, Benhamou D, Vicaut E, Duranteau J (2007) Induced mild hypothermia reduces mortality during acute inflammation in rats. Acta Anaesthesiol Scand 51:1211–6
61. Amiry-Moghaddam M, Otsuka T, Hurn PD et al (2003) An alpha-syntrophin-dependent pool of AQP4 in astroglial end-feet confers bidirectional water flow between blood and brain. Proc Natl Acad Sci USA 100:2106–11
62. Xiao F, Arnold TC, Zhang S, Brown C, Alexander JS, Carden DL, Conrad SA (2004) Cerebral cortical aquaporin-4 expression in brain edema following cardiac arrest in rats. Acad Emerg Med 11:1001–7
63. Weng Y, Sun S (2012) Therapeutic hypothermia after cardiac arrest in adults: mechanism of neuroprotection, phases of hypothermia, and methods of cooling. Crit Care Clin 28:231–43
64. Hypothermia after Cardiac Arrest Study Group (2002) Mild therapeutic hypothermia to improve the neurologic outcome after cardiac arrest. N Engl J Med 346:549–56
65. Bernard SA, Gray TW, Buist MD, Jones BM, Silvester W, Gutteridge G, Smith K (2002) Treatment of comatose survivors of out-of-hospital cardiac arrest with induced hypothermia. N Engl J Med 346:557–63
66. Arrich J, Holzer M, Havel C, Müllner M, Herkner H (2012) Hypothermia for neuroprotection in adults after cardiopulmonary resuscitation. Cochrane Database Syst Rev 9, CD004128
67. Rothman SM, Olney JW (1986) Glutamate and the pathophysiology of hypoxic–ischemic brain damage. Ann Neurol 19:105–11
68. Vaagenes P, Ginsberg M, Ebmeyer U et al (1996) Cerebral resuscitation from cardiac arrest: pathophysiologic mechanisms. Crit Care Med 24:S57–68
69. Belousov AB (2012) Novel model for the mechanisms of glutamate-dependent excitotoxicity: role of neuronal gap junctions. Brain Res 1487:123–30
70. Lipton SA, Rosenberg PA (1994) Excitatory amino acids as a final common pathway for neurologic disorders. N Engl J Med 330:613–22
71. Gundersen RY, Vaagenes P, Breivik T, Fonnum F, Opstad PK (2005) Glycine—an important neurotransmitter and cytoprotective agent. Acta Anaesthesiol Scand 49:1108–16
72. Globus M, Ginsberg M, Busto R (1991) Excitotoxic index—a biochemical marker of selective vulnerability. Neurosci Lett 127:39–42
73. Kaindl AM, Degos V, Peineau S et al (2012) Activation of microglial N-methyl-D-aspartate receptors triggers inflammation and neuronal cell death in the developing and mature brain. Ann Neurol 72:536–49
74. Zipfel G, Babcock J, Lee J, Choi D (2000) Neuronal apoptosis after CNS injury: the roles of glutamate and calcium. J Neurotrauma 17:857–69
75. Choi D (1994) Calcium and excitotoxic neuronal injury. Ann N Y Acad Sci 747:162–71
76. Moore LE, Traystman RJ (1994) Role of oxygen free radicals and lipid peroxidation in cerebral reperfusion injury. Adv Pharmacol 31:565–76
77. Yang J, Ahn H-N, Chang M, Narasimhan P, Chan PH, Song YS (2013) Complement component 3 inhibition by an antioxidant is neuroprotective after cerebral ischemia and reperfusion in mice. J Neurochem 124(4):523–35

78. Van Beek J, Elward K, Gasque P (2003) Activation of complement in the central nervous system: roles in neurodegeneration and neuroprotection. Ann N Y Acad Sci 992:56–71
79. Donadello K, Favory R, Salgado-Ribeiro D, Vincent J-L, Gottin L, Scolletta S, Creteur J, De Backer D, Taccone FS (2011) Sublingual and muscular microcirculatory alterations after cardiac arrest: a pilot study. Resuscitation 82:690–5
80. Lo Pizzo M, Schiera G, Di Liegro I, Di Liegro CM, Pál J, Czeiter E, Sulyok E, Dóczi T (2012) Aquaporin-4 distribution in control and stressed astrocytes in culture and in the cerebrospinal fluid of patients with traumatic brain injuries. Neurol Sci. doi:10.1007/s10072-012-1233-4
81. Suehiro E, Fujisawa H, Akimura T, Ishihara H, Kajiwara K, Kato S, Fujii M, Yamashita S, Maekawa T, Suzuki M (2004) Increased matrix metalloproteinase-9 in blood in association with activation of interleukin-6 after traumatic brain injury: influence of hypothermic therapy. J Neurotrauma 21:1706–11
82. Soukup J, Zauner A, Doppenberg EMR, Menzel M, Gilman C, Bullock R, Young HF (2002) Relationship between brain temperature, brain chemistry and oxygen delivery after severe human head injury: the effect of mild hypothermia. Neurol Res 24:161–8
83. Vespa P, Bergsneider M, Hattori N, Wu H-M, Huang S-C, Martin NA, Glenn TC, McArthur DL, Hovda DA (2005) Metabolic crisis without brain ischemia is common after traumatic brain injury: a combined microdialysis and positron emission tomography study. J Cereb Blood Flow Metab 25:763–74
84. Glenn TC, Kelly DF, Boscardin WJ, McArthur DL, Vespa P, Oertel M, Hovda DA, Bergsneider M, Hillered L, Martin NA (2003) Energy dysfunction as a predictor of outcome after moderate or severe head injury: indices of oxygen, glucose, and lactate metabolism. J Cereb Blood Flow Metab 23:1239–50
85. Bacher A, Illievich UM, Fitzgerald R, Ihra G, Spiss CK (1997) Changes in oxygenation variables during progressive hypothermia in anesthetized patients. J Neurosurg Anesthesiol 9:205–10
86. Sydenham E, Roberts I, Alderson P (2009) Hypothermia for traumatic head injury (2009 review). Cochrane Database Syst Rev 1–48
87. Rivera-Lara L, Zhang J, Muehlschlegel S (2012) Therapeutic hypothermia for acute neurological injuries. Neurotherapeutics 9:73–86
88. Choi HA, Badjatia N, Mayer SA (2012) Hypothermia for acute brain injury—mechanisms and practical aspects. Nat Rev Neurol 8:214–22
89. Sadaka F, Veremakis C (2012) Therapeutic hypothermia for the management of intracranial hypertension in severe traumatic brain injury: a systematic review. Brain Inj 26:899–908
90. Jiang J-Y, Xu W, Li W-P, Gao G-Y, Bao Y-H, Liang Y-M, Luo Q-Z (2006) Effect of long-term mild hypothermia or short-term mild hypothermia on outcome of patients with severe traumatic brain injury. J Cereb Blood Flow Metab 26:771–6
91. Bratton SL, Chestnut RM, Ghajar J et al (2007) Guidelines for the management of severe traumatic brain injury. III. Prophylactic hypothermia. J Neurotrauma 24(Suppl 1):S21–5
92. Nunnally ME, Jaeschke R, Bellingan GJ et al (2011) Targeted temperature management in critical care: a report and recommendations from five professional societies. Crit Care Med 39:1113–25
93. Andrews PJ, Sinclair HL, Battison CG et al (2011) European society of intensive care medicine study of therapeutic hypothermia (32–35 °C) for intracranial pressure reduction after traumatic brain injury (the Eurotherm3235Trial). Trials 12:8
94. Greenhalgh AD, Brough D, Robinson EM, Girard S, Rothwell NJ, Allan SM (2012) Interleukin-1 receptor antagonist is beneficial after subarachnoid haemorrhage in rat by blocking haem-driven inflammatory pathology. Dis Model Mech 5:823–33
95. Chou SH-Y, Feske SK, Atherton J, Konigsberg RG, De Jager PL, Du R, Ogilvy CS, Lo EH, Ning M (2012) Early elevation of serum tumor necrosis factor-α is associated with poor outcome in subarachnoid hemorrhage. J Investig Med 60:1054–8
96. Maddahi A, Povlsen G, Edvinsson L (2012) Regulation of enhanced cerebrovascular expression of proinflammatory mediators in experimental subarachnoid hemorrhage via the

mitogen-activated protein kinase kinase/extracellular signal-regulated kinase pathway. J Neuroinflammation 9:274
97. Rodling-Wahlström M, Olivecrona M, Koskinen L-OD, Naredi S, Hultin M (2012) Subarachnoid haemorrhage induces an inflammatory response followed by a delayed persisting increase in asymmetric dimethylarginine. Scand J Clin Lab Invest 72:484–9
98. Török E, Klopotowski M, Trabold R, Thal SC, Plesnila N, Schöller K (2009) Mild hypothermia (33 degrees C) reduces intracranial hypertension and improves functional outcome after subarachnoid hemorrhage in rats. Neurosurgery 65:352–9, discussion 359
99. Kawamura Y, Yamada K, Masago A, Katano H, Matsumoto T, Mase M (2000) Hypothermia modulates induction of hsp70 and c-jun mRNA in the rat brain after subarachnoid hemorrhage. J Neurotrauma 17:243–50
100. Wang Z, Chen H, Wang F-X (2011) Influence of plasma and cerebrospinal fluid levels of endothelin-1 and NO in reducing cerebral vasospasm after subarachnoid hemorrhage during treatment with mild hypothermia, in a dog model. Cell Biochem Biophys 61:137–43
101. Schubert GA, Poli S, Schilling L, Heiland S, Thomé C (2008) Hypothermia reduces cytotoxic edema and metabolic alterations during the acute phase of massive SAH: a diffusion-weighted imaging and spectroscopy study in rats. J Neurotrauma 25:841–52
102. Schubert GA, Poli S, Mendelowitsch A, Schilling L, Thomé C (2008) Hypothermia reduces early hypoperfusion and metabolic alterations during the acute phase of massive subarachnoid hemorrhage: a laser-Doppler-flowmetry and microdialysis study in rats. J Neurotrauma 25:539–48
103. Todd MM, Hindman BJ, Clarke WR, Torner JC (2005) Mild intraoperative hypothermia during surgery for intracranial aneurysm. N Engl J Med 352:135–45
104. Anderson SW, Todd MM, Hindman BJ, Clarke WR, Torner JC, Tranel D, Yoo B, Weeks J, Manzel KW, Samra S (2006) Effects of intraoperative hypothermia on neuropsychological outcomes after intracranial aneurysm surgery. Ann Neurol 60:518–27
105. Hindman BJ, Bayman EO, Pfisterer WK, Torner JC, Todd MM (2010) No association between intraoperative hypothermia or supplemental protective drug and neurologic outcomes in patients undergoing temporary clipping during cerebral aneurysm surgery. Anesthesiology 112:86–101
106. Gasser S, Khan N, Yonekawa Y, Imhof H-G, Keller E (2003) Long-term hypothermia in patients with severe brain edema after poor-grade subarachnoid hemorrhage: feasibility and intensive care complications. J Neurosurg Anesthesiol 15:240–8
107. Connolly ES, Rabinstein AA, Carhuapoma JR et al (2012) Guidelines for the management of aneurysmal subarachnoid hemorrhage: a guideline for healthcare professionals from the American Heart Association/american Stroke Association. Stroke 43:1711–37
108. Nguyen H, Zaroff J, Bayman E, Gelb A, Todd M, Hindman B (2010) Perioperative hypothermia (33 °C) does not increase the occurrence of cardiovascular events in patients undergoing cerebral aneurysm surgery: findings. Anesthesiology 113:327–42
109. Badjatia N, Fernandez L, Schmidt JM, Lee K, Claassen J, Connolly ES, Mayer SA (2010) Impact of induced normothermia on outcome after subarachnoid hemorrhage: a case–control study. Neurosurgery 66:696–700, discussion 700–701
110. Kawanishi M, Kawai N, Nakamura T, Luo C, Tamiya T, Nagao S (2008) Effect of delayed mild brain hypothermia on edema formation after intracerebral hemorrhage in rats. J Stroke Cerebrovasc Dis 17:187–95
111. Fingas M, Penner M, Silasi G, Colbourne F (2009) Treatment of intracerebral hemorrhage in rats with 12 h, 3 days and 6 days of selective brain hypothermia. Exp Neurol 219:156–62
112. MacLellan CL, Davies LM, Fingas MS, Colbourne F (2006) The influence of hypothermia on outcome after intracerebral hemorrhage in rats. Stroke 37:1266–70
113. Morgenstern LB, Hemphill JC, Anderson C et al (2010) Guidelines for the management of spontaneous intracerebral hemorrhage: a guideline for healthcare professionals from the American Heart Association/American Stroke Association. Stroke 41:2108–29
114. Staykov D, Wagner I, Volbers B, Doerfler A, Schwab S, Kollmar R (2013) Mild prolonged hypothermia for large intracerebral hemorrhage. Neurocrit Care 18:178–83. doi:10.1007/s12028-012-9762-5

115. Kollmar R, Juettler E, Huttner HB, Dörfler A, Staykov D, Kallmuenzer B, Schmutzhard E, Schwab S, Broessner G (2012) Cooling in intracerebral hemorrhage (CINCH) trial: protocol of a randomized German-Austrian clinical trial. Int J Stroke 7:168–72
116. Vezzani A, Moneta D, Richichi C, Aliprandi M, Burrows SJ, Ravizza T, Perego C, De Simoni MG (2002) Functional role of inflammatory cytokines and antiinflammatory molecules in seizures and epileptogenesis. Epilepsia 43(Suppl 5):30–5
117. Marchi N, Fan Q, Ghosh C et al (2009) Antagonism of peripheral inflammation reduces the severity of status epilepticus. Neurobiol Dis 33:171–81
118. Hsieh PF, Hou C-W, Yao P-W, Wu S-P, Peng Y-F, Shen M-L, Lin C-H, Chao Y-Y, Chang M-H, Jeng K-C (2011) Sesamin ameliorates oxidative stress and mortality in kainic acid-induced status epilepticus by inhibition of MAPK and COX-2 activation. J Neuroinflammation 8:57
119. Zhou Y, Wang Y, Shao X, Chen L (2012) Effects of hypothermia on brain injury induced by status epilepticus. Front Biosci 17:1882–90
120. Wang Y, Liu P-P, Li L-Y, Zhang H-M, Li T (2011) Hypothermia reduces brain edema, spontaneous recurrent seizure attack, and learning memory deficits in the kainic acid treated rats. CNS Neurosci Ther 17:271–80
121. Shorvon S, Ferlisi M (2012) The outcome of therapies in refractory and super-refractory convulsive status epilepticus and recommendations for therapy. Brain 135:2314–28
122. Orlowski J, Erenberg G, Lueders H, Cruse R (1984) Hypothermia and barbiturate coma for refractory status epilepticus. Crit Care Med 12:367–72
123. Robakis T, Hirsch L (2006) Literature review, case report, and expert discussion of prolonged refractory status epilepticus. Neurocrit Care 04:35–46
124. Fernandez A, Claassen J (2012) Refractory status epilepticus. Curr Opin Crit Care 18: 127–31
125. Fisher A, Wang X, Cock HR, Thom M, Patsalos PN, Walker MC (2004) Synergism between topiramate and budipine in refractory status epilepticus in the rat. Epilepsia 45:1300–7
126. Turtle L, Griffiths MJ, Solomon T (2012) Encephalitis caused by flaviviruses. QJM 105: 219–23
127. Wang T, Rumbaugh JA, Nath A (2006) Viruses and the brain: from inflammation to dementia. Clin Sci (Lond) 110:393–407
128. Scheld WM, Koedel U, Nathan B, Pfister H-W (2002) Pathophysiology of bacterial meningitis: mechanism(s) of neuronal injury. J Infect Dis 186(Suppl):S225–33
129. Rowin M, Xue V, Irazuzta J (2001) Hypothermia attenuates beta 1 integrin expression on extravasated neutrophils in an animal model of meningitis. Inflammation 25:137–44
130. Koedel U, Paul R, Winkler F, Kastenbauer S, Huang P, Pfister W (2001) Lack of endothelial nitric oxide synthase aggravates murine pneumococcal meningitis. J Neuropathol Exp Neurol 60:1041–50
131. Humphries HE, Triantafilou M, Makepeace BL, Heckels JE, Triantafilou K, Christodoulides M (2005) Activation of human meningeal cells is modulated by lipopolysaccharide (LPS) and non-LPS components of Neisseria meningitidis and is independent of Toll-like receptor (TLR)4 and TLR2 signalling. Cell Microbiol 7:415–30
132. Stevens NT, Sadovskaya I, Jabbouri S, Sattar T, O'Gara JP, Humphreys H, Greene CM (2009) Staphylococcus epidermidis polysaccharide intercellular adhesin induces IL-8 expression in human astrocytes via a mechanism involving TLR2. Cell Microbiol 11:421–32
133. Irazuzta JE, Pretzlaff R, Rowin M, Milam K, Zemlan FP, Zingarelli B (2000) Hypothermia as an adjunctive treatment for severe bacterial meningitis. Brain Res 881:88–97
134. Mpabanzi L, Jalan R (2012) Neurological complications of acute liver failure: pathophysiological basis of current management and emerging therapies. Neurochem Int 60:736–42
135. Stravitz RT, Larsen FS (2009) Therapeutic hypothermia for acute liver failure. Crit Care Med 37:S258–64
136. Jiang W, Desjardins P, Butterworth RF (2009) Hypothermia attenuates oxidative/nitrosative stress, encephalopathy and brain edema in acute (ischemic) liver failure. Neurochem Int 55: 124–8

137. Gregorios JB, Mozes LW, Norenberg LO, Norenberg MD (1985) Morphologic effects of ammonia on primary astrocyte cultures. I. Light microscopic studies. J Neuropathol Exp Neurol 44:397–403
138. Lavoie J, Giguère JF, Layrargues GP, Butterworth RF (1987) Amino acid changes in autopsied brain tissue from cirrhotic patients with hepatic encephalopathy. J Neurochem 49:692–7
139. Lavoie J, Giguère JF, Layrargues GP, Butterworth RF (1987) Activities of neuronal and astrocytic marker enzymes in autopsied brain tissue from patients with hepatic encephalopathy. Metab Brain Dis 2:283–90
140. Clemmesen JO, Larsen FS, Kondrup J, Hansen BA, Ott P (1999) Cerebral herniation in patients with acute liver failure is correlated with arterial ammonia concentration. Hepatology 29:648–53
141. Hindfelt B, Plum F, Duffy TE (1977) Effect of acute ammonia intoxication on cerebral metabolism in rats with portacaval shunts. J Clin Invest 59:386–96
142. Lai JC, Cooper AJ (1986) Brain a-ketoglutarate dehydrogenase complex: kinetic properties, regional distribution, and effects of inhibitors. J Neurochem 47:1376–86
143. Zwingmann C, Chatauret N, Leibfritz D, Butterworth RF (2003) Selective increase of brain lactate synthesis in experimental acute liver failure: results of a [H-C] nuclear magnetic resonance study. Hepatology 37:420–8
144. Chatauret N, Zwingmann C, Rose C, Leibfritz D, Butterworth RF (2003) Effects of hypothermia on brain glucose metabolism in acute liver failure: a 1H/13C-nuclear magnetic resonance study. Gastroenterology 125:815–24
145. Murthy C, Rama Rao K, Bai G, Norenberg MD (2001) Ammonia-induced production of free radicals in primary cultures of rat astrocytes. J Neurosci Res 288:282–8
146. Rao VL, Audet RM, Butterworth RF (1997) Increased neuronal nitric oxide synthase expression in brain following portacaval anastomosis. Brain Res 765:169–72
147. Schliess F, Görg B, Fischer R, Desjardins P, Bidmon HJ, Herrmann A, Butterworth RF, Zilles K, Häussinger D (2002) Ammonia induces MK-801-sensitive nitration and phosphorylation of protein tyrosine residues in rat astrocytes. FASEB J 16:739–41
148. Horowitz ME, Schafer DF, Molnar P, Jones EA, Blasberg RG, Patlak CS, Waggoner J, Fenstermacher JD (1983) Increased blood–brain transfer in a rabbit model of acute liver failure. Gastroenterology 84:1003–11
149. Skowrońska M, Zielińska M, Wójcik-Stanaszek L, Ruszkiewicz J, Milatovic D, Aschner M, Albrecht J (2012) Ammonia increases paracellular permeability of rat brain endothelial cells by a mechanism encompassing oxidative/nitrosative stress and activation of matrix metalloproteinases. J Neurochem 121:125–34
150. Wright G, Shawcross D, Olde Damink SWM, Jalan R (2007) Brain cytokine flux in acute liver failure and its relationship with intracranial hypertension. Metab Brain Dis 22:375–88
151. Sawara K, Desjardins P, Chatauret N, Kato A, Suzuki K, Butterworth RF (2009) Alterations in expression of genes coding for proteins of the neurovascular unit in ischemic liver failure. Neurochem Int 55:119–23
152. Butterworth RF (2011) Neuroinflammation in acute liver failure: mechanisms and novel therapeutic targets. Neurochem Int 59:830–6
153. Bémeur C, Butterworth RF (2012) Liver-brain proinflammatory signalling in acute liver failure: role in the pathogenesis of hepatic encephalopathy and brain edema. Metab Brain Dis 28:145–50. doi:10.1007/s11011-012-9361-3
154. Chastre A, Bélanger M, Beauchesne E, Nguyen BN, Desjardins P, Butterworth RF (2012) Inflammatory cascades driven by tumor necrosis factor-alpha play a major role in the progression of acute liver failure and its neurological complications. PLoS One 7:e49670
155. Jiang W, Desjardins P, Butterworth RF (2009) Direct evidence for central proinflammatory mechanisms in rats with experimental acute liver failure: protective effect of hypothermia. J Cereb Blood Flow Metab 29:944–52
156. Barba I, Chatauret N, García-Dorado D, Córdoba J (2008) A 1H nuclear magnetic resonance-based metabonomic approach for grading hepatic encephalopathy and monitoring the effects of therapeutic hypothermia in rats. Liver Int 28:1141–8

157. Jalan R, Damink SW, Deutz NE, Lee A, Hayes PC (1999) Early reports moderate hypothermia for uncontrolled intracranial hypertension in acute liver failure. Lancet 354:1164–8
158. Jalan R, Olde Damink SWM, Deutz NEP, Davies NA, Garden OJ, Madhavan KK, Hayes PC, Lee A (2003) Moderate hypothermia prevents cerebral hyperemia and increase in intracranial pressure in patients undergoing liver transplantation for acute liver failure. Transplantation 75:2034–9
159. Jalan R, Olde Damink SWM, Deutz NEP, Hayes PC, Lee A (2004) Moderate hypothermia in patients with acute liver failure and uncontrolled intracranial hypertension. Gastroenterology 127:1338–46
160. Vaquero J (2012) Therapeutic hypothermia in the management of acute liver failure. Neurochem Int 60:723–35
161. Nakamura K (2011) Central circuitries for body temperature regulation and fever. Am J Physiol Regul Integr Comp Physiol 301:R1207–28
162. Ogoina D (2011) Fever, fever patterns and diseases called "fever"—a review. J Infect Public Health 4:108–24
163. Zhang Y, Calderwood SK (2011) Autophagy, protein aggregation and hyperthermia: a mini-review. Int J Hyperthermia 27:409–14
164. Bisschops LLA, Van der Hoeven JG, Hoedemaekers CWE, Van Der Hoeven JG (2012) Effects of prolonged mild hypothermia on cerebral blood flow after cardiac arrest. Crit Care Med 40:2362–7
165. Krüger A, Ošťádal P, Vondráková D, Janotka M, Herget J (2012) Nitrotyrosine and nitrate/nitrite levels in cardiac arrest survivors treated with endovascular hypothermia. Physiol Res 61:425–30
166. Matsui T, Motoki Y, Inomoto T, Miura D, Kato Y, Suenaga H, Hino K, Nojima J (2012) Temperature-related effects of adenosine triphosphate-activated microglia on pro-inflammatory factors. Neurocrit Care 17:293–300
167. Lee H-C, Chuang H-C, Cho D-Y, Cheng K-F, Lin P-H, Chen C-C (2010) Applying cerebral hypothermia and brain oxygen monitoring in treating severe traumatic brain injury. World Neurosurg 74:654–60
168. Baumann E, Preston E, Slinn J, Stanimirovic D (2009) Post-ischemic hypothermia attenuates loss of the vascular basement membrane proteins, agrin and SPARC, and the blood–brain barrier disruption after global cerebral ischemia. Brain Res 1269:185–97
169. Polderman KH, Herold I (2009) Therapeutic hypothermia and controlled normothermia in the intensive care unit: practical considerations, side effects, and cooling methods. Crit Care Med 37:1101–20
170. Badjatia N (2009) Hyperthermia and fever control in brain injury. Crit Care Med 37:S250–7
171. Axelrod YK, Diringer MN (2008) Temperature management in acute neurologic disorders. Neurol Clin 26:585–603, xi
172. Fernandez A, Schmidt JM, Claassen J et al (2007) Fever after subarachnoid hemorrhage: risk factors and impact on outcome. Neurology 68:1013–9
173. Todd MM, Hindman BJ, Clarke WR, Torner JC, Weeks JB, Bayman EO, Shi Q, Spofford CM (2009) Perioperative fever and outcome in surgical patients with aneurysmal subarachnoid hemorrhage. Neurosurgery 64:897–908, discussion 908

Index

J. Chen et al. (eds.), *Immunological Mechanisms and Therapies in Brain Injuries and Stroke*, Springer Series in Translational Stroke Research, DOI 10.1007/978-1-4614-8915-3,

MIX
Papier aus verantwortungsvollen Quellen
Paper from responsible sources
FSC® C105338

If you have any concerns about our products,
you can contact us on
ProductSafety@springernature.com

In case Publisher is established outside the EU,
the EU authorized representative is:
Springer Nature Customer Service Center GmbH
Europaplatz 3, 69115 Heidelberg, Germany

Printed by Libri Plureos GmbH
in Hamburg, Germany